Studies in Big Data

Volume 33

Series editor

Janusz Kacprzyk, Polish Academy of Sciences, Warsaw, Poland
e-mail: kacprzyk@ibspan.waw.pl

The series "Studies in Big Data" (SBD) publishes new developments and advances in the various areas of Big Data- quickly and with a high quality. The intent is to cover the theory, research, development, and applications of Big Data, as embedded in the fields of engineering, computer science, physics, economics and life sciences. The books of the series refer to the analysis and understanding of large, complex, and/or distributed data sets generated from recent digital sources coming from sensors or other physical instruments as well as simulations, crowd sourcing, social networks or other internet transactions, such as emails or video click streams and other. The series contains monographs, lecture notes and edited volumes in Big Data spanning the areas of computational intelligence incl. neural networks, evolutionary computation, soft computing, fuzzy systems, as well as artificial intelligence, data mining, modern statistics and Operations research, as well as self-organizing systems. Of particular value to both the contributors and the readership are the short publication timeframe and the world-wide distribution, which enable both wide and rapid dissemination of research output.

More information about this series at http://www.springer.com/series/11970

Aboul Ella Hassanien · Mohamed Elhoseny
Janusz Kacprzyk
Editors

Quantum Computing: An Environment for Intelligent Large Scale Real Application

Editors
Aboul Ella Hassanien
Faculty of Computers and Information
Cairo University
Cairo
Egypt

Janusz Kacprzyk
Systems Research Institute
Polish Academy of Sciences
Warsaw
Poland

Mohamed Elhoseny
Faculty of Computer and Information Sciences
Mansoura University
Mansoura
Egypt

ISSN 2197-6503 ISSN 2197-6511 (electronic)
Studies in Big Data
ISBN 978-3-319-87602-3 ISBN 978-3-319-63639-9 (eBook)
https://doi.org/10.1007/978-3-319-63639-9

Printed on acid-free paper

This Springer imprint is published by Springer Nature
The registered company is Springer International Publishing AG
The registered company address is: Gewerbestrasse 11, 6330 Cham, Switzerland

Preface

From being functionally monolithic and binary data based in the past, the real-life applications had been transferred to being quantum big data processing systems that adopt the traditional binary data processing approach. Openness, heterogeneity, adaptivity, and dynamism of work environments are becoming the main characteristics of modern real-life applications and it becomes obvious that they are becoming increasingly interconnected and more difficult to maintain. Due to the increase in these applications size, complexity, and the number of components, it is no longer practical to anticipate and model all possible interactions and data processing in these applications using the traditional data processing model. The emerging of new engineering research areas are a clear evidence of the emergence of new demands and requirements of modern real-life applications to be more intelligent, and quantum computing based. Recently, the intelligent systems gained a great attention in all real-life applications. It is being promoted by the software engineering community to use such systems as the adequate solution to handle the current requirements of complex big data processing problems that demanding distribution, flexibility, and robustness. In addition, the increasing usage of wide area networks such as the Internet as a communication medium for such emerging commercial applications made them more overloaded with big amount of data. Thus, we believe that the near-future scientific research will focus quantum computing and big data processing in intelligent systems.

This book aims to provide a collection of high-quality research works that address broad challenges in both theoretical and application aspects of big data processing for intelligent real-life applications in quantum computing environment. This book contains a set of book chapters that stimulates the continuing effort on the application of the intelligent systems that leads to solve the problem of big data processing in quantum-based environment.

In this book, we present the concepts associated with Quantum Computing applications in three distinct parts.

1. **Part I** discusses the issues surrounding Quantum in Network Cryptology. This part aims to provide a clear view about the fundamental concepts for building a secure data communication in a quantum environment. It contains nine chapters.
2. **Part II** discusses different topics related to the applications of Quantum in Physics. This part includes four chapters.
3. **Part III** discusses the applications of Quantum Computing in intelligent environments. Different topics are covered in that part and are well presented in six chapters to ensure the context is simple for understanding.

Cairo, Egypt — Aboul Ella Hassanien
Mansoura, Egypt — Mohamed Elhoseny
Warsaw, Poland — Janusz Kacprzyk

Contents

Part I
Quantum in Network Cryptology

Quantum Information Protocols for Cryptography

Bassem Abd-El-Atty, Salvador E. Venegas-Andraca and Ahmed A. Abd El-Latif

Abstract Quantum cryptography is a robust field of quantum computation and quantum information that focuses on protecting data secrecy by using properties of quantum-mechanical systems. Over the last few years, quantum cryptography has evolved into an emergent high-tech market with companies capable of delivering off-the-shelf products. This chapter introduces a succinct overview of some fundamental concepts of quantum computation, quantum information protocols and their use on the development of quantum cryptography protocols. Key concepts include quantum key distribution, quantum secret sharing, quantum secure direct communication, and deterministic secure quantum communication.

1 Introduction to Quantum Cryptography

Nowadays, computers are key resources in all branches of science, engineering, commerce, and business in general. Indeed, computer science and computer engineering have pervaded every aspect of modern society. The complex relationship between computer science and computer engineering includes the following aspects:

- Computer science provides the fundamental mathematical structures required to build both specific- and general-purpose computers (automata theory) as well as to quantify the amount of resources needed to execute an algorithm (complexity theory). Computer science serves as a guide to understand and estimate the capacities and limits of practical computers and algorithm design.

B. Abd-El-Atty · A.A. Abd El-Latif (✉)
Faculty of Science, Department of Mathematics, Menoufia University,
Menofia Governorate 32511, Egypt
e-mail: a.rahiem@gmail.com; ahmed_rahiem@yahoo.com

S.E. Venegas-Andraca
Escuela de Ingenieria y Ciencias, Tecnologico de Monterrey,
Ave. Eugenio Garza Sada 2501, 64849 Monterrey, N.L., México, Mexico
e-mail: salvador.venegas-andraca@keble.oxon.org

A.E. Hassanien et al. (eds.), *Quantum Computing: An Environment for Intelligent Large Scale Real Application*, Studies in Big Data 33,
https://doi.org/10.1007/978-3-319-63639-9_1

- Open problems and engineering challenges faced at the development of computer technology usually become a decisive resource to produce new scientific questions, among them the need to develop new abstract models of computers.
- Ultimately, information is processed, stored and transmitted by physical systems. Indeed, theoretical computer science is a branch of mathematics but the actual manufacture and behavior of tangible computers and communications systems is fundamentally described by the laws of physics.
- Since the 1950s, the development of computer industry is based on transistor technology whose behavior, due to transistor size, has been largely described by classical physics.

The birth and evolution of quantum computation, the interdisciplinary scientific and engineering field devoted to build information processing systems using the quantum mechanical properties of Nature, is an example of the mutual fertilization of computer science and computer technology described above:

- On the one hand, the idea of using quantum mechanical systems in order to simulate other quantum mechanical systems is the very theoretical foundation of quantum computers as introduced by Feynman [1].
- On the other hand, current semiconductor-based transistors for computer technology are now at the nano scale level. The state-of-the-art of transistor technology has two important consequences for quantum computation: (a) the description and dynamics of nano-transistors require quantum mechanics [2] and (b) the end of Moore's law beseeches new technologies for the computer industry, being quantum computation a leading and most promising paradigm [3].

Physics and computer science have cross-fertilized each other for long time. For example, in trying to quantify the minimum amount of energy required to perform a computation, von Neumann [4] showed that "a minimum amount of energy required per elementary decision of a two-way alternative and the elementary transmittal of one unit of information" was close to kT, where k is Boltzmann's constant and T is the temperature of the system. Later on, Landauer studied the relationship between energy consumption and reversible computation [5] and Bennet presented a formal reversible version of a Universal Turing Machine [6]. Another famous side of this cross-fertilization is the field of algorithms inspired in the mathematical description of physical phenomena, e.g., simulated annealing [7] and percolation theory [8].

Quantum computation and quantum information are scientific and engineering disciplines devoted to the development of novel quantum algorithms and quantum information processing protocols and devices. Some key results in quantum computation include the discovery of quantum teleportation [9, 10] which consists of transmitting quantum information (originally contained in a qubit) via a quantum channel, quantum dense coding [11, 12] which allows sending two classical bits using only one qubit, quantum steganography [13–15] and quantum cryptography [16–44].

Quantum cryptography is a robust field of science and engineering that evolved in an emergent high-tech market with companies capable of delivering off-the-shelf products [45, 46]. Quantum cryptography protocols can be categorized into quantum key distribution protocols (QKD) [16–18], quantum secret sharing protocols (QSS) [19–22], quantum secure direct communication protocols (QSDC) [23–38] and deterministic secure quantum communication protocols (DSQC) [39–44].

Let us now provide a succinct introduction to some fundamental concepts of quantum computation needed to formally describe quantum cryptography protocols.

1.1 Concise Introduction to Preliminary Mathematics and Fundamentals of Quantum Computation

Section 1.1 is based on [47, 48].

1.1.1 The Qubit

In classical computation, information is stored and manipulated in the form of bits. The mathematical structure of a classical bit is rather simple. It suffices to define two logical values, traditionally labelled as $\{0, 1\}$ (i.e. a classical bit lives in a scalar space) and to relate those values to two different and mutually exclusive outcomes of a classical measurement.

In quantum computation, information is stored, manipulated and measured in the form of qubits. A qubit is a physical entity described by the laws of quantum mechanics. A qubit may be mathematically represented as a unit vector in a two-dimensional Hilbert space $|\psi\rangle \in \mathscr{H}^2$ (for the purposes of this chapter, a Hilbert space is a complex inner-product vector space, e.g., $\mathbb{C}^n$). A qubit $|\psi\rangle$ may be written in general form as

$$|\psi\rangle = \alpha|p\rangle + \beta|q\rangle \tag{1}$$

where $\alpha, \beta \in \mathbb{C}$, $|\alpha|^2 + |\beta|^2 = 1$ and $\{|p\rangle, |q\rangle\}$ is an arbitrary basis spanning $\mathscr{H}^2$. Note that a most important consequence of the vectorial nature of a qubit is the possibility of writing it as a linear combination of elements of any basis (this is known as the principle of superposition.)

The choice of $\{|p\rangle, |q\rangle\}$ is often *the orthonormal* basis

$$\{|0\rangle, |1\rangle\}$$

known as **the computational basis**. In addition to the computational basis $\{|0\rangle, |1\rangle\}$, it is customary in quantum cryptography protocols to use **the diagonal basis**

$$\{|+\rangle, |-\rangle\}$$

defined as $|+\rangle = (|0\rangle + |1\rangle)/\sqrt{2}$ and $|-\rangle = (|0\rangle - |1\rangle)/\sqrt{2}$.

1.1.2 Dirac Notation

The symbol $|\cdot\rangle$ is known as a ket. We can always represent kets $|\psi\rangle$ of $\mathscr{H}^2$ as column vectors by choosing a basis for $\mathscr{H}^2$. For example, let $\mathscr{H}^2 = \mathbb{C}^2$ be a 2-dimensional Hilbert space and let us choose the vector basis $\{|0\rangle, |1\rangle\}$, where

$$|0\rangle = \begin{pmatrix} 1 \\ 0 \end{pmatrix} \text{ and } |1\rangle = \begin{pmatrix} 0 \\ 1 \end{pmatrix} \tag{2}$$

Then, every element $|\psi\rangle \in \mathscr{H}^2$ can be written as

$$|\psi\rangle = \alpha|0\rangle + \beta|1\rangle = \alpha \begin{pmatrix} 1 \\ 0 \end{pmatrix} + \beta \begin{pmatrix} 0 \\ 1 \end{pmatrix} = \begin{pmatrix} \alpha \\ \beta \end{pmatrix}, \alpha, \beta \in \mathbb{C}.$$

We can also define bras, which formally speaking are functionals (i.e. functions of vector spaces into corresponding fields) and in practice can be thought of as row vectors:

$$|\psi\rangle = \alpha|0\rangle + \beta|1\rangle \text{ if and only if } \langle\psi| = \alpha^*\langle 0| + \beta^*\langle 1|$$

For example, let

$$|\psi\rangle = \frac{i}{\sqrt{2}}\begin{pmatrix} 1 \\ 0 \end{pmatrix} + \frac{1}{\sqrt{2}}\begin{pmatrix} 0 \\ 1 \end{pmatrix} = \begin{pmatrix} \frac{i}{\sqrt{2}} \\ \frac{1}{\sqrt{2}} \end{pmatrix} = \frac{i}{\sqrt{2}}|0\rangle + \frac{1}{\sqrt{2}}|1\rangle$$

The corresponding bra $\langle\psi|$ is

$$\langle\psi| = \frac{-i}{\sqrt{2}}(1,\ 0) + \frac{1}{\sqrt{2}}(0,\ 1) = (\frac{-i}{\sqrt{2}}, \frac{1}{\sqrt{2}}) = \frac{-i}{\sqrt{2}}\langle 0| + \frac{1}{\sqrt{2}}\langle 1|$$

Thus, if $\mathscr{H}$ is an n-dimensional Hilbert space then the ket $|\psi\rangle \in \mathscr{H}$ can be represented as an n-dimensional column vector, and its corresponding bra $\langle\psi| \in \mathscr{H}^*$ can be seen as an n-dimensional row vector; $|\psi\rangle \leftrightarrow \langle\psi|$ corresponds to transposition and conjunction. Kets and bras are succinctly referred to as **Dirac notation**.

1.1.3 Inner and Outer Products

We can use Dirac notation to make calculations. For example, $\langle\phi|\psi\rangle$ is the usual row-column matrix operator that computes the *inner product* in finite dimensional vector spaces.

Let us take the representations of $|0\rangle$ and $|1\rangle$ given in Eq. (2). Note that $|0\rangle \perp |1\rangle$ as well as the fact that both vectors have unitary norm. Consequently, the inner product of $|0\rangle$ and $|1\rangle$ must be zero and the inner product of each vector with itself must be equal to one:

$$\langle 0|1\rangle = (1,\ 0)\begin{pmatrix}0\\1\end{pmatrix} = (1\times 0+0\times 1) = 0 = (0\times 1+1\times 0) = (0,\ 1)\begin{pmatrix}1\\0\end{pmatrix} = \langle 1|0\rangle$$

Moreover

$$\langle 0|0\rangle = (1,\ 0)\begin{pmatrix}1\\0\end{pmatrix} = (1\times 1+0\times 0) = 1 = (0\times 0+1\times 1) = (0,\ 1)\begin{pmatrix}0\\1\end{pmatrix} = \langle 1|1\rangle$$

We may also use kets and bras to create linear operators (i.e. linear functions between vector spaces). Let $|\psi\rangle, |a\rangle \in \mathscr{H}_1$ and $|\phi\rangle \in \mathscr{H}_2$ then the *outer product* is the linear operator from $\mathscr{H}_1$ to $\mathscr{H}_2$ defined by

$$(|\phi\rangle\langle\psi|)|a\rangle \equiv (\langle\psi|a\rangle)|\phi\rangle$$

As it may be expected, the summation of outer products is also a linear operator. For example, let us define the Hadamard operator

$$\hat{H} = \frac{1}{\sqrt{2}}(|0\rangle\langle 0| + |0\rangle\langle 1| + |1\rangle\langle 0| - |1\rangle\langle 1|)$$

The action of $\hat{H}$ on the ket $|0\rangle$ is given by

$$\begin{aligned}\hat{H}|\psi\rangle &= (\tfrac{1}{\sqrt{2}}|0\rangle\langle 0| + \tfrac{1}{\sqrt{2}}|0\rangle\langle 1| + \tfrac{1}{\sqrt{2}}|1\rangle\langle 0| - \tfrac{1}{\sqrt{2}}|1\rangle\langle 1|)|0\rangle\\ &= \tfrac{\langle 0|0\rangle}{\sqrt{2}}|0\rangle + \tfrac{\langle 1|0\rangle}{\sqrt{2}}|0\rangle + \tfrac{\langle 0|0\rangle}{\sqrt{2}}|1\rangle - \tfrac{\langle 1|0\rangle}{\sqrt{2}}|1\rangle\\ &= \tfrac{1}{\sqrt{2}}|0\rangle + \tfrac{1}{\sqrt{2}}|1\rangle\end{aligned}$$

1.1.4 Quantum Measurements

In quantum mechanics, measurement is a non-trivial and highly counter-intuitive process. Firstly, because measurement outcomes are probabilistic: regardless of the carefulness in the preparation of a measurement procedure, the possible outcomes of such measurement will be distributed according to a certain probability distribution. Secondly, once a measurement has been performed, a quantum system is unavoidably altered due to the interaction with the measurement apparatus. Consequently, for an arbitrary quantum system, pre-measurement and post-measurement quantum states are different in general. Thirdly, in order to perform a measurement it is needed to define a set of measurement operators. This set of operators must fulfill a number of rules that allows one to compute the actual probability distribution as well as post-measurement quantum states.

Quantum measurements are described by a set of measurement operators $\{\hat{M}_m\}$ where index m labels the different measurement outcomes. Measurement outcomes correspond to values of *observable*, such as position, energy and momentum.

Let $|\psi\rangle$ be the state of the quantum system immediately before the measurement. Then, the probability that result m occurs is given by

$$p(m) = \langle\psi|\hat{M}_m^\dagger \hat{M}_m|\psi\rangle \tag{3}$$

and the post-measurement quantum state is

$$|\psi\rangle_{pm} = \frac{\hat{M}_m|\psi\rangle}{\sqrt{\langle\psi|\hat{M}_m^\dagger \hat{M}_m|\psi\rangle}}. \tag{4}$$

Let us work out a simple example. Assume we have a polarized photon with associated polarization orientations 'horizontal' and 'vertical'. The horizontal polarization direction is denoted by $|0\rangle$ and the vertical polarization direction is denoted by $|1\rangle$.

Thus, an arbitrary initial state for our photon can be described by the quantum state $|\psi\rangle = \alpha|0\rangle + \beta|1\rangle$, where α and β are complex numbers constrained by the normalization condition $|\alpha|^2 + |\beta|^2 = 1$ and $\{|0\rangle, |1\rangle\}$ is the computational basis spanning $\mathscr{H}^2$.

Now, we construct two measurement operators $\hat{M}_0 = |0\rangle\langle 0|$ and $\hat{M}_1 = |1\rangle\langle 1|$ and two measurement outcomes a_0, a_1. Then, the full *observable* used for measurement in this experiment is $\hat{M} = a_0|0\rangle\langle 0| + a_1|1\rangle\langle 1|$. According to Eq. (3), the probabilities of obtaining outcome a_0 or outcome a_1 are given by $p(a_0) = |\alpha|^2$ and $p(a_1) = |\beta|^2$. Corresponding post-measurement quantum states (Eq. (4)) are as follows: if outcome $= a_0$ then $|\psi\rangle_{pm} = |0\rangle$; if outcome $= a_1$ then $|\psi\rangle_{pm} = |1\rangle$.

The following results shall be used in the next section:

1.1.5 Entanglement

Entanglement is a unique type of correlation shared between components of a quantum system. Entangled quantum systems are often best used collectively, i.e. an optimal use of entangled quantum systems for information storage and retrieval must manipulate and measure those systems as a whole, rather than on an individual basis.

The concept of correlation is deeply rooted in every branch of science. A typical and simple example is the following experiment: let us suppose we have two balls, one white and one black, as well as two boxes. If we randomly put a ball in each box and then close both boxes, we need to perform only one experiment, that is, to open one box, in order *to know which of the balls is in each box*. In other words, by means of one measurement, namely opening one box and seeing which ball was stored in it, we obtain two pieces of information, namely the colour of the ball stored in both boxes.

The former experiment is an example of classical correlation. Quantum entanglement is also a kind of correlation, but one that has been detected only in quantum phenomena so far.

For example, consider the following 2-particle state:

$$|\Psi^-\rangle = \frac{|01\rangle - |10\rangle}{\sqrt{2}} \tag{5}$$

Clearly, $|\Psi^-\rangle$ lives in a four-dimensional Hilbert space. It can be seen, after some calculations, that it is impossible to find quantum states $|a\rangle, |b\rangle \in \mathscr{H}^2$ such that $|a\rangle \otimes |b\rangle = |\Psi^-\rangle$, that is, $|\Psi^-\rangle$ is not a product state of $|a\rangle$ and $|b\rangle$. This is indeed a criterion to determine whether a quantum state is entangled or not, whether it is possible to express such a composite quantum state as a simple tensor product of quantum subsystems.

Another example is the tripartite entangled GHZ state

$$|GHZ\rangle = \frac{|000\rangle + |111\rangle}{\sqrt{2}} \tag{6}$$

Again, it is not possible to find three quantum states $|a\rangle, |b\rangle, |c\rangle \in \mathscr{H}^2$ such that $|a\rangle \otimes |b\rangle \otimes |c\rangle = |GHZ\rangle$.

2 Quantum Key Distribution Protocols

The quantum key distribution protocols we shall review in this chapter are methodologies designed and implemented to produce and distribute private keys. The importance of (quantum) private keys is described in the following lines.

Cryptography is the branch of science and engineering devoted to the design and implementation of techniques for secure communication, under the assumption that a third party is interested in reading our messages. In other words, cryptography is the science of keeping information secure [49]. Encryption is the process of transforming plaintext (i.e. the actual message) into a ciphertext (i.e. a coded message) by using a cryptosystem. Decryption is the inverse process, i.e. transforming a ciphertext into a plain message. Cryptanalysis is the field focused on recovering the plain text breaking the encryption method used to produce the ciphertext [49].

Modern cryptography has two types of encryption: symmetric key and asymmetric key cryptosystems. Let us briefly describe each type.

- Symmetric key cryptosystems, also known as private key cryptosystems, are encryption algorithms that use a single key for both encryption and decryption of the ciphertext. In this field, algorithms for encryption and decryption are known while the key must remain private, i.e. secret [50].

- Asymmetric key cryptosystems use two keys: a public key and a private key, the public key is used for encryption and the private key for decryption. These cryptosystems work under the rationale of a safe with two keys: the key to lock the safe is known by everybody but the key for opening is available to only one person [49, 50].

Private key cryptosystems can be very powerful. For example, the cryptosystem known as the one-time pad (also known as the Vernam cipher) can be proved to be perfectly secured as long as the key is truly random, the length of the key is equal to the size of the message and the key is only used once [50]. Other private key cryptosystems include the Data Encryption Standard (DES) and the Advanced Encryption Standard (AES) [49, 50].

One of the most sensible issues with private key cryptosystems is safe key distribution, i.e. the process of establishing a private key between two users who cannot use a perfectly secure communication channel [50]. For example, this situation may arise if two users, who do not share secrete information a priori, need to contact each other urgently. The issue is the fact that, if the channel used to distribute a private key is governed by the laws of classical physics, then in principle any (classical) key can be passively eavesdropped, i.e. without users becoming aware of this vulnerability [50].

Quantum key distribution protocols use properties of quantum mechanical systems in order to significantly enhance the security of key distribution channels. Let us now introduce the BB84 and E91 quantum key distribution protocols.

2.1 BB84

In 1984, Charles Bennett and Gilles Brassard [16] proposed a quantum key distribution protocol known as BB84. The key idea of BB84 is to produce a key by encoding bits in qubits that are taken from a set composed by the union of the computational and diagonal bases, i.e. a set of four *non-orthogonal* quantum states. Proofs of the security of BB84 against attacks of eavesdropping strategies have been presented in [51, 52].

Let us suppose the following setting: two people, usually referred to as Alice and Bob, want to create a key for encoding a message. The key, to be shared by Alice and Bob, must be random and secret. There is a third character in this scenario: Eve, an eavesdropper. The conflict in this story is: Alice and Bob's key need their key to be secret and Eve will do anything in her power to reveal at least some portions of the key. Note that the purpose of BB84 is to produce a key, i.e. actual encoded data transmission is *not* part of the BB84.

In the following lines, we present the BB84 protocol step by step. Hereinafter, we assume that:
(a) Alice has a source of individual photons as well as the experimental equipment and expertise to manipulate the polarization of her photons. Moreover, she has access to a source of random bits.
(b) There is a channel available to Alice and Bob that they can use to send quantum states (e.g., an optical fiber.)
(c) Bob has the experimental facilities and expertise required to measure quantum states using different bases.
(d) Finally, Alice and Bob have a classical channel (e.g., a telephone line) that may or may not be a secure line, it does not really matter.

The BB84 protocol is composed of the following steps [53]:

1. Alice starts by generating two sets of *random* bits $A = \{a_1, a_2, \dots a_n\}$ and $B = \{b_1, b_2, \dots b_n\}$
2. Alice uses the set of random bits A to select the vector basis with which she will prepare the initial polarization state of her photons, according to the following criterion: Alice will read one bit at a time from set A and, depending on her reading (either 0 or 1), she will choose the computational basis (+) or the diagonal basis (×) to prepare her photons. In computational terms, this step would be more or less equivalent to variable initialization.
   For instance, let us suppose that $A = \{0, 1, 0, 0, 1, 0, 1, 1, 1, 0\}$ and that Alice has *chosen* the following criterion: if 0 then use the computational basis; if 1 then use the diagonal basis. Then, Alice would initialize her qubits using the following list: (+, ×, +, +, ×, +, ×, ×, ×, +).
3. So far, Alice has a random list of vector space bases (+, ×) to initialize her photons. Now, Alice must *randomly* choose one vector from each basis to prepare the polarization initial state of each photon. To do so, Alice will use the elements of set B according to the following function:

$$\text{initial polarization state} = \begin{cases} |0\rangle \text{ (or } |1\rangle) & \text{if } (+, 0) \\ |1\rangle (\text{ or } |0\rangle) & \text{if } (+, 1) \\ |+\rangle (\text{ or } |-\rangle) & \text{if } (\times, 0) \\ |-\rangle (\text{ or } |+\rangle) & \text{if } (\times, 1) \end{cases}$$

   Following the example presented in the previous step, let us suppose that $B = \{1, 0, 1, 0, 1, 0, 0, 1, 0, 1\}$ and that Alice's list for preparing the initial polarization state of her photons is (+, ×, +, +, ×, +, ×, ×, ×, +). Moreover, let us suppose that she has decided to prepare her photons according to the following

$$\text{initial polarization state} = \begin{cases} |0\rangle & \text{if } +, 1) \\ |1\rangle & \text{if } +, 0) \\ |+\rangle & \text{if } \times, 0) \\ |-\rangle & \text{if } (\times, 1) \end{cases}$$

Then, her photons will have the following initial polarization states:

$$\{|0\rangle, |+\rangle, |0\rangle, |1\rangle, |-\rangle, |1\rangle, |+\rangle, |-\rangle, |+\rangle, |0\rangle\}$$

4. Alice sends her qubit sequence (step 3) to Bob via the quantum channel they both have access to. This step is Eve's chance to extract information from the quantum channel. We now describe two strategies that Eve may try to follow, together with an explanation of why such strategies would fail:

   a. To *perfectly* copy photon polarization states on the fly. Eve may try to design a device $D_{\hat{U}}(|\psi\rangle_i|0\rangle) = |\psi\rangle_i|\psi\rangle$
   to allow her to perform the following operation:

   $$\hat{U}(|\psi\rangle_i|0\rangle) = |\psi\rangle_i|\psi\rangle$$

   where $\hat{U}$ must be a unitary operator (this is a requisite imposed by the mathematical structure of quantum mechanics), $|\psi\rangle_i$ is the polarization state of photon i and $|0\rangle$ is the initial state of a qubit (e.g., the polarization state of a photon contained in the device). In this strategy, Eve intends to copy quantum states as photon travel along the quantum channel, it is *not* her intention to perturb the photons sent by Alice in any sense.
   The no-cloning theorem [54, 55], a most remarkable and counter-intuitive result of quantum mechanics, prevents the realization of a device that makes *clones*, i.e. perfect copies, of arbitrary quantum states. In other words, the no-cloning theorem makes Eve's device D_U impossible to create.

   b. To measure photon polarization states followed by preparing new photons for Bob. In this case, Eve intercepts photon i and measures its polarization state. Moreover and based on the information extracted from photon i, Eve prepares a new photon and sends it to Bob.
   In order to measure photon i, Eve must choose a basis to build the corresponding measurement operators. Eve may know that Alice has prepared her qubits using the computational and diagonal bases, but she does not know the order in which bases were picked, hence she does not know for sure which measurement operators she must use in order to extract information from the photon sequence; in fact, Eve can only *guess* which basis she should use for measuring each photon.
   Let us then suppose that Eve randomly chooses to use the computational basis or the diagonal basis to measure photon polarization states. Of course,

Table 1 Probabilities and post-measurement quantum states $|0\rangle, |1\rangle, |+\rangle, |-\rangle$ and measurement operators $|0\rangle\langle 0|, |1\rangle\langle 1|$

Qubit	Measurement operator	$p(0)$	$|\psi\rangle_{pm}^{0}$	Measurement operator	$p(1)$	$|\psi\rangle_{pm}^{1}$
$|0\rangle$	$\hat{M}_0 = |0\rangle\langle 0|$	1	$|0\rangle$	$\hat{M}_1 = |1\rangle\langle 1|$	0	$\nexists$
$|1\rangle$	$\hat{M}_0 = |0\rangle\langle 0|$	0	$\nexists$	$\hat{M}_1 = |1\rangle\langle 1|$	1	$|1\rangle$
$|+\rangle$	$\hat{M}_0 = |0\rangle\langle 0|$	1/2	$|0\rangle$	$\hat{M}_1 = |1\rangle\langle 1|$	1/2	$|1\rangle$
$|-\rangle$	$\hat{M}_0 = |0\rangle\langle 0|$	1/2	$|0\rangle$	$\hat{M}_1 = |1\rangle\langle 1|$	1/2	$|1\rangle$

Table 2 Probabilities and post-measurement quantum states $|0\rangle, |1\rangle, |+\rangle, |-\rangle$ and measurement operators $|+\rangle\langle +|, |-\rangle\langle -|$

Qubit	Measurement operator	$p(+)$	$|\psi\rangle_{pm}^{0}$	Measurement operator	$p(-)$	$|\psi\rangle_{pm}^{1}$
$|0\rangle$	$\hat{M}_+ = |+\rangle\langle +|$	1/2	$|+\rangle$	$\hat{M}_- = |-\rangle\langle -|$	1/2	$|-\rangle$
$|1\rangle$	$\hat{M}_+ = |+\rangle\langle +|$	1/2	$|+\rangle$	$\hat{M}_- = |-\rangle\langle -|$	1/2	$|-\rangle$
$|+\rangle$	$\hat{M}_+ = |+\rangle\langle +|$	1	$|+\rangle$	$\hat{M}_- = |-\rangle\langle -|$	0	$\nexists$
$|-\rangle$	$\hat{M}_+ = |+\rangle\langle +|$	0	$\nexists$	$\hat{M}_- = |-\rangle\langle -|$	1	$|-\rangle$

Eve will make right choices sometimes and will be wrong some other times. Let us analyze both cases:

i. Right choice of measurement operators. For example, let us suppose that the polarization state of photon i is $|1\rangle$ and that Eve has chosen the computational basis $\{|0\rangle, |1\rangle\}$ to produce $\{\hat{M}_0, \hat{M}_1\}$. Hence, the probability of measuring outcome 0 is 1, i.e. $p(0) = 1$, and the corresponding post-measurement state is $|0\rangle$, i.e. $|\psi\rangle_{pm}^{0} = |0\rangle$, (please see Table 1). After learning that the polarization of photon i is $|0\rangle$, Eve prepares a new photon i' and sends it to Bob. Since Eve has sent to Bob the same quantum state that Alice prepared, her eavesdropping will go unnoticed (Table 2).
ii. Wrong choice of measurement operators. For example, Eve chooses the diagonal basis $\{|+\rangle, |-\rangle\}$ to measure photon j whose polarization state is $|0\rangle$. Then, according to the rules of quantum mechanics, the probability of having $|+\rangle$ as post-measurement quantum state is $1/2$ and, correspondingly, the probability of having $|-\rangle$ as post-measurement quantum state is also $1/2$. Here, Eve faces two problems: she has not extracted any information from photon j and the photon that she will prepare and send to Bob will be different from the photon sent by Alice, hence there is room for detecting her eavesdropping activity (remember that BB84 is about safely creating a private key, not about transmitting any message between Alice and Bob, so detecting Eve's activity does not jeopardize the message itself, it only renders the key unusable.)

So, if Eve randomly chooses between the computational basis and the diagonal basis to measure photon polarization states, it is reasonable to expect that she will be successful only half of the time, hence her chances of making a mistake are high (50%), an unacceptable scenario for a professional spy.

5. Let us now describe Bob's activities. Bob knows that Alice has prepared her qubits using either the computational basis or the diagonal basis but, just like Eve, he does not know the order in which Alice chose between those vector bases. So, Bob randomly selects the vector basis he will use to produce the corresponding measurement operators (i.e. he randomly chooses $\{|0\rangle, |1\rangle\}$ or $\{|+\rangle, |-\rangle\}$) and measures the polarization states of the photons he receives at his end. The result of this procedure will be a list C composed of ordered pairs (*basis,outcome*) where *basis* is either + or × and *outcome* is either 0 or 1.
6. Once the full photon sequence has been processed as described above, Alice and Bob use a classical channel (a telephone line or e-mail, for instance) to tell each other the sequence of basis they use to initialize and measure photon polarization states, respectively. Alice and Bob will discard those outcomes that correspond to disagreements of initialization and measurement bases (for instance, if Alice used $\{|0\rangle, |1\rangle\}$ to prepare photon i and Bob used basis $\{|+\rangle, |-\rangle\}$ to measure the same photon i) and will keep the remaining outcomes. Let us label the remaining bit sequences os Alice and Bob as R_A and R_B, respectively.
Assuming no interference on the quantum channel, R_A and R_B must be identical, that is, Alice and Bob have succeeded at producing a private key $K(= R_A = R_B.)$ However, it is reasonable to assume some discrepancies between R_A and R_B due to errors in transmission and/or Eve's activity. In this case, Alice and Bob may follow error correction and privacy amplification procedures in order to produce two identical bit strings R'_A and R'_B, that is, the private key $K(= R_A = R_B)$ they needed to generate.

A succinct example of BB84 is presented on Table 3.

Table 3 Example of the BB84 protocol in action

Alice's bits	1	1	0	1	0	0	1	1	1	0	0	1
Alice's bases	+	+	×	+	×	+	×	+	×	+	×	×
Alice's qubits	$\|0\rangle$	$\|0\rangle$	$\|+\rangle$	$\|0\rangle$	$\|+\rangle$	$\|1\rangle$	$\|-\rangle$	$\|0\rangle$	$\|-\rangle$	$\|1\rangle$	$\|+\rangle$	$\|-\rangle$
Eve's bases	+	×	+	+	×	×	+	×	+	+	+	+
Eve's measurements	$\|0\rangle$	$\|+\rangle$	$\|1\rangle$	$\|0\rangle$	$\|+\rangle$	$\|0\rangle$	$\|0\rangle$	$\|+\rangle$	$\|1\rangle$	$\|1\rangle$	$\|1\rangle$	$\|0\rangle$
Bob's bases	×	+	+	+	×	×	×	+	+	+	+	×
Bob's measurements	$\|+\rangle$	$\|0\rangle$	$\|1\rangle$	$\|0\rangle$	$\|+\rangle$	$\|-\rangle$	$\|+\rangle$	$\|1\rangle$	$\|1\rangle$	$\|1\rangle$	$\|1\rangle$	$\|+\rangle$
Bob's bits	1	1	0	1	0	1	0	0	0	0	0	0
Selection		•		•	•		•	•		•		•

2.2 E91

In 1991, Ekert [17] proposed a quantum key distribution protocol based on Bell states and known as E91. Bell states, presented in Eqs. (7a–7d), constitute a widely used set of entangled states

$$|\Phi^+\rangle = \frac{1}{\sqrt{2}}(|00\rangle + |11\rangle) \tag{7a}$$

$$|\Phi^-\rangle = \frac{1}{\sqrt{2}}(|00\rangle - |11\rangle) \tag{7b}$$

$$|\Psi^+\rangle = \frac{1}{\sqrt{2}}(|01\rangle + |10\rangle) \tag{7c}$$

$$|\Psi^-\rangle = \frac{1}{\sqrt{2}}(|01\rangle - |10\rangle) \tag{7d}$$

When Alice holds the first particle and Bob holds the second particle, this implies that the measurement results of Alice and Bob are completely correlated with each other when they used one of two bases ($+, \times$) to measure the state. Eve has no information about the particles (qubits) hold by Alice and Bob, due to the two particles are in entangled state and Eve cannot reach the two particles in the same time. To transmit the whole key, Alice and Bob generate a large numbers of Bell states and measure each particle in separate way. E91 is composed of the following steps:

1. Alice generates a sequence of Bell states.
2. Alice randomly selects a subset of this sequence and applies the Hadamard transformation on the first particle for each pair when the corresponding bit string is '1'.
3. Alice sends the sequence of second qubits to Bob and holds the sequence of first qubits.
4. Alice tells Bob the bits of string and which qubits are selected.
5. Bob applies the Hadamard transformation on the selected qubits where the bit string is '1'.
6. Both Alice and Bob measure the selected qubits with same bases and delete the bits which are of different bases and the remaining bits are called shifted key. Then they check the exiting of Eve by estimate the error-rate of choosing a subset of the sifted key. If the bits are differing, they abort the protocol.
7. Finally, by performing error correction Alice and Bob obtain the secret key.

Without performing Hadamard transformation on the first qubit (which is equivalent to transforming $+$ into $\times$, Eve could easily attack the protocol via intercept-resend attack.

3 Quantum Secret Sharing Protocols (QSS)

Secret sharing, as one of the most important branches in cryptography, is a technique developed to distribute a secret message to several parties so that no participant can access and read the secret message without the collaboration of other parties. To further illustrate the observations, imagine Alice holding a secret message and want to gain access for both Bob and Charlie. She knows that one of them is dishonest and she does not know which the honest one is. So, Alice cannot send the secret message directly to both of them, because the dishonest one will steal the information, but she knows that if Bob and Charlie carry out it together, the honest one will keep the dishonest one from doing any damage.

Quantum secret sharing (QSS) firstly proposed by Hillery et al. [19] in 1999, namely Hillery's protocol. It's based on three-particle quantum entangled states and the classical secret sharing protocol presented by Shamir [56] in 1979. Then, Anders Karlsson et al. [20] presented a new QSS protocol based on entangled two-photon states. Thereafter, several QSS protocols have been proposed, for example, but not limited, [19, 20].

Herein, we shed the light to QSS protocol in [19]. This protocol uses Greenberger-Horne-Zeilinger (GHZ) three-particle states. Supposing that there are three participants Alice, Bob and Charlie, each of them in possession of one particle of the GHZ state presented in Eq. (8)

$$|\Psi\rangle = \frac{|000\rangle + |111\rangle}{\sqrt{2}} \tag{8}$$

All participants randomly choose to measure their own particles using directions x or y. Then they communicate with each other to declare which direction they used in measurement process, without announcing their measurement results. By combining the measurement results of Alice and Bob, they can determine the measurement result of Charlie. This allows Charlie to build up a joint key with Alice and Bob, which can be used to send the secret message. The directions of particles x and y are defined as follows:

$$|x^{\pm}\rangle = \frac{|0\rangle \pm |1\rangle}{\sqrt{2}} \tag{9a}$$

$$|y^{\pm}\rangle = \frac{|0\rangle \pm i|1\rangle}{\sqrt{2}} \tag{9b}$$

If Alice and Bob perform their measurement in the x direction and they want to get information about Charlie's particle. The GHZ state can be written as

$$|\Psi\rangle = \frac{(|x^+\rangle_a|x^+\rangle_b + |x^-\rangle_a|x^-\rangle_b)(|0\rangle_c + |1\rangle_c) + (|x^+\rangle_a|x^-\rangle_b + |x^-\rangle_a|x^+\rangle_b)(|0\rangle_c - |1\rangle_c)}{2\sqrt{2}} \tag{10}$$

From Eq. (10), if Alice and Bob have the same result of measurement, then they conclude that Charlie have the state $\frac{|0\rangle_c+|1\rangle_c}{\sqrt{2}}$, otherwise he have the state $\frac{|0\rangle_c-|1\rangle_c}{\sqrt{2}}$. Similarly, Bob and Charlie can determine the quantum state of Alice by combining their results of measurement. Generally, no one of the three participants can determine the quantum state of each other without combining their results.

Now, what happens if the participants choose to measure their own particles in any direction without announcing their measurement basis. In other words, how to determine whether the participant's measurements are correlated or not. The answer is, all participant's particles must be measured in the same direction. For example, if both Bob and Charlie measure their particles in the x direction. Alice must also measure her particle in x direction. Otherwise, she gains no information from her result. Therefore, all the three participants must announce their measurements basis to each other to decide whether measurement basis to keep or which to discard.

4 Quantum Secure Direct Communication Protocols (QSDC)

Quantum secure direct communication (QSDC) protocols differ from QKD, QSS and DSQC protocols as a secret message in QSDC protocols is transmitted by the quantum channel directly without having to share a private key between two legitimate users beforehand. There are three requirements should be satisfied in any secure QSDC protocol:

1. Before the sender (Alice) and the receiver (Bob) communicate to encode their secret message on the quantum states, they can detect the exiting of the eavesdropper (Eve).
2. Bob can read the secret message directly without establishing additional classical channels with Alice to exchange the secret key and ensure the security of the protocol.
3. In any type of attacks performed by Eve, no any useful information stolen about the transmitted secret message.

In 2002, Beige et al. [23] proposed the first QSDC protocol based on the exchange of single photons and each photon transmits one bit of the secret message. In the same year, Boström et al. [24] presented a ping-pong QSDC scheme based on EPR pairs. In 2003, Deng and Long [25] introduced a two-step QSDC scheme based on dense coding operations. In 2004, Deng and Long [26] introduced another QSDC protocol based on a sequence of single photons. Then, several QSDC protocols are proposed to carry the secret message based on single photons and entangled particles through a quantum channel. In the literature, there are several QSDC protocols utilized entangled particles in their internal structure such as EPR pairs [27, 28], Greenberger-Horne-Zeilinger (GHZ) [29], cluster states [30], W states [31] and χ-type states [32]. EPR pairs are more easily prepared and therefore widely used. In fact, single photons have several advantages that lead to the development of quantum communication protocols such as flexible implementation, high efficiency as well as simple operations. In 2004, Deng and Long [26] developed a QSDC scheme by using batches of single photons, which serves as quantum one-time pad cryptosystem. In 2006, Wang et al. [33] proposed a QSDC scheme based on order rearrangement of single photons, in which all single photons are used to encode the secret message except those used for eavesdropping check. In the same year, Li et al. [34] presented a deterministic QSDC scheme by using a sequence of single photons. In 2010, Quan et al. [35] proposed a one-way DQSDC scheme by using a sequence of single photons. In 2013, Chang et al. [36] proposed quantum secure direct communication and authentication protocol with single photons and XOR operation. In 2015, Zhao [37] proposed two quantum secure communication protocols based on single photon sequence and the XOR operation. Also, in the same year, Xin et al. [38] proposed a quantum authentication protocol based on Hash function and Bell states. Hereinafter, we introduce examples of these protocols.

4.1 QSDC Deng-Long Protocol

As mentioned above, quantum communication protocols based on single photons are easier to implement than quantum protocols based on entangled states. Therefore, Deng and Long presented a QSDC protocol [26] based on QKD idea. In this protocol, Alice shares with Bob a sequence of single photon states, then Alice encodes the secret message and transmits the photon states to Bob. The detailed steps of QSDC Deng-Long protocol is given below:

1. Bob generates a sequence S of single photons in one of the four polarized states $|0\rangle, |1\rangle, |+\rangle = \frac{1}{\sqrt{2}}(|0\rangle + |1\rangle), |-\rangle = \frac{1}{\sqrt{2}}(|0\rangle - |1\rangle)$ which the polarized states $|0\rangle$ and $|+\rangle$ represent the binary value 0, the polarized states $|1\rangle$ and $|-\rangle$ represent the binary value 1 and sends these sequence S to Alice. To check the security of the protocol, Bob selects a subset group of photons in sequence S and tells Alice the positions of the selected group. Alice measures the photons in the selected group using the same basis and compares the results of measurement with Bob. If their results are the same, there is no attack and the quantum channel is secure, otherwise, the connection is not secure and they abort from the quantum channel.
2. To encode the secret message, Alice performs the unitary transformation U_0 and U_1 which represent binary values 0 and 1, respectively on each photon in the selected positions according to the secret message. where

$$U_0 = |0\rangle\langle 0| + |0\rangle\langle 0| = \begin{bmatrix} 1 & 0 \\ 0 & 1 \end{bmatrix} = I \tag{11a}$$

$$U_1 = |0\rangle\langle 1| - |1\rangle\langle 0| = \begin{bmatrix} 0 & 1 \\ -1 & 0 \end{bmatrix} = i\sigma_Y \tag{11b}$$

3. Alice sends the result of the sequence S to Bob. Bob knows the initial state of photon polarization and the selected positions, due to the sequence is generated by her. Then Bob measure each photon in the selected positions by the original basis and deduce the secret message from the unitary transformation Alice applied on the selected photons.

4.2 QSDC Deng Protocol

Deng et al. [25] presented a two-step QSDC protocol, which the first QSDC protocol based on Bell states. In this protocol, Alice generates a sequence of entangled particles all in $|\Phi\rangle^+$ state. Alice separates the sequence of particles into two subsequences. The first subsequence called the message-carrier and labels it as SA. The second subsequence called the checking and labels it as SB. At first Alice and Bob agree that classical bits 00, 01, 10 and 11 correspond to the four Bell states $|\Psi\rangle^+, |\Psi\rangle^-, |\Phi\rangle^+, |\Phi\rangle^-$, respectively. Then, Alice sends the sequence SB to Bob. To check the security of the established quantum channel, Bob selects randomly a subset of photons in the sequence SB and measures this selected group by one of the two basis (Z and X). Thus, Bob communicates with Alice via a classical channel to tells here the positions of photons in the selected group. Then, Alice measures the corresponding photons in sequence SA using the same basis. Alice's result is then compared with Bob's result via the classical channel. If two results are completely opposite, there is no attack performed by Eve, otherwise Eve performed attack and they must abort the connection. After that Alice encodes her secret message by applying unitary transformations $(\sigma_I, \sigma_z, \sigma_x, \sigma_{iy})$ on their own particles in the selected positions to transform the Bell state to another Bell state (see Table 4) according to the encoded secret message and sends the result sequence SA to Bob. Then Bob extract the secret message from the Bell states in the selected position utilizing Bell measurements.

where,

$$\sigma_I = |0\rangle\langle 0| + |1\rangle\langle 1| = \begin{bmatrix} 1 & 0 \\ 0 & 1 \end{bmatrix} \tag{12a}$$

Table 4 The relationship between the initial Bell states and the final Bell states

	σ_I	σ_X	σ_Z	$i\sigma_Y$
$\|\Phi^+\rangle$	$\|\Phi^+\rangle$	$\|\Psi^+\rangle$	$\|\Phi^-\rangle$	$\|\Psi^-\rangle$
$\|\Phi^-\rangle$	$\|\Phi^-\rangle$	$\|\Psi^-\rangle$	$\|\Phi^+\rangle$	$\|\Psi^+\rangle$
$\|\Psi^+\rangle$	$\|\Psi^+\rangle$	$\|\Phi^+\rangle$	$\|\Psi^-\rangle$	$\|\Phi^-\rangle$
$\|\Psi^-\rangle$	$\|\Psi^-\rangle$	$\|\Phi^-\rangle$	$\|\Psi^+\rangle$	$\|\Phi^+\rangle$

$$\sigma_X = |0\rangle\langle 1| + |1\rangle\langle 0| = \begin{bmatrix} 0 & 1 \\ 1 & 0 \end{bmatrix} \tag{12b}$$

$$\sigma_Z = |0\rangle\langle 0| - |1\rangle\langle 1| = \begin{bmatrix} 1 & 0 \\ 0 & -1 \end{bmatrix} \tag{12c}$$

$$i\sigma_Y = |0\rangle\langle 1| - |1\rangle\langle 0| = \begin{bmatrix} 0 & 1 \\ -1 & 0 \end{bmatrix} = \sigma_Z\sigma_X \tag{12d}$$

5 Deterministic Secure Quantum Communication Protocols (DSQC)

The deterministic secure quantum communication (DSQC) protocols are designed to obtain deterministic information, not random information. It is similar to QKD protocols which at first generate a key to encrypt the secret message. In DSQC, Bob can extract the secret message by using for each transmitted qubit at least an additional classical bit. So the classical channels are needed besides the DSQC protocols. In 2004, Cai and Li [56] presented a DSQC protocol based on single qubit in a mixed state and its security based on the security of BB84 protocol. In 2005, Gao et al. [39] proposed a DSQC protocol based on GHZ states and entanglement swapping. In 2006, Shaari et al. [40] presented a two-way deterministic protocol using six mutually unbiased states in the Poincare sphere which the information are encoded by not-flip or flip operations on the states. Also, in the same year, Li et al. [41] presented two DSQC protocols, one based on d-dimensional single-photon states and the other based on pure entangled states which single-photon measurements are only used for the two participants in these two protocols. In 2012, Huang et al. [42] proposed two DSQC protocols with collective detection, one is a DSQC network and the other is a two-party DSQC scheme. Finally, in 2015, Yan et al. [43] presented a controlled DSQC protocol based on three-particle GHZ state in X-basis.

Let us now describe one of these DSQC protocols in details. Huang et al. [42] presented a two-party DSQC protocol using single photons with collective detection. In this protocol, Alice and Bob perform single-photon measurements on their photons and communicate via classical channel to exchange a certain classical information bits. This protocol based on collective detection strategy, which is used to detect any attack on the quantum protocol after the whole process of qubit transmission and reduces the cost of protocol realization.

At first in this protocol, Alice and Bob share a sequence of single photons, and then Alice encodes the message using two unitary operations then encrypts the encoded states using a tilt-adjustable phase plate. After receiving Bob the sequence of photons from Alice via quantum channel, he can decode the message with the help of transmitted classical information bits via classical channel. The steps of this protocol can be described as follows:

1. Bob prepares a sequence of single photons all in the state $|+\rangle$ and sends the sequence to Alice.
2. After receiving the sequence, Alice generates a random string M with length equal to the number of elements in Bob's sequence of states S.
3. Alice apply on the photon sequence S a phase shift $\psi_i \in \{0, \pi/2, \pi, 3\pi/2\}$ to encode 00, 01, 10 and 11, respectively.
4. Alice randomly chooses some polarized photons in sequence S as decoy states and performs unitary transformation σ_i and σ_z on the polarized photons in selected positions, then Alice sends the sequence to Bob.
5. To check eavesdropping via classical channel:
    - At first Alice tells Bob the positions of decoy photons and the corresponding values of M for each decoy particle.
    - Then Bob measures each decoy photon using $\{|+\rangle, |-\rangle, |s\rangle, |t\rangle\}$ to obtain corresponding values, where $|s\rangle = \frac{|0\rangle + i|1\rangle}{\sqrt{2}}$ and $|t\rangle = \frac{|0\rangle - i|1\rangle}{\sqrt{2}}$
    - Bob estimates the error rate and decides whether they abort the protocol; otherwise, they go to the next step.
6. Bob measures each photon in the sequence with measurement basis $\{|+\rangle, |-\rangle, |s\rangle, |t\rangle\}$ to reveal the encoded secret message M.

The main advantage of this protocol is that Alice and Bob can utilize all transferred polarized photons to transmit the secret message, except for the ones used for eavesdropping check.

References

1. Feynman, R.P.: Simulating physics with computers. Int. J. Theor. Phys. **21**(6/7), 467–488 (1982)
2. Svizhenko, A., Anantram, M.P., Govindan, T.R., Biegel, B., Venugopal, R.: Two-dimensional quantum mechanical modeling of nanotransistors. J. Appl. Phys. **91**(4), 2343–2354 (2002)
3. IEEE. Rebooting Computing Initiative. http://rebootingcomputing.ieee.org/
4. von Neumann, J.: Fourth University of Illinois Lecture (Theory of self-reproducing Automata). University of Illinois Press (1966)
5. Landauer, R.: Irreversibility and heat generation in the computing process. IBM J. Res. Dev. **5**(3), 183–191 (1961)
6. Landauer, R.: Logical reversibility of computation. IBM J. Res. Dev. **17**(6), 525–532 (1973)
7. Kirkpatrick, S., Gelatt, C.D. Jr., Vecchi, M.P.: Optimization by simulated annealing. Science **220**(4598), 671–680 (1983)
8. Mertens, S., Moore, C.: Continuum percolation thresholds in two dimensions. Phys. Rev. E **86**, 061109 (2012)
9. Bennett, H.C., Brassard, G., Crpeau, C., Jozsa, R., Peres, A., Wootters, W.K.: Teleporting an unknown quantum state via dual classical and Einstein-Podolsky-Rosen channels. Phys. Rev. Lett. **70**(13), 1895–1899 (1993)
10. Lee, J., Kim, M.S.: Entanglement teleportation via werner states. Phys. Rev. Lett. **84**(18), 4236–4239 (2000)
11. Bennett, C.H., Wiesner, S.J.: Communication via one- and two-particle operators on Einstein-Podolsky-Rosen states. Phys. Rev. Lett. **69**(20), 2881–2884 (1992)

12. Mermin, N.D.: Deconstructing dense coding. Phys. Rev. A. **66**, 032308 (2002)
13. Abd-El-Atty, B., Abd El-Latif, A.A., Amin, M.: New quantum image steganography scheme with Hadamard transformation. In: International Conference on Advanced Intelligent Systems and Informatics, pp. 342–352, Springer International Publishing (2016)
14. Jiang, N., Zhao, N., Wang, L.: LSB based quantum image steganography algorithm. Int. J. Theor. Phys. **55**(1), 107–123 (2015)
15. Wang, S., Sang, J., Song, X., Niu, X.: Least significant qubit (LSQb) information hiding algorithm for quantum images. Measurement **73**, 352–359 (2015)
16. Bennett, C.H., Brassard, G.: Quantum cryptography: public key distribution and coin tossing. In: Proceedings of the IEEE International Conference on Computers, Systems and Signal Processing, pp. 175–179, Bangalore, India (1984)
17. Ekert, A.K.: Quantum cryptography based on Bell's theorem. Phys. Rev. Lett. **67**(6), 661–663 (1991)
18. Bennett, C.H., Brassard, G., Mermin, N.D.: Quantum cryptography without Bell's theorem. Phys. Rev. Lett. **68**(5), 557–559 (1992)
19. Hillery, M., Bužek, V., Berthiaume, A.: Quantum secret sharing. Phys. Rev. A **59**(3), 1829–1834 (1999)
20. Karlsson, A., Koashi, M., Imoto, N.: Quantum entanglement for secret sharing and secret splitting. Phys. Rev. A **59**(1), 162–168 (1999)
21. Xiao, L., Long, G.L., Deng, F.G., Pan, J.W.: Efficient multiparty quantum secret-sharing schemes. Phys. Rev. A **69**, 052307 (2004)
22. Chen, P., Long, G.L., Deng, F.G.: High-dimension multiparty quantum secret sharing scheme with Einstein-Podolsky-Rosen pairs. Chin. Phys. B **15**, 2228–2235 (2006)
23. Beige, A., Englert, B.G., Urtsiefer, C.K., Weinfurter, H.: Secure communication with a publicly known key. Acta Physica Polonica A **101**(3), 357–368 (2002)
24. Bostrom, K., Felbinger, T.: Deterministic secure direct communication using entanglement. Phys. Rev. Lett. **89**, 187902 (2002)
25. Deng, F.G., Long, G.L., Liu, X.S.: Two-step quantum direct communication protocol using the Einstein-Podolsky-Rosen pair block. Phys. Rev. A **68**, 042317 (2003)
26. Deng, F.G., Long, G.L.: Secure direct communication with a quantum one-time pad. Phys. Rev. A **69**, 052319 (2004)
27. Yin, X., Ma, W., Shen, D., Hao, C.: Efficient three-party quantum secure direct communication with EPR pairs. Quantum Inf. Sci. **3**, 1–5 (2013)
28. Zhang, C., Long, G.F.: Quantum secure direct dialogue using Einstein-Podolsky-Rosen pairs. Sci. Chin. Phys. Mech. Astron. **57**(7), 1238 (2014)
29. Man, Z.X., Xia, Y.J., An, N.B.: Quantum secure direct communication by using GHZ states and entanglement swapping. J. Phys. B **39**(18), 3855–3863 (2006)
30. Chang, Y., Zhang, S.B., Yan, L.L.: A bidirectional quantum secure direct communication protocol based on five-particle cluster state. Chin. Phys. Lett. **30**, 090301 (2013)
31. Cao, H.J., Song, H.S.: Quantum secure direct communication with W state. Chin. Phys. Lett. **23**, 290–292 (2006)
32. Lin, S., Wen, Q.Y., Gao, F., Zhu, F.C.: Quantum secure direct communication with χ-type entangled states. Phys. Rev. A **78**, 064304 (2008)
33. Wang, J., Zhang, Q., Tang, C.J.: Quantum secure direct communication based on order rearrangement of single photons. Phys. Lett. A **358**, 256–258 (2006)
34. Li, X.H., Deng, F.G., Li, C.Y., Liang, Y.J., Zhou, P., Zhou, H.Y.: Deterministic secure quantum communication without maximally entangled states. J. Korean Phys. Soc. **49**, 1354–1359 (2006)
35. Quan, D.X., Pei, C.X., Liu, D., Nan, Z.: One-way deterministic secure quantum communication protocol based on single photons. Acta. Phys. Sin. **59**, 2493–2497 (2010)
36. Chang, Y., Xu, C.X., Zhang, S.B., Yan, L.L.: Quantum secure direct communication and authentication protocol with single photons. Chin. Sci. Bull. **58**(36), 4571–4576 (2013)
37. Zhao, G.: Quantum secure communication protocol based on single-photon. Int. J. Sec. Appl. **9**(3), 267–274 (2015)

38. Xin, X., Hua, X., Song, J., Li, F.: Quantum authentication protocol for classical messages based on bell states and hash function. Int. J. Sec. Appl. **9**(7), 285–292 (2015)
39. Cai, Q.Y., Li, B.W.: Deterministic secure communication without using entanglement. Chin. Phys. Lett. **21**(4), 601 (2004)
40. Gao, T., Yan, F.L., Wang, Z.X.: Deterministic secure direct communication using GHZ states and swapping quantum entanglement. J. Phys. A Math. Gen. **38**(25), 5761 (2005)
41. Shaari, J.S., Lucamarini, M., Wahiddin, M.R.B.: Deterministic six states protocol for quantum communication. Phys. Lett. A **358**(2), 85–90 (2006)
42. Li, X., Deng, F.G., Li, C.Y., Liang, Y.J., Zhou, P., Zhou, H.Y.: Quantum secure direct communication without maximally entangled states. J. Korean Phys. Soc. **49**, 1354–1359 (2006)
43. Huang, W., Wen, Q.Y., Liu, B., Gao, F., Chen, H.: Deterministic secure quantum communication with collective detection using single photons. Int. J. Theor. Phys. **51**(9), 2787–2797 (2012)
44. Yan, C., Shi-Bin, Z., Li-Li, Y., Gui-Hua, H.: Controlled deterministic secure quantum communication protocol based on three-particle GHZ states in X-basis. Commun. Theor. Phys. **63**(3), 285–290 (2015)
45. IID Quantique. http://www.idquantique.com/
46. IID Quantique. http://www.nucrypt.net/
47. Nielsen, M.A., Chuang, I.L.: Quantum computation and quantum information, Cambridge University Press (2000)
48. Venegas-Andraca, S.E.: Quantum walks and quantum image processing. DPhil Thesis, The University of Oxford (2005)
49. Andress, J.: The Basics of Information Security: Understanding the Fundamentals of InfoSec in Theory and Practice, 2nd edn. Elsevier (2014)
50. Bouwmeester, D.: The Basics of Information Security: Understanding the Fundamentals of InfoSec in Theory and Practice, 2nd edn. Elsevier (2014)
51. Mayers, D.: Unconditional security in quantum cryptography. J. ACM (JACM) **48**(3), 351–406
52. Shor, P.W., Preskill, J.: Simple proof of security of the BB84 quantum key distribution protocol. Phys. Rev. Lett. **85**(2), 441–444 (2000)
53. Loepp, S., Wootters, W., Zurek, W.: Protecting information: from classical error correction to quantum cryptography. Cambridge University Press (2006)
54. Wootters, W., Zurek, W.: A single quantum cannot be cloned. Nature **299**, 802–803 (1982)
55. Dieks, D.: Communication by EPR devices. Phys. Lett. A **92**(6), 271–272 (1982)
56. Shamir, A.: How to share a secret. Commun. ACM **22**(11), 612–613 (1979)

Applications of Quantum Mechanics in Secure Communication

Mosayeb Naseri, Negin Fatahi, Ahmed Farouk, O. Tarawneh and M. Elhoseny

Abstract Over the last half century, the components of computers have become smaller by a factor of two every 18 months, a phenomenon known as Moore's law. In state-of-the-art computers, the smallest wires and transistors are approaching 100 nm feature size, which is approximately 1000x the diameter of an atom. Quantum mechanics is the theory of physics that describes the behavior of matter and energy in extreme conditions, such as short times and tiny distances. As transistors and wires become smaller and smaller, they inevitably begin to behave in intrinsically quantum mechanical ways. In this chapter it will be shown how it can be possible by using simple principles of quantum mechanics to reach a new field of communication science, named quantum communication. Also, the most recent development in quantum secure communication will be introduced and finally, the new method of secure dialogue between two agents (Alice, Bob), with the help of measurement concept in quantum mechanics will be presented.

Keywords Quantum effect · Entanglement · No cloning · Quantum cryptography · Quantum teleportation

---

M. Naseri (✉) · N. Fatahi
Department of Physics, Islamic Azad University,
Kermanshah Branch, Kermanshah, Iran
e-mail: Sepehr1976@yahoo.com

A. Farouk · M. Elhoseny
Faculty of Computer and Information Sciences, Mansoura University,
Mansoura, Egypt

A. Farouk · O. Tarawneh
Information Technology Department, Al-Zahra College for Women,
3365, Muscat, Oman

A.E. Hassanien et al. (eds.), *Quantum Computing: An Environment for Intelligent Large Scale Real Application*, Studies in Big Data 33,
https://doi.org/10.1007/978-3-319-63639-9_2

1 Introduction

Quantum mechanics is probably the most successful physical theory of this century. It pro-vides powerful tools which form one of the cornerstones of scientific progress, and which are indispensable for the understanding of omnipresent technical devices such as the transistor, semiconductor chips and the laser. The most important areas where these devices are used are modern communication and information- processing technologies. But quantum mechanics, until now, has only been used to construct these devices and quantum effects are absolutely avoided in the representation and manipulation of information. Rather than using single pho-tons, we still use strong light pulses to send information along optical high-speed connections, and we rely on electrical currents in semiconductor logic chips instead of applying single electrons as signal carriers.

Quantum Communication is the art of transferring a message from one place to another by using the quantum state as a message carrier, traditionally, the sender is named Alice and the receiver Bob. Quantum communication methods utilize fundamental properties of quantum mechanics to enhance the power and potential of today's communication systems. Quantum information processing and communication theory is a broad field, including quantum teleportation, quantum cryptography, quantum dense coding. By way of 2017, the improvement and growth of a real quantum computer is still in early stages but many poetical and theoretical experimentations were implemented by many research groups [1–22].

In this chapter there is a brief introduction, in Sect. 2 we present the necessary quantum mechanical back-grounds for investigation of quantum communication in their simplest forms and some pure quantum mechanical phenomena are discussed. Section 3 describes the fundamentals of quantum communication including the concepts of quantum teleportation and quantum cryptography. In Sect. 4 a brief history of research on quantum secure communication is presented and finally, in the last section we give a brief summary [23–38].

2 Quantum Mechanics

Quantum mechanics arose from the need to understand the thermal properties of radiation and the discrete spectral features of atoms. From this, the present understanding of the non-classical behavior of the fundamental units of matter and radiation was developed. Quantum theory has turned out to be the most universally successful theory of physics. From its start in atomic spectroscopy, it has developed to predict structures of molecules, nuclei, and even the large-scale structures of the universe. Much of our electronics industry today utilizes quantum phenomena in an essential manner. Without the understanding offered by quantum theory, our ability to build integrated circuits and communication devices would not have emerged. In these areas the basic theoretical progress took place in the middle of the twentieth century; the engineers who plan electronics devices need hardly worry about the problems still lingering on our interpretation of quantum theory.

2.1 States Space and Measurement

In quantum mechanics a physical state for example, a silver atom with a definite spin orientation is represented by state vector in a complex vector space. We call such a vector a ket and denote it by $|\alpha\rangle$, this state ket is postulated to contain complete information about system, everything we ask about the state is contained in the ket. Any ket $|\alpha\rangle$ can be written as [39],

$$|\alpha\rangle = \sum_{a'} c_{a'} |a'\rangle. \tag{1}$$

With $a', a'', \ldots$ up to a^N and $c_{a'}$ is a complex coefficient. In quantum mechanics each observable, such as momentum and spin components are represented by operators that act on kets.

When measurement is performed, the system is "thrown into" one of the eignestates, say $|a'\rangle$ of observable A [39],

$$|\alpha\rangle \rightarrow |a'\rangle, \tag{2}$$

we do not know in advance into which of the various states the system will be thrown as the result of measurement. we do postulate, however, that the probability for jumping into some particular state is given by;

$$P_{a'} = |\langle a'|\alpha\rangle|^2. \tag{3}$$

So quantum physics establishes a set of negative rules stating things that cannot be done.

(1) One cannot take a measurement without perturbing the system.
(2) One cannot draw pictures of individual quantum processes.
(3) One cannot duplicate an unknown quantum state.

This negative viewpoint of quantum physics, due to its contrast with classical physics, has recently been turned positive, and quantum information is one of the best illustrations of this psychological revolution. We present two novel and typical quantum computation phenomena. It is useful to encounter them early in the study of quantum computation.

2.2 Composite Quantum Systems and Entanglement, Hidden Variables and Bell Inequalities

Consider a two-electron system in a state;

$$|\psi^{\pm}\rangle_{12} = \frac{1}{\sqrt{2}}(|0\rangle_1|1\rangle_2 \pm |1\rangle_1|0\rangle_2) \tag{4}$$

Suppose we make a measurement on the spin component of one of the electrons, clearly, there is a 50–50% chance of getting either spin-up or spin-down, because the composite system may be in $|0\rangle_1|1\rangle_2$ or $|1\rangle_1|0\rangle_2$ with equal probabilities. But if one of the components is shown to be in the spin-up state, the other is necessarily in the spin-down state, and vice versa. In other words, when the spin component of electron 1 is shown to be up, the measure-ment apparatus has selected the first term $|0\rangle_1|1\rangle_2$ of (4). It is remarkable that this kind of correlation can persist even if the two particles are well separated and have ceased to interact provided that they fly apart!. The above states together with;

$$|\psi^{\pm}\rangle_{12} = \frac{1}{\sqrt{2}}(|0\rangle_1|0\rangle_2 \pm |1\rangle_1|1\rangle_2) \tag{5}$$

are called Entangled states or EPR states [39–41].

Some have argued that the difficulties encountered here are inherent in the probabilistic interpretation of quantum mechanics and that the dynamic behavior at the microscopic level appears probabilistic only because some yet unknown parameters, so-called hidden variables, have not been specified.

In 1964, John Bell proposed a mechanism to test for the existence of these hidden variables, and he developed his inequality principle as the basis for such a test. He showed that if the inequality was ever not satisfied, then there were no such hidden variables [42].

2.3 No Cloning

It can be proved that it is impossible to copy an unknown quantum state perfectly. This feature is a direct consequence of the linearity of the Schrodinger equation. This pure quantum effect is known as “No Cloning” theorem [43].

3 Using Quantum Effects in Secure Communication

The goal of quantum communication is to transmit an unknown quantum state from one person to another one at a distant location. This can be obtained either by direct transmission of the state [44], or by disembodied transport, i.e., quantum teleportation [45]. Here we briefly introduce two pillars of quantum communication science, quantum teleportation and quantum cryptography.

3.1 Quantum Teleportation

Quantum teleportation is a process that enables the transmission of an unknown quantum state via a previously shared EPR pair with the help of only two classical bits transmitted on a classical channel. The No-cloning theorem forbids a perfect copy of an arbitrary unknown quantum state.

Suppose Alice and a remote Bob share an EPR pair in the state;

$$|\phi^+\rangle_{12} = \frac{1}{\sqrt{2}}(|0\rangle_1|0\rangle_2 \pm |1\rangle_1|1\rangle_2) \tag{6}$$

and she has a qubit that she want to send to Bob. Suppose that the state of the qubit is;

$$|\psi\rangle = \alpha|0\rangle + \beta|1\rangle \tag{7}$$

So in the beginning of teleportation, Alice has the following state of the above diagram [43]:

$$\begin{aligned} |\psi\rangle|\phi^+\rangle &= \alpha|0\rangle + \beta|1\rangle\frac{1}{\sqrt{2}}(|0\rangle_1|0\rangle_2 + |1\rangle_1|1\rangle_2) \\ &= \frac{1}{\sqrt{2}}(\alpha|0\rangle 0\rangle 0\rangle + \alpha|0\rangle 0\rangle 1\rangle + \beta|1\rangle 1\rangle 1\rangle) \end{aligned} \tag{8}$$

Suppose that Alice applies a joint measurement on her two qubits, she first applies a CNOT quantum gate which transforms the state to:

$$\frac{1}{\sqrt{2}}(\alpha|0\rangle 0\rangle 0\rangle + \alpha|0\rangle 1\rangle 1\rangle + \beta|1\rangle 1\rangle 0\rangle + \beta|1\rangle 0\rangle 1\rangle) \tag{9}$$

where the first qubit is a control qubit and the second one is a target qubit. Next, the Hadamard transform is applied, which transforms the state to;

$$
\begin{aligned}
&\frac{1}{2}(\alpha|0\rangle 0\rangle 0\rangle + \alpha|1\rangle 0\rangle 0\rangle + \alpha|0\rangle 1\rangle 1\rangle + \alpha|1\rangle 1\rangle 1\rangle) \\
&\quad + \frac{1}{2}(\beta|0\rangle 1\rangle 0\rangle + \beta|1\rangle 0\rangle 0\rangle + \beta|1\rangle 1\rangle 0\rangle + \beta|1\rangle 1\rangle 1\rangle) \\
&= \frac{1}{2}|0\rangle|0\rangle(\alpha|0\rangle + \beta|1\rangle) + \frac{1}{2}|0\rangle|1\rangle(\alpha|1\rangle + \beta|0\rangle) \\
&\quad + \frac{1}{2}|1\rangle|0\rangle(\alpha|0\rangle - \beta|1\rangle) + \frac{1}{2}|1\rangle|1\rangle(\alpha|1\rangle - \beta|0\rangle)
\end{aligned} \tag{10}
$$

A measurement by Alice of her particles produces two classical bits. These bits specify one of four possible results (0, 0), (0, 1), (1, 0), (1, 1). So if Alice sends these two bits to Bob over a classical channel, this allows him to choose one of the following rotations to apply to his particle and transform it into the initial transformed state, so the minimal resources required for faithful teleportation are one EPR pair.

3.2 *Quantum Cryptography*

It is clear that traditional cryptosystems are breakable given enough computing time, quantum cryptography offers the promise of unconditional security without face-to-face exchanges. Rather than relying on problems believed to be computationally "difficult," quantum cryptography uses basic physical laws to provide provable unconditional security.

BB84 Quantum Cryptography Protocol

The first quantum cryptographic communication protocol, called BB84, was invented in 1984 by Bennett and Brassard [46]. This protocol has been experimentally demonstrated to work for a transmission over 30 km of fiber optic cable [47, 48]. In this section we describe the BB84 protocol. The basic tools are a quantum channel connecting Alice and Bob and a public classical channel, where Eve is allowed to listen passively, but not allowed to change the transmitted message. For the quantum channel, we use four signal states. For simplicity, let us for now regard the signals as realized by single photons in the polarization degree of freedom. Consider two sets of orthogonal signals, one formed by a horizontal and a vertical polarized photon, and the other formed by a 45° and a 135° polarized photon. These four polarized states are non-orthogonal. The overlap probability between signals from two different sets is one half. Bob has two measurement devices at his hand, one in the rectilinear (i.e., vertical/horizontal) basis and one in the diagonal (i.e., 45°/135°) basis. Notice that Bob's two measurements do not commute. To assure the detection of Eve's eavesdropping, Bennett and Brassard require Alice and Bob to communicate in two phases; at the first phase Alice and Bob are communicating over a one-way quantum channel as follows: (a) Alice sends a sequence of signals, each randomly chosen from one of the above four polarizations. (b) For each signal, Bob randomly chooses one of the two measurements, rectilinear or diagonal basis devices to perform a measurement. (c) Bob confirms that he has received and measured all signals. In the first phase, Alice is required, each time she transmits a single bit, to use randomly with equal probability one of the two orthogonal alphabets; it follows from the Heisenberg uncertainty principle that no one, not even Bob or Eve, can receive Alice's transmission with an accuracy greater than 75% [49].

For each bit transmitted by Alice, we assume that Eve performs one of two actions, opaque eavesdropping with probability λ, or no eavesdropping with probability $1 - \lambda$. Thus, Eve is eavesdropping on each transmitted bit or Eve is not eavesdropping at all. Because Bob's and Eve's choice of measurement operators are stochastically independent of each other and of Alice's choice of alphabet, Eve's eavesdropping has an immediate and detectable impact on Bob's received bits. Eve's eavesdropping causes Bob's error rate to jump from 25% to;

$$\frac{1}{4} \cdot (1 - \lambda) + \frac{1}{2} \cdot \frac{1}{4} \cdot \lambda = \frac{1}{4} + \frac{\lambda}{8} \tag{11}$$

Thus, if Eve eavesdrops on every bit, i.e., if $\lambda = 1$, then Bob's error rate jumps from 25 to 37.5%, a 50% increase.

Phase 2 is dedicated to eliminating the bit locations (and hence the bits at these locations) at which error could have occurred without Eve's eavesdropping. Bob begins by publicly communicating to Alice which measurement operators he used for each of the received bits. Alice then in turn publicly communicates to Bob which of his measurement operator choices were correct. After this two way communication, Alice and Bob delete the bits corresponding to the incompatible measurement choices to produce shorter sequences of bits which we call respectively Alice's raw key and Bob's raw key. If there is no intrusion, then Alice's and Bob's raw keys will be in total agreement. However, if Eve has been at work, then corresponding bits of Alice's and Bob's raw keys will not agree with the probability;

$$0 \cdot (1 - \lambda) + \frac{1}{4} \cdot \lambda = \frac{\lambda}{4} \tag{12}$$

In the absence of noise, any discrepancy between Alice's and Bob's raw keys is proof of Eve's intrusion. So to detect Eve, Alice and Bob select a publicly agreed upon random subset of m bit locations in the raw key, and publicly compare corresponding bits, making sure to discard from the raw key, each bit as it is revealed. The other technique is to select a publicly agreed upon random subset of n bit locations in the canceled bits, making sure that these bits will violate a Bell inequality. The amount by which a Bell inequality is violated is thus an ideal measure of the security of the key.

4 Secure Quantum Communication

Since BB84 quantum key distribution has developed quickly. Quantum communication holds secret promise for transmission of secure message via quantum cryptography, distribution of quantum information, and distributing protocol and quantum teleportation. Many attempts have been made to apply these methods in design-ing communication protocols [50]. In 1999, Shimizo and Imoto proposed a DSQC protocol using entangled photon pairs [51]. In their scheme the ciphertext is encoded in the state of the entangled pairs, and they are transmitted from Alice to Bob. Bob performs a Bell-basis measurement to read out the partial information. Full information of the ciphertext is read out after Alice notifies him of the encoding basis through a classical communication. Beige et al. proposed a Deterministic Secure Quantum Communication (DSQC) scheme based on a single photon two-qubit state in 2002; in this scheme, the message can be read out only after transmission of additional classical information for each qubit [52]. In 2002, Bostrom and Felbinger presented a scheme for quasi-secure direct communication with EPR pairs [53], this scheme was based on quantum dense coding and the protocol called the ping-pong scheme [54]. Also in 2002, Long and Liu proposed a two-step highly efficient QKD protocol [55]. In 2003, the formal procedure to use protocol for quantum secure direct communication (QSDC) was given [56]. Today there are many people in the world studying the subject.

5 Quantum Key Distribution in Satellite Communication

It is clear that traditional cryptosystems are breakable given enough computing time, quantum cryptography offers the promise of unconditional security without face-to-face exchanges. Rather than relying on problems believed to be computationally "difficult," quantum cryptography uses basic quantum physics laws to provide provable unconditional security. The main principles which are used for quantum cryptography are the following:

(a) It is not possible to determine simultaneously the position and the momentum of a particle with arbitrary high accuracy (Heisenbergs uncertainty principle).
(b) It is not possible to measure the polarization of a photon in the vertical-horizontal basis and simultaneously in the diagonal basis.
(c) Each measurement of the quantum state perturbs the quantum state.
(d) It is not possible to copy quantum states (No-cloning-theorem).

5.1 BB84 Quantum Cryptography Protocol

The first quantum cryptographic communication protocol, called BB84, was invented in 1984 by Bennett and Brassard [46]. This protocol has been experimentally demonstrated to work for a transmission over 30 km of fiber optic cable [47, 48]. In this section we describe the BB84 protocol. The basic tools are a quantum channel connecting Alice and Bob and a pub-lic classical channel, where Eve is allowed to listen passively, but not allowed to change the transmitted message. For the quantum channel, we use four signal states. For simplicity, let us for now regard the signals as realized by single photons in the polarization degree of freedom. Consider two sets of orthogonal signals, one formed by a horizontal and a vertical polarized photon, and the other formed by a 45° and 135° polarized photon. These four polarized states are non-orthogonal. The overlap probability between signals from two differ-ent sets is one half. Bob has two measurement devices at his hand, one in the rectilinear (i.e., vertical/horizontal) basis and one in the diagonal (i.e., 45°/135°) basis. Notice that Bob's two measurements do not commute. To assure the detection of Eve's eavesdrop-ping, Bennett and Brassard require Alice and Bob to communicate in two phases; in the first phase Alice and Bob communicate over a one-way quantum channel as follows: (a) Alice sends a sequence of signals, each randomly chosen from one of the above four polarizations. (b) For each signal, Bob randomly chooses one of the two measurements, rectilinear or di-agonal basis devices to perform a measurement. (c) Bob confirms that he has received and measured all signals. In the first phase, Alice is required, each time she transmits a single bit, to use randomly with equal probability one of the two orthogonal alphabets it follows from Heisenberg's uncertainty principle that no one, not even Bob or Eve, can receive Alice's transmission with an accuracy greater than 75% [49].

For each bit transmitted by Alice, we assume that Eve performs one of two actions, opaque eavesdropping with probability λ, or no eavesdropping with probability $1 - \lambda$. Thus, Eve is eavesdropping on each transmitted bit or Eve is not eavesdropping at all. Because Bob's and Eve's choice of measurement operators are stochastically independent of each other and of Alice's choice of alphabet, Eve's eavesdropping has an immediate and detectable impact on Bob's received bits. Eve's eavesdropping causes Bob's error rate to jump from 25% to;

$$\frac{1}{4}\cdot(1-\lambda)+\frac{1}{2}\cdot\frac{1}{4}\cdot\lambda=\frac{1}{4}+\frac{\lambda}{8} \tag{13}$$

Thus, if Eve eavesdrops on every bit, i.e., if $\lambda = 1$, then Bob's error rate jumps from 25 to 37.5%, a 50% increase. So to intercept and gain information on the key, an eavesdropper must make measurements on some or all of the sent pulses. An eavesdropper can intercept, measure and resend every pulse but has to guess the random basis. This results in a 25% error rate in the key established between sender and receiver. The sender and receiver can monitor for eavesdropping by monitoring the error rate of their system. Any increase of the error rate above a threshold value can be interpreted as an insecure line.

Phase 2 is dedicated to eliminating the bit locations (and hence the bits at these locations) at which error could have occurred without Eve's eavesdropping. Bob begins by publicly communicating to Alice which measurement operators he used for each of the received bits. Alice then in turn publicly communicates to Bob which of his measurement operator choices were correct. After this two way communication, Alice and Bob delete the bits corresponding to the incompatible measurement choices to produce shorter sequences of bits which we call respectively Alice's raw key and Bob's raw key. If there is no intrusion, then Alice's and Bob's raw keys will be in total agreement. However, if Eve has been at work, then corresponding bits of Alice's and Bob's raw keys will not agree with probability;

$$0 \cdot (1-\lambda) + \frac{1}{4} \cdot \lambda = \frac{\lambda}{4} \tag{14}$$

In the absence of noise, any discrepancy between Alice's and Bob's raw keys is proof of Eve's intrusion. So to detect Eve, Alice and Bob select a publicly agreed upon random subset of m bit locations in the raw key, and publicly compare corresponding bits, making sure to discard from the raw key each bit as it is revealed. The other technique is to select a publicly agreed upon random subset of n bit locations in the cancelled bits, making sure that these bits will violate a Bell inequality. The amount by which a Bell inequality is violated is thus an ideal measure of the security of the key.

5.2 Entangled Photon Based Quantum Cryptography Protocol

Einstein, Podolsky, and Rosen (EPR) in their famous 1935 paper [42] challenged the foundations of quantum mechanics by pointing out a "paradox." There exist spatially separated pairs of particles, henceforth called EPR pairs, whose states are correlated in such a way that the measurement of a chosen observable A of one, automatically determines the result of the measurement of A of the other. Since EPR pairs can be pairs of particles separated at great distances, this leads to what appears to be a paradoxical "action at a distance."

For example, it is possible to create a pair of photons (each of which we label below with the subscripts 1 and 2, respectively) with correlated linear polarizations. An example of such an entangled state is given by;

$$|\Omega_0\rangle = \frac{1}{\sqrt{2}}\left(|0\rangle_1\left|\frac{\pi}{2}\right\rangle_2 + \left|\frac{\pi}{2}\right\rangle_1|0\rangle_2\right) \tag{15}$$

where state $|0\rangle$ is a vertical linear polarization photon and state $|1\rangle$ is a horizontal linear polarization photon [42].

Einstein et al. [42] then state that such quantum correlation phenomena could be a strong indication that quantum mechanics is incomplete, and that there exist "hidden variables," inaccessible to experiments, which explain such "action at a distance".

In 1964, Bell [45] gave a means for actually testing for locally hidden variable (LHV) theories. He proved that all such LHV theories must satisfy the Bell inequality. Quantum mechanics has been shown to violate the inequality.

In 1991, Ekert has devised a quantum protocol based on the properties of quantum-correlated particles [43].

The EPR quantum protocol is a 3-state protocol that uses Bell's inequality to detect the presence or absence of Eve as a hidden variable. Consider three possible polarization states of EPR pair,

$$|\Omega_0\rangle = \frac{1}{\sqrt{2}}\left(|0\rangle_1\left|\frac{3\pi}{6}\right\rangle_2 + \left|\frac{3\pi}{6}\right\rangle_1|0\rangle_2\right) \tag{16}$$

$$|\Omega_1\rangle = \frac{1}{\sqrt{2}}\left(\left|\frac{\pi}{6}\right\rangle_1\left|\frac{4\pi}{6}\right\rangle_2 + \left|\frac{4\pi}{6}\right\rangle_1\left|\frac{3\pi}{6}\right\rangle_2\right) \tag{17}$$

$$|\Omega_2\rangle = \frac{1}{\sqrt{2}}\left(\left|\frac{2\pi}{6}\right\rangle_1\left|\frac{5\pi}{6}\right\rangle_2 + \left|\frac{5\pi}{6}\right\rangle_1\left|\frac{2\pi}{6}\right\rangle_2\right) \tag{18}$$

For each of these states, we choose the following corresponding operators M_0, M_1, and M_2, given as the following:

$$M_0 = |0\rangle\langle 0|, \; M_1 = \left|\frac{\pi}{6}\right\rangle\left\langle\frac{\pi}{6}\right|, \; M_2 = \left|\frac{2\pi}{6}\right\rangle\left\langle\frac{2\pi}{6}\right|.$$

there are two stages to the EPR protocol, the first stage over a quantum channel, the second over a public channel.

In the first stage for each time slot, a state $|\Omega_j\rangle$ is randomly selected with equal probability from the set of states $\{|\Omega_1\rangle, |\Omega_2\rangle, |\Omega_3\rangle\}$.

One photon of this EPR pair is sent to Alice, the other to Bob. Alice and Bob at random with equal probability separately and independently select one of the three measurement operators M_0, M_1, and M_2, and accordingly measure their respective photons. Alice records her measured bit. On the other hand, Bob records the complement of his measured bit. This procedure is repeated for as many time slots as needed.

In stage 2, Alice and Bob communicate over a public channel. At first Alice and Bob carry on a discussion over a public channel to determine those bit slots at which they used the same measurement operators. They each then separate their respective bit sequences into two subsequences. One subsequence, called the raw key, consists of those bit slots at which they used the same measurement operators. The other subsequence, called the rejected key, consists of all the remaining bit slots. Now Alice and Bob now carry on a discussion over a public channel comparing their respective rejected keys to determine whether or not Bell's inequality is satisfied. If it is, Eve's presence is detected. If not, then Eve is absent.

Finally in the presence of noise, the remaining phase of the EPR protocol is the reconciliation phase [44].

6 Free-Space Quantum Cryptography

There are two applications which require free-space quantum cryptography rather than fibre based one. The first is short distance communication up to several kilometers, mainly in urban areas, where a fibre based connection is too expensive to deploy. The second is secure satellite communication, where a fibre link is not possible.

The success of QKD over free-space optical paths depends on the transmission and de-tection of single photons against a high background through a turbulent medium. Although this problem is difficult, a combination of sub-nanosecond timing, narrow filters [45, 57], spatial filtering [46] and adaptive optics [47] can render the transmission and detection problems tractable. Furthermore, the essentially non-birefringent nature of the atmosphere at optical wavelengths allows the faithful transmission of the single-photon polarization states used in the free-space QKD protocol.

From 1991, when the free-space QKD was first introduced over an optical path of about 30 cm several demonstrations (indoor optical paths of 205 m and outdoor optical paths of 75 m) increased the utility of QKD by extending it to line-of-site laser communications systems. There are certain key distribution problems in this category for which free-space QKD would have definite practical advantages (as for example, it is impractical to send a courier to a satellite). In 1998 a research group at Los Alamos National Laboratory, New Mexico, USA developed a free-space QKD over outdoor optical paths of up to 950 m under night-time conditions [48]. Four years later, in 2002 the same laboratory have demonstrated that free-space QKD is possible in daylight or at night, protected against intercept/resend, beamsplitting and unambiguous state discrimination (USD) eavesdropping, and even photon number splitting (PNS) eavesdropping at night, over a 10 km, 1-airmass path, which is representative of poten-tial ground-to-ground applications and is several times longer than any previously reported results [49].

7 Free-Space Quantum Cryptography in Satellite Communication

With the exponential expansion of electronic commerce the need for global protection of data is paramount. Conventional key exchange methods generally utilize public key methods and rely on computational complexity as proof against tampering and eavesdropping. Satellite systems thus require future-proofing against the rapid improvements in computational power that may occur during their operational lifetime (many years). In this chapter we examine how can we use the free-space quantum channel in the future years of satellite telecommunication.

The quantum computing algorithms can be used to affirm our communication in the following four ways [58, 59]:

(1) Open-air communication (horizontal telecommunication, below 100 km, instead of optical cable, using the twisted surface of Earth).
(2) Satellite communications (between 300 and 800 km altitude, signal encoding and decoding). Quantum error correction allows quantum computation in a noisy environment. Quantum computation of any length can be created as accurately as desired, as long as the noise is below a certain threshold, e.g. $P < 10^{-4}$.
(3) Satellite broadcast (our broadcast satellite orbit at 36,000 km, using 27 MHz signal) [50]. In quadrature phase shift keying (QPSK) every symbol contains two bits, this is why the bit speed is 55 Mbs. Half the bits is for error-coding, in the best case we only have 38 Mbs, but in common solutions there is only 27–28 Mbs, in which 5–6 TV-channels can be stored with a bandwidth of 2–5 Mbs each. The quantum algorithms can prove the effective bandwidth to better fill the brand as in the traditional case.
(4) Satellite-satellite communication (between broadcast or other satellites, using free-space, for signal coding and encoding, super density coding etc.).

So by placing a source of single photons and entangled photons on satellites we can propose a satellite based global key exchange system for key exchange between any two arbitrary points on the globe. This system would work by first exchanging keys between one ground station and the satellite. The satellite would then have to store the key securely until the second ground station came into view (up to several hours later). Exchanging the key with this second ground station would allow the first key to be sent down using an absolutely secure one-time-pad encoding scheme. The global reach of the system may be what drives the development but it will probably cost well in excess of ten million Euros (dollars) to build and fly [51].

8 Conclusions

In this chapter we survey some results in quantum communication. A brief introduction to some principles of quantum mechanics that are essential to understanding quantum communication is presented, and the main connections between quantum mechanics and secure information transfer have been discussed. It has been seen that the laws of quantum physics guarantee the security of sharing keys between two parties based on quantum cryptography, and provide a mechanism by which any attempt at eavesdropping can be detected immediately.

So, although quantum cryptography is not so practical right now, it is still worthy of study for several reasons. Unlike public-key cryptosystems, currently it works only over short distances. Also, with sufficient technical improvements, it might be possible in the future to implement quantum cryptography over long distances.

In this chapter, some results of the application of quantum cryptography in satellite communication has been presented. A brief introduction to quantum cryptography that is essential to understanding quantum satellite communication is presented. Although quantum cryptography is not so practical right now, it is still worthy of study for several reasons. Unlike public-key cryptosystems, currently it works only over short distances. Also, with sufficient technical improvements, it might be possible in the future to implement quantum cryptography over long distances. So, it will be possible by placing a source of single photons and entangled photons on satellites, to design global secure quantum communication networks.

References

1. Metwaly, A., Rashad, M.Z., Omara, F.A., Megahed, A.A.: Architecture of point to multipoint QKD communication systems (QKDP2MP). In: 8th International Conference on Informatics and Systems (INFOS), Cairo, pp. NW 25–31. IEEE (2012)
2. Farouk, A., Omara, F., Zakria, M., Megahed, A.: Secured IPsec multicast architecture based on quantum key distribution. In: The International Conference on Electrical and Bio-medical Engineering, Clean Energy and Green Computing, pp. 38–47. The Society of Digital Information and Wireless Communication (2015)
3. Farouk, A., Zakaria, M., Megahed, A., Omara, F.A.: A generalized architecture of quantum secure direct communication for N disjointed users with authentication. Sci. Rep. **5**, 16080–16080 (2014)
4. Wang, M.M., Wang, W., Chen, J.G., Farouk, A.: Secret sharing of a known arbitrary quantum state with noisy environment. Quantum Inf. Process. **14**(11), 4211–4224 (2015)
5. Naseri, M., Heidari, S., Batle, J., Baghfalaki, M., Gheibi, R., Farouk, A., Habibi, A.: A new secure quantum watermarking scheme. Optik-Int. J. Light Electron Opt. **139**, 77–86 (2017)
6. Batle, J., Ciftja, O., Naseri, M., Ghoranneviss, M., Farouk, A., Elhoseny, M.: Equilibrium and uniform charge distribution of a classical two-dimensional system of point charges with hard-wall confinement. Phys. Scr. **92**(5), 055801 (2017)

7. Geurdes, H., Nagata, K., Nakamura, T., Farouk, A.: A note on the possibility of incomplete theory (2017). arXiv:1704.00005
8. Batle, J., Farouk, A., Alkhambashi, M., Abdalla, S.: Multipartite correlation degradation in amplitude-damping quantum channels. J. Korean Phys. Soc. **70**(7), 666–672 (2017)
9. Batle, J., Naseri, M., Ghoranneviss, M., Farouk, A., Alkhambashi, M., Elhoseny, M.: Shareability of correlations in multiqubit states: Optimization of nonlocal monogamy inequalities. Phys. Rev. A **95**(3), 032123 (2017)
10. Batle, J., Farouk, A., Alkhambashi, M., Abdalla, S.: Entanglement in the linear-chain Heisenberg antiferromagnet Cu (C 4 H 4 N 2) (NO 3) 2. Eur. Phys. J. B **90**, 1–5 (2017)
11. Batle, J., Alkhambashi, M., Farouk, A., Naseri, M., Ghoranneviss, M.: Multipartite non-locality and entanglement signatures of a field-induced quantum phase transition. Eur. Phys. J. B **90**(2), 31 (2017)
12. Nagata, K., Nakamura, T., Batle, J., Abdalla, S., Farouk, A.: Boolean approach to dichotomic quantum measurement theories. J. Korean Phys. Soc. **70**(3), 229–235 (2017)
13. Abdolmaleky, M., Naseri, M., Batle, J., Farouk, A., Gong, L.H.: Red-Green-Blue multi-channel quantum representation of digital images. Optik-Int. J. Light Electron Opt. **128**, 121–132 (2017)
14. Farouk, A., Elhoseny, M., Batle, J., Naseri, M., Hassanien, A.E.: A proposed architecture for key management schema in centralized quantum network. In: Handbook of Research on Machine Learning Innovations and Trends, pp. 997–1021. IGI Global (2017)
15. Zhou, N.R., Li, J.F., Yu, Z.B., Gong, L.H., Farouk, A.: New quantum dialogue protocol based on continuous-variable two-mode squeezed vacuum states. Quantum Inf. Process. **16**(1), 4 (2017)
16. Batle, J., Abutalib, M., Abdalla, S., Farouk, A.: Persistence of quantum correlations in a XY spin-chain environment. Eur. Phys. J. B **89**(11), 247 (2016)
17. Batle, J., Abutalib, M., Abdalla, S., Farouk, A.: Revival of Bell nonlocality across a quantum spin chain. Int. J. Quantum Inf. **14**(07), 1650037 (2016)
18. Batle, J., Ooi, C.R., Farouk, A., Abutalib, M., Abdalla, S.: Do multipartite correlations speed up adiabatic quantum computation or quantum annealing? Quantum Inf. Process. **15**(8), 3081–3099 (2016)
19. Batle, J., Bagdasaryan, A., Farouk, A., Abutalib, M., Abdalla, S.: Quantum correlations in two coupled superconducting charge qubits. Int. J. Mod. Phys. B **30**(19), 1650123 (2016)
20. Batle, J., Ooi, C.R., Abutalib, M., Farouk, A., Abdalla, S.: Quantum information approach to the azurite mineral frustrated quantum magnet. Quantum Inf. Process. **15**(7), 2839–2850 (2016)
21. Batle, J., Ooi, C.R., Farouk, A., Abdalla, S.: Nonlocality in pure and mixed n-qubit X states. Quantum Inf. Process. **15**(4), 1553–1567 (2016)
22. Metwaly, A.F., Rashad, M.Z., Omara, F.A., Megahed, A.A.: Architecture of Multicast Network Based on Quantum Secret Sharing and Measurement (2015)
23. Bennett, C.H., Brassard, G.: Quantum cryptography: public key distribution and coin tossing. In: Proceedings of the International Conference on Computer Systems and Signal Processing, vol. 175, Bangalore, 1984
24. Ekert, A.: Phys. Rev. Lett. **67**, 661 (1991)
25. Shor, P.W., Preskill, J.: Phys. Rev. Lett. **85**, 441 (2000)
26. Lutkenhaus, N.: Phys. Rev. A **61**, 052304 (2000)
27. Einstein, A., Podolsky, B., Rosen, N.: Can quantum, mechanical description of physical reality be considered complete? Phys. Rev. **47**, 777 (1935)
28. Bohm, D.: Quantum Theory. Prentice-Hall, Englewood Cliffs (1951)
29. Ekert, A.K.: Quantum cryptography based on Bell's theorem. Phys. Rev. Lett. **67**(6), 661–663 (1991)
30. Walker, J.G., Seward, S.F., Rarity, J.G., Tapster, P.R.: Quantum Opt. **1**,75–82 (1989)
31. Seward, S.F., Tapster, P.R., Walker, J.G., Rarity, J.G.: Quantum Opt. **3**, 201–207 (1991)
32. Buttler, W.T., et al.: Phys. Rev. A **57**, 2379–2382 (1998)
33. Primmerman, C.A., et al.: Nature **353**, 141–143 (1991)

34. Bennett, C.H., DiVincenzo, D.P., Smolin, J.A.: Capacities of quantum erasure channels. arXiv:quant-ph/9701015
35. Buttler, W.T., Hughes, R.J., Kwiat, P.G., Lamoreaux, S.K., Luther, G.G., Morgan, G.L., Nordholt, J.E., Peterson, C.G., Simmons, C.M.: Practical free-space quantum key distribution over 1 km. arXiv:quant-ph/9805071
36. Bacsardi, L.: Using quantum computing algorithms in future satellite communication. Acta Astronautica **57**(28), 224229 (2005)
37. Bacsardi, L.: Satellite communication over quantum channel. Acta Astronautica **61**, 151–159 (2007). Gschwindt, A.: Satellite broadcast, in Hungarian, Muszaki Konyvkiado, Budapest (1997)
38. Rarity, J.G., Tapster, P.R., Gorman, P.M., Knight, P.: Ground to satellite secure key exchange using quantum cryptography. New J. Phys. **4**, 82 (2002)
39. Sakurai, J.J.: Modern Quantum Mechanics. Addison-Wesley Publication Company (1985)
40. Nielson Michael, A., Chuang Hsaac, L.: Quantum Information and Computation. Cambridge University Press (2000)
41. Deng, F.G., Li, X.H., Li, C.Y., Zhou, P., Zhou, H.Y.: Phys. Rev. A **72**, 044301 (2005)
42. Bell, J.S.: Physics **1**, 195–200 (1964)
43. Johns, W.: Quantum Teleportation; Deutsch's Algorithm. Lecture Notes. University of Calgary (2006)
44. Cirac, J.I. et al.: Phys. Rev. Lett. **78**, 3221 (1997); Van Enk, S.J., Cirac, J.I., Zoller, P.: Science **279**, 205 (1998)
45. Bennett, C.H., et al.: Phys. Rev. Lett. **70**, 1895 (1993)
46. Bennett, Charles H., Gilles, B.: Quantum cryptography: public key distribution and coin tossing. In: International Conference on Computers, Systems and Signal Processing, Bangalore, India, pp. 175–179, December 10–12, 1984
47. Phoenix, S.J., Townsend, P.D.: Quantum cryptography: how to beat the code breakers using quantum mechanics. Contemp. Phys. **36**(3), 165–195 (1995)
48. Townsend, P.D.: Secure key distribution system based on quantum cryptography. Electron Lett. **30**(10), 809–811 (1994)
49. Lomonaco, Jr. S.J.: A Quick Glance at Quantum Cryptography. arXiv:quant-ph/9811056
50. Gui-In, L., et al.: Front. Phys. China **2**(3): 252–273 (2007)
51. Shimizu, K., Imoto, N.: Phys. Rev. A **60**, 157 (1999)
52. Beige, A., Englert, B.G., Kurtsiefer, C.: Acta Phys. Pol. **101**(3), 357 (2002)
53. Bostrom, K., Felbinger, T.: Phys. Rev. Lett. **89**, 187902 (2002)
54. Wojcik, A.: Phys. Rev. Lett. **90**, 157901 (2003)
55. Long, G.l., Liv, X.S.: Phys. Rev. A **65**, 032302 (2002)
56. Deng, F.G., Long, G.l., Liv, X.S.: Phys. Rev. A **68**, 042317 (2003)
57. Shannon, C.E., Weaver, W.: A more complete analysis of the communication problem can be found. In: The Mathematical Theory of Communication. University of Illinois Press, Chicago (1963)
58. Fu-Guo, D., Xi-Han, L., Chun-Yan, L., Ping, Z., Hong-Yu, Z.: Phys. Scr. **76**, 25–30 (2007)
59. Deng, F.G., Long, G.L., Liu, X.S.: Phys. Rev. A **68**, 042317 (2003)

Different Architectures of Quantum Key Distribution Network

Ahmed Farouk, O. Tarawneh, Mohamed Elhoseny, J. Batle, Mosayeb Naseri, Aboul Ella Hassanien and Muzaffar Lone

Abstract Most existing realizations of quantum key distribution (QKD) are point-to-point systems with one source transferring to only one destination. Growth of these single-receiver systems has now achieved a reasonably sophisticated point. However, many communication systems operate in a point-to-multi-point (Multicast) configuration rather than in point-to-point mode, so it is crucial to demonstrate compatibility with this type of network in order to maximize the application range for QKD. The researchers have proposed several approaches for Quantum Key Distribution Network. In this chapter we will discuss these various architectures.

Keywords Quantum key distribution · Quantum cryptography · DARPA network · Deterministic secure direct communication

---

A. Farouk (✉) · M. Elhoseny
Faculty of Computer and Information Sciences, Mansoura University, Mansoura, Egypt
e-mail: dr.ahmedfarouk85@yahoo.com

A. Farouk
University of Science and Technology, Zewail City of Science and Technology, Giza, Egypt

O. Tarawneh
Information Technology Department, Al-Zahra College for Women, 3365, Muscat, Oman

J. Batle
Departament de Física, Universitat de Les Illes Balears, 07122 Palma de Mallorca, Balearic Islands, Spain

M. Naseri
Department of Physics, Kermanshah Branch, Islamic Azad University, Kermanshah, Iran

A.E. Hassanien
Faculty of Computers and Information, Cairo University, Giza, Egypt

M. Lone
Department of Physics, University of Kashmir, Srinagar 190006, India

A.E. Hassanien et al. (eds.), *Quantum Computing: An Environment for Intelligent Large Scale Real Application*, Studies in Big Data 33,
https://doi.org/10.1007/978-3-319-63639-9_3

1 Secure Communication Based on Quantum Cryptography (SECOQC) Network

The majority of the current applications of Quantum cryptography are point-to-point approaches with one source transfers messages to only one destination which delivers a secured key. There are number of approaches and prototypes for the exploitation of *QKD* to secure communications. In 2003, the European project (SECOQC) is started by effort of 41 research and modern groups from the European Union, Switzerland and Russia. SECOQC aspires to improve the point-to-point Quantum cryptography as long distance key distribution. Symmetric Cryptography is applied when the two communicating parties share a key before any encryption and decryption are done. The secret key should be circulated before the transmission between communicated parties [1–11]. The architecture of SECOQC *QKD* network and its components are shown in Fig. 1.

The structure of SECOQC *QKD* network consists of two main divisions. The first division is a trusted private network which is established between end points and Quantum Back Bone "QBB". The second division is a quantum system which made up of QBB nodes. QBB is contained with a number of *QKD* devices that are attached with other *QKD* devices in one-to-one connection as shown in Fig. 1 [12, 13]. The whole architecture delivers a backbone network for key circulation and management. Quantum Back Bone nodes and links are like gateways and unicast communication links respectively in a classical network. Quantum Back Bone nodes are utilized for connecting the communicated users to private and secure network. Furthermore, the communicated users can be connected to a specific access point with restricted routing responsibilities called Quantum Access Node (QAN) [1–11].

The architecture of SECOQC *QKD* is physically partitioned into numerous routing zones. Each one consists of a cluster of Quantum Back Bone nodes which are given a single IP addresses from a particular pool of the obtainable address space. There're three main responsibilities of Quantum Back Bone node are [1–11]. 1. Manage and generate the keys. 2. Distribute the keys over the quantum network among the communicated peers. 3. Act as access points for generated key clients' functions. The Geographical Division of the SECOQC *QKD* Network into Several Routing Areas based on Open Shortest Path First protocol is shown in Fig. 2.

In Fig. 3, the structure of SECOQC Quantum Back Bone Node and Quantum Routing Protocol are represented. The main elements of quantum Back Bone node are: (1) numerous occurrences of unicast quantum protocol (Q3P); (2) a routing element which is in charge of preserving the routing information and make it up-to-date; (3) path determination and route decision mechanism. The unicast quantum protocol (Q3P) is responsible for administering the communication process across the classical channel. The main functions of Q3P are verification, encoding/decoding, segmentation/assembling, flow control and connection management among nearby quantum backbone nodes [1–11].

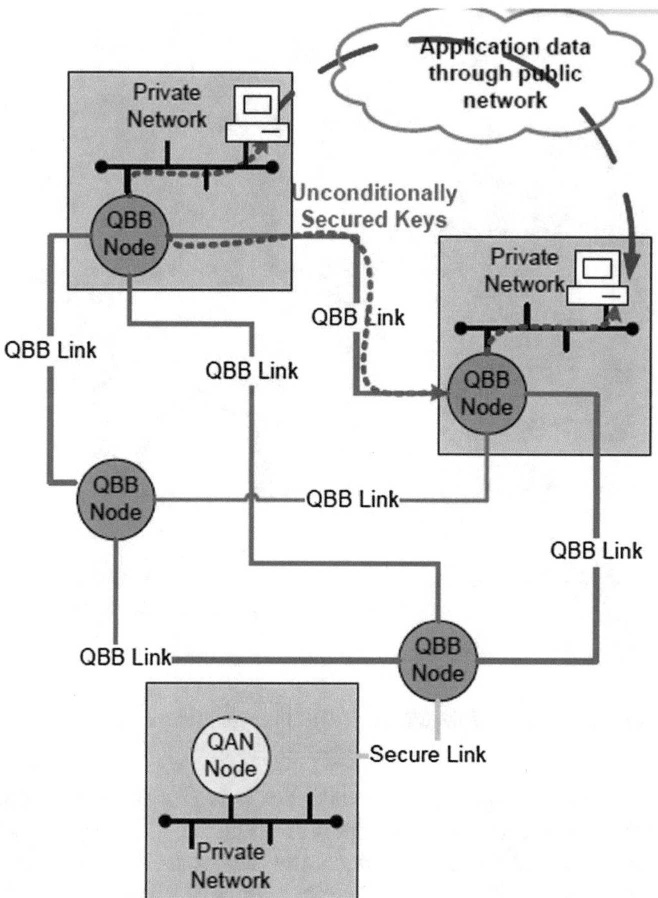

Fig. 1 SECOQC QKD network structure [1]

2 Defense Advanced Research Projects Agency (DARPA) Network

In 2003, DARPA and Bolt, Beranek and Newman (BBN) Technologies have constructed the first completely functioning quantum cryptography network in the world. Initially, DARPA—BBN project consists of 6 nodes, but now total nodes is 10 among Harvard University, Boston University, and BBN. All nodes are successively running on a high-speed fiber optic telecommunications infrastructure. DARPA—BBN project goal is to distribute the security keys from node-to-node using Quantum Key Distribution, as well as, protect message movement among communicating nodes using IPsec- VPN technology. DARPA—BBN network maintains a diversity of QKD technologies fiber modulated phase, entanglement

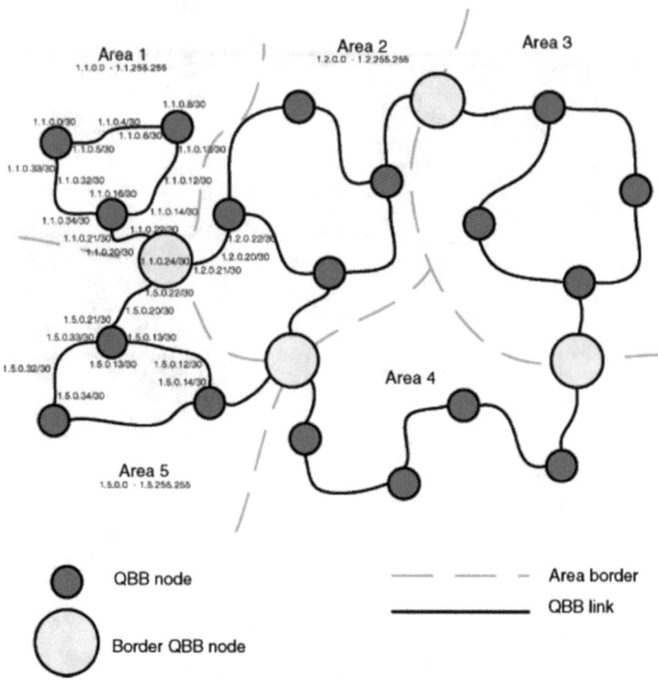

Fig. 2 Geographical division of the SECOQC QKD network into several routing areas [33]

over fiber and wireless quantum network, but it is still used for point to point communication system as shown in Fig. 4 [1–11].

The transmission of signals between the sender and receiver is based on weak-coherent link. As, the sender transmits a series of single photons by using an extremely weakened pulse at wavelength equal to 1550 nm. The transmitted single photon will be adapted to one of four phases by passing it via a Mach-Zehnder interferometer. Therefore, the transmitted photon will be encoded according to its adopted phase and basis. Afterwards, the encoded photon is transmitted to the receiver. The receiver performs the inverse operation as he has another Mach-Zehnder interferometer which randomly adopted the encoded photon to the transmitted value as shown in Fig. 5 [1–11]. By way of 2017, the improvement and growth of a real quantum computer is still in early stages but many poetical and theoretical experimentations were implemented by many research groups [12–34].

In Fig. 6, the structure of DARPA *QKD* Protocol Stack is represented. The architecture of DARPA *QKD* consists of four main processes which are Sifting, Error Correction, Privacy Amplification and Authentication. Sifting is the procedure for examine missing all the obviously unsuccessful transmitted quantum bits between the sender and receiver. The unsuccessful transmitted quantum bits can be resulted from sender's laser never transfer pulses, receiver's detectors failed or photons were missing during transmission. Error Correction is the process for

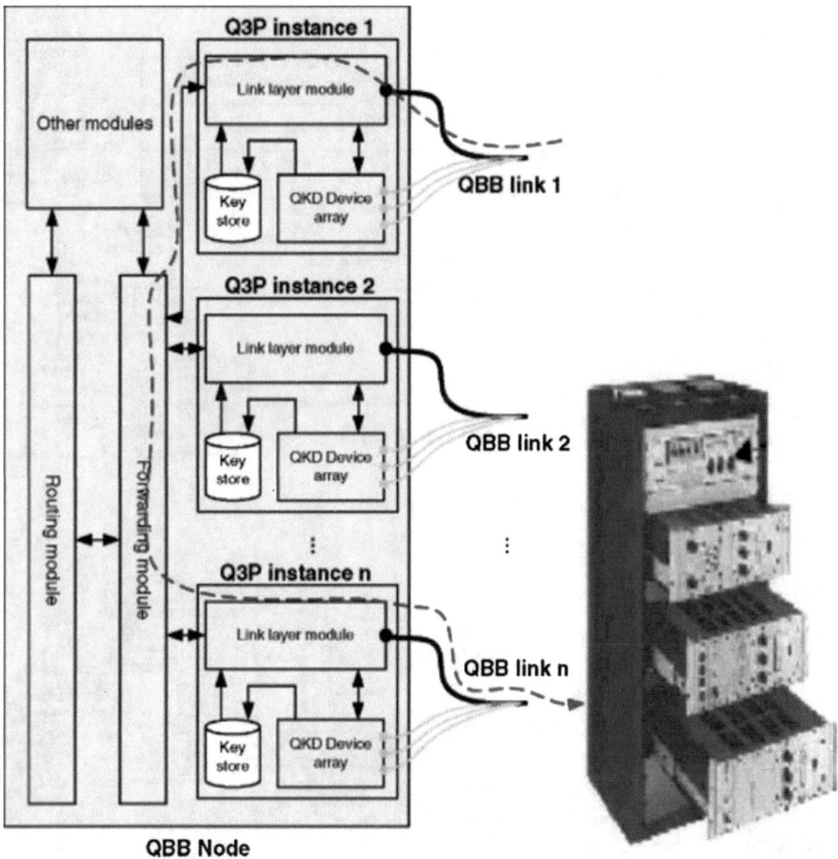

Fig. 3 SECOQC quantum back bone node structure [3]

detecting and correcting all the fault quantum bits and inaccuracy rate. The quantum bit error can be resulted as either that sender transmits as zeros but receiver receives it as ones or vice versa. Furthermore, it can be caused by intruder on channel noisy. In Privacy amplification phase, both sender and a receiver use several bases for measurements purpose. The objective of this step is to recognize and exclude those bit positions where communicators use different bases. These positions are then discarded by both communicators over a public channel. Furthermore, improve the security of the key string by correcting resulted errors during transmission and eavesdropper detection phases. Authentication is the process for verifying the identity of both sender and receiver which allows the communication process to be protected against many types of attacks. The authentication process must be executed on a continuing base for all key interchange controlling stages since the eavesdropper can enclose himself at any stage of the *QKD* process [1–11].

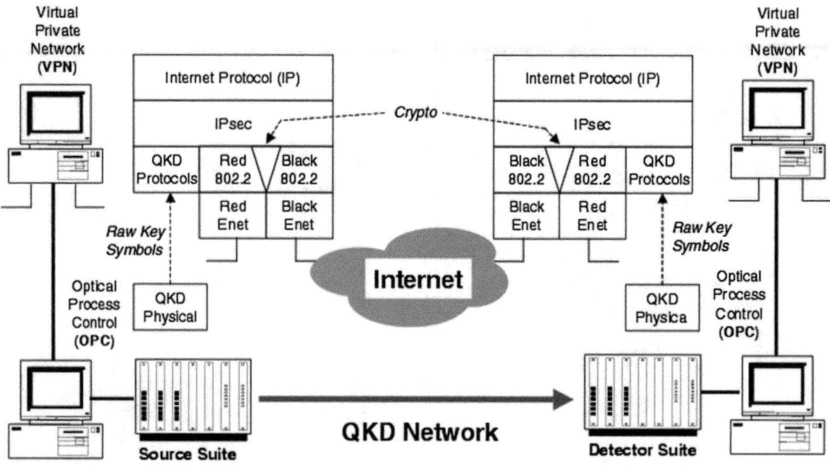

Fig. 4 Architecture of DARPA QKD IP_{sec}–VPN network [3]

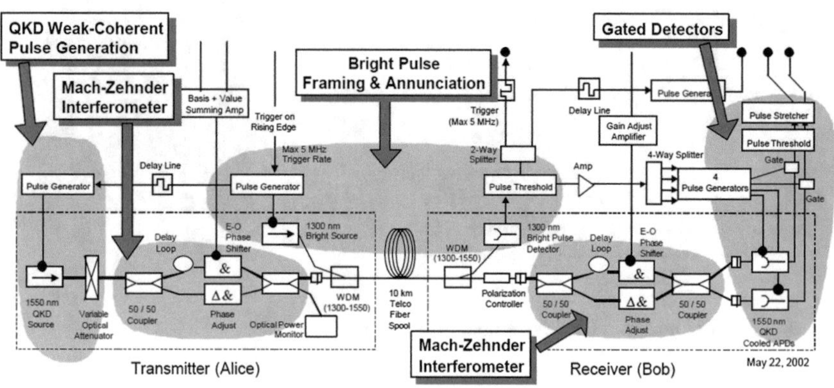

Fig. 5 DARPA QKD network components [34]

DARPA and BBN Technologies have been proposed an integrated architecture for IPsec-based Virtual Private Networks (see Fig. 7). One of the main components of IPsec is the generated keys which are used for encoding and authentication of successive transmitted messages by agreed IPsec Security Association (SA) among the communicated peers. The most important role of the proposed architecture is to define the generated keys for IPsec Security Associations which come from quantum key distribution protocols. The duration for utilizing the generated keys is managed by defined key lifetime. The key lifetime can be represented either in seconds or in kilobytes according to the configuration parameters in Security Policy Database (SPD). When the key lifetime expired, a fresh SA has to be negotiated. Consequently, the old key will be replaced by new one. A Virtual Private Network

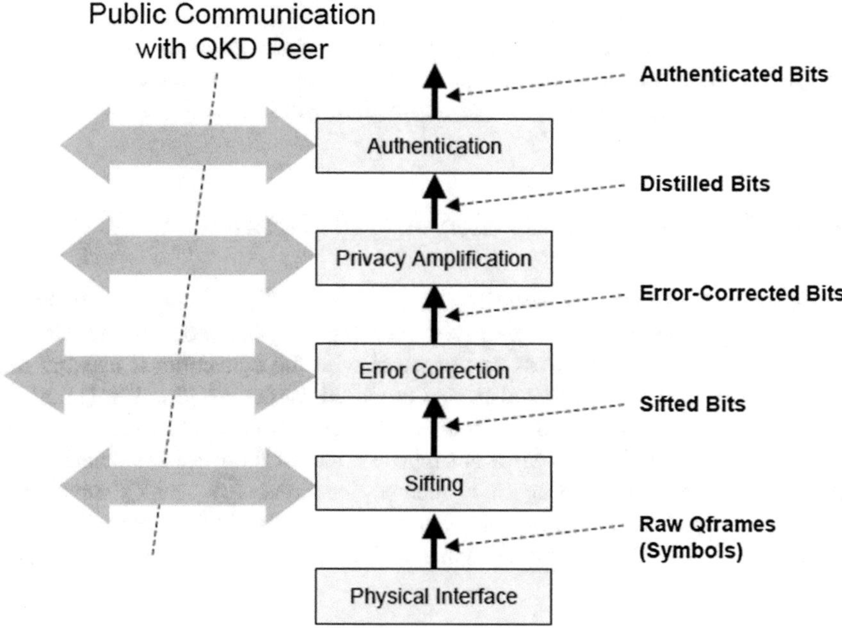

Fig. 6 The full DARPA QKD protocol stack [34]

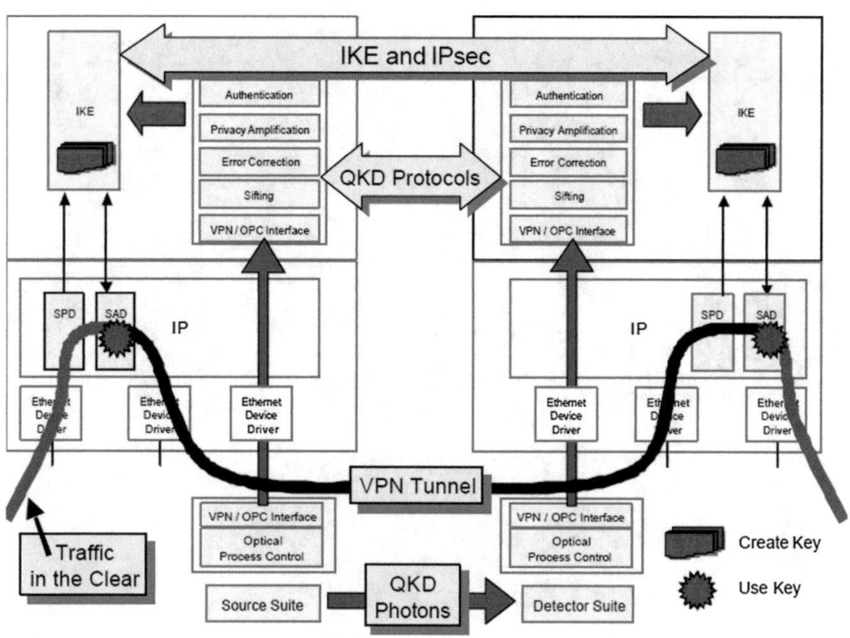

Fig. 7 The upper layers of the IPsec-VPN DARPA QKD protocol stack [3]

(VPN) is used for securing the transmitted traffic with quantum cryptography. The present implementation of the proposed architecture supports both classical encryption algorithms and one-time pads. The choice of used protocol depends on the sensitivity level of the transmitted information within a particular VPN [1–11].

3 Grid Security Model Based on *QKD* Network

In [35], an architecture for security enhancement of Grid Computing based on Quantum Key Distribution is proposed (see Fig. 8). The proposed architecture combines the Quantum network and protocol with the conventional network and protocol. The principal characteristics of proposed model are [36, 37]: (1) all the communicated parties of grid are connected over two channels which Quantum channel and Public channel. (2) It is supposed that all the members connected to quantum channel are existing in a secured locations. (3) The Quantum Key Distribution delivers unconditional security for key distribution, controlling and authentication. (4) The communication link between the two users has the same

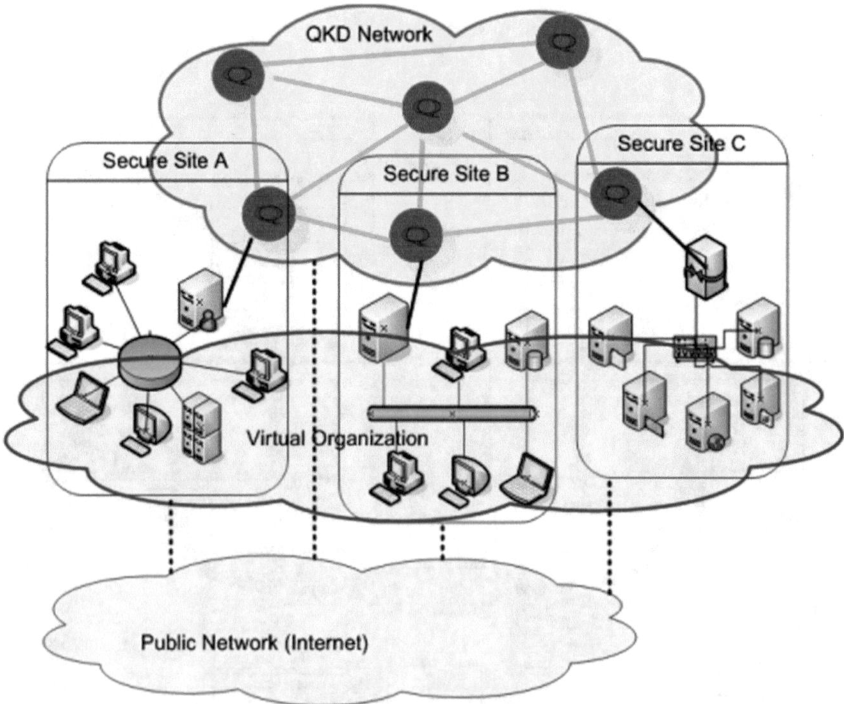

Fig. 8 Theoretical model of grid computing based on QKD network [36]

features and functions of SECOQC *QKD* network. (5) The Grid Security Structure consists of public key encryption, X.509 PKI certificates and SSL/TLS protocol to offer message security, verification, allocation and authorization.

4 Deterministic Secure Direct Communication Using Entanglement

In [38], an innovative protocol for securing Direct Communication Deterministically Using Entangled *EPR* pairs is proposed. The proposed protocol permits transmission of information in a deterministic way by utilizing the property of quantum key distribution and secured pseudo direct communication. The information can be encrypted through the transmission process. So, the protocol has proved its unconditional security against many types of attacks. In case of attacker spy on the communication channel to achieve maximum transmitted information, the attacker will be detected since an error rate equivalent to 50% will be resulted. The protocol has two communication modes which are message and control respectively. By default, the communicated peers are always in message mode. In messaged mode, the sender will prepare an entangled *EPR* pairs of polarized photons using two Particles (see Fig. 9). Afterwards, he will keep one particle for himself and transmit the other one to the receiver. Then, the receiver encodes the transmitted particle using corresponding Pauli gates. Consequently, the receiver

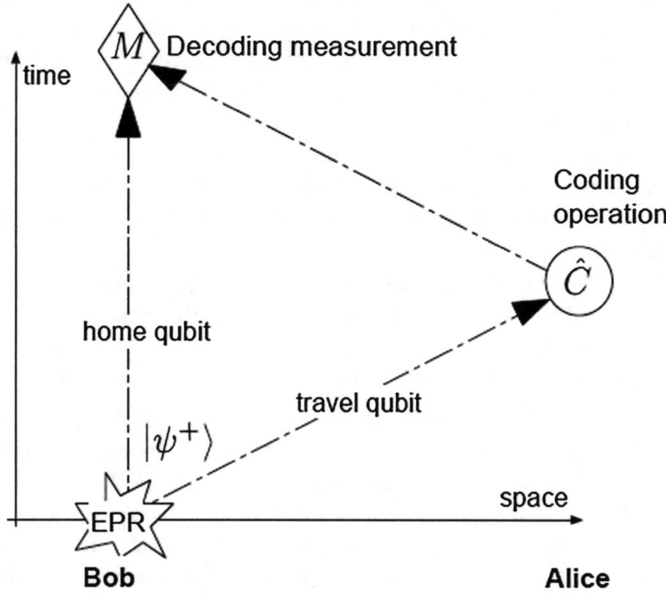

Fig. 9 Message mode. *Dashed lines* represent quantum bits transfer [38]

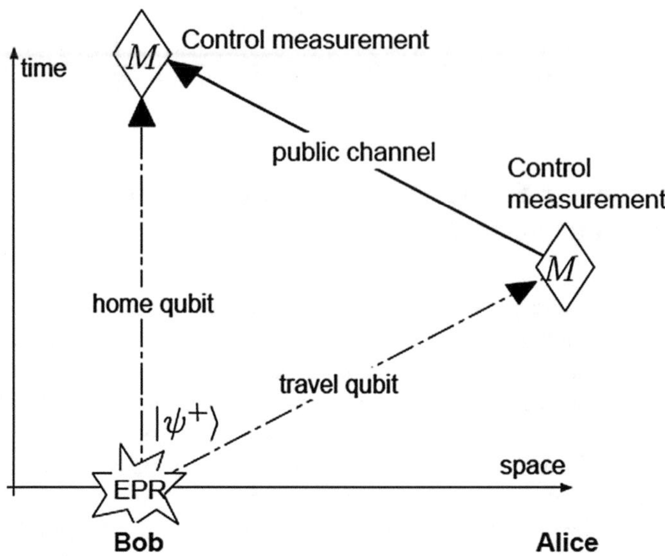

Fig. 10 Control mode. *Solid lines* represent classical bits transfer [38]

will send the encoded quantum bit to the sender. In control mode, rather than the receiver encodes the transmitted particle using corresponding Pauli gates, he applies a measurement in the basis of σ_z. Then, the receiver transmits the resulted measurement over a public channel. Therefore, the sender swaps to the control mode and applies his measurement in the basis of σ_z. According to the result of the measurement, the sender will decide to continue or terminate the communication process (see Fig. 10).

If the measurement result of both sender and receiver are the same then the communication process will continue otherwise it will be terminated. If the measurement result is not matched, the sender will conclude that an eavesdropper spy over the channel.

Since the eavesdropper don't have an access to the sender particle, he only works on the receiver particle. The eavesdropper will listen to the transmitted information to discover which Pauli transformation the receiver applied. So, as to obtain any information about applied receiver's Pauli operation, the eavesdropper will apply his unitary Pauli operation $\hat{\mathrm{E}}$ on the transmitted particle from the sender to the receiver. Subsequently, the receiver will apply his unitary Pauli operation $\hat{\mathrm{C}}$ on the transmitted particle. Lastly, the receiver performs his measurement on the composed system. As a possible control measurement by the receiver would be happened before the eavesdropper's decisive measurement, the receiver hasn't inspiration on the discovery possibility for eavesdropper's attack (see Fig. 11).

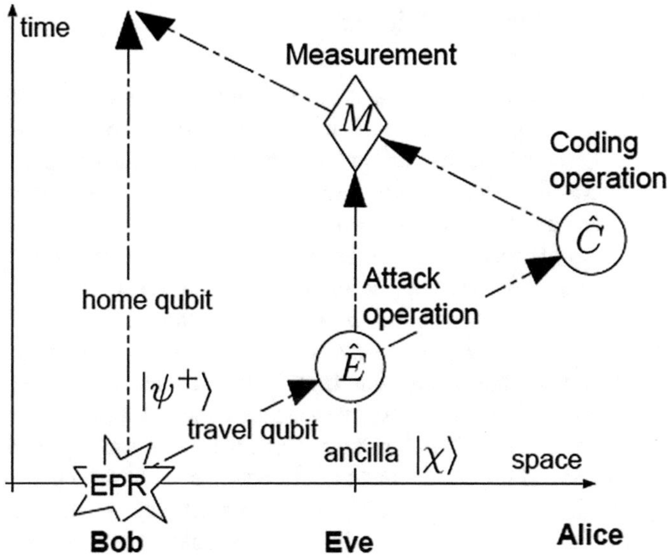

Fig. 11 A general eavesdropping attack [38]

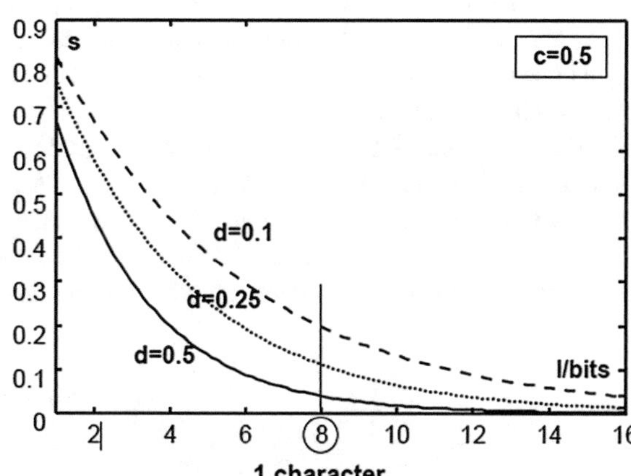

Fig. 12 Eavesdropping success probability as a function of the maximal eavesdropped information [41]

The eavesdropping achievement possibility as a function of the obtained information, $c = 0.5$ and for various discovery possibilities d that eavesdropper can select is charted (see Fig. 12). By analyzing the discovery possibility (d) for the eavesdropper operation Ê, the eavesdropper can achieve only a portion of the transmitted message and he doesn't know which portion of the message is [38–40].

5 Secret Key Sharing with *GHZ*

In [42], an approach has been proposed for dividing the transmitted quantum state into two segments. Therefore, the communicated participants are essential to use the two segments for retrieving the original quantum state. The proposed approach distributes a Secret Key sharing with three participants and extended to four participants. The implementation of the proposed approach is based on *GHZ* states and measurement. The proposed protocol demonstrated it is imaginable to merge the quantum cryptography with secret sharing technique for verification if the attacker is active throughout the communication process of secret sharing or not. Suppose the quantum communication system consists of three users which are $User_1, User_2$ and $User_3$ and each one has one particle from a *GHZ*. triplet states. Every user will select a randomly measure basis direction to measure the status of his particle which in x or y direction (see Fig. 13). Afterward, each user will publicly publish which direction he will use. If the arrangement of directions is authorized and acceptable, then $User_1$ and $User_2$ will be able to collaborate to achieve the measurement outcome of $User_3$. From the Table 1 it is shown that if $User_3$ distinguishes the correct measurements of both $User_1$ and $User_2$, then he can conclude whether their outcomes are identical or different. Consequently, $User_3$ can't achieve any actual information about their results. By the same way, $User_1$ can't distinguish the result of user's measurement without the help of $User_3$.

As each user will select a randomly measure basis direction to measure the status of his particle which in x or y direction, so only half of the *GHZ* triplet states will provide convenient results. For instance if, both $User_1$ and $User_2$ measurements are in x direction, then $User_3$ has to measure his particle in x direction to achieve the desirable information. However, if $User_3$ measured his particle in y direction, then he won't achieve any information. Since $User_3$ selects his measurement direction randomly, he will select accurately measurement only half the time. This is the reason why all users have to publish which direction they will use in a public way.

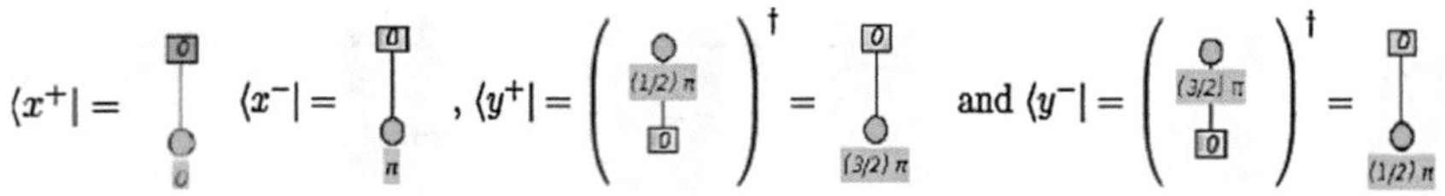

Fig. 13 Measurement outcome represented by ZX—calculus [42]

Table 1 Secret key sharing with GHZ measurement outcome

	$User_2$				
$User_1$		x^+	x^-	y^+	y^-
	x^+	$\|0\rangle + \|1\rangle$	$\|0\rangle - \|1\rangle$	$\|0\rangle - i\|1\rangle$	$\|0\rangle + i\|1\rangle$
	x^-	$\|0\rangle - \|1\rangle$	$\|0\rangle + \|1\rangle$	$\|0\rangle + i\|1\rangle$	$\|0\rangle - i\|1\rangle$
	y^+	$\|0\rangle - i\|1\rangle$	$\|0\rangle + i\|1\rangle$	$\|0\rangle - \|1\rangle$	$\|0\rangle + \|1\rangle$
	y^-	$\|0\rangle + i\|1\rangle$	$\|0\rangle - i\|1\rangle$	$\|0\rangle + \|1\rangle$	$\|0\rangle - \|1\rangle$

Therefore, they have the ability to decide whether to save or ignore the outcomes of *GHZ* states. The publication should be performed in the following way; both $User_1$ and $User_3$ transmit to $User_2$ the direction of their measurements. Subsequently, all users transmit their measurement direction to $User_1$ and $User_3$ (see Table 1) [42–44].

In case of implementation of Quantum Secret Sharing with four users. Therefore, the quantum communication system consists of four users which are $User_1, User_2, User_3$ and $User_4$ and each has one particle from a four *GHZ* states. Every user will select a randomly measure basis direction to measure the status of his particle which in x or y direction. Afterward, each user will publicly publish which direction he will use with $User_1$. Subsequently, $User_1$ determines whether the whole basis is functional and distributes all the selected bases to the other users. When three users measure the transmitted particle in the same basis and only one user is different, then is called invalid basis. However, if all users measure the transmitted particle in the same basis, it is called valid basis. Then $User_2, User_3$ and $User_4$ can collaborate to retrieve the original quantum state sent by $User_1$. By way of three measurements will be enough to conclude the definitive result of the transmitted quantum state [42–44].

6 Quantum Teleportation of *EPR* Pair by Three-Particle Entanglement

In [45, 46], a procedure for teleportation of *EPR* pair by three-particle entanglement is proposed. This procedure involves a sender can transmit an *EPR* pair for two disjoint receivers. Transmitted *EPR* pair will be teleported by utilizing the property of *GHZ* entanglement. It requires a maximal entanglement of *GHZ* and *EPR* states. The sender will perform *GHZ* joint measurement based on the generated *GHZ* state. Both receivers can retrieve the original unidentified *EPR* pair according to the outcome of measurement. However, both receivers have to cooperate together for successfully retrieving *EPR* pair as only one receiver can't retrieve it alone. The proposed approach requires five quantum bits; two for *EPR* pair and three for *GHZ* states. The scheme teleport *EPR* pair which consists of quantum bits 1 and 2 with the utilization of maximally entangled *GHZ* state which consists of quantum bits 3, 4 and 5 (see Fig. 14) and (see Eq. (1))

$$|\psi>_{Teleporated} = |\psi>_{EPR} \otimes |\psi>_{GHZ} \tag{1}$$

In Fig. 14, there are three users; one sender $User_A$ and two receivers $User_B$ and $User_C$. The three users share a maximally *GHZ* entangled particles 3, 4 and 5. $User_A$ would like to send unknown *EPR* pair to $User_B$ and $User_C$. $User_A$ performs a *GHZ* joint measurement on quantum bits 1, 2 and 3. Then, $User_A$ transmits the result of measurement to both $User_B$ and $User_C$ over a public channel. The most

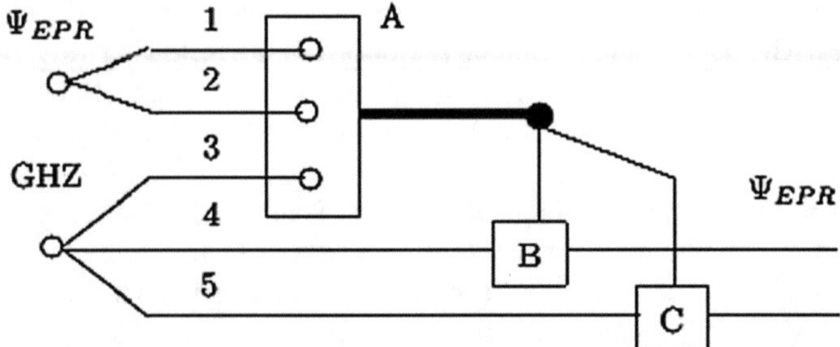

Fig. 14 Teleportation of EPR pair of quantum bits 1 and 2 using GHZ triplet. $User_A$, $User_B$, and $User_C$ share quantum bits 3, 4 and 5 of GHZ. $User_A$. sends outcome of a joint measurement to $User_B$ and $User_C$ who recover an unknown *EPR*-state [45]

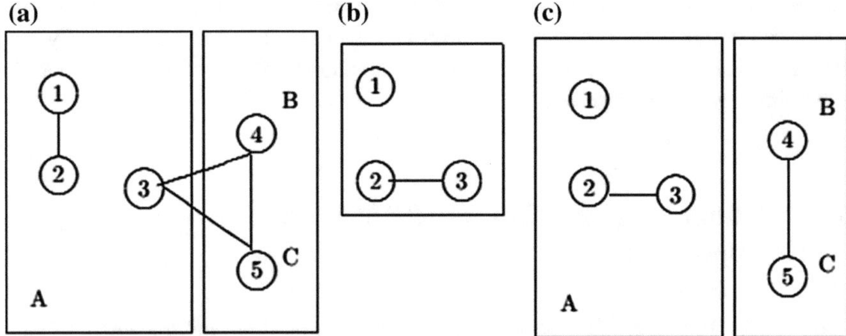

Fig. 15 The entanglement structure of the states. **a** The initial state, where quantum bits of EPR pair 1, 2 and quantum bits 3, 4, 5 of GHZ are entangled. **b** Projection basis (4). **c** The state after measuring [46]

important aspect is the way that performs the joint measurement which requires eight projection operators.

The protocol for teleportation works as follows; $User_A$ performs a joint measurement on quantum bits 1, 2 and 3 according to the selected appropriate base and transmit the result to both $User_B$ and $User_C$. Both $User_B$ and $User_C$ retrieve the original *EPR* according to the result of measurement. If the value of measurement is equal to 0, 1, 2 or 3, then the $User_B$ will rotate his quantum bit by Pauli operations $\sigma_x, i\sigma_y, -i\sigma_y - \sigma_x$ respectively and $User_C$. will not do anything. But if the value of measurement is equal to 4, 5, 6 or 7, then the $User_C$ will rotate his quantum bit by Pauli operations $\sigma_x, i\sigma_y, -i\sigma_y - \sigma_x$ respectively and $User_B$ will not do anything (see Fig. 15).

The teleportation network for *EPR* state consists of *CNOT* and *Hadamard* gates. The *CNOT* gate is used for generating entangled *EPR* pair on quantum bits 1 and 2. As *CNOT* gate changes the value of target quantum bit if the control quantum bit equal to one. The division of *GHZ* prepares a maximal entangled *GHZ* state by

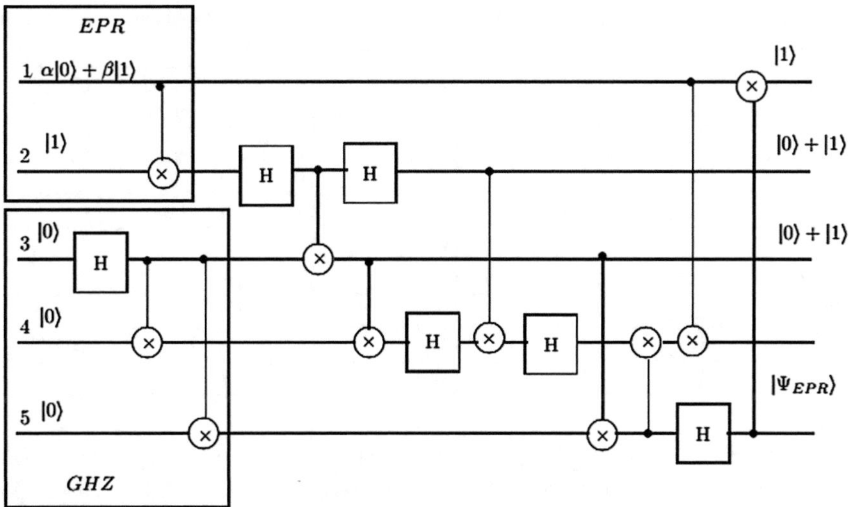

Fig. 16 Circuit for Teleportation of *EPR* pair [46]

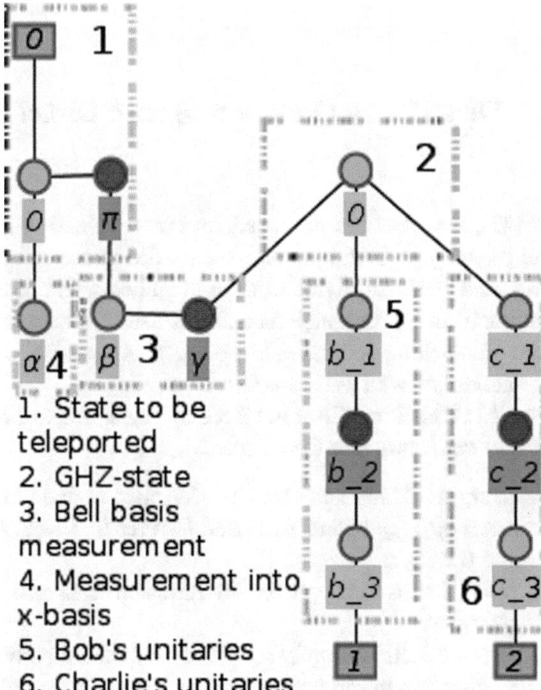

Fig. 17 Graphical representation of the set of instructions of teleportation of $\mathbf{a}|01\rangle + \mathrm{b}\ |10\rangle$ through GHZ protocol [45]

Table 2 The corresponding local operations that $user_b$ and $user_c$ perform, given a measurement outcome of $User_A$. additionally, the corresponding phases are displayed

$User_A$ measurement	$User_A$ direction	$User_B$ measurement	$User_C$ measurement	α	B	γ	$User_B$ operation	$User_C$ operation
$\lvert\phi^+\rangle$	$\lvert x^+\rangle$	σ_x	I	0	0	0	σ_x	I
$\lvert\phi^-\rangle$	$\lvert x^+\rangle$	$i\sigma_y$	I	0	π	0	$i\sigma_y$	I
$\lvert\phi^+\rangle$	$\lvert x^-\rangle$	$-i\sigma_y$	I	π	0	0	$-i\sigma_y$	I
$\lvert\phi^-\rangle$	$\lvert x^-\rangle$	$-\sigma_x$	I	π	π	0	$-\sigma_x$	I
$\lvert\Psi^+\rangle$	$\lvert x^+\rangle$	I	σ_x	0	0	π	I	σ_x
$\lvert\Psi^-\rangle$	$\lvert x^+\rangle$	I	$-i\sigma_y$	0	π	π	I	$-i\sigma_y$
$\lvert\Psi^+\rangle$	$\lvert x^-\rangle$	I	$i\sigma_y$	π	0	π	I	$i\sigma_y$
$\lvert\Psi^-\rangle$	$\lvert x^-\rangle$	I	$-\sigma_x$	π	π	π	I	$-\sigma_x$

applying the *Hadamard* gate on quantum bit three and applying two *CNOT* operations. The first *CNOT* operation is applied on quantum bits three and four. The second *CNOT* operation is applied on quantum bits three and five. Consequently, the value of quantum bits four and five at the end side constitute the transmitted *EPR* pair which being independent of others (see Figs. 16 and 17) and (see Table 2).

7 Three-Party Quantum Secure Direct Communication with *GHZ*

In [47], a protocol is proposed for communication among three users by utilizing the property of both quantum secure direct communications and *GHZ* states. The proposed protocol divided into two stages which are channel checking and direct communication. Channel checking is used for assuring that the channel is safe and free of attacking and eavesdropping. Then, the real communication process among the communicated parties could be begin.

Stage 1: Checking Channel Free of Error while $User_A$, $User_B$ and $User_C$ would like to communicate (see Table 3).

1. $User_A$ randomly generates *N GHZ* state in one of the various defined eight *GHZ* states $|\psi_i\rangle_{ABC}$ where A, B and C refer to $User_A$, $User_B$ and $User_C$ respectively and $0 \le i \le 7$.
2. $User_A$ keeps one particle for him and transmits the other two for $User_B$ and $User_C$.
3. $User_B$ selects randomly subset of transmitted particles and measures them in z or x direction at random way. Therefore, $User_B$ will transmit the outcome of his measurement to both $User_A$ and $User_C$ over classical channel.

Table 3 Three-party quantum secure direct communication with GHZ Stage 1 Operation with corresponding graphical representation

#	Binary	Standard Form (SF)	Unitaries	Graphical Representation
0	000	$\frac{1}{\sqrt{2}}(\|000\rangle + \|111\rangle)$	$I \otimes I$	
1	001	$\frac{1}{\sqrt{2}}(\|001\rangle + \|110\rangle)$	$I \otimes \sigma_x$	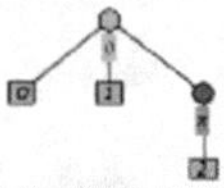
2	010	$\frac{1}{\sqrt{2}}(\|010\rangle + \|101\rangle)$	$\sigma_x \otimes I$	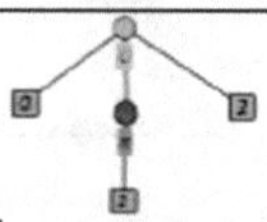
3	011	$\frac{1}{\sqrt{2}}(\|011\rangle + \|100\rangle)$	$\sigma_x \otimes \sigma_x$	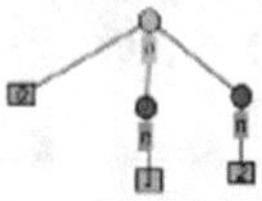
4	100	$\frac{1}{\sqrt{2}}(\|000\rangle - \|111\rangle)$	$\sigma_z \otimes I$	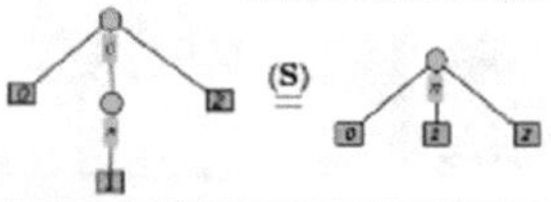
5	101	$\frac{1}{\sqrt{2}}(\|001\rangle - \|110\rangle)$	$\sigma_z \otimes \sigma_x$	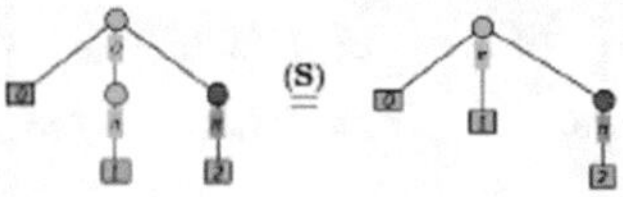
6	110	$\frac{1}{\sqrt{2}}(\|010\rangle - \|101\rangle)$	$i\sigma_y \otimes I$	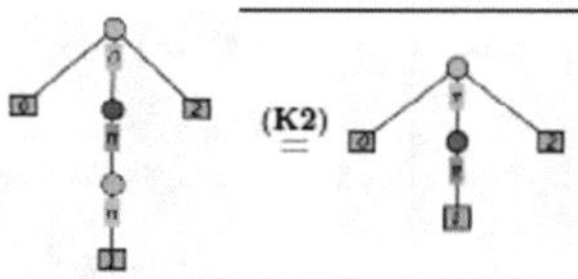
7	111	$\frac{1}{\sqrt{2}}(\|011\rangle - \|100\rangle)$	$i\sigma_y \otimes \sigma_x$	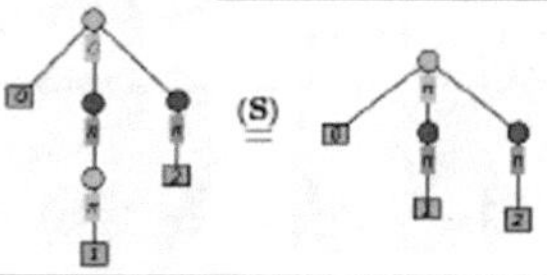

4. $User_A$ and $User_C$ measure their corresponding particles in the same direction. Afterward, $User_C$ transmits the outcome of his measurement to $User_A$ over a classical channel.
5. $User_A$ will detect if there's attacker on the communication channel by utilizing both the result of measurement and the succession of prepared *GHZ* states.
6. If $User_A$ suspicious an attacking on the communication channel, then the communication process will be terminated as the quantum communication channel is unsafe. Otherwise, the communication process will continue and communicated parties will proceed for stage 2.

Table 4 Three-party quantum secure direct communication with GHZ stage 2 operation with corresponding graphical representation

#	Binary	Alternative Form (AF)	Unitaries	Graphical Representation
0	000	$\frac{1}{\sqrt{2}}(\|111\rangle + \|000\rangle)$	$\sigma_z \otimes \sigma_z$	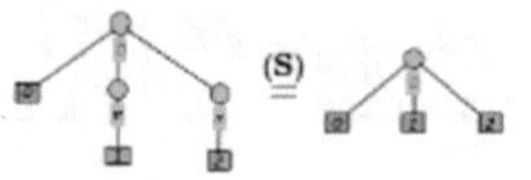 $\overset{(S)}{=}$
1	001	$\frac{1}{\sqrt{2}}(\|110\rangle + \|001\rangle)$	$\sigma_z \otimes i\sigma_y$	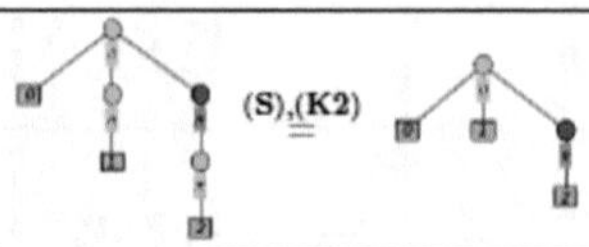 $\overset{(S),(K2)}{=}$
2	010	$\frac{1}{\sqrt{2}}(\|101\rangle + \|010\rangle)$	$i\sigma_y \otimes \sigma_z$	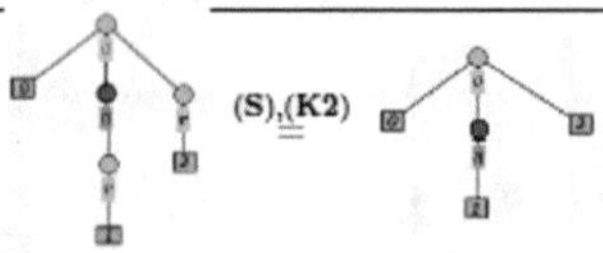 $\overset{(S),(K2)}{=}$
3	011	$\frac{1}{\sqrt{2}}(\|100\rangle + \|011\rangle)$	$i\sigma_y \otimes i\sigma_y$	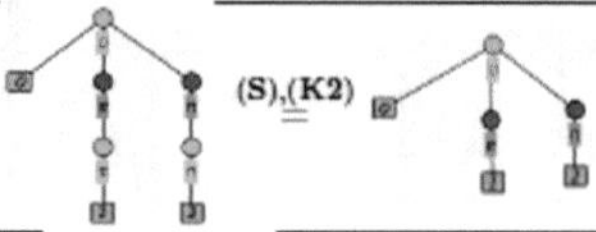 $\overset{(S),(K2)}{=}$
4	100	$\frac{1}{\sqrt{2}}(\|111\rangle - \|000\rangle)$	$I \otimes \sigma_z$	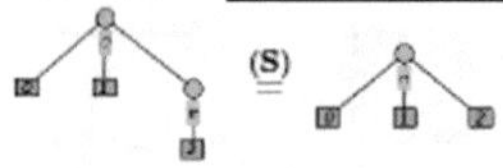 $\overset{(S)}{=}$
5	101	$\frac{1}{\sqrt{2}}(\|100\rangle - \|001\rangle)$	$I \otimes i\sigma_y$	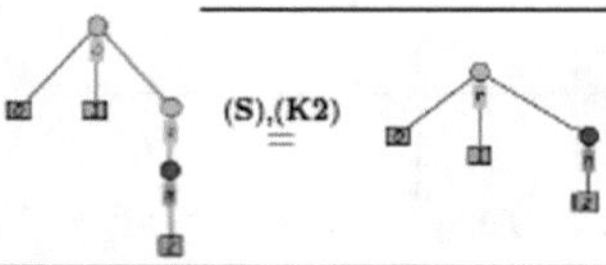 $\overset{(S),(K2)}{=}$
6	110	$\frac{1}{\sqrt{2}}(\|101\rangle - \|010\rangle)$	$\sigma_x \otimes \sigma_z$	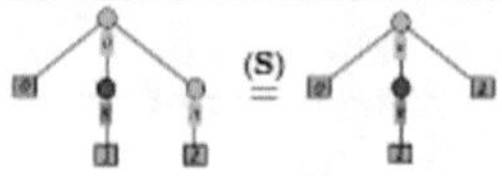 $\overset{(S)}{=}$
7	111	$\frac{1}{\sqrt{2}}(\|100\rangle - \|011\rangle)$	$\sigma_x \otimes i\sigma_y$	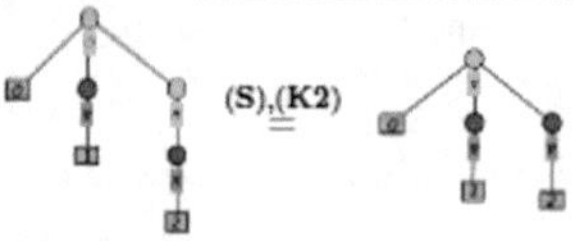 $\overset{(S),(K2)}{=}$

Stage 2: **Direct Communication among communicated parties** (see Table 4).

1. All three users will work on the outstanding *GHZ* states after performing stage 1.
2. $User_B$ and $User_C$ encode the transmitted bits by utilizing the property of Pauli transformation. If the transmitted bit equal to zero, then they won't do any thing which means applying *I* operation else they will perform *Y* operation on their quantum bits.

3. Afterward, both $User_B$ and $User_C$ transmit the transformed quantum bits to $User_A$ as two classical bits.
4. $User_A$ performs its local Pauli operation on his quantum bits according to the received classical values. If the received two classical bits equal to 00, 01, 10 and 11, then he will perform, Y and Z on his quantum bits respectively.
5. Then, $User_A$ measures the whole GHZ state and announces the result of his measurement and corresponding original GHZ state.
6. According to the distributed information and Pauli transformations, the users can retrieve the original secret message.

References

1. Poppe, A., Peev, M., Maurhart, O.: Outline of the SECOQC quantum-key-distribution network in Vienna. Int. J. Quantum Inf. **6**(02), 209–218 (2008)
2. Peev, M., Pacher, C., Alléaume, R., Barreiro, C., Bouda, J., Boxleitner, W., Tualle-Brouri, R.: The SECOQC Quantum Key Distribution Network in Vienna. New J. Phys. **11**(7), 075001 (2009)
3. Elliott, C.: Building the Quantum Network. New J. Phys. **4**(1), 46 (2002)
4. Elliott, C., Colvin, A., Pearson, D., Pikalo, O., Schlafer, J., Yeh, H.: Current status of the DARPA quantum network. In: Defense and Security, International Society for Optics and Photonics, pp. 138–149, May 2005
5. Metwaly, A.F., Mastorakis, N.E.: Architecture of decentralized multicast network using quantum key distribution and hybrid WDM-TDM. In: Proceedings of the 9th International Conference on Computer Engineering and Applications (CEA'15), Advances in Information Science And Computer Engineering, pp. 504–518 (2015)
6. Alleaume, R., Riguidel, M., Weinfurter, H., Gisin, N., Grangier, P., Dianati, M., Godfrey, M., et al.: SECOQC White Paper on Quantum Key Distribution and Cryptography. No. quant-ph/0701168 (2007)
7. Dianati, M., Alléaume, R., Gagnaire, M., Shen, X.S.: Architecture and protocols of the future european quantum key distribution network. Secur. Commun. Netw. **1**(1), 57–74 (2008)
8. Khan, M.M., Hyder, S., Pathan, M.K., Sheikh, K.H.: A Quantum key distribution network through single mode optical fiber. In: 2006 International Symposium on Collaborative Technologies and Systems, CTS 2006, pp. 386–391. IEEE, May 2006
9. Le, Q.C., Bellot, P.: Enhancement of AGT telecommunication security using quantum cryptography. In: 2006 International Conference on Research, Innovation and Vision for the Future, pp. 7–16. IEEE (2006, February)
10. Kimble, H.: The quantum internet. Nature **453**(7198), 1023–1030 (2008)
11. Dianati, M., Alléaume, R.: Architecture of the Secoqc Quantum Key Distribution nNetwork. arXiv: preprint quant-ph/0610202 (2006)
12. Farouk, A., Omara, F., Zakria, M., Megahed, A.: Secured IPsec multicast architecture based on quantum key distribution. In: The International Conference on Electrical and Bio-medical Engineering, Clean Energy and Green Computing, pp. 38–47. The Society of Digital Information and Wireless Communication (2015)
13. Farouk, A., Zakaria, M., Megahed, A., Omara, F.A.: A generalized architecture of quantum secure direct communication for N disjointed users with authentication. Sci. Rep. **5**, 16080–16080 (2014)

14. Metwaly, A.F., Rashad, M.Z., Omara, F.A., Megahed, A.A.: Architecture of multicast centralized key management scheme using quantum key distribution and classical symmetric encryption. Eur. Phys. J. Spec. Top. **223**(8), 1711–1728 (2014)
15. Metwaly, A., Rashad, M.Z., Omara, F.A., Megahed, A.A.: Architecture of point to multipoint QKD communication systems (QKDP2MP). In: 8th International Conference on Informatics and Systems (INFOS), Cairo, (pp. NW 25–31). IEEE, May 2012
16. Wang, M.M., Wang, W., Chen, J.G., Farouk, A.: Secret sharing of a known arbitrary quantum state with noisy environment. Quantum Inf. Process. **14**(11), 4211–4224 (2015)
17. Naseri, M., Heidari, S., Batle, J., Baghfalaki, M., Gheibi, R., Farouk, A., Habibi, A.: A new secure quantum watermarking scheme. Optik-Int. J. Light Electron Opt. **139**, 77–86 (2017)
18. Batle, J., Ciftja, O., Naseri, M., Ghoranneviss, M., Farouk, A., Elhoseny, M.: Equilibrium and uniform charge distribution of a classical two-dimensional system of point charges with hard-wall confinement. Phys. Scr. **92**(5), 055801 (2017)
19. Geurdes, H., Nagata, K., Nakamura, T., Farouk, A.: A note on the possibility of incomplete theory (2017). arXiv preprint arXiv:1704.00005
20. Batle, J., Farouk, A., Alkhambashi, M., Abdalla, S.: Multipartite correlation degradation in amplitude-damping quantum channels. J. Korean Phys. Soc. **70**(7), 666–672 (2017)
21. Batle, J., Naseri, M., Ghoranneviss, M., Farouk, A., Alkhambashi, M., Elhoseny, M.: Shareability of correlations in multiqubit states: optimization of nonlocal monogamy inequalities. Phys. Rev. A **95**(3), 032123 (2017)
22. Batle, J., Farouk, A., Alkhambashi, M., Abdalla, S.: Entanglement in the linear-chain Heisenberg antiferromagnet Cu (C 4 H 4 N 2) (NO 3) 2. Eur. Phys. J. B **90**, 1–5 (2017)
23. Batle, J., Alkhambashi, M., Farouk, A., Naseri, M., Ghoranneviss, M.: Multipartite non-locality and entanglement signatures of a field-induced quantum phase transition. Eur. Phys. J. B **90**(2), 31 (2017)
24. Nagata, K., Nakamura, T., Batle, J., Abdalla, S., Farouk, A.: Boolean approach to dichotomic quantum measurement theories. J. Korean Phys. Soc. **70**(3), 229–235 (2017)
25. Abdolmaleky, M., Naseri, M., Batle, J., Farouk, A., Gong, L.H.: Red-green-blue multi-channel quantum representation of digital images. Optik-Int. J. Light Electron Opt. **128**, 121–132 (2017)
26. Farouk, A., Elhoseny, M., Batle, J., Naseri, M., Hassanien, A.E.: A proposed architecture for key management schema in centralized quantum network. In: Handbook of Research on Machine Learning Innovations and Trends, pp. 997–1021. IGI Global (2017)
27. Zhou, N.R., Li, J.F., Yu, Z.B., Gong, L.H., Farouk, A.: New quantum dialogue protocol based on continuous-variable two-mode squeezed vacuum states. Quantum Inf. Process. **16**(1), 4 (2017)
28. Batle, J., Abutalib, M., Abdalla, S., Farouk, A.: Persistence of quantum correlations in a XY spin-chain environment. Eur. Phys. J B **89**(11), 247 (2016)
29. Batle, J., Abutalib, M., Abdalla, S., Farouk, A.: Revival of Bell nonlocality across a quantum spin chain. Int. J. Quantum Inf. **14**(07), 1650037 (2016)
30. Batle, J., Ooi, C.R., Farouk, A., Abutalib, M., Abdalla, S.: Do multipartite correlations speed up adiabatic quantum computation or quantum annealing? Quantum Inf. Process. **15**(8), 3081–3099 (2016)
31. Batle, J., Bagdasaryan, A., Farouk, A., Abutalib, M., Abdalla, S.: Quantum correlations in two coupled superconducting charge qubits. Int. J. Mod. Phys. B **30**(19), 1650123 (2016)
32. Batle, J., Ooi, C.R., Abutalib, M., Farouk, A., Abdalla, S.: Quantum information approach to the azurite mineral frustrated quantum magnet. Quantum Inf. Process. **15**(7), 2839–2850 (2016)
33. Batle, J., Ooi, C.R., Farouk, A., Abdalla, S.: Nonlocality in pure and mixed n-qubit X states. Quantum Inf. Process. **15**(4), 1553–1567 (2016)
34. Metwaly, A.F., Rashad, M.Z., Omara, F.A., Megahed, A.A.: Architecture of Multicast Network Based on Quantum Secret Sharing and Measurement (2015)
35. Khan, M.M., Xu, J.: Enhancing grid security using quantum key distribution. Int. J. Secur. Appl. **6**(4) (2012)

36. Chakrabarti, A., Damodaran, A., Sengupta, S.: Grid computing security: a taxonomy. IEEE Secur. Priv. **1**, 44–51 (2008)
37. Zhao, S., Aggarwal, A., Kent, R.D.: PKI-based authentication mechanisms in grid systems. In: 2007 International Conference on Networking, Architecture, and Storage, NAS 2007, pp. 83–90. IEEE, July 2007
38. Boström, K., Felbinger, T.: Deterministic secure direct communication using entanglement. Phys. Rev. Lett. **89**(18), 187902 (2002)
39. Beige, A., Englert, B.G., Kurtsiefer, C., Weinfurter, H.: Secure communication with single-photon two-qubit states. J. Phys. A Math. Gen. **35**(28), L407 (2002)
40. Deng, F.G., Long, G.L.: Secure direct communication with a quantum one-time pad. Phys. Rev. A **69**(5), 052319 (2004)
41. Batle, J., Ciftja, O., Abdalla, S., Elhoseny, M., Alkhambashi, M., Farouk, A.: Equilibrium-charge distribution on a finite straight one-dimensional wire. Eur. J. Phys. **38**(5),(2017). http://iopscience.iop.org/article/10.1088/1361-6404/aa78bb/meta
42. Hillery, M., Bužek, V., Berthiaume, A.: Quantum secret sharing. Phys. Rev. A **59**(3), 1829 (1999)
43. Man, Z.X., Xia, Y.J., An, N.B.: quantum secure direct communication by using GHZ states and entanglement swapping. J. Phys. B At. Mol. Opt. Phys. **39**(18), 3855 (2006)
44. Hillebrand, A.: Superdense coding with gHZ and quantum key distribution with W in the ZX-calculus. In: EPTCS, vol. 95, pp. 103–121
45. Gorbachev, V.N., Trubilko, A.I.: Quantum teleportation of EPR pair by three-particle entanglement. JETP Lett. **91**(quant-ph/9906110), 894–898 (2000)
46. Joo, J., Park, Y.J., Oh, S., Kim, J.: Quantum teleportation via a W state. New J. Phys. **5**(1), 136 (2003)
47. Jin, X.R., Ji, X., Zhang, Y.Q., Zhang, S., Hong, S.K., Yeon, K.H., Um, C.I.: Three-party quantum secure direct communication based on GHZ states. Phys. Lett. A **354**(1), 67–70 (2006)

Quantum Computing and Cryptography: An Overview

Ahmed Farouk, O. Tarawneh, Mohamed Elhoseny, J. Batle, Mosayeb Naseri, Aboul Ella Hassanien and M. Abedl-Aty

Abstract In this chapter the principles of quantum computing and communications has been proposed.

Keywords Quantum key distribution · Quantum-Back-Bone · Quantum data link layer

1 Introduction

The principle of quantum computing and communications has been initiated in 1970 when Stephan Wiesner has proposed a discussion paper about the basis of several improvements of quantum computing and cryptography area. Stephan Wiesner introduced the basic idea of no-cloning theory which declares that the

A. Farouk · M. Elhoseny
Faculty of Computer and Information Sciences,
Mansoura University, Mansoura, Egypt

A. Farouk (✉) · M. Abedl-Aty
University of Science and Technology,
Zewail City of Science and Technology, Giza, Egypt
e-mail: dr.ahmedfarouk85@yahoo.com

O. Tarawneh
Information Technology Department, Al-Zahra College for Women,
3365, Muscat, Oman

J. Batle
Department de Física, Universitat de Les Illes Balears, 07122 Palma
de Mallorca, Balearic Islands, Spain

M. Naseri
Department of Physics, Kermanshah Branch, Islamic Azad University,
Kermanshah, Iran

A.E. Hassanien
Faculty of Computers and Information, Cairo University, Giza, Egypt

A.E. Hassanien et al. (eds.), *Quantum Computing: An Environment for Intelligent Large Scale Real Application,* Studies in Big Data 33,
https://doi.org/10.1007/978-3-319-63639-9_4

quantum status can't be copied [1]. In 1973, Charles Bennett has introduced a changeable classical Turing machine which leads to the invention of quantum Turing machine [2]. From 1980–1982, the first concept of quantum Turing machine is developed by Paul Benioff. The structure of proposed model is based on quantum mechanical Hamiltonian [3–6]. In 1982, Richard Feynman demonstrated that quantum mechanical schemes can't be reproduced effectively and proficiently by classical one. Furthermore, Feynman recommended that the only method for building the structure of quantum systems is based only on quantum laws and principles [7]. In 1982 Wootters and Zurek [8] have identified the no-cloning theorem and its deep effects in the quantum computing and associated areas. The no-cloning theorem is a consequence of quantum system which prevents the formation of duplicate copies of an unidentified random quantum bit. In 1984, Bennett and Brassard have achieved that the transfer of quantum states is very essential for securing communication system. Consequently, they developed the first quantum key distribution method which latterly is known as BB84 protocol. BB84 protocol was applied practically in 1989 [9]. In 1985, David Deutsch has proposed an idea of using the quantum systems for resolving the computational problems quicker and more efficient than the classical systems [10]. In 1992, Charles Bennett and Stephan Wiesner have pioneered the principle of superdense coding [11]. In Dense coding, the classical information can be encoded and transmitted between distant parties based on both one quantum bit and maximally shared quantum entanglement among the distant parties as each quantum bit can transmit two classical bits [11, 12]. In 1993, the principle of quantum teleportation is introduced by Charles Bennett et al. [13]. In quantum teleportation, the quantum information can be transmitted between distant parties based on both classical communication and maximally shared quantum entanglement among the distant parties [13–16]. Unlike the classical computer, the quantum computer can improve the massive processing power and accomplish many tasks by utilizing all potential transformations concurrently. The quantum computers are dissimilar from classical computers depending on transistors. While a classical computer involves data to be converted into bits (0 or 1), the quantum computer involves quantum bits (qubits) can be in superposition state. The superposition state means that the quantum state can be 0, 1 or in both states at the same time. Quantum computers share hypothetical relationships with non-deterministic and probabilistic computers [17–19]. By way of 2017, the improvement and growth of a real quantum computer is still in early stages but many poetical and theoretical experimentations were implemented by many research groups [20–43].

2 Classical and Quantum Bit

The classical bit is the fundamental element of information. It is used to represent information by computers. Nevertheless of its physical realization, a classical bit has two possible states, 0 and 1. On the other hands, it is recognized that the quantum state is a fundamental concept in quantum mechanics. Actually, the quantum bit is the same as the quantum state. The quantum bit can be represented and measured using two states $|0>$ and $|1>$ which well known as Dirac notation [18, 44]. In classical computer, information is expressed in terms of classical bit which can be either 0 or 1 at any time. On the other hand, the quantum computer uses the quantum bit rather than a bit. It can be in a state of 0 or 1, also there is usage of a form of linear combinations of state called superposition state. Quantum bit can take the properties of 0 and 1 simultaneously at any one moment [45]. Quantum bit definition is described as follow: Definition: A quantum bit, or qubit for short, is a 2 dimensional Hilbert space H_2. An orthonormal basis of H_2 is specified by $\{|0>, |1>\}$. The state of the qubit is an associated unit length vector in H_2. If a state is equal to a basis vector then we say it is a pure state. If a state is any other linear combination of the basis vectors, we say it is a mixed state, or that the state is a superposition of $|0>$ and $|1>$ [45, 46]. In general, the state of a quantum bit is described by Eq. (1) Where $|\psi>$ is a quantum state, α and β are complex numbers:

$$\Psi = \alpha|0> + \beta|1> \tag{1}$$

The quantum bit can be measured in the traditional basis where it is equal to the probability of effect for α^2 in $|0>$ direction and the probability of effect for β^2 in $|1>$ direction where α and β must be constrained by Eq. (2) and Fig. 1 [18, 44]

$$\alpha^2 + \beta^2 = 1 \tag{2}$$

3 Quantum Bit Transformation

Quantum computers can manipulate quantum information where the quantum state can be transformed from a pure or mixed state to another pure or mixed state correspondingly. The transformation can be achieved by applying a unitary linear operation $\breve{U}$, where $\breve{U}\breve{U}\dagger = \breve{U}\dagger\breve{U} = I$ if the quantum state is a single qubit. If we have a pure quantum state $|\psi>$ can be transformed into another pure state $\breve{U}|\psi>$,

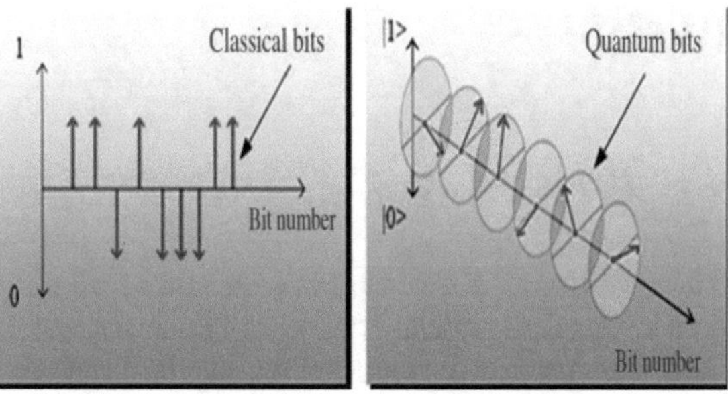

Fig. 1 Classical and quantum bits

as well if we have a mixed quantum state ρ can be transformed into another mixed state $\breve{U}\rho\breve{U}\dagger$ [12, 14, 17, 18, 23]. The unitary transformation operations are defined by (Eq. (3)). X, Y, Z can be used instead of σ_x, $i\sigma_y$, σ_z respectively. They're used to transform *GHZ* state at sender(s) side into unreadable form according to the generated original classical message before transmitting it to the receiver.

$$\begin{aligned} I &= |0\rangle\langle 0| + |1\rangle\langle 1| \\ X &= |0\rangle\langle 1| + |1\rangle\langle 0| \\ Y &= |0\rangle\langle 1| - |1\rangle\langle 0| \\ Z &= |0\rangle\langle 0| - |1\rangle\langle 1| \end{aligned} \tag{3}$$

In case of the quantum state is a multi qubit, the transformation can be achieved by applying controlled quantum gates as *CNOT* (*Controlled-NOT*), *FREDKIN* (*Controlled-NOT*), and *TOFFOLI* (*Controlled-NOT*). *CNOT* has two qubits as input, it transforms the computational basis states by flipping the state of the second qubit only when the first qubit is 1, otherwise the quantum state remain unchanged [12, 14, 17, 18, 47]. To achieve such kind of changes, a unitary transformation operation is required. Mostly, the quantum transformation concept can be stated as follows; let $|A>$ and $|B>$ to be a random qubit with $|A>$ as input and $|B>$ as output then a random quantum transformation is defined as $|B> = F|A>$, where F is a random matrix for a quantum transformation operation and its size be contingent with the dimensions of both $|A>$ and $|B>$.

4 Single Quantum Bit Gates

A single quantum bit state $|\psi>$ can be transformed into another quantum state $\breve{U}|\psi>$ by employing a single quantum gate. The commonly representation of a single gate is 2 × 2 unitary matrices. Each unitary transformation operation ($\breve{U}$) is an authoritative single quantum bit gate. We can generate unlimited number of single quantum bit gate by utilizing unlimited number of 2 × 2 unitary matrices. Conversely, in the traditional classical system, just two single bit logic gates are workable, specifically the Uniqueness gate and the logical *NOT* (Invertible) gate [17, 48–51].

4.1 Pauli X Gate

The operation of X gate (σ_X) in a quantum system is the same as operation of *NOT* gate in classical system. Unlike the classical system *NOT* gate accepts only 0 or 1, the X gate accepts a superposition quantum state as input. X gate changes the state

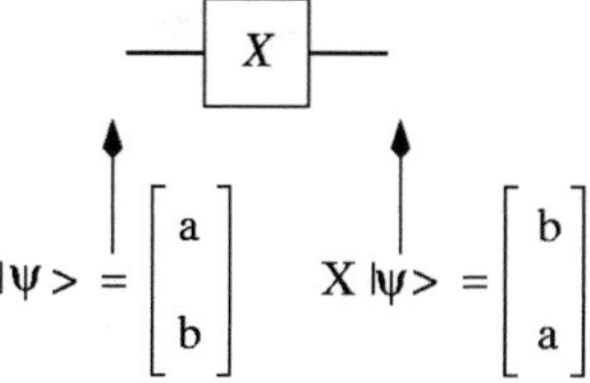

Fig. 2 *X* Gate operation [17]

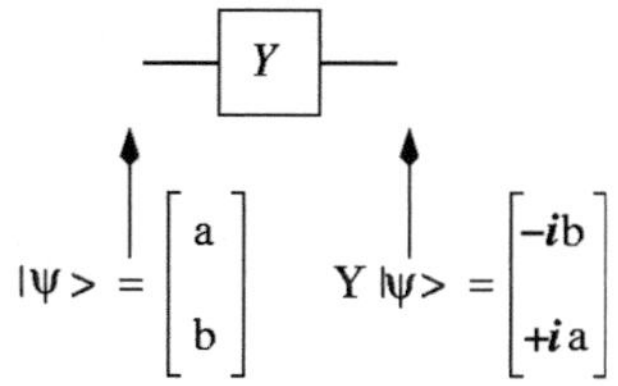

Fig. 3 Y Gate operation [49]

of quantum bit. Let $|\psi> = \alpha|0> + \beta|1>$ be a random quantum state, then the *X* gate changes it to $|\psi> = \alpha|1> + \beta|0>$. *X* gate transforms $|0>$ to $|1>$ and $|1>$ to $|0>$ (see Eq. (4)) and Fig. 2 [17, 48–51].

$$X|\psi> = \begin{bmatrix} 0 & 1 \\ 1 & 0 \end{bmatrix} \begin{bmatrix} \alpha \\ \beta \end{bmatrix} = \begin{bmatrix} \beta \\ \alpha \end{bmatrix} = \alpha|1> + \beta|0> \tag{4}$$

4.2 Pauli Y Gate

Y gate involves an imaginary component *i*. The operation of *Y* gate (σ_Y) is to flip the quantum bit. Let $|\psi> = \alpha|0> + \beta|1>$ be a random quantum state, then the *Y* gate changes it to $|\psi> = e^{i\pi/2}(-\beta|0> + \alpha|1>$ (see Eq. (5)) and Fig. 3 [17, 48–51].

$$Y|\psi> = \begin{bmatrix} 0 & -i \\ i & 0 \end{bmatrix} \begin{bmatrix} \alpha \\ \beta \end{bmatrix} = e^{\frac{i\pi}{2}}(-\beta|0> + \alpha|0>) \tag{5}$$

4.3 Pauli Z Gate

The operation of *Z* gate (σ_Z) is to rotate $\pi/2$ around $|1>$ axis. Let $|\psi> = \alpha|0> + \beta|1>$ be a random quantum state, then the *Z* gate changes it to $|\psi> = \alpha|0> - \beta|1>$ (see Eq. (6)) and Fig. 4 [17, 48–51].

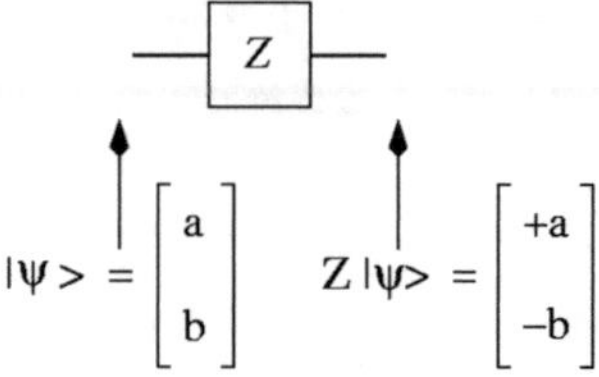

Fig. 4 Z Gate operation [50]

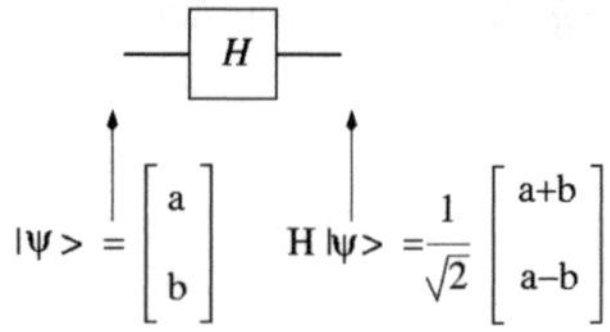

Fig. 5 **Hadamard** Gate operation [48]

$$Z|\psi\rangle = \begin{bmatrix} 1 & 0 \\ 0 & -1 \end{bmatrix}\begin{bmatrix} \alpha \\ \beta \end{bmatrix} = \begin{bmatrix} \alpha \\ -\beta \end{bmatrix} = \alpha|0> - \beta|1> \tag{6}$$

4.4 Hadamard Gate

The *Hadamard* (*H*) gate has a relation with Pauli gates, as the operation of *Hadamard* gate is equal to both *X* and *Z* gates. Let $|\psi> = \alpha|0> + \beta|1>$ be a random quantum state, then the *H* gate changes it to $\frac{(\alpha+\beta)}{\sqrt{2}}|0> - \frac{(\alpha+\beta)}{\sqrt{2}}|1>$ (see Eqs. (1.7 – 1.10)) and Fig. 5 [17, 48–51].

$$H = \begin{bmatrix} 1 & 1 \\ 1 & -1 \end{bmatrix} \tag{7}$$

$$H|0> \to |+> = \frac{(0+1)}{\sqrt{2}} \tag{8}$$

$$H|1> \to |-> = \frac{(0-1)}{\sqrt{2}} \tag{9}$$

$$H|\psi> \to \alpha|+> + \beta|-\rangle = \frac{(\alpha+\beta)}{\sqrt{2}}|0> - \frac{(\alpha+\beta)}{\sqrt{2}}|1> \tag{10}$$

5 Two Quantum Bits Gates

5.1 Controlled-NOT

In the case of the quantum state is a multi-quantum bits, so the transformation can be achieved by applying controlled quantum gates as *CNOT* (*Controlled-NOT*), *FREDKIN* (*Controlled-SWAP*), and *TOFFOLI* (*Controlled-Contolled-NOT*). *CNOT* has two quantum bits as input, it transforms the computational basis states by flipping the state of the second qubit only when the first quantum bit is 1, otherwise the quantum state remain unchanged [17, 48–51]. The circuit for *CNOT* is shown in Fig. 6 where the upper line denotes the control quantum bit and the lowest line represents the target quantum bit. So, if the input for *CNOT* circuit is 00, 01, 10 and 11, therefore the output is 00, 01, 11 and 10 respectively (see Eq. (11)) and Fig. 7

$$CNOT = \begin{bmatrix} 1 & 0 & 0 & 0 \\ 0 & 1 & 0 & 0 \\ 0 & 0 & 0 & 1 \\ 0 & 0 & 1 & 0 \end{bmatrix} \tag{11}$$

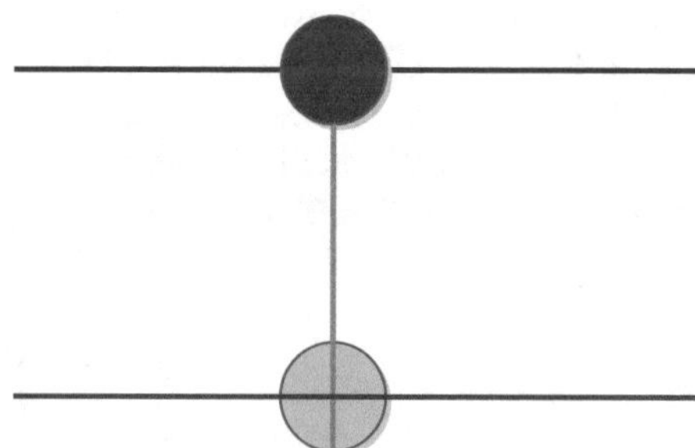

Fig. 6 *Controlled-NoT* Gate operation [51]

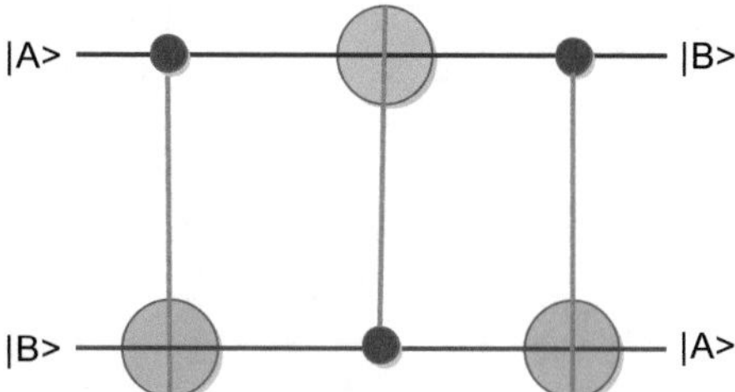

Fig. 7 **SWAP** Gate operation [51]

5.2 *SWAP Gate*

SWAP gate has two quantum bits as input. It swaps the states of the two quantum bits; thus it maps $|ab> \rightarrow |ba>$. So, if the input for *Controlled-SWAP* circuit is 00, 01, 10 and 11 therefore, the output is 00, 10, 01 and 11 respectively (see Eq. (12)) and Fig. 7 [17, 48–51].

$$SWAP = \begin{bmatrix} 1 & 0 & 0 & 0 \\ 0 & 0 & 1 & 0 \\ 0 & 1 & 0 & 0 \\ 0 & 0 & 0 & 1 \end{bmatrix} \tag{12}$$

6 Three Quantum Bits Gates

6.1 *TOFFOLI (Controlled-Controlled-NOT) Gate*

TOFFOLI gate has three quantum bits as input. It transforms the computational basis states by flipping the state of the third quantum bit only when the first two quantum bits are 1, otherwise the quantum state remain unchanged [17, 48–51]. The circuit for *TOFFOLI* is shown in Fig. 8. Where the upper line denotes the target quantum bit and the lowest two line represents the control quantum bits. So, if the input for *TOFFOLI* circuit is 000, 001, 010, 011, 100, 101, 110 and 111 therefore the output is 000, 001, 010, 011, 100, 101, 111 and 110 respectively (see Eq. (13))

$$TOFFOLI = \begin{bmatrix} 1 & 0 & 0 & 0 & 0 & 0 & 0 & 0 \\ 0 & 1 & 0 & 0 & 0 & 0 & 0 & 0 \\ 0 & 0 & 1 & 0 & 0 & 0 & 0 & 0 \\ 0 & 0 & 0 & 1 & 0 & 0 & 0 & 0 \\ 0 & 0 & 0 & 0 & 1 & 0 & 0 & 0 \\ 0 & 0 & 0 & 0 & 0 & 1 & 0 & 0 \\ 0 & 0 & 0 & 0 & 0 & 0 & 0 & 1 \\ 0 & 0 & 0 & 0 & 0 & 0 & 1 & 0 \end{bmatrix} \tag{13}$$

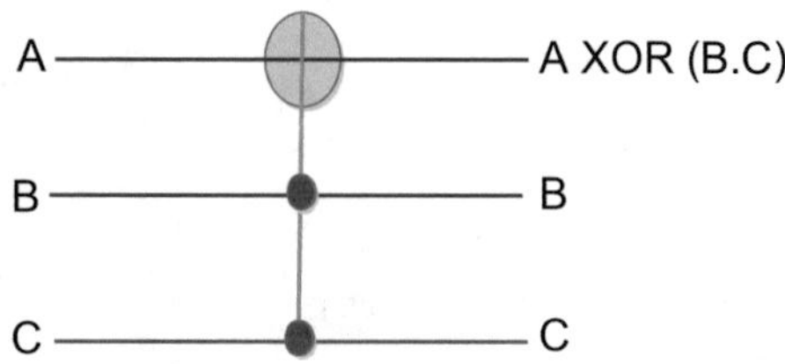

Fig. 8 **TOFFOLI** Gate operation [48]

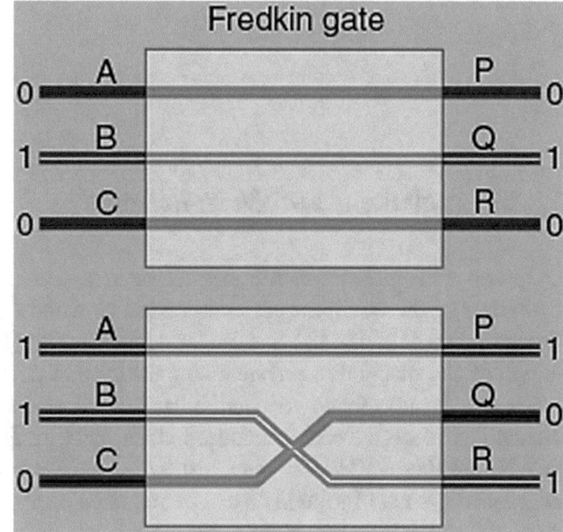

Fig. 9 *FREDKIN* Gate operation [48]

6.2 FREDKIN (Controlled-SWAP)

FREDKIN gate has three quantum bits as input. It swaps the states of the second and third quantum bit only if the first quantum bit is equal to 1 as shown in Fig. 9. So, if the input for *Controlled-Swap* circuit is 000, 001, 010, 011, 100, 101, 110 and 111 therefore, the output is 000, 001, 010, 011, 100, 110, 101 and 111 respectively (see Eq. (14)) [17, 48–51]

$$FREDKIN = \begin{bmatrix} 1 & 0 & 0 & 0 & 0 & 0 & 0 & 0 \\ 0 & 1 & 0 & 0 & 0 & 0 & 0 & 0 \\ 0 & 0 & 1 & 0 & 0 & 0 & 0 & 0 \\ 0 & 0 & 0 & 1 & 0 & 0 & 0 & 0 \\ 0 & 0 & 0 & 0 & 1 & 0 & 0 & 0 \\ 0 & 0 & 0 & 0 & 0 & 0 & 1 & 0 \\ 0 & 0 & 0 & 0 & 0 & 1 & 0 & 0 \\ 0 & 0 & 0 & 0 & 0 & 0 & 0 & 1 \end{bmatrix} \quad (14)$$

7 Entanglement and Measurement

Quantum entanglement is a constructive area which can be used to measure the operation of the quantum communication as Shor's algorithm [52–54], Quantum Teleportation [13, 55–58], and Super dense Coding [11, 12, 59, 60]. Entanglement is one of the distinctive physical singularity that defines the way of how the particles can be correlated to each other regardless of the distance. Entanglement means that if we have two quantum states $|\psi>_A \in H_A$ and $|\psi>_B \in H_B$ where H_A and H_B are the two Hilbert space for $|\psi>_A$ and $|\psi>_B$ respectively. Therefore, the composite system of joint state can be represented by $|\psi>_{AB} \in H_A \otimes H_B$. The joint state can or can't be represented by the tensor product of the Hilbert space, as the two entangled particles will have interrelated physical characteristics even though they're disjointed by distance. If the composite system state can be written as Eq. (15), then the composite state is distinguishable [16–18, 56].

$$|\psi>_{AB} = |\psi>_A \otimes |\psi>_B \quad (15)$$

Otherwise, the composite system state is entangled if it meets the condition for entanglement (see Eq. (16))

$$|\psi>_{AB} \neq |\psi>_A \otimes |\psi>_B \quad (16)$$

$|00> = |0> \otimes |0>, 1/\sqrt{2}(|01> \pm |11>) = 1/\sqrt{2}(|0> \pm |1>) \otimes |1>$ are examples of distinguishable states. If the quantum state is a mixed one, then the composite system state can be characterized by the density matrix.

7.1 Bell States

The Bell states are one of the main theories in quantum information processing which denote the entanglement concept [61–64]. Bell states are certain extremely entangled quantum states of two particles denoted by *EPR*. As the two entangled particles will have interrelated physical characteristics even though they're disjointed by distance. Bell states are entitled in many applications but the most useful examples are quantum teleportation and dense coding [17, 18]. The really essential phase for quantum teleportation and dense coding is Bell measurement. The

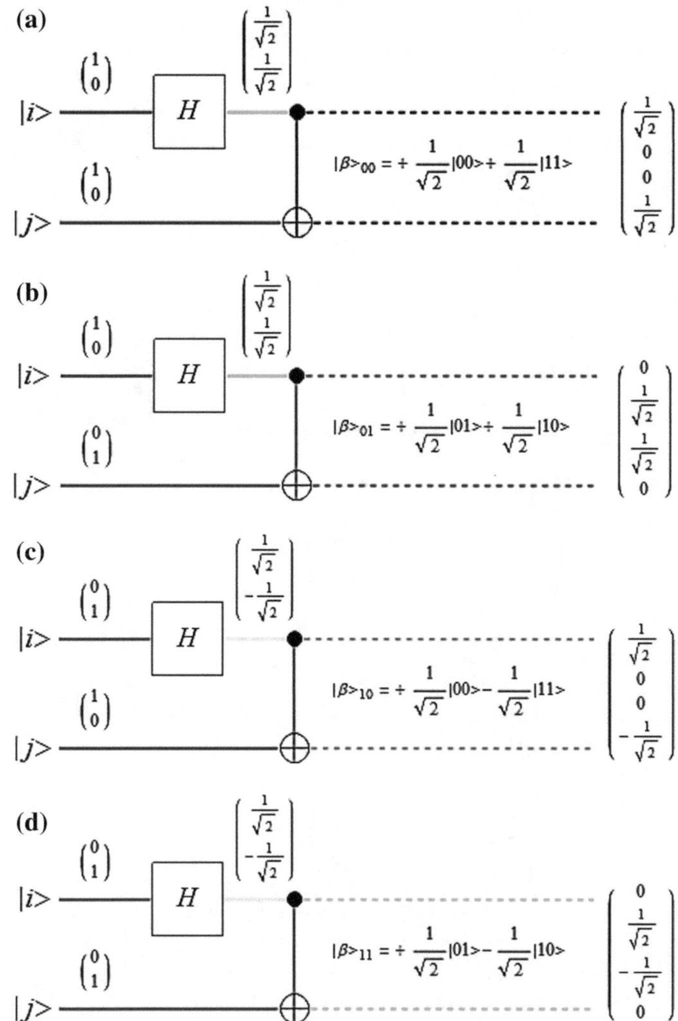

Fig. 10 Bell state generation **a** Generation of bell state $|\Phi^+>$ while **i** and **j** equal to 00, **b** Generation of bell state $|\psi^+>$ while **i** and **j** equal to 01, **c** Generation of bell state $|\Phi^->$ while **i** and **j** equal to 10, **d** Generation of bell state $|\psi^->$ while **i** and **j** equal to 11 [12, 18]

outcome of Bell measurement is a couple of classical bits, which can be used for retrieve the original state [12, 65]. Bell measurement is used in Communication Process for determining which unitary operation is used to transform the original classical message so, the receiver can retrieve it. The four Bell states (*EPR* pairs) are defined by (Eq. (17))

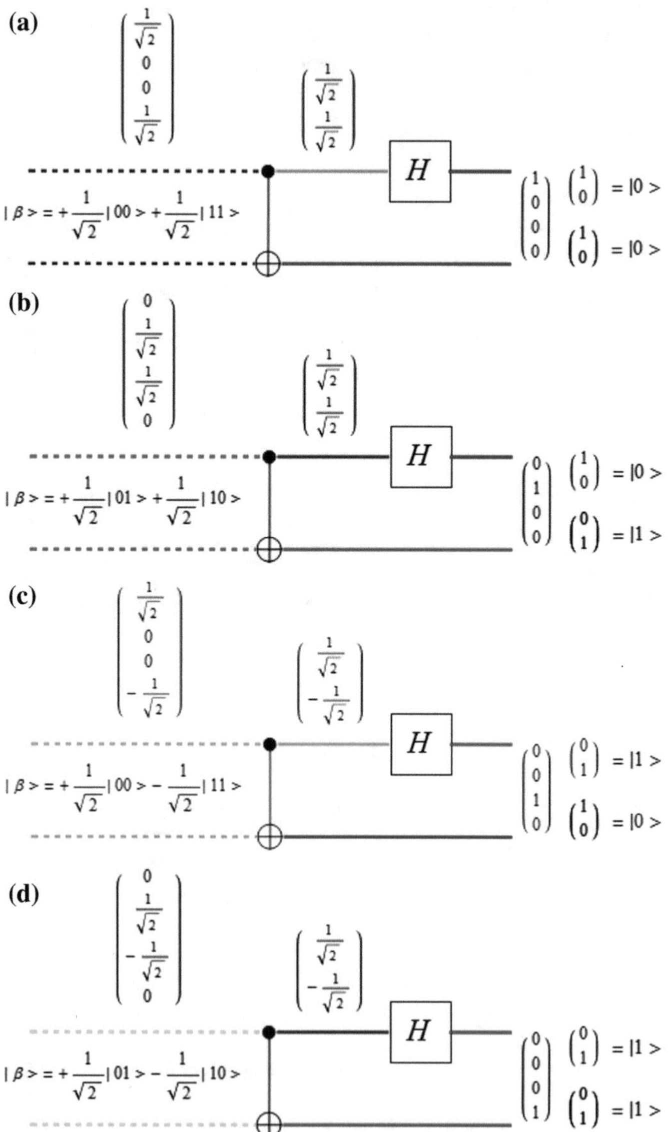

Fig. 11 Bell state measurement **a** Measurement of bell state $|\boldsymbol{\Phi}^+>$ yields to 00, **b** Measurement of Bell State $|\psi^+>$ yields to 01, **c** Measurement of Bell State $|\boldsymbol{\Phi}^->$ yields to 10, **d** Measurement of Bell State $|\psi^->$ yields to 11 [12, 18]

$$|\Phi^{\pm}> = \frac{1}{\sqrt{2}}(|00> \pm |11>) \quad |\psi^{\pm}> = \frac{1}{\sqrt{2}}(|01> \pm |10>) \tag{17}$$

Bell states can be generated by utilizing the properties of *Hadamard* gate and *Controlled-NOT* gate. Figure 10 shows the four possibilities of Bell states (EPR) according to the input bits. While the input bits are 00, 01, 10 and 11, then the generated EPR states are $|\Phi^+> = \frac{1}{\sqrt{2}}(|00> + |11>)$ $|\psi^+> = \frac{1}{\sqrt{2}}(|01> + |10>)$, $|\Phi^-> = \frac{1}{\sqrt{2}}(|00> - |11>)$ $|\psi^-> = \frac{1}{\sqrt{2}}(|01> - |10>)$.

Bell measurement refers to apply a projective measurement according to bell basis. Bell measurement plays an important role in quantum teleportation and cryptography. As shown in Fig. 11, Bell measurement can be achieved by choosing one of four maximally entangled Bell states and applying the measured quantum circuit.

The quantum circuit performs a *Controlled-NOT* gate on the entangled pair of quantum state, afterward a *Hadamard* gate on the upper quantum state. This produces a pair of classical bits, which can then be measured to detect which of the four Bell entangled states were input to the circuit [12, 17–19, 44].

7.2 GHZ *States, Measurement and Source*

GHZ states are certain maximally entangled quantum states which include at least three qubits (particles). It was first examined by Greenberger, Horne, and Zeilinger in 1989 [66]. The standard *GHZ* state was defined for $qubits = 3$ but for $qubits > 3$, the *GHZ* state is defined by (Eq. (18))

$$|GHZ> = \frac{|0>^{\otimes \text{qubits}} + |1>^{\otimes \text{qubits}}}{\sqrt{2}} \tag{18}$$

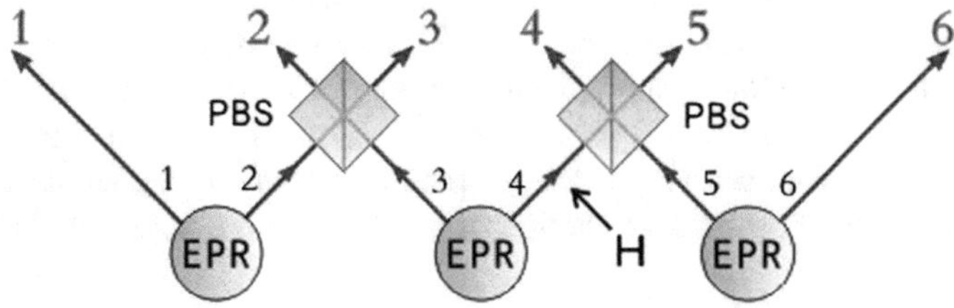

Fig. 12 Experimental entanglement of six photons *GHZ* state [67]

GHZ measurement is used by receiver or quantum server according to which type of used cooperation during the communication among the users. Consistent with *GHZ* measurement result, the receiver can conclude which unitary operations are used by senders for transforming *GHZ* state according to the generated original classical message before transmit it to the receiver. By obtaining the applied unitary operations, the receiver can retrieve the sent original message [12, 17–19, 44, 67]. The Eight *GHZ* States are defined by (Eq. (19))

$$\begin{aligned} |\Psi^{\pm}> &= \frac{1}{\sqrt{2}}(|000> \pm |111>), \quad |\psi^{\pm}> = \frac{1}{\sqrt{2}}(|011> \pm |100>) \\ |\phi^{\pm}> &= \frac{1}{\sqrt{2}}(|010> \pm |101>), \quad |\varphi^{\pm}> = \frac{1}{\sqrt{2}}(|001> \pm |110>) \end{aligned} \tag{19}$$

In [67], an Experimental entanglement of six photons *GHZ* state and cluster state in graph states has been proposed. The generation of six-photon *GHZ* states and cluster state is achieved by Einstein Podolsky-Rosen (*EPR*) entangled photon pairs (see Fig. 12).

8 Quantum Cryptography

The pioneering work of Bennett and Brassard has been developed for the purpose of quantum cryptography [9]. Quantum cryptography is one of the most significant prospects associated with laws of quantum mechanics in order to ensure unconditional security [16–19, 47]. The quantum cryptography proves unconditional security characteristic through no cloning theory as the transmitted quantum bit can't be replicated or copied but its state can be teleported [8]. The most used quantum principles are quantum teleportation and dense coding. In quantum teleportation, the quantum information can be transmitted between distant parties based on both classical communication and maximally shared quantum entanglement among the distant parties [13–16]. In Dense coding the classical information can be encoded and transmitted between distant parties based on both one quantum bit and maximally shared quantum entanglement among the distant parties as each quantum bit can transmit two classical bits [11, 12]. There are number of approaches and prototypes for the exploitation of quantum principles' to secure the communication between two parties and multi-parties [68–72]. While these approaches used different techniques for achieving a private communication among authorized users, but most of them depend on generation of a secret random keys [73, 74].

8.1 Security Requirements of Communication

The performance is an indispensable concern for designing a secured classical or quantum communication system. The only way to ensure the optimum performance of the communication system is the Confidentiality, Integrity and Availability. Confidentiality means the transmitting message over the communication channel should be secured against intruder, so the intruder can't gain any useful information from it. The Confidentiality goal is achieved through encryption and authentication principles. Encryption assures that the transmitted message is converted to unreadable format. Authentication verifies the identity of communicators and incorporation of the transferred message among them. So, copy or fake of the transmitted message is unachievable or may be avoided which confirm that the message has not been changed during the transmission by intruder. Integrity refers to the credibility and securing information against incorrectly changed by illegal participants during the communication process. The integrity goal is achieved by cryptography which acts as encryption and authentication to assure data integrity. Availability denotes that information can be retrieved by authorized users at any given instant of time. Availability can be reached through a copy of data may be stored by offline or online backup sites. With the intention of explore properties of the private communication [111, 112], it is encouraged for developing a private communication model [16–19, 44]. In 1949, Shannon exhibited a mathematically classical private communication model, afterwards a quantum private communication system is structured accordingly with the same idea of classical one. In a private communication system, there are normally three correspondents (sender, receiver and eavesdropper) participate in the communication process. The eavesdropper will attempt to spy on the communication channel, therefore he can monitor and steal a valuable information from the transmitted messages. There're two types of attack approaches; passive and active. Passive attack is only spy and disturbing the confidentiality of the transmitted messages without alteration it. On the other side, the active attack includes the alteration of transmitted messages and attempts to gain unauthorized access to the communication systems. In support of protecting the transmitted messages with respect of confidentiality and authentication, a private communication model should be established [44, 75, 76].

8.2 Quantum Private Communication Model

At present, there're two approaches of quantum private communication. One is a hybrid of classical cryptosystem and quantum key distribution. The other one applies a completely quantum cryptosystem with natural quantum physics laws. In first approach, the employed encoding and decoding algorithms come from classical. Whilst the generated keys for message encoding and decoding which act as significant role in the cryptosystem derives from a distinguished quantum key

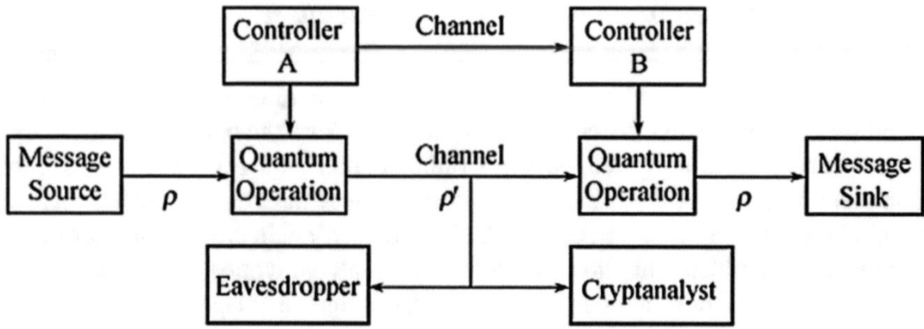

Fig. 13 Quantum private communication model [44]

distribution scheme. In second approach, the encoding and decoding algorithms are quantum one and the keys for message encoding and decoding derives from a distinguished quantum key distribution scheme. The quantum communication system can be described using the same way of classical model. The messages in quantum system represented by quantum state which can be pure or mixed [16–19, 44]. In Fig. 13, ρ and $\rho^{'}$ represent the quantum states of original message and cipher message respectively. The controllers A and B denote the employed encryption and decryption quantum keys. The generated quantum keys have to follow the steps of quantum key distribution technique. The controllers A and B infer random numbers for controlling the selection of appropriate measurements bases by the communicators. The quantum operation indicates the used encryption and decryption algorithm which can be classical or quantum. The most three principal components for designing a quantum communication system are cryptosystem, authentication and key management system. All included processes in these components may be classical or quantum but in any case at minimum one of these components has to apply a quantum features and laws [16–19, 44].

8.3 No Cloning Theorem

In 1982, Wootters et al. [8, 18, 44] have identified the no-cloning theorem and its deep effects in the quantum computing and associated areas. The no-cloning theorem is a consequence of quantum system which prevents the formation of duplicate copies of an unidentified random quantum bit. The property of quantum no-cloning is a fundamental component in the quantum cryptography, as it prevents eavesdroppers from creating copies of a transmitted quantum cryptographic key.

Quantum no-cloning theorem proves that there is no such operation U that can perform the cloning operation for any arbitrary quantum state which means an arbitrary, unknown quantum state cannot be copied exactly without altering the original state in any way. Accordingly, the no-cloning theorem forbids intruders or eavesdroppers to create identical replicas of unidentified random quantum state and forwards it without change as *QKD*, which has been invented and plays an important role in quantum cryptography and quantum private communication [8, 18, 44].

This characteristic leads to the particular copy of quantum state which is very simple in the classic world to be unachievable. Furthermore, the no-cloning theory inhibits us from consuming classical error correction methods on quantum states. Consequently, no one will be able to generate alternate replicas of a quantum state during the quantum computation [8, 18, 44].

Theorem In a complex Hilbert space H, there does not exist a unitary transformation $U: H \otimes H \rightarrow H \otimes H$ such that there exists a state $|s> \in H$ satisfying the following condition in Eq. (20)

$$U(|\psi>|s>) = |\psi>|\psi>, \quad \forall|\psi> \in H \tag{20}$$

Proof Assume that such a unitary operation U exists, then we can have Eq. (21) and Eq. (22)

$$U(|\psi>|s>) = |\psi>|\psi> \tag{21}$$

$$U(|\phi>|s>) = |\phi>|\phi> \tag{22}$$

Now combining Eq. (21) and Eq. (22) and using the property of unitary operator, we may obtain Eq. (23)

$$<\psi|\phi> = <\psi| <s|s> \phi> = <\psi, s|U^{\dagger}U|s, \phi> = <\psi, \psi|\phi, \phi> = (<\psi|\phi>)^2 \tag{23}$$

Equation (23) can be satisfied only for $<\psi|\phi> = 0$ or $<\psi|\phi> = 1$. Consequently, our initial assumption is wrong and we cannot copy.

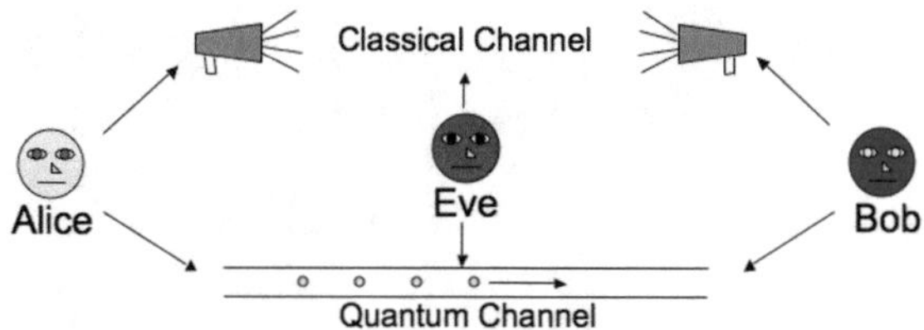

Fig. 14 Quantum key distribution [18]

8.4 Quantum Key Distribution

Innovative communication systems depend on cryptographic methods to guarantee confidentiality and integrity of transmitted traffic among communicated peers over the communication network. Basically, cryptographic methods rely on the generated and distributed secret keys for encryption, decryption and authentication process. According to basic characteristics of quantum physics, a new model for key generation is initiated and known as Quantum Key Distribution "*QKD*" [18, 19, 44, 77].

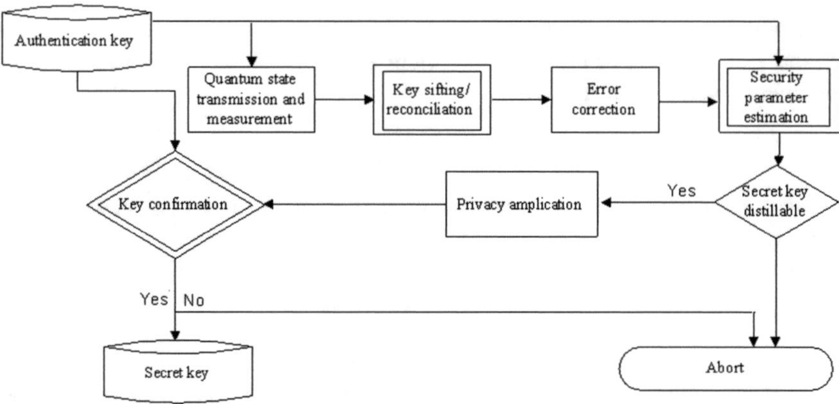

Fig. 15 Basic stages of quantum key distribution [44]

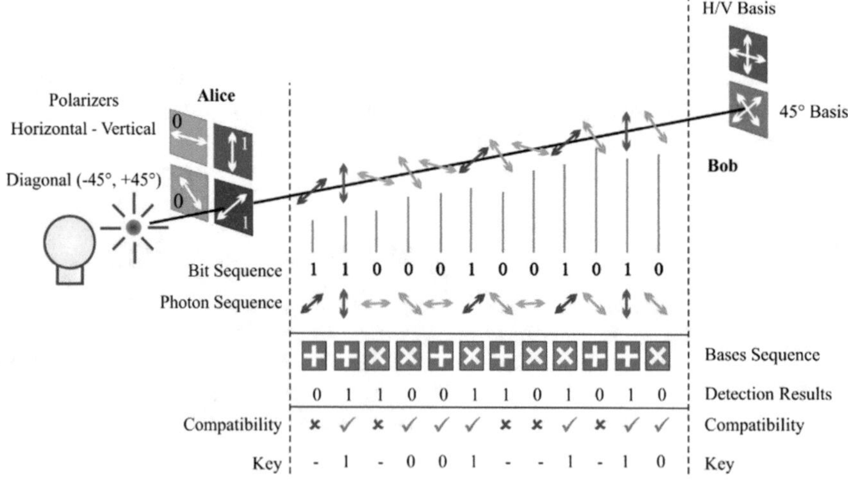

Fig. 16 Example of *QKD* between Alice and Bob [78]

QKD uses two channels one is quantum channel and other is classical channel. The quantum channel is used for transmission of quantum keys through light pulses. The classical channel is used for transmission of cryptographic protocols, ciphered traffic and key agreement protocols. By the nature of quantum physics, any attacker that spies on the quantum channel will produce a measurable interruption to the movement of single and continues fire of photons (see Fig. 14) [18, 19, 44, 77].

8.5 Basic Stages of Quantum Key Distribution

According to Figs. (15 and 16), a *QKD* scheme consists of three stages are [44, 78, 79]: First stage; Quantum Coding and Transmission. Second stage; Raw Key Generation. Third stage; Eavesdropper Detection. First stage runs through a quantum channel. The sender generates a random bit string and encodes each with a quantum source. After encoding, quantum bits are transmitted actually from the sender to the receiver over a transmission channel. Clearly, a *QKD* system requires a transmission channel so that encoding quantum bits are transmitted from one communicator to another communicator through quantum carriers. Two popular types of transmission channels are optical fiber used for telecommunication networks and open air for satellite communications.

The receiver generates measurements on received encoded qubits by selecting basis on his realized [44, 78, 79]. In second stage, during transmission, communicators use different bases for measurements purpose. The objective of this step is to recognize and exclude those bit positions where communicators use different bases. These positions are then discarded by both communicators over a public channel. [44, 78, 79]. In third stage; during the transmission between communicators; eavesdropper might spy on quantum channel and retrieve potential secret key bits. Figure 16 illustrates the steps for quantum key distribution between Alice and Bob with final shared sifted key equal to 1001110.

Using quantum laws, eavesdropper operation on quantum channel can be detected. Eavesdropping is discovered as follows; An arbitrary subset of the raw key is agreed by communicators, and those bits are evaluated openly. If whichever two agreeing bits vary, this specifies the existence of an eavesdropper and so communicators go back to stage 1. Otherwise, the exchanged bits will be abandoned and the remainder of the raw key is used as the final secret key [44, 78, 79].

8.6 Quantum Teleportation and Super-Dense Coding

The quantum cryptography proves unconditional security characteristic through no cloning theory as the transmitted quantum bit can't be replicated or copied but its state can be teleported [8]. The most used quantum principles are quantum teleportation and dense coding. In quantum teleportation, the quantum information can be transmitted between distant parties based on both classical communication and maximally shared quantum entanglement among the distant parties [13–16]. In Dense coding, the classical information can be encoded and transmitted between distant parties based on both one quantum bit and maximally shared quantum entanglement among the distant parties as each quantum bit can transmit two classical bits [11, 12]. There are number of approaches and prototypes for the exploitation of quantum principles to secure the communication between two parties and multi-parties [68–72]. While these approaches used different techniques for achieving a private communication among authorized users, but most of them depend on the generation of secret random keys [73, 74].

In quantum teleportation, supposes that there're two spatially disjointed communicators and sender would like to transmit unidentified quantum state $\Psi = \alpha|0> + \beta|1>$ to the receiver on the other side. The sender can't send the state completely over the quantum channel since the time he will attempt for measuring the value of transmitted state, it will be collapsed either to $|0>$ or $|1>$. The stated problem is solved by employing quantum teleportation characteristic. Therefore, the sender can transmit the quantum state by both classical communication and maximally shared quantum entanglement state. The maximally shared quantum entanglement state represents the quantum channel between the communicators. Teleportation can be accomplished by applying various kinds of entangled states as a quantum channel. The quantum state will be destroyed at the sender side and rematerialized at the receiver side. In [13], the novel teleportation scheme has been proposed and followed by an enormous number of teleportation approaches and their applications. The teleportation approaches can be classified either perfect or probabilistic.

Perfect teleportation signifies that the achievement degree is unity. It involves a maximally entangled quantum channel. However, after the innovative realization of Bennett et al. it was attained that teleportation can be achieved even if the employed quantum channel is non-maximally entangled. Therefore the achievement degree of the receiver will be a probabilistic one that is called probabilistic teleportation [12, 14, 17–19, 44, 80].

Teleportation approaches are not restricted to two-communicator teleportation, but also generalized to several communicators quantum teleportation. One of the most used multi-communicators quantum teleportation approaches is controlled teleportation (CT). In this approach, the sender shares previous entanglement with the receiver and as a minimum one trusted center (TC). No cloning theory proves that the sender can't teleport duplicates of the transmitted unknown quantum state to both the receiver and trusted center. Subsequently, if the sender succeeds for teleporting the unknown quantum state to both the receiver and trusted center, afterward only one of them can create a copy of the transmitted unknown quantum state with the support of the other. As a consequence, the transmitted information is fragmented between the sender and the trusted center, so both will cooperated together for retrieving the transmitted unknown state by the sender. Meanwhile, the trusted center control the whole teleportation process. This protocol is denoted as controlled teleportation (CT) [12, 14, 17–19, 44, 80].

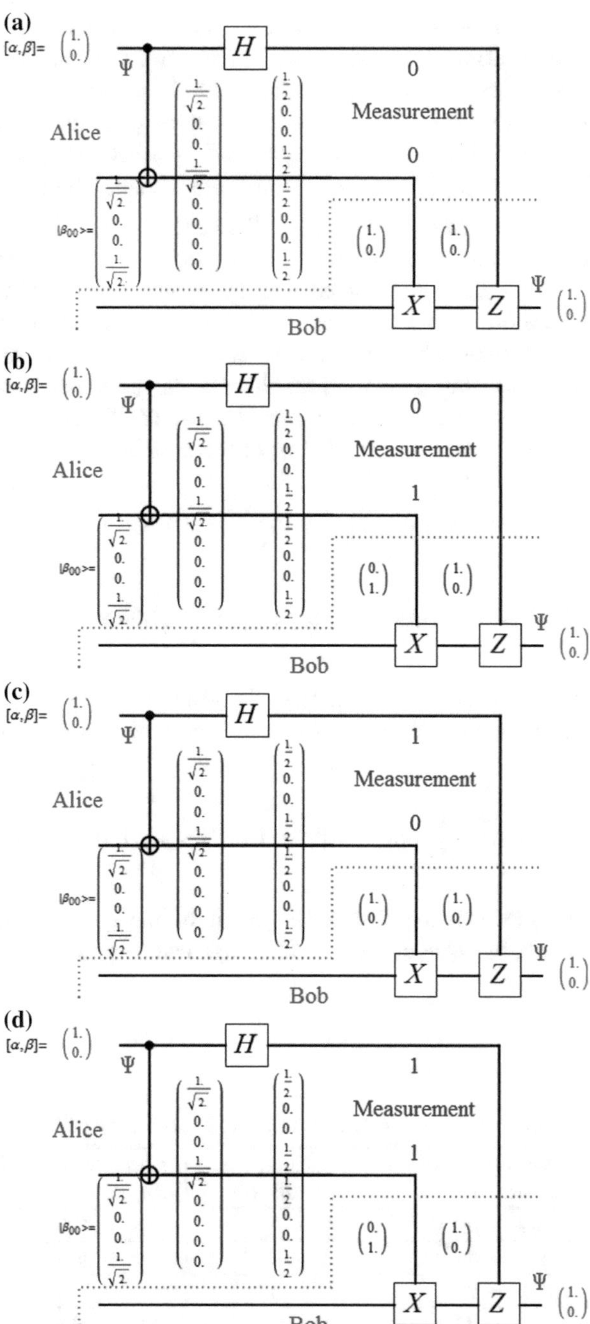

Fig. 17 Quantum teleportation **a** Measurement of bell state $|\Phi^+>$ yields to 00, **b** Measurement of bell state $|\psi^+>$ yields to 01, **c** Measurement of bell state $|\Phi^->$ yields to 10, **d** Measurement of Bell State $|\psi^->$ yields to 11 [18, 44]

The most common maximally entangled quantum state intended for implementing a quantum channel for two-communicator teleportation scheme is EPR pairs. Nevertheless, *GHZ* state, *GHZ*-like state and *W* state are commonly used as a quantum channel for applying several communicators quantum teleportation. Figure 17 shows an illustrative example of perfect teleportation where $|\psi^+> = \frac{1}{\sqrt{2}}|00> + |11>$ is used as a quantum channel between sender and receiver. Wherever the first quantum state of the EPR pair is retrieved by the sender and the second quantum state of EPR is retrieved by the receiver. The Maximal EPR entangled pair can be generated by quantum circuit consisting of a *Hadamard* gate followed by a *Controlled-NOT* gate. Currently, the sender would like to transmit the unknown quantum state $|\psi> = \alpha|0> + \beta|1>$ to the receiver. The unknown state will move through the teleportation circuit with a *Controlled-NOT* gate and a *Hadamard* gate [12, 14, 17–19, 44, 80].

With the unknown state, the initial state of the system is defined by (Eq. (24))

$$|\psi>_1 = \alpha|0> + \beta|1> \otimes \frac{1}{\sqrt{2}}(|00> + |11>)$$

$$= \left(\alpha|0> \frac{1}{\sqrt{2}}(|00> + |11>) + \beta|1> \frac{1}{\sqrt{2}}(|00> + |11>\right) \tag{24}$$

Subsequently, the *CNOT* gate transform the state as in (Eq. (25)) according to using sender quantum bit as the control one and receiver quantum bit as the target one.

$$|\psi>_2 = \left(\alpha|0> \frac{1}{\sqrt{2}}(|00> + |11>) + \beta|1> \frac{1}{\sqrt{2}}(|10> + |01>\right) \tag{25}$$

Since sender transmits the first quantum bit of the quantum state over the *Hadamard* gate. So, the state of the overall system can be transformed as shown in (Eq. (26))

Table 1 Relationship between sender measurement and receiver's operation

Sender measurement	Status of receiver's quantum bit	Receiver's operation	Status of receiver quantum bit after pauli operation
00	$\alpha|0> + \beta|1>$	I	$\alpha|0> + \beta|1>$
01	$\alpha|1> + \beta|0>$	X	$\alpha|1> + \beta|0>$
10	$\alpha|0> - \beta|1>$	Z	$\alpha|0> - \beta|1>$
11	$\alpha|1> - \beta|0>$	$ZX = iY$	$\alpha|1> - \beta|0>$

Table 2 Relationship between sender message, receiver's operation and receiver's state

Transmitted classical bits at sender side	Pauli operation at receiver side	Final state at receiver side
00	I	$\frac{(\|00>+\|11>)}{\sqrt{2}}$
01	X	$\frac{(\|10>+\|01>)}{\sqrt{2}}$
10	iY	$\frac{(\|01>-\|10>)}{\sqrt{2}}$
11	Z	$\frac{(\|00>-\|11>)}{\sqrt{2}}$

$$
\begin{aligned}
|\psi\rangle_3 &= \begin{pmatrix} \alpha\frac{(|0>+|1>)}{\sqrt{2}}\frac{1}{\sqrt{2}}(|00>+|11>) \\ +\beta\frac{(|0>+|1>)}{\sqrt{2}}\frac{1}{\sqrt{2}}(|10>+|01>) \end{pmatrix} \\
&= \frac{1}{2}\begin{pmatrix} |00>(\alpha|0>+\beta|1>)+|01>(\alpha|1>+\beta|0>) \\ +|01>(\alpha|0>+\beta|1>)+|11>(\alpha|1>+\beta|0>) \end{pmatrix}
\end{aligned}
\tag{26}
$$

Afterward, the sender computes the first two quantum bits and publishes the result of his measurement through the classical channel. When the receiver receives the two classical bits, he will conclude which unitary operation should be applied for restructuring the transmitted original unknown quantum state sent by the sender as shown in Table 1 and Fig. 17.

In Dense coding the classical information can be encoded and transmitted between distant parties based on both one quantum bit and maximally shared quantum entanglement among the distant parties as each quantum bit can transmit two classical bits [11, 12]. The most common application of dense coding is transmitting two classical bits by utilizing only one quantum bit. Firstly, the sender and receiver share a maximal entangled EPR pair $|\psi^+> = \frac{1}{\sqrt{2}}(|00>+|11>)$. The sender reserved the first qubit for himself and transmitted the other qubit to the receiver. The sender would like to transmit a message consists of serial of classical bits, so the sender encrypts each two classical bits by applying appropriate unitary transformations. When the sender would like to send 00 or 01 or 10 or 11, he requires no less than four different unitary transformations. Therefore, he can encrypt the transmitted message 00, 01, 10 and 11 by applying U_{00}, U_{01}, U_{10} and U_{11} respectively.

Table 2 presents an illustrative example as the sender will apply $U_{00}=I$, $U_{01}=X$, $U_{10}=ZX=iY$ and $U_{11}=Z$ for encrypting 00, 01, 10 and 11 respectively. For transmitting a classical message consists of 00, the sender will not do anything as he applies I operation. If the transmitted classical message consists of 01, then he applies X gate on his first qubit. So, the original state will be transformed to $|\Phi^+> = \frac{(|10>+|01>)}{\sqrt{2}}$. Correspondingly, when the transmitted bits are 10, then he applies Y gate on his qubit and that transforms the original state to be $|\Phi^-> = \frac{(|01>-|10>)}{\sqrt{2}}$. Finally, when the transmitted bits are 11, then he applies

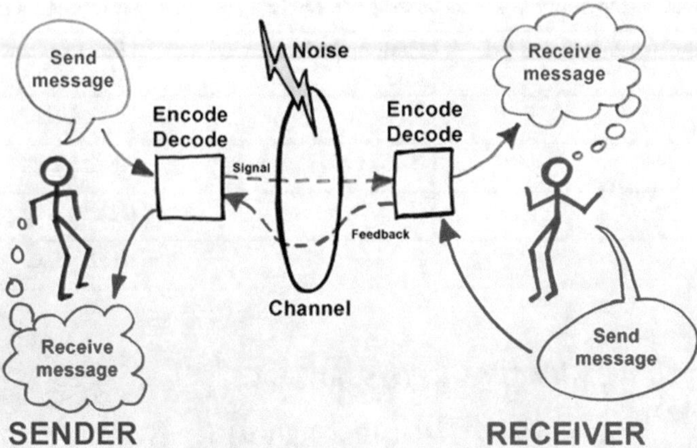

Fig. 18 Shannon communication model [75]

Z gate on his qubit and that transforms the original state to be $|\psi^- > = \frac{(|00> - |11>)}{\sqrt{2}}$. When the receiver receives the encoded state, he performs EPR measurement on the received encoded state. According to the result of measurement, the receiver will know which classical bits and applied unitary operation at the sender side. So, he can retrieve the original classical message which has been sent by the sender [11, 12, 14, 17–19, 44, 80].

9 Classical Cryptography and Multicast

9.1 Classical Private Communication Model

Shannon communication model is appropriate for both classical and quantum communication systems [75]. Classical private communication model consists of six components; source, encoding algorithm, communication channel, decoding algorithm, attacker, and receiver. The source of communication system creates consecutive series of messages and a transducer that transforms the message into an electrical signal. Message is a series of bits transmitted from source to destination or a group of destinations. The transmitted messages are encoded by cryptographic algorithms, so the transmitted messages are converted to ciphertext (see Fig. 18).

The function of the transmission process is to send and circulate an analogue or digital information signal through communication channel. A communication channel transfers the transmitted signal from one or several senders to one or several receivers. The communication channel can be guided or unguided media. Guided media includes twisted pair, coaxial cable, fiber optic which the electromagnetic waves are guided along a physical route. Unguided media includes air, vacuum, water which called wireless the transmitting electromagnetic waves which are not guided along with a physical path. The communication channel has a specified capacity for transferring information usually called bandwidth and measured in bits per seconds (bps) or in Hertz (Hz) [75, 80–84].

The capacity of the communication channel defines the maximum amount of information can be transmitted without an error and can be calculated using Shannon's channel capacity theorem by providing the following relationship for maximum channel capacity (bits per second) denoted as C, in terms of bandwidth W, S is the average received signal power over the bandwidth (in case of a modulated signal, often denoted C, i.e. modulated carrier), measured in watts (or volts squared); N is the average noise or interference power over the bandwidth, measured in watts (or volts squared); and signal—to—noise ratio $\frac{S}{N}$ is given by (Eq. (27))

$$\frac{C}{W} = \log_2\left(1 + \frac{S}{N}\right) \tag{27}$$

The attacker will try to disrupt the communication by utilizing the gained information from the channel and some previous knowledge so that he may achieved some useful information. The noise is an error, superfluous or disruption of the transmitted original message from the sender to the receive. If the transmitted has an error rate higher than the agreed threshold, then the message will be discarded and not delivered by the receiver. The receiver is the user(s) at the other end of the communication channel. He receives the encrypted message and decrypts it to retrieve the original message which sent by the sender. If the encoding and decoding keys are the same then the used cryptosystem called symmetric. Otherwise it is called asymmetric [75, 80–84].

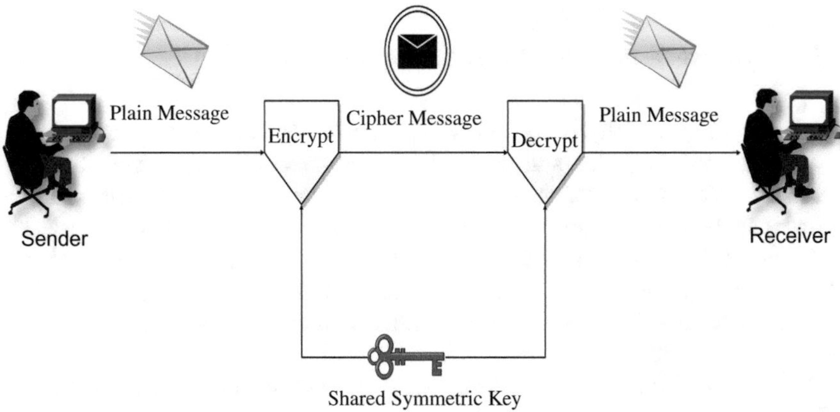

Fig. 19 Symmetric cryptography process

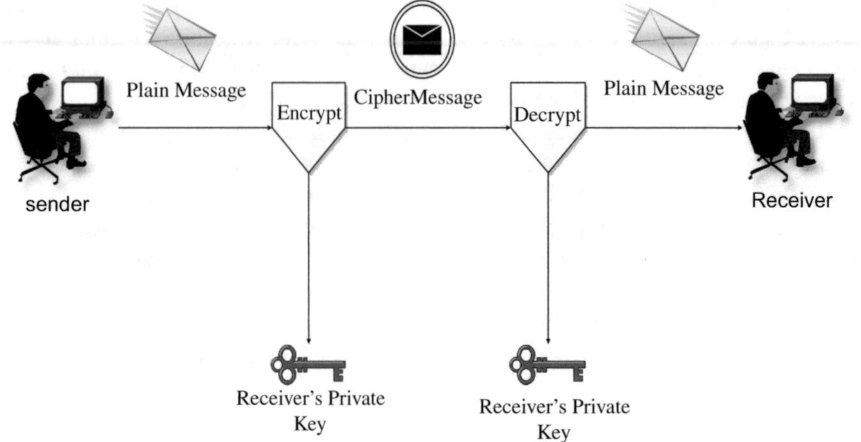

Fig. 20 A symmetric cryptography process

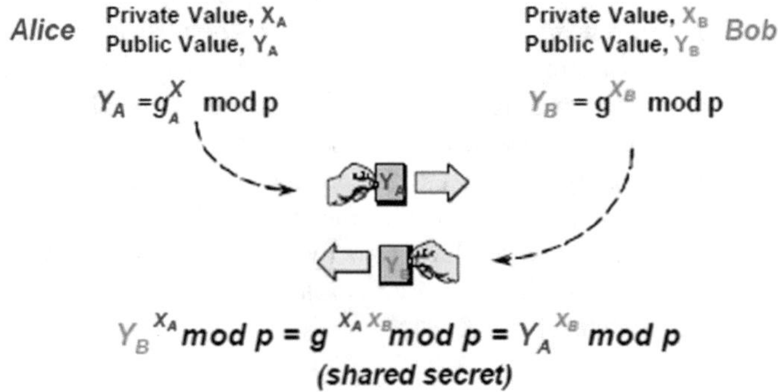

Fig. 21 Diffie Hellman summarized operation [90]

9.2 Symmetric and Asymmetric Cryptography

Symmetric Cryptography is applied when the two communicating parties share a key before any encryption and decryption to be done. The secret key should be circulated before transmission between communicated parties. The same key has the advantage of encryption and decryption data (see (Fig. 19.) Symmetric key performance depends on size of used key. The longer used key, the harder to break. Most common symmetric cryptography algorithms are Data Encryption Standard (DES), Triple Data Encryption Standard (3DES), and Advanced Encryption Standard (AES) [78, 80–86].

Asymmetric Cryptography is used when the two communicating parties use different keys for encryption and decryption as shown in Fig. 20. Two separate keys are used; private and public keys. Public key is publicly available and used for

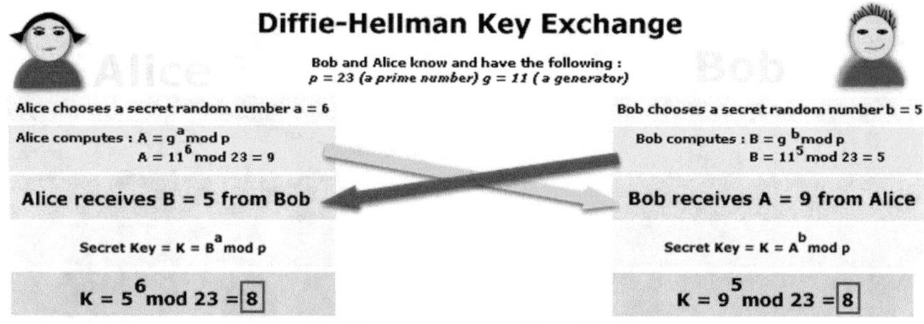

Fig. 22 Diffie Hellman example [90]

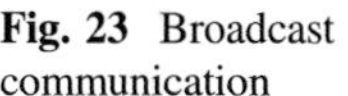

Fig. 23 Broadcast communication

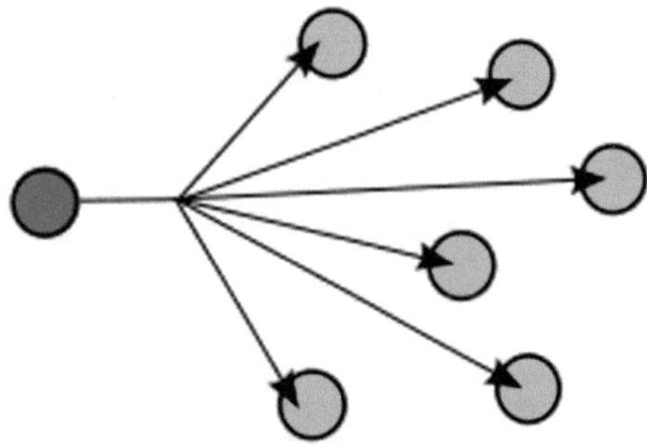

encryption. Private Key is known only to the user and used for decryption. Neither key can perform both functions by itself. In contrast with symmetric key, there is no need for distributing keys before transmission between communicated parties. Most common asymmetric cryptography algorithms are Digital Signature Algorithm and RSA [78, 80–86].

9.3 Diffie-Hellman Algorithm

In 1976, Whitfield Diffie and Martin Hellman have introduced Diffie-Hellman algorithm [87]. The Diffie-Hellman algorithm is used for securing key transmission over unsafe media. Furthermore, is used excessively in present key management for delivery keying information for cryptographic algorithms as *RSA*, 3DES or keyed-MD5 (*HMAC*). It obtains its protection from the complexity of computing the discrete logarithms for each large numbers [88–90]. Figure 21 summarizes the operation of Diffie-Hellman algorithm between two participants called Alice and Bob. The two participants agreed on two random numbers one has a small number called generator (g) and other has a large number called modulus (p). Every

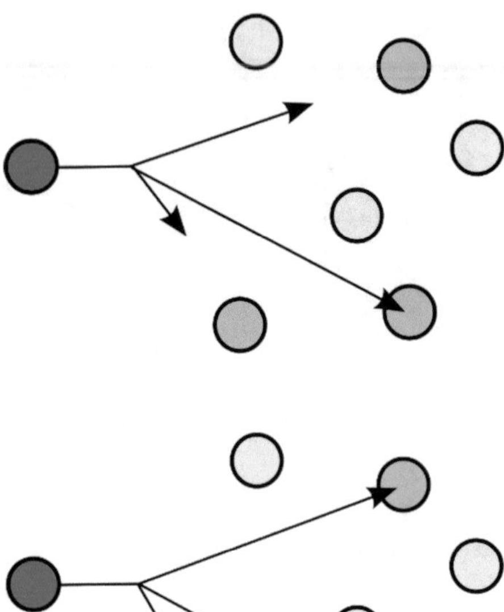

Fig. 24 Anycast communication

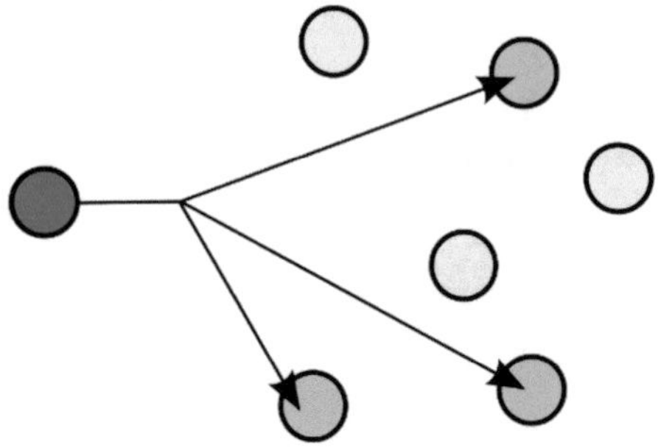

Fig. 25 Multicast communication

participant randomly create secret (x). According to g, p and x, each participant creates and transmits public values.

Figure 22 demonstrates an example of Diffie Hellman operation with p = 23 and g = 11. The result of common secret key is 8.

9.4 Types of Multiparty Communication

Broadcast Communication denotes transmission of messages from one sender to all participants concurrently over the network (see Fig. 23). The most applications of broadcasting are radio and television. The eavesdropper can misuse the broadcasting communication by transmitting a series of fake messages which affects the utilization of computational resources, such as bandwidth, memory, disk space, or processor time. There is no constraint regarding to the group of destinations, everybody is received the messages whether they need it or not [78, 80–86].

Anycast communication provides available services from one sender (Server) to a group of destinations (Clients) (see Fig. 24). However, the group of destinations is commonly a group of servers offering a specific service. The distinction between anycast and multicast is that the anycast communications is not used for the definite transmission of data as per in multicast, it places an obtainable server from a

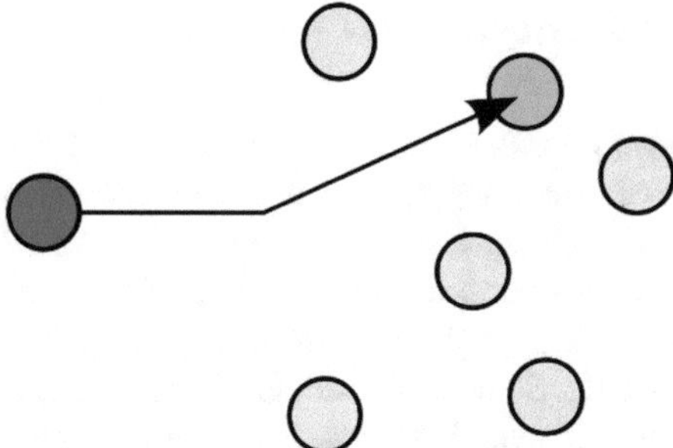

Fig. 26 Unicast communication

group. Afterward server localization, the transmission keep on a conventional unicast client-server operation [78, 80–86].

Multicast communication signifies transmission of messages from one sender to multiple (group) concurrently by utilizing only one transmission (see Fig. (25)). The main concepts of multicast communication are multicast group and characteristics. The multicast group is the collection of receivers. The characteristics of multicast group are Candidness to new members and senders, Dynamic Lifetime of keys and security. The Multicast Communication Operations are Joining and Leaving. Therefore, Group members have the ability to join any interested group but there not always have the motivation to leave [78, 80–86].

Unicast communication means transmission of series of messages from one sender to one receiver over the network. Each destination address is recognized by a unique address (see Fig. 26). [78, 80–86].

10 Group Key Management

Key Management is the suite of procedures and methods which providing the establishment and distribution of a secured among the participated communicators. Furthermore, the maintenance of continuing keying association between communicated parties containing previous with novel keys as essential. There're two principal difficulties with group key management which group rekeying and group key establishment. Key establishment provides a secured key agreement for authentication and communication processes among the communicated parties. Group rekeying denotes replacement of the present used group key as soon as it is supposed vulnerable. Rekeying is happened either if the group status has been changed by joining and leaving member or regularly updating the group key [91–96].

10.1 Group Key Management Categories

Group key management protocols can be almost organized into three classifications, specifically centralized, distributed, and decentralized. Centralized Group Key Management "CGKM" systems require a single centralized confidence group controller which is responsible to manage the secured transmitted messages among group members', as well as, synchronize member joining/leaving and rekeying the messages [91, 92, 97–101]. So, a single member is responsible to manage the entire group and for re-keying, calculation and distributing group key to all group members. CGKM suffers from many serious problems as single point of failure which means that by failing CGKM, the whole multicast network stops working, The other problem is the performance bottleneck where CGKM receives requests from multiple members of the whole group concurrently which leads to transmission suspension or network breakdown [97–100]. Centralized protocols are categorized into three approaches; secure locks, Pairwise key and Hierarchy of keys [91–96].

In Distributed group key protocols, group members themselves participate to establish a group or session key. These members are similarly in charge of the re-keying and distribution of group keys. Distributed protocols are categorized into three approaches which are Ring based cooperation, Hierarchical-based cooperation and Broadcast-based cooperation [91–96]. In Decentralized group key protocols, the secured multicast group is divided into smaller groups or clusters; each sub-group is assigned by a local controller. Each local controller is accountable for security controlling of members and its subgroup. Decentralized protocols are categorized into two approaches which are distinguished as static and dynamic schemes [91–96].

10.2 Group Key Management Responsibilities

Group Key management acts as a principal entity for designing a secured group communications by administering access control on the group key. It maintains the process for generation and backup of key among communicated participants consistent with the specified security policy [91–96]. The main responsibilities of Group Key Management are:

1. Identification and verification of group members for preventing an eavesdropper from copying and spying on the transmitted messages over the group network.
2. After the group member verified, it's enroll and join up should be authorized. The process for authorization group members beforehand providing them access to group transmission called Access control
3. Generation, distribution and maintenance of key information during the active group communication process. The distributed key has to be changed on a

periodic time to preserve its confidentiality. The distributed new key must be totally different from any earlier used and future keys.

4. Confidentiality of key when the membership of the group member is changed. This can be achieved through backward and forward secrecy. The method for inhibiting a new joined member for decrypting exchanged messages before he joined the group is called backward secrecy. The process for preventing a leaving member from a group to stay has the access of group's communication is called forward secrecy [91–96].

10.3 Group Key Security Requirements

The requirements of Group Key Security would be summarized as follows [91–96]:

1. Forward secrecy guarantees a user won't be able to encode transmitted messages after he left the group. Furthermore, he should has no access to any forthcoming secured group key. The best solution to ensure forward secrecy is performing a

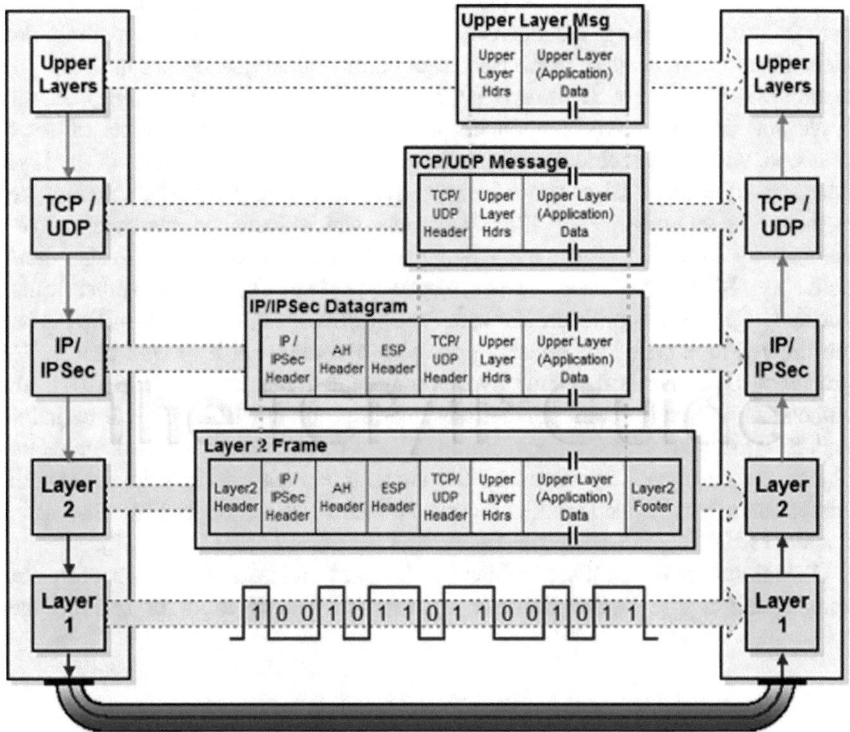

Fig. 27 IPsec encapsulation and decapsulation processes [105]

re-key approach of the group with a new secured key once the user left from the group.

2. The ability for a new joined user should has no access to any used key is called backward secrecy. Backward secrecy ensures that a user won't be able to encode transmitted messages which sent before he joined the group. The best solution to ensure forward secrecy is performing a re-key approach of the group with a new secured key once a new user join the group.
3. When a key confession does not compromise any other keys then the key management protocol is called independence.
4. The key management protocol should minimize confidence in a high amount of communicated parties. Otherwise, the effectual implementation and utilization of key management system would not be easy.

10.4 IPsec

IPsec delivers security for IP communications at the internet layer of the TCP/IP Model. Sensitive information which is sent over the internet can be encrypted via IPsec to maintain the secrecy of the information. IPsec encrypts data at layer 3 IP packet layer proposing exhaustively secured solution through providing data authentication, anti-replay protection, data confidentiality, and data integrity protection [88, 89]. Figure 27 illustrates the encapsulation a de-capsulation process of IPsec and its supportive components over OSI layers. IPsec consists of set of protocols with each protocol concentrates on particular characteristics of the IPsec purposes to protect IP communications over untrusted networks. Internet key exchange is an example of IPsec protocols which focus on message delivery authenticity as well Encapsulating Security Payload which focus on data confidentiality. IPsec uses encapsulated security payload (ESP) and authentication header (AH) to accomplish desired security objectives [85, 88, 89, 102–105]. AH is constructed to enhance the security principles by protection IP packet header. The protection objective is delivered by cryptographic authentication [106, 107]. The authentication service confirms that any interference with IP traffic will be identified. ESP provides protection and confidentiality for IP packet data. This protection objective is achieved by encoding the content of data packet using symmetric encryption algorithm as Data Encryption Standard (DES), Triple Data Encryption Standard (3DES), and Advanced Encryption Standard (AES) [104–108].

Traditional multicast IPsec architecture is based on classical key generation and authentication. The Diffie -Hellman algorithm is mainly used for securing key transmission over unsafe media. Furthermore, is used excessively in present key management for delivery keying information for IPsec [87–89]. Key generation and management based on Diffie-Hellman algorithm impaired from many weaknesses. Firstly, the frequency of changing the distributed keys between the communicated IPsec peers is limited. Secondly, the generated keys are conditional security which means intruder can spy on the communication channel and copy keys without any given warning for the communicated IPsec peers. Lastly, the whole architecture of traditional multicast IPsec deteriorates from man in the middle attack [87, 102, 103]. IPsec delivers security for IP communications at the internet layer of TCP/IP model as well it depends on supportive components for securing transmitted traffic among communicated peers.

IPsec has two modes transport and tunnel. IPsec uses transport and tunnel mode to establish a secure communication channel between network nodes. In transport mode only transport layer protocols which communicate between two IPsec hosts together encapsulated using AH/ESP. In tunnel mode the entire IP packet which communicates between two security gateways encapsulated using AH/ESP and a new IP packet will be generated. Security associations (SAs) are essential to IPsec. An SA is a suite of components including the protocols (AH, ESP or both), encapsulation mode (transport mode or tunnel mode), encryption algorithm (DES, 3DES, or AES), shared key used for flood protection and key lifetime. An SA can be created manually or with IPsec supportive components [85, 106, 107, 109, 110].

References

1. Wiesner, S.: Conj. Coding. SIGACT News **15**(1), 78–88 (1983). doi:10.1145/1008908.1008920
2. Bennett, C.: Logical reversibility of computation. IBM J. Res. Dev. **17**(6), 525–532 (1973). doi:10.1147/rd.176.0525
3. Benioff, P.: The computer as a physical system: a microscopic quantum mechanical hamiltonian model of computers as represented by turing machines. J. Stat. Phys. **22**(5), 563–591 (1980). doi:10.1007/bf01011339
4. Benioff, P.: Quantum mechanical models of turing machines that dissipate no energy. Phys. Rev. Lett. **48**(23), 1581–1585 (1982). doi:10.1103/physrevlett.48.1581
5. Benioff, P.: Quantum mechanical hamiltonian models of turing machines. J. Stat. Phys. **29** (3), 515–546 (1982). doi:10.1007/bf01342185
6. Benioff, P.: Quantum mechanical hamiltonian models of discrete processes that erase their own histories: application to turing machines. Int. J. Theor. Phys. **21**(3–4), 177–201 (1982). doi:10.1007/bf01857725
7. Feynman, R.: Simulating physics with computers. Int. J. Theor. Phys. **21**(6–7), 467–488 (1982). doi:10.1007/bf02650179
8. Wootters, W., Zurek, W.: A single quantum cannot be cloned. Nature **299**(5886), 802–803 (1982). doi:10.1038/299802a0
9. Bennett, C., Brassard, G.: Quantum cryptography: public key distribution and coin tossing. Theor. Comput. Sci. **560**, 7–11 (2014). doi:10.1016/j.tcs.2014.05.025

10. Deutsch, D.: Quantum theory, the church-turing principle and the universal quantum computer. Proc. Royal Soc. A Math. Phys. Eng. Sci. **400**(1818), 97–117 (1985). doi:10.1098/rspa.1985.0070
11. Bennett, C., Wiesner, S.: Communication via one- and two-particle operators on Einstein-Podolsky-Rosen states. Phys. Rev. Lett. **69**(20), 2881–2884 (1992). doi:10.1103/physrevlett.69.2881
12. Kaye, P., Laflamme, R.: An Introduction to Quantum Computing. Oxford University Press (2007)
13. Bennett, C., Brassard, G., Crépeau, C., Jozsa, R., Peres, A., Wootters, W.: Teleporting an unknown quantum state via dual classical and Einstein-Podolsky-Rosen channels. Phys. Rev. Lett. **70**(13), 1895–1899 (1993). doi:10.1103/physrevlett.70.1895
14. Stinson, D.: Cryptography. CRC Press, Boca Raton (1995)
15. Chakrabarty, I.: Teleportation via a mixture of a two qubit subsystem of a N-qubit W and GHZ state. Eur. Phys. J. D **57**(2), 265–269 (2010). doi:10.1140/epjd/e2010-00017-8
16. Liang, H., Liu, J., Feng, S., Chen, J.: Quantum teleportation with partially entangled states via noisy channels. Quant. Inf. Process. **12**(8), 2671–2687 (2013). doi:10.1007/s11128-013-0555-3
17. Nielsen, M., Chuang, I.: Quantum Computation and Quantum Information. Cambridge University Press, Cambridge (2000)
18. Zeng, G.H.: Quantum Cryptology: Science Press (2006)
19. Van Assche, G.: Quantum Cryptography and Secret-key Distillation. Cambridge University Press, Cambridge (2006)
20. Metwaly, A.F., Rashad, M.Z., Omara, F.A., Megahed, A.A.: Architecture of multicast centralized key management scheme using quantum key distribution and classical symmetric encryption. Eur. Phys. J. Special Topics **223**(8), 1711–1728 (2014)
21. Metwaly, A., Rashad, M.Z., Omara, F.A., Megahed, A.A.: Architecture of Point to Multipoint QKD Communication Systems (QKDP2MP). In: 8th International Conference on Informatics and Systems (INFOS), Cairo, IEEE pp. NW 25–31. (2012)
22. Farouk, A., Omara, F., Zakria, M., Megahed, A.: Secured IPsec multicast architecture based on quantum key distribution. In: The International Conference on Electrical and Bio-medical Engineering, Clean Energy and Green Computing. The Society of Digital Information and Wireless Communication, pp. 38–47 (2015)
23. Farouk, A., Zakaria, M., Megahed, A., Omara, F.A.: A generalized architecture of quantum secure direct communication for N disjointed users with authentication. Sci. Rep. **5**, 16080–16080
24. Wang, M.M., Wang, W., Chen, J.G., Farouk, A.: Secret sharing of a known arbitrary quantum state with noisy environment. Quant. Inf. Process. **14**(11), 4211–4224 (2015)
25. Naseri, M., Heidari, S., Batle, J., Baghfalaki, M., Gheibi, R., Farouk, A., Habibi, A.: A new secure quantum watermarking scheme. Optik Int. J. Light Electron Optics **139**, 77–86 (2017)
26. Batle, J., Ciftja, O., Naseri, M., Ghoranneviss, M., Farouk, A., Elhoseny, M.: Equilibrium and uniform charge distribution of a classical two-dimensional system of point charges with hard-wall confinement. Phys. Scr. **92**(5), 055801 (2017)
27. Geurdes, H., Nagata, K., Nakamura, T., Farouk, A.: A note on the possibility of incomplete theory. arXiv preprint (2017) arXiv:1704.00005
28. Batle, J., Farouk, A., Alkhambashi, M., Abdalla, S.: Multipartite correlation degradation in amplitude-damping quantum channels. J. Korean Phys. Soc. **70**(7), 666–672 (2017)
29. Batle, J., Naseri, M., Ghoranneviss, M., Farouk, A., Alkhambashi, M., Elhoseny, M.: Shareability of correlations in multiqubit states: optimization of nonlocal monogamy inequalities. Phys. Rev. A **95**(3), 032123 (2017)
30. Batle, J., Farouk, A., Alkhambashi, M., Abdalla, S.: Entanglement in the linear-chain Heisenberg antiferromagnet Cu (C_4 H_4 N_2) (NO_3) 2. Eur. Phys. J. B **90**, 1–5 (2017)
31. Batle, J., Alkhambashi, M., Farouk, A., Naseri, M., Ghoranneviss, M.: Multipartite non-locality and entanglement signatures of a field-induced quantum phase transition. Eur. Phys. J. B **90**(2), 31 (2017)

32. Nagata, K., Nakamura, T., Batle, J., Abdalla, S., Farouk, A.: Boolean approach to dichotomic quantum measurement theories. J. Korean Phys. Soc. **70**(3), 229–235 (2017)
33. Abdolmaleky, M., Naseri, M., Batle, J., Farouk, A., Gong, L.H.: Red-Green-Blue multi-channel quantum representation of digital images. Optik Int. J. Light Elect. Opt. **128**, 121–132 (2017)
34. Farouk, A., Elhoseny, M., Batle, J., Naseri, M., Hassanien, A.E.: A proposed architecture for key management schema in centralized quantum network. In: Handbook of Research on Machine Learning Innovations and Trends IGI Global, pp. 997–1021
35. Zhou, N.R., Li, J.F., Yu, Z.B., Gong, L.H., Farouk, A.: New quantum dialogue protocol based on continuous-variable two-mode squeezed vacuum states. Quant. Inf. Process. **16**(1), 4 (2017)
36. Batle, J., Abutalib, M., Abdalla, S., Farouk, A.: Persistence of quantum correlations in a XY spin-chain environment. Eur. Phys. J. B **89**(11), 247 (2016)
37. Batle, J., Abutalib, M., Abdalla, S., Farouk, A.: Revival of bell nonlocality across a quantum spin chain. Int. J. Quant. Inf. **14**(07), 1650037 (2016)
38. Batle, J., Ooi, C.R., Farouk, A., Abutalib, M., Abdalla, S.: Do multipartite correlations speed up adiabatic quantum computation or quantum annealing? Quant. Inf. Process. **15**(8), 3081–3099 (2016)
39. Batle, J., Bagdasaryan, A., Farouk, A., Abutalib, M., Abdalla, S.: Quantum correlations in two coupled superconducting charge qubits. Int. J. Mod. Phys. B **30**(19), 1650123 (2016)
40. Batle, J., Ooi, C.R., Abutalib, M., Farouk, A., Abdalla, S.: Quantum information approach to the azurite mineral frustrated quantum magnet. Quant. Inf. Process. **15**(7), 2839–2850 (2016)
41. Batle, J., Ooi, C.R., Farouk, A., Abdalla, S.: Nonlocality in pure and mixed n-qubit X states. Quant. Inf. Process. **15**(4), 1553–1567 (2016)
42. Metwaly, A.F., Mastorakis, N.E.: Architecture of decentralized multicast network using quantum key distribution and hybrid WDM-TDM. In: Advances in Information Science and Computer Engineering, 504–518 (2015)
43. Metwaly, A.F., Rashad, M.Z., Omara, F.A., Megahed, A.A.: Architecture of Multicast Network Based on Quantum Secret Sharing and Measurement (2015)
44. Zeng, G.: Quantum Private Communication. Higher Education Press, Beijing (2010)
45. Barenco, A., Bennett, C., Cleve, R., DiVincenzo, D., Margolus, N., Shor, P., et al.: Elementary gates for quantum computation. Phys. Rev. A **52**(5), 3457–3467 (1995). doi:10.1103/physreva.52.3457
46. Hirvensalo, M.: Quantum Computing. Springer, Berlin (2001)
47. Sharbaf, M.S.: Quantum cryptography: a new generation of information technology security system. In: Information Technology: New Generations, 2009. ITNG'09. Sixth International Conference on IEEE pp. 1644–1648 (April, 2009)
48. Aharonov, D.: A Simple Proof that Toffoli and Hadamard are Quantum Universal. arXiv preprint quant-ph/0301040 (2003)
49. Williams, C.P., Clearwater, S.H.: Explorations in Quantum Computing, vol. 1. Springer, New York (1998)
50. Mohammadi, M., Eshghi, M.: On figures of merit in reversible and quantum logic designs. Quant. Inf. Process. **8**(4), 297–318 (2009)
51. Haghparast, M., Mohammadi, M., Navi, K., Eshghi, M.: Optimized reversible multiplier circuit. J. Circ. Syst. Comp. **18**(02), 311–323 (2009)
52. Shor, P.W.: Scheme for reducing decoherence in quantum computer memory. Phys. Rev. A **52**(4), R2493 (1995)
53. Martín-López, E., Laing, A., Lawson, T., Alvarez, R., Zhou, X.Q., O'Brien, J.L.: Experimental realization of shor's quantum factoring algorithm using Qubit recycling. Nat. Photon. **6**(11), 773–776 (2012)
54. Politi, A., Matthews, J.C., O'Brien, J.L.: Shor's Quantum factoring algorithm on a photonic chip. Science, **325**(5945), 1221–1221 (2009)
55. Jin, X.M., Ren, J.G., Yang, B., Yi, Z.H., Zhou, F., Xu, X.F., Pan, J.W.: Experimental Free-space Quantum Teleportation. Nat. Photonics **4**(6), 376–381 (2010)

56. Yin, J., Ren, J.G., Lu, H., Cao, Y., Yong, H.L., Wu, Y.P., Pan, J.W.: Quantum Teleportation and Entanglement Distribution over 100-kilometre Free-space Channels. Nature **488**(7410), 185–188 (2012)
57. Zhang, Q., Goebel, A., Wagenknecht, C., Chen, Y.A., Zhao, B., Yang, T., Pan, J.W.: Experimental Quantum Teleportation of a Two-qubit Composite System. Nat. Phys. **2**(10), 678–682 (2006)
58. Huang, Y.F., Ren, X.F., Zhang, Y.S., Duan, L.M., Guo, G.C.: Experimental Teleportation of a Quantum Controlled-NOT Gate. Phys. Rev. Lett. **93**(24), 240501 (2004)
59. Fang, X., Zhu, X., Feng, M., Mao, X., Du, F.: Experimental implementation of dense coding using nuclear magnetic resonance. Phys. Rev. A, **61**(2), (2000) doi:10.1103/physreva.61.022307
60. Mattle, K., Weinfurter, H., Kwiat, P.G., Zeilinger, A.: Dense Coding in Experimental Quantum Communication. Phys. Rev. Lett. **76**(25), 4656 (1996)
61. Bell, J.S.: On the Einstein-Podolsky-Rosen Paradox. Physics **1**(3), 195–200 (1964)
62. Aspect, A., Dalibard, J., Roger, G.: Experimental Test of Bell's Inequalities using Time-varying Analyzers. Phys. Rev. Lett. **49**(25), 1804 (1982)
63. Shimizu, K., Imoto, N.: Communication channels secured from eavesdropping via transmission of photonic bell states. Phys. Rev. A, **60**(1), 157 (1999)
64. Einstein, A., Podolsky, B., Rosen, N.: Can Quantum-mechanical Description of Physical Reality be Considered Complete? Phys. Rev. **47**(10), 777 (1935)
65. He, G., Zhu, J., Zeng, G.: Quantum Secure Communication using Continuous Variable Einstein-Podolsky-Rosen Correlations. Phys. Rev. A, **73**(1), 012314 (2006)
66. Greenberger, D.M., Horne, M., Zeilinger, A.: Bell's Theorem, Quantum Theory, and Conceptions of the Universe, Ed. Kafatos, M. (1989)
67. Lu, C.Y., Zhou, X.Q., Gühne, O., Gao, W.B., Zhang, J., Yuan, Z.S., Pan, J.W.: Experimental Entanglement of Six Photons in Graph States. Nat. Phys. **3**(2), 91–95 (2007)
68. Poppe, A., Peev, M., Maurhart, O.: Outline of the SECOQC Quantum-key-distribution Network in Vienna. Int. J. Quant. Inf. **6**(02), 209–218 (2008)
69. Peev, M., Pacher, C., Alléaume, R., Barreiro, C., Bouda, J., Boxleitner, W., Tualle-Brouri, R.: The SECOQC Quantum Key Distribution Network in Vienna. New J. Phys. **11**(7), 075001 (2009)
70. Elliott, C.: Building the Quantum Network. New J. Phys. **4**(1), 46 (2002)
71. Elliott, C., Colvin, A., Pearson, D., Pikalo, O., Schlafer, J., Yeh, H.: Current status of the DARPA quantum network. In: Defense and Security. International Society for Optics and Photonics pp. 138–149 (May 2005)
72. Metwaly, A.F., Mastorakis, N.E.: Architecture of Decentralized Multicast Network Using Quantum Key Distribution and Hybrid WDM-TDM. Proceedings of the 9th International Conference on Computer Engineering and Applications (CEA '15). Advances in Information Science And Computer Engineering, 504–518 (2015)
73. Gisin, N., Ribordy, G., Tittel, W., Zbinden, H.: Quantum Cryptography. Rev. Mod. Phys. **74**(1), 145 (2002)
74. Beige, A., Englert, B.G., Kurtsiefer, C., Weinfurter, H.: Secure communication with Single-photon two-qubit states. J. Phys. A: Math. Gen. **35**(28), L407 (2002)
75. Shannon, C.E.: A Mathematical Theory of Communication. ACM SIGMOBILE Mobile Computing and Communications Review **5**(1), 3–55 (2001)
76. Shannon, C.E.: Communication Theory of Secrecy Systems*. Bell Syst. Tech. J. **28**(4), 656–715 (1949)
77. Shields, A., Zhiliang, Y.: Key to the Quantum Industry. Phys. World **20**(3), 24–29 (2007)
78. Kumar, Y., Munjal, R., Sharma, H.: Comparison of Symmetric and Asymmetric Cryptography with Existing Vulnerabilities and Countermeasures. Int. J. Comp. Sci. Manag. Studies, **11**(03) (2011)
79. Ansari, H., Parameswaran, A., Antani, L., Aditya, B., Taly, A., Kumar, L.: Quantum Cryptography and Quantum Computation. IIT, Bombay

80. Pathak, A.: Elements of Quantum Computation and Quantum Communication. Taylor & Francis (2013)
81. Forouzan, A.B.: Data Communications & Networking (sie). Tata McGraw-Hill Education (2006)
82. Friend, G.: Understanding Data Communications. Texas Instruments, Dallas, Tx. (1984)
83. Hughes, L.: Data communications. McGraw-Hill, New York (1992)
84. Stallings, W.: Data and Computer Communications. Pearson/Prentice Hall (2007)
85. Ferguson, N., Schneier, B.: Practical Cryptography. Indianapolis, IN [etc.]: Wiley (2003)
86. Van Lint, J.H.: Introduction to Coding Theory, vol. 86. Springer Science & Business Media (1999)
87. Diffie, W., Hellman, M.E.: New Directions in Cryptography. Information Theory, IEEE Transactions on **22**(6), 644–654 (1976)
88. Bellovin, S.M.: Problem areas for the IP security protocols. In: Proceedings of the 6th conference on USENIX Security Symposium, Focusing on Applications of Cryptography vol. 6, pp. 21–21. USENIX Association (1996)
89. Paterson, K.G., Yau, A.K.: Cryptography in theory and practice: the case of encryption in IPsec. In: Advances in Cryptology-EUROCRYPT pp. 12–29. Springer, Berlin Heidelberg (2006)
90. Schneier, B.: Applied Cryptography: Protocols, Algorithms, and Source Code in C. Wiley (2007)
91. Rafaeli, S., Hutchison, D.: A survey of key management for secure group communication. ACM Comput. Surveys (CSUR) **35**(3), 309–329 (2003)
92. Bandara, H.D., Jayasumana, A.P.: Collaborative applications over peer-to-peer systems-challenges and solutions. Peer Peer Network. Appl. **6**(3), 257–276 (2013)
93. Guo, C.J., Huang, Y.M.: Residency-based distributed collaborative key agreement for dynamic peer groups. Int. J. Innov. Comput. Inform. Control **8**(8), 5523–5542 (2012)
94. Siramdasu, H., Krishna, H.: Communication in vibrant peer groups for cluster key management. Int. J. Eng. Trends Technol. **4**(5), 1367–1373 (2013)
95. SuganyaDevi, D., Padmavathi, G.: Secure Multicast Key Distribution for Mobile Ad Hoc Networks. arXiv preprint (2010)arXiv:1003.1799
96. Devaraju, S., Ganapathi, P.: Dynamic clustering for QoS based secure multicast key distribution in mobile ad hoc networks. IJCSI Int. J. Comp. Sci. **7**(1–2), 30–37 (2010)
97. Canetti, R., Garay, J., Itkis, G., Micciancio, D., Naor, M., Pinkas, B.: Multicast security: a taxonomy and some efficient constructions. In: INFOCOM'99. Eighteenth Annual Joint Conference of the IEEE Computer and Communications Societies. Proceedings. IEEE vol. 2, pp. 708–716. IEEE (1999)
98. Canetti, R., Malkin, T., Nissim, K.: Efficient communication-storage tradeoffs for multicast encryption. In: Advances in Cryptology—EUROCRYPT'99 Springer Berlin Heidelberg, pp. 459–474 (1999)
99. Caronni, G., Waldvogel, M., Sun, D., Plattner, B.: Efficient Security for large and dynamic multicast groups. In: Enabling Technologies: Infrastructure for Collaborative Enterprises, 1998. (WET ICE'98) Proceedings on Seventh IEEE International Workshops on, pp. 376–383. IEEE (1998)
100. Wallner, D., Harder, E., Agee, R.: Key Management for Multicast: Issues and Architectures. RFC 2627 (1999)
101. Wong, C.K., Gouda, M., Lam, S.S.: Secure group communications using key graphs. Network. IEEE/ACM Trans. **8**(1), 16–30 (2000)
102. Degabriele, J.P., Paterson, K.G.: Attacking the IPsec Standards in Encryption-only Configurations. In: IEEE Symposium on Security and Privacy vol. 161, pp. 335–349 (2007)
103. Aiello, W., Bellovin, S.M., Blaze, M., Canetti, R., Ioannidis, J., Keromytis, A.D., Reingold, O.: Just fast keying: key agreement in a hostile internet. ACM Trans. Inform. Syst. Secur. (TISSEC) **7**(2), 242–273 (2004)
104. Kent, S., Atkinson, R.: RFC 2401: Security Architecture for the Internet Protocol (1998)
105. Kent, S., Seo, K.: RFC 4301: Security Architecture for the Internet Protocol (2005)

106. Atkinson, R., Header, I.A.: RFC 1826. Naval Research Laboratory (1995)
107. Kent, S., Header, I.A.: RFC 4302. IETF, December (2005)
108. Kent, S., Atkinson, R.: RFC 2402: IP Authentication Header (1998)
109. Atkinson, R.: RFC 1827. IP Encapsulating Security Payload (ESP) (1995)
110. Errata, K.S.: IP Encapsulating Security Payload. RFC 4303 (2005)
111. Elhoseny, M., El-Minir,R.A/., Yuan, X.: A secure data routing schema for WSN using elliptic curve cryptography and homomorphic encryption. J. King Saud Univ. Comp. Inform. Sci, **28**(3): 262–275 (2016)
112. Elhoseny, M., Yuan, X., El-Minir, H., Riad, A.: An energy efficient encryption method for secure dynamic WSN. Sec. Commun. Networks, (9):2024–2031 (2016)

Quantum Key Distribution Over Multi-point Communication System: An Overview

Ahmed Farouk, O. Tarawneh, Mohamed Elhoseny, J. Batle, Mosayeb Naseri, Aboul Ella Hassanien and M. Abedl-Aty

Abstract Most existing realizations of quantum key distribution (QKD) are point-to-point systems with one source transferring to only one destination. Growth of these single-receiver systems has now achieved a reasonably sophisticated point. However, many communication systems operate in a point-to-multi-point (Multi-cast) configuration rather than in point-to-point mode, so it is crucial to demonstrate compatibility with this type of network in order to maximize the application range for QKD.

Keywords Quantum key distribution · Quantum-back-bone · Quantum data link layer

A. Farouk (✉) · M. Elhoseny
Faculty of Computer and Information Sciences, Mansoura University, Mansoura, Egypt
e-mail: dr.ahmedfarouk85@yahoo.com

A. Farouk · M. Abedl-Aty
Zewail City of Science and Technology, University of Science and Technology, Giza, Egypt

O. Tarawneh
Information Technology Department, Al-Zahra College for Women, P.O. Box 3365, Muscat, Oman

J. Batle
Departament de Física, Universitat de les Illes Balears, 07122 Palma de Mallorca, Balearic Islands, Spain

M. Naseri
Department of Physics, Kermanshah Branch, Islamic Azad University, Kermanshah, Iran

A.E. Hassanien
Faculty of Computers and Information, Cairo University, Giza, Egypt

A.E. Hassanien et al. (eds.), *Quantum Computing: An Environment for Intelligent Large Scale Real Application*, Studies in Big Data 33,
https://doi.org/10.1007/978-3-319-63639-9_5

1 Introduction

In this chapter we provide an overview about the architecture of Point-to-Multipoint QKD (QKDP2MP) systems and how it will be presented in terms of Quantum Back Bone "QBB" and Quantum layered architecture. The architecture shows the components of quantum communication layer and relation with its perspective classical one. Furthermore, a general architecture of Point-to-Multipoint Quantum Key Distribution based on combined classical and quantum components is shown. Additionally, the architecture of routing component and transportation of messages from one node to another node through the quantum network is proposed. In 1984 [1], the concept of utilizing the properties of quantum physics for key generation and distribution through the network was developed. Consequently, there're many research centers and individuals which effectively proved and established quantum key distribution techniques in a real world [2]. Unfortunately, most of experimental implementations of quantum key distribution are focused on point-to-point communication link which intended transmission from one sender to one receiver. The growth of these unicast procedures has now achieved a reasonably sophisticated phase. Nevertheless, most modern networking systems exploit multicast transmission technology. The fundamental benefit of multicast is that a sender can transmit messages to multiple receivers simultaneously [3, 4]. Currently, most of recent networks rely on classical cryptographic protocols to assure the confidentiality and integrity of transmitted messages. The most important aspect for designing a secured multicast network is key generation and distribution. The extremely two used cryptographic approaches are symmetric and asymmetric. Asymmetric Cryptography is used when the two communicating parties use different keys for encryption and decryption. Two separate keys are used; private and public keys. A public key is publicly available and used for encryption. Private Key is known only to the user and used for decryption. Neither key can perform both functions by itself. In reality, best communication systems employ Asymmetric Cryptography approach, as one key used for identity verification of the user and the other key used for securing the transmitted information over the classical network [5–14].

Nowadays, transition from classical world to quantum one has been achieved advanced stages. Most of modem systems depend on quantum key distribution protocols instead of classical one as it proved its security against many types of attacks. Quantum key distribution requires two transmission channels between the communicated parties which are classical and quantum (see Fig. 1). The quantum channel is used for transferring of generated photon pulses, in other words, the key information between the communicated participants. The key information is produced according to generated photons, key agreement process, sifting, privacy amplifications and eavesdropper detection. Afterwards, the agreed generated key is used for classical cryptographic protocols. The classical channel transmits the flow of messages, encryption/decryption algorithms and secret keys. The most important thing is the combination and compatibility of quantum and classical world. By integrating quantum keys with classical cryptographic protocols, any eavesdropper

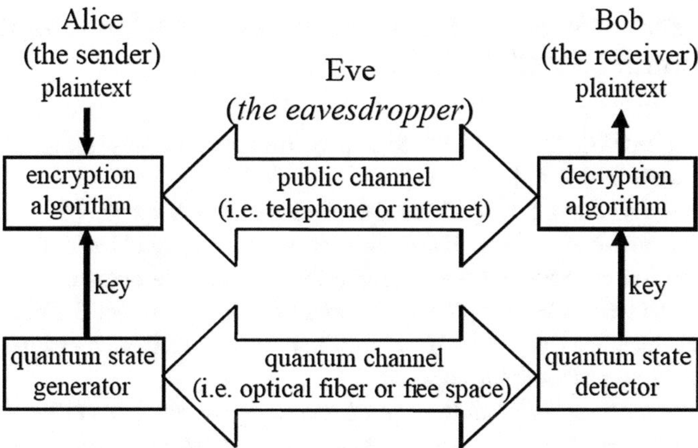

Fig. 1 Quantum key distribution

spy on the communication channel will be detected as he will produce a computable disruption of the movement of photons [5, 13, 15–28].

Several various theoretical and experimental quantum network approaches and protocols have been started and developed. Really, many Quantum Cryptography protocols have been developed, and particular protocols can transmit the generated keys over tens of kilometers. The used infrastructure for building quantum networks are optical fiber and free space which used for cabling and wireless technology respectively [13, 29–33].

Currently, multicast classical communication systems deteriorate from many security aspects. Firstly, the used processes for key agreement and distribution which are commonly employed in modern network system security architecture can be broken. Therefore, the communicated parties will lose their privilege to communicate securely across the network. Secondly, the communication process susceptible to many attacking approaches and especially man-in-the-middle attack. Therefore, the movement of key material can be disrupted and the control of communication process will be switched in the hand of the attacker. The most major compensations of proposed architecture are security of generated keys, authentication and robustness. The security of generated keys comes from the principles of quantum laws like superposition and entanglement. The generated keys prove its unconditional security through no-cloning theory. Therefore, the intruded can't copy or access the transmitted quantum state. In case of the attacker intercepts the communication process, the transmission process will be erroneous with at least probability 50%. Quantum cryptography proves the identity of the communicators. So, the generated keys will forward to the correct person [5, 34–41]. Furthermore, various applications of quantum have be proposed in [42–60].

2 The Basic Phases of Constructing Point-to-Multipoint Quantum Key Distribution

The basic phases which could be used to construct the point-to-Multipoint quantum key distribution are [13, 34–41, 61]:

The sender Quantum Key Distribution endpoint will create connections with every Receiver Quantum Key Distribution endpoints through fiber optic technology for the quantum channel and over the internet for classical channel.

The Sender and Receiver prepare communication and select the cryptographic collection which will be used for controlling the transmitted messages over the Point-to-Multipoint Quantum network.

When the receiver recipient the agreed message, the receiver verifies its identity to the receiver. Furthermore, in some cases, the sender also requires to verify its identity to the receiver. Asymmetric cartography approach is the most used for developing the authentication process between the sender and multiple receipts. The accurate technique utilized for authentication process is governed by the exchanged cryptographic collection agreement.

Subsequently, both sender and receivers exchange generated random numbers which are used for establishing a secured key. The secured key is coming from the well-known quantum key distribution protocol and used for cryptographic collection.

3 General Architecture of Point to Multipoint *QKD* (*QKDP2MP*)

General architecture of Point-to-Point Quantum Key Distribution based on combined classical and quantum components is shown in Fig. 2.

Furthermore, general architecture of Point-to-Multipoint Quantum Key Distribution based on combined classical and quantum components is shown in Fig. 3. In Fig. 3, the sender and receivers can transmit secret messages over the classical channel through quantum key distribution link. The quantum key distribution link connects one private communicators and multiple private communicators which represent the sender and multiple receipts simultaneously. The sender and receivers shared two types of communication channels which are classical and quantum. The quantum channel is used for generating and distributing the agreed quantum keys which employ for securing the transmitted traffic. While the classical channel carries the transmitted encrypted traffic over the proposed architecture [13, 34–41, 61].

The sender and receivers side consist of one or more local fiber optic system or free space system. The nature of communication infrastructure will determine which system will be used. If the multicast network infrastructure's is cabling one then the best choice is fiber optic. On the other side, if network infrastructure's is wireless then the best choice is free space. The proposed architecture is popular

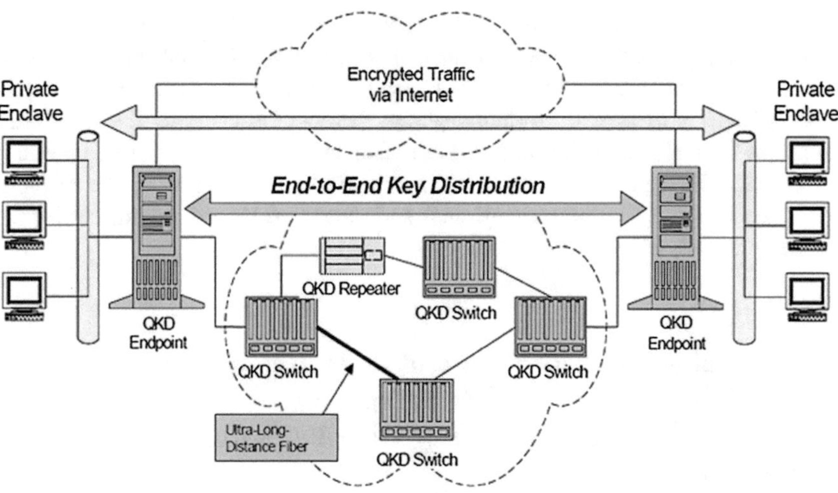

Fig. 2 General architecture of point to point *QKD* (*QKDP*2*P*)

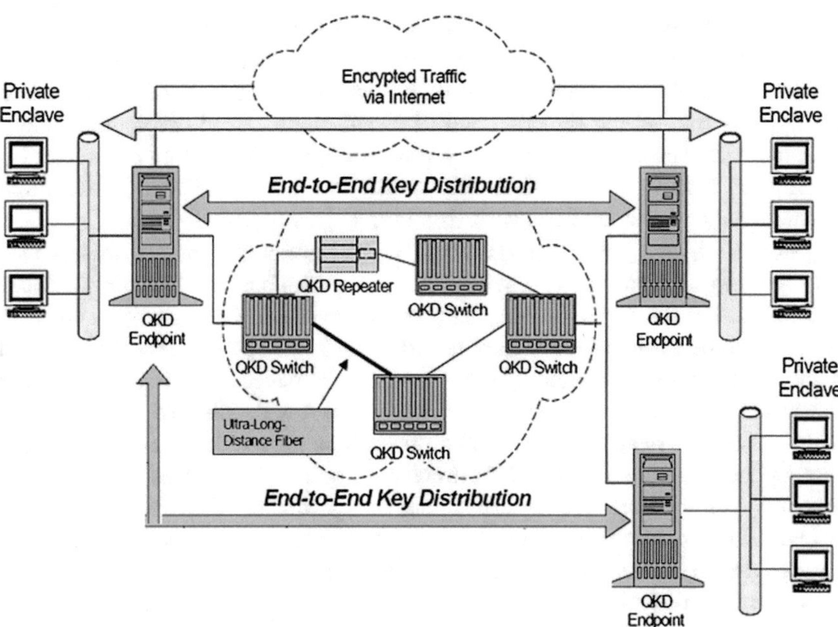

Fig. 3 General architecture of point to multipoint *QKD* (*QKDP*2*MP*)

used in modern networks across various applications for securing the transmitted messages. The most example of proposed architecture is connecting a headquarter office "which represents sender" and its associated branch offices "which represents

Receivers". Such examples are frequently secured the communication process by implementing particular devices for instance IP Security—Virtual Private Network "IPsec-VPN" technology. IPsec-VPN is used for establishing and monitoring the security aspects of private communications. IPsec-VPN encrypts the transmitted messages on the sender side before inserted it to the classical channel by encapsulating it with agreed security associations. At the receiver side, the receipted encrypted message will be authenticated and decrypted before sending it ahead to the receiver [13, 61].

4 General Architecture of Point to Multipoint Quantum Key Distribution Network Based on Quantum-Back-Bone Key Distribution Link Interface

Here, we will present a simplified, general block diagram of a point-to-multipoint QKD using Quantum–Back–Bone Key Distribution Link Interface "$QBB-KDL$". A $QBB-KDL$ communicates and connects two nodes which are sender and intended receiver through quantum and classical channels. $QBB-KDL$ uses the quantum channel for transmitting a series of quantum photons. Furthermore, it uses the classical channel for signaling and distribution the agreed cryptographic keys and protocols. The combination of key generation, key distribution and secret key traffic from many $QBB-KDL$ constitute the backbone of Point to Multipoint Quantum Key Distribution Network (see Fig. 4) [13, 61].

$QBB-KDL$ and communicated nodes are appeared jointly from resource sharing viewpoint, so they can deliver many facilities and services for communicated users. The most important aspect is transportation of messages from one node to another node through the quantum network. This concept with classical networks is referred to routing. The routing concept will be achieved by three steps. The first one is the transmitted messages from many users will collected to the nearest $QBB-KDL$. Afterward, the collected traffic will be moved across router-by-router through the quantum network infrastructure. The most implemented devices of quantum infrastructure are optical router and switches. Finally, the intended recipient will receive the transmitted messages. The key generation rate of quantum network can be affected in case of a lot of users join the network. So, while we have to implement this type of network, the number of deployed quantum devices and channels have to be taken into consideration. The best solution is adopting numerous quantum devices and parallel quantum channels for enhancement the key generation rate among the communicated nodes over $QBB-KDL$ (see Fig. 5) [13, 61].

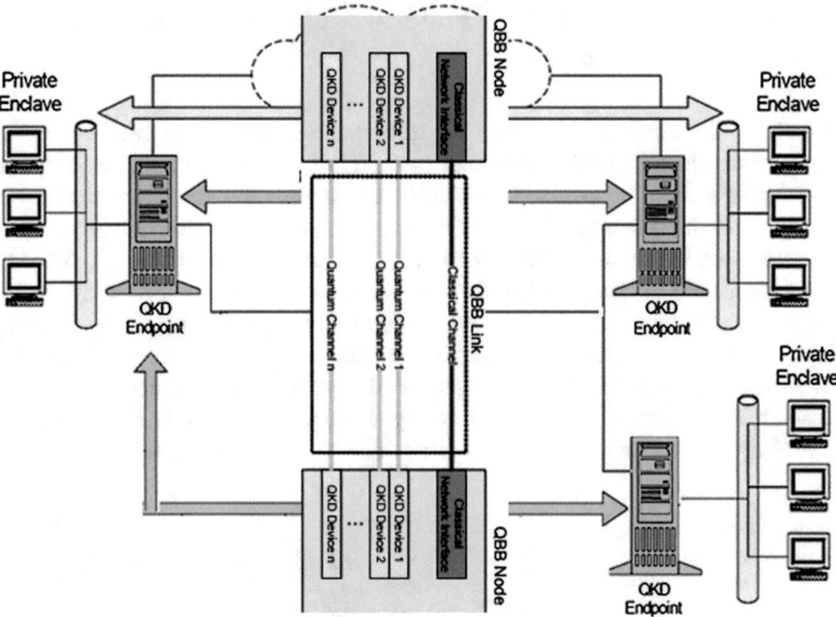

Fig. 4 General architecture of point to multipoint *QKD* using quantum–back–bone link interface

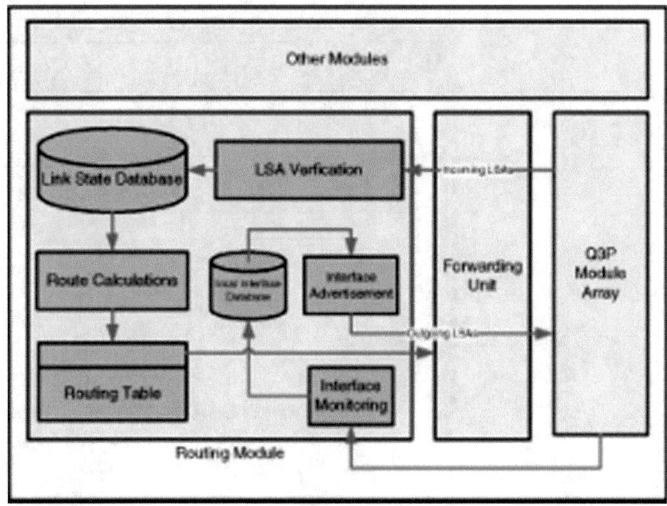

Fig. 5 The routing frame work of quantum node

5 General Architecture of Point to Multipoint *QKD* Using a Layered Architecture

In classical communication system the transmitted messages are encapsulated by moving through a layered architecture (see Fig. 6). Each layer adds its important information which necessary for delivery the transmitted messages totally and accurately to the desired receiver. By analogy, in quantum communication system, a layered architecture is essential for transmitted the quantum messages. The layered architecture of quantum key distribution network consists of four layers. The first layer is Quantum Key Distribution Application Layer "*QKDAL*". The Quantum Key Distribution Application Layer is interface between the user and quantum network. Furthermore, it utilizes the generated keys and became it accessible by the quantum key management layer. The key management layer is responsible for generating the quantum keys, as well many encoding devices have been

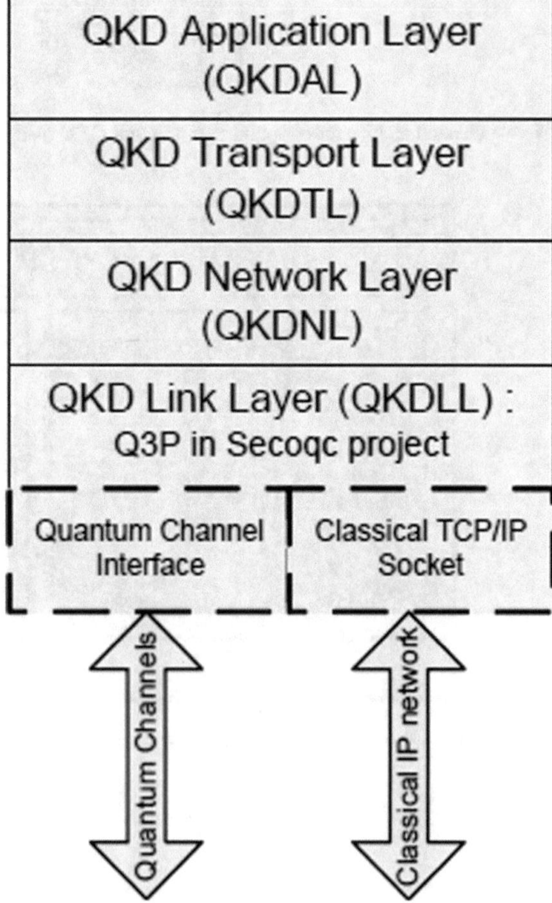

Fig. 6 A layered architecture for the communications protocol stack of the *QKD* network

implemented for using keys delivered by it. The second layer is Quantum Key Distribution Transport Layer "*QKATL*". The Quantum Key Distribution Transport Layer is responsible for linking the applications who might use the quantum network, recovery from network disasters and fragmentation and reassembly of application data. Fragmentation means break down the application data into suitable packet size and it is done at the sender side. Reassembly means the fragments data will be collected at the receiver side to reflect what was transmitted [13, 61].

The third layer is Quantum Key Distribution Network Layer "*QKDNL*". The Quantum Key Distribution Network Layer is responsible for movement the quantum messages from the sender to the receiver, route determination, switching and congestion criticism. Route determination indicates the taken path of transmitted packets from the sender and the receiver. Route determination can be achieved by using classical routing protocols and hop-by-hop principle. Quantum Back Bone works as intermediate routers between the sender and receiver. The switching provides the ability for moving the packets from incoming port to the appropriate outgoing port.

Furthermore, Quantum Key Distribution Network Layer permits forwarding of quantum keys over the quantum network. The last layer is Quantum Key Distribution Data Link Layer "*QKDLL*". Quantum Key Distribution Data Link Layer is the boundary between the Quantum Back Bone and the upper layers. Quantum Key Distribution Data Link Layer consists of two interface which one for quantum

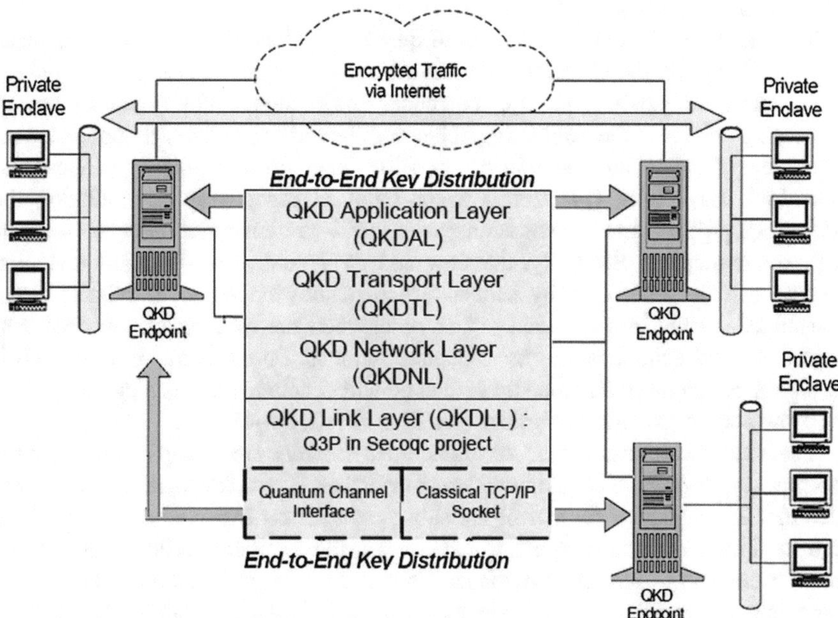

Fig. 7 General architecture of point to multipoint *QKD* using layered architecture

channel and the other for classical one. *QKDLL* is responsible for error handling and flow regulation. Error handling is assured the transmitted messages are received completely, accurately and in same order they sent (see Fig. 7) [13, 61].

6 Quantum Key Distribution Transport Layer Logical Architecture

The logical architecture of Quantum Key Distribution Transport Layer consists of three sub-layers which are packaging and transmission, admission control and connection manager. The first step for initiating the communication process is sending a request to Quantum Key Distribution Application Layer. Consequently, the request is forwarded to admission control sub-layer which decide to accept or refuse the request. The acceptance or refusing of the request decided according to the condition of the network. In case of acceptance of the communication request, the connection manager will start the negotiation process for establishing a secure Quantum Key Distribution Transport Layer connection session between the corresponding quantum back bone nodes. The movement of cryptographic keys between incoming and outgoing quantum back bone nodes will be controlled by established secured *QKDTL* connection. The responsibilities of the Transport Layer are achieved by exchanging Quantum Key Distribution Transport Layer packets (see Fig. 8) [13, 34–41, 61].

The structure of the transport layer of quantum system is similar to the classical system. Therefore, the Quantum Key Distribution Transport Layer packet consists of six main components. Firstly, the sender and receiver identify each other by addressing scheme. The addressing scheme differs at each layer, as port address, mac address and IP address will be used for Quantum transport layer, Quantum Data Link Layer and Quantum Transport Layer respectively. By utilizing the addressing property the transmission packets will be transmitted between the correct communicators. Secondly, the flow and error control of the communication process can be controlled by sequence and acknowledgement numbers. If the transmitted packets are received out of order, then the receiver can reorder it to mirror original sent message by sequence number. Furthermore, if the receiver receipts a packet more than one time, then he will conclude there is a duplication by its corresponding sequence number (see Fig. 9) [13, 34–41, 61].

The sender can assure that the transmitted packets are receipted by intended receiver and correctly by acknowledgment number. The acknowledgment number identifies the sequence number of the subsequent packet expected by the receiver. In case of transmitted message corrupted or lost, the receiver will inform the sender through the acknowledgment number. When the sender receives the acknowledgment number, he will retransmit the erroneous or lost packets again to the receiver. Thirdly, the size of transmitted packet header can be specified by data offset part. Furthermore, it specifies the starting of actual transmitted data [13, 34–41, 61].

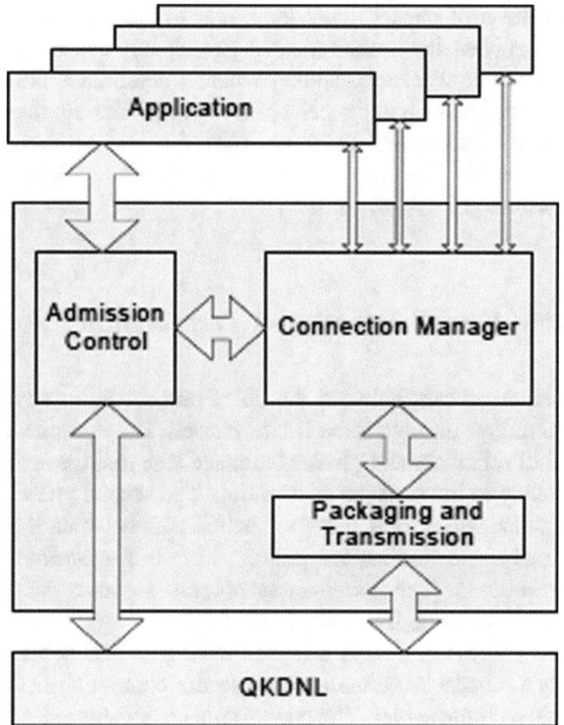

Fig. 8 The logical components of the *QKDTL*

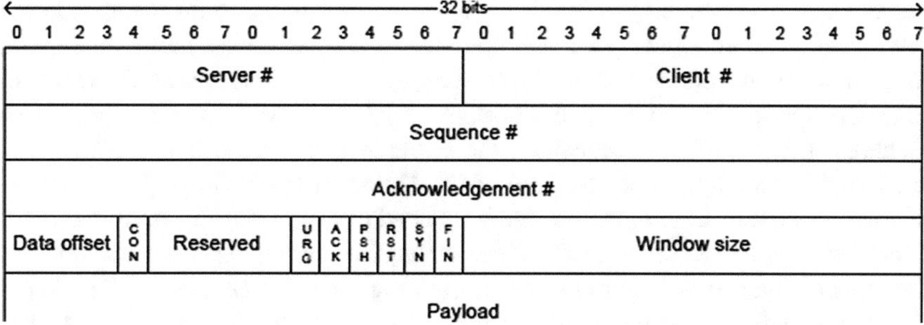

Fig. 9 *QKDTL* packet

Fourthly, the nature of transmitted packets can be determined by flag bits. There're six different types of flag bits which are *URG*, *ACK*, *PSH*, *RST*, *SYN* and *FIN*. *RST* is used for resetting the communication process. *FIN* is used for terminating the communication process. *SYN* is used for synchronize the communication process by coordinating the sequence number at establishment phase. *ACK* indicates the type of transmitted packet is acknowledgement one. *PSH* is used to force the transmission of packets from the buffer to the receiver. Fifthly, the flow control process between the communicated participants can be governed by utilizing the sliding window feature. Sliding window determines the amount of data which the sender can transmit and receiver can receive. Finally, the error correction can be achieved by employing checksum operation on the transmitted packets. Checksum has the ability for error detection and correction while the transmission or storage of packets [13, 34–41, 61].

7 Connection Establishment and Termination

The communication process between the client and server or any two communicated parties should go through three basics phases. The first one is establishment phase. The goal of establishment phase is to make sure that the sender and receiver are online and ready to transmit the information. The second phase is the exchange phase which is responsible for starting the actual data between the communicated participants. Finally, the termination phase which is responsible for ending the communication process. The termination process signifies neither sender nor receiver has more data to be transmitted [13, 61].

Establishment process is always achieved through a 3-way handshake process. The client sends a request for establishing a secure communication process over a classical network with the server. The client starts its registration with the Quantum Key Distribution network by establishing a TCP connection with the nearest Quantum Back Bone. Subsequently, the client will be assigned IP address if he registered successfully with its perspective Quantum Back Bone. The client IP address consists of both Quantum Back Bone address and a 16-bit unique address. Afterward, the composed packet will be moved across router-by-router through the quantum network infrastructure. The server starts its registration with the Quantum Key Distribution network by establishing a TCP connection with the nearest Quantum Back Bone. Subsequently, the server will be assigned IP address if he registered successfully with its perspective Quantum Back Bone. The server IP address consists of both Quantum Back Bone address and a 16-bit unique address. Consequently, the server and the client successfully established a Quantum Key Distribution Transport Layer connection (see Figs. 10, 11, 12 and 13) [13, 61].

After successfully established the communication process between the communicated participants, both clients and server starting the negotiation of generation a secure keys. The client sent a key request for its corresponding connection manager. The connection manager generates random bit string and encodes each with a quantum source. After encoding, quantum bits are transmitted actually from the sender to the receiver over a transmission channel. Clearly, a *QKD* system requires a transmission channel so that encoding quantum bits are transmitted from one communicator to another communicator through quantum carriers. Two popular types of transmission channels are optical fiber and open air often used for telecommunication networks and satellite communications respectively.

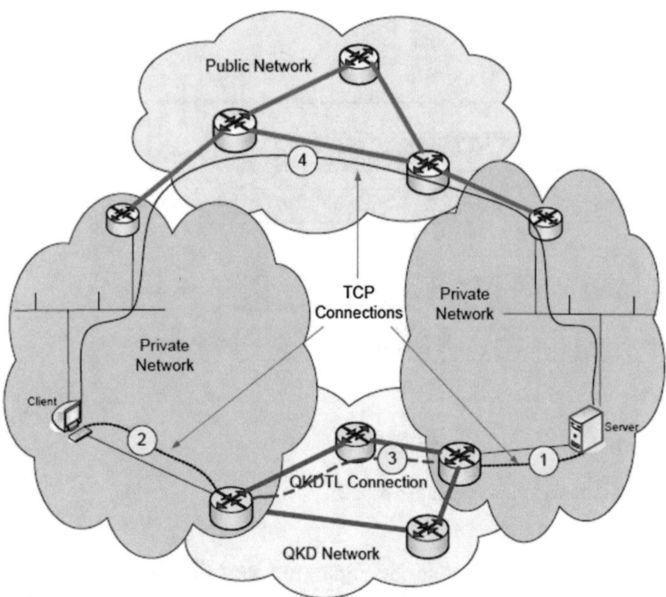

Fig. 10 *QKDTL* connection establishment and termination

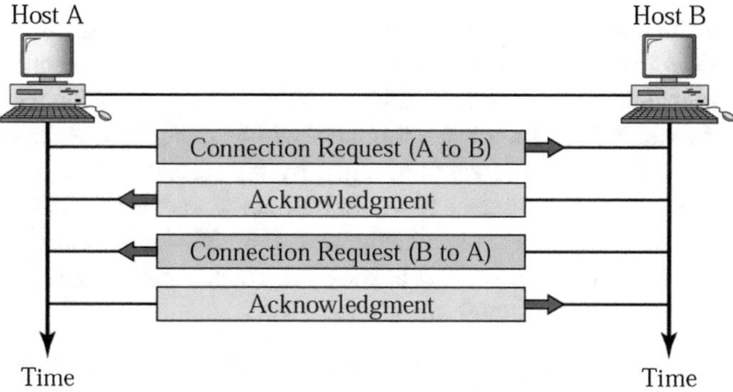

Fig. 11 *QKDTL* connection establishment process

If the transmitted quantum packet is successfully verified, then the connection manager at the recipient side transfers the message to its perspective application. Subsequently, the receiver will send an acknowledgment message to the sender for confirming the receiving of message correctly without any corruption. The receiver generates measurements on received encoded quantum bits by selecting basis on his realized. Subsequently, during transmission the communicators may use different

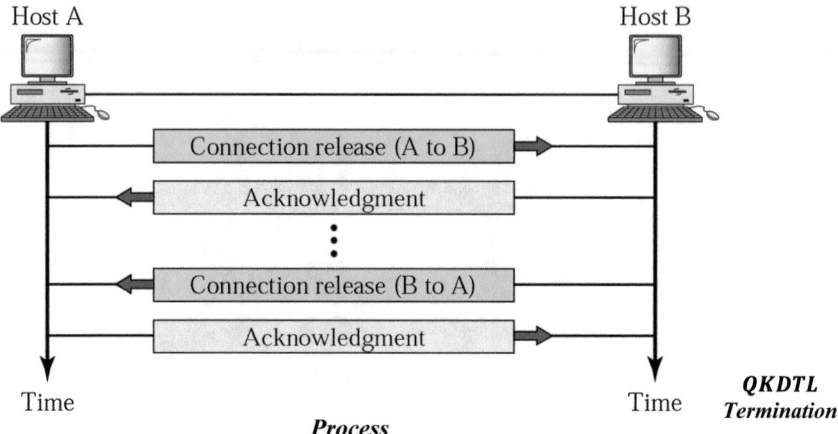

Fig. 12 *QKDTL* connection termination process

bases for measurements purpose. So, quantum key distribution protocol will recognize and exclude those bit positions where communicators use different bases. These positions are then discarded by both communicators over a public channel [13, 61].

Finally, during the transmission between communicators; eavesdropper might spy on quantum channel and retrieve potential secret key bits. The responsibility of *QKD* is to detect if any attacker or eavesdropper intercepts the transmitted keys. If the size of transmitted packet is not suitable for the network, then the packet will be divided into suitable fragments. Each fragment will be assigned a sequence number. Every transmitted fragment is acknowledged by the receiver. Acknowledgements are cumulative which indicate multiple correctly received packets can be acknowledged by only one acknowledgment [13, 34–41, 61]. This procedure reduces the overhead of transmitted acknowledgements packet and improves the efficiency of the network through three basic steps:

1. If the sender does not receive *ACK* within a specified amount of time, the sender retransmits the data.
2. Accepts out of order but does not send negative acknowledgements,
3. If a segment is not acknowledged before time-out, it is considered to be either corrupted or lost and the sender will retransmit the segment only when it times-out

The termination of established communication between the client and server can be achieved through three phases. Firstly, the Quantum Key Distribution Transport Layer "*QKATL*" between the client and server must be ended. In this case either client or server can request termination of connection. Secondly, the Quantum Key

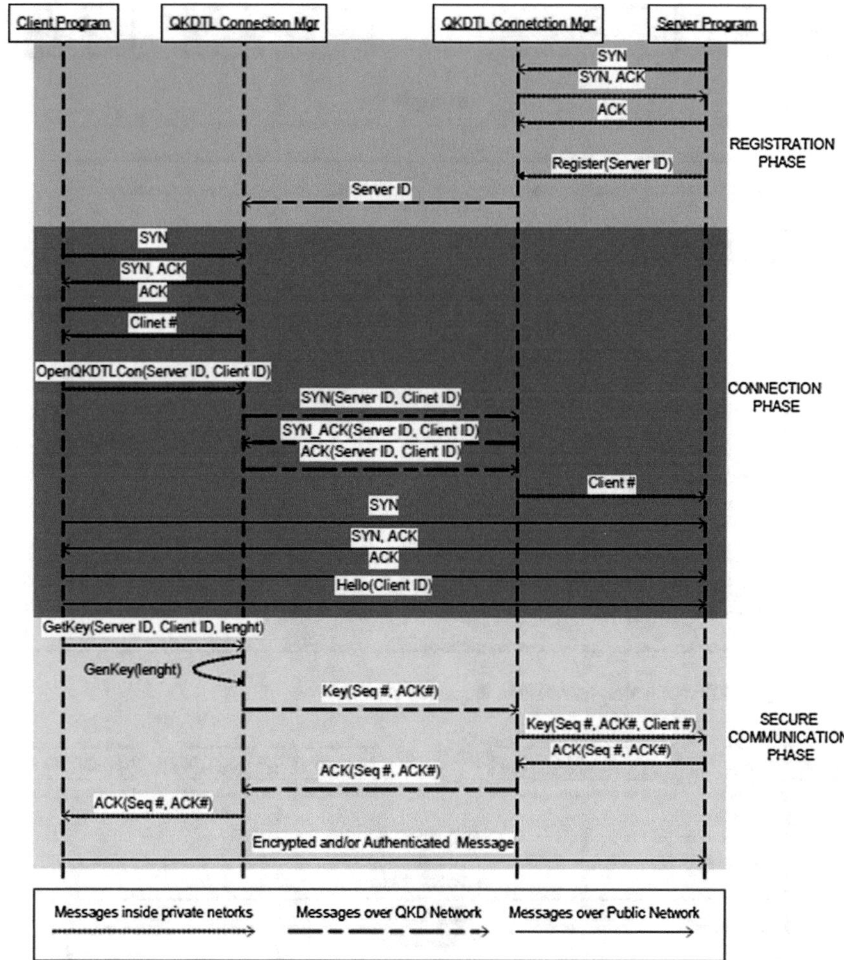

Fig. 13 *QKDTL* communication process steps in details

Distribution Data Link Layer connection between the incoming and outgoing Quantum Back Bone nodes must be ended. Lastly, the Quantum Key Distribution Transport Layer connection between the client and the perspective Quantum Back Bone node must be terminated. In Figs. 14 and 15 show the sequence of termination connection steps which initiated by the client and server respectively. Figs. 16 and 17 show the state diagram of Quantum Key Distribution Data Link Layer connection for client and server respectively [13, 34–41, 61].

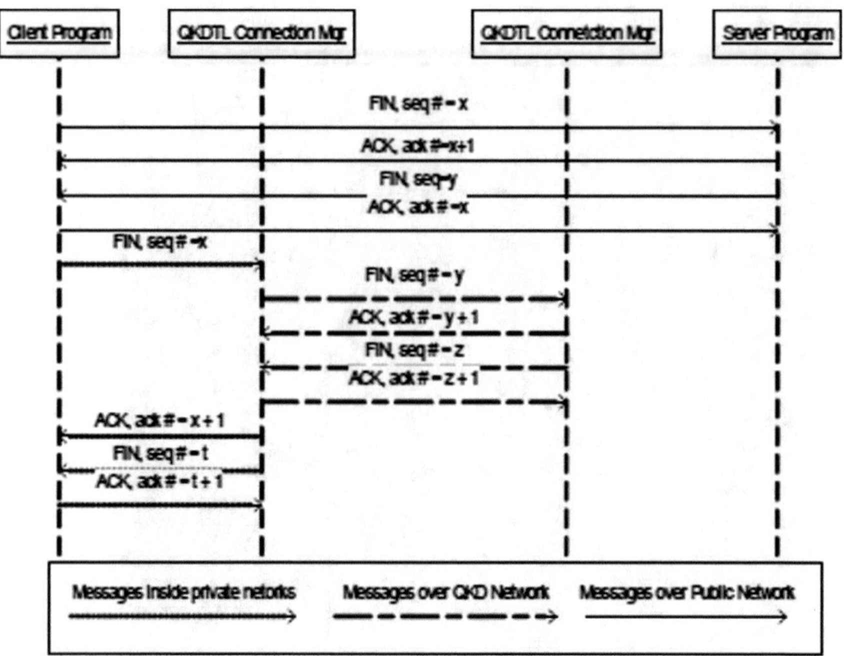

Fig. 14 *QKDTL* connection termination sequence at client

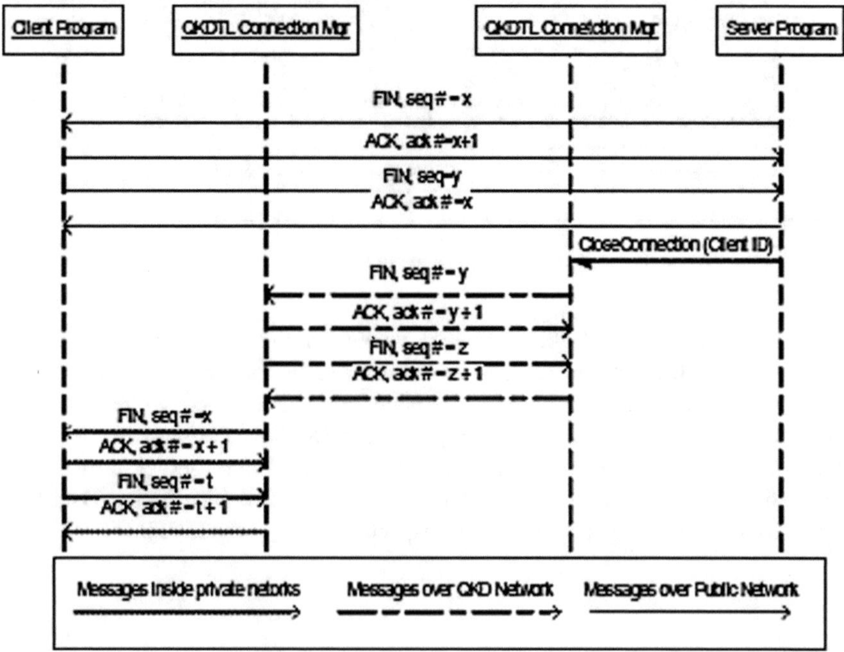

Fig. 15 *QKDTL* connection termination sequence at server

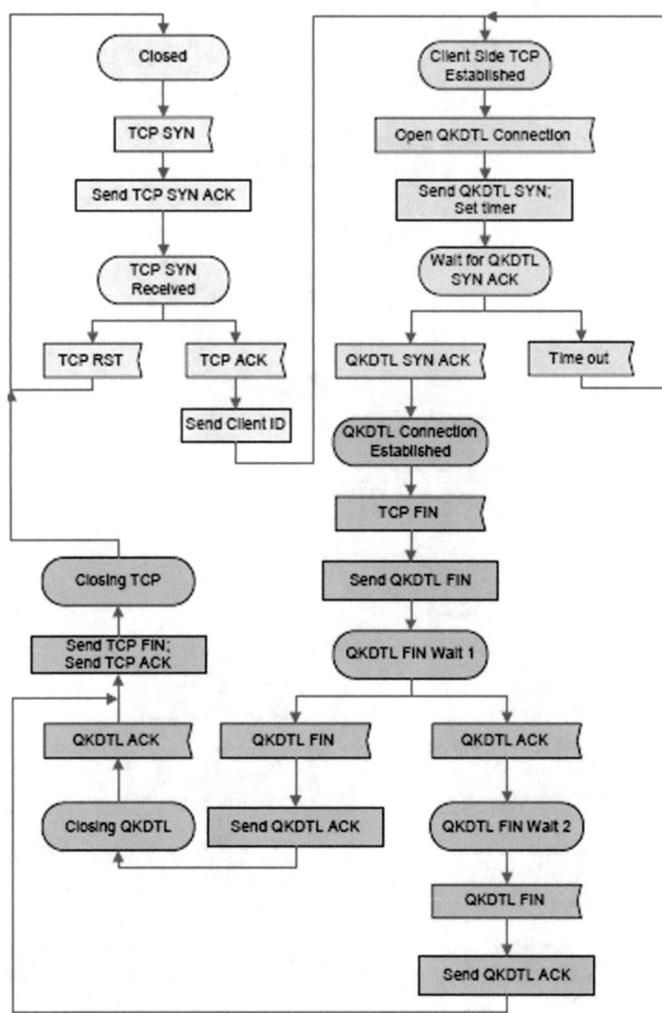

Fig. 16 *QKDTL* connection termination state diagram when client initiates the process

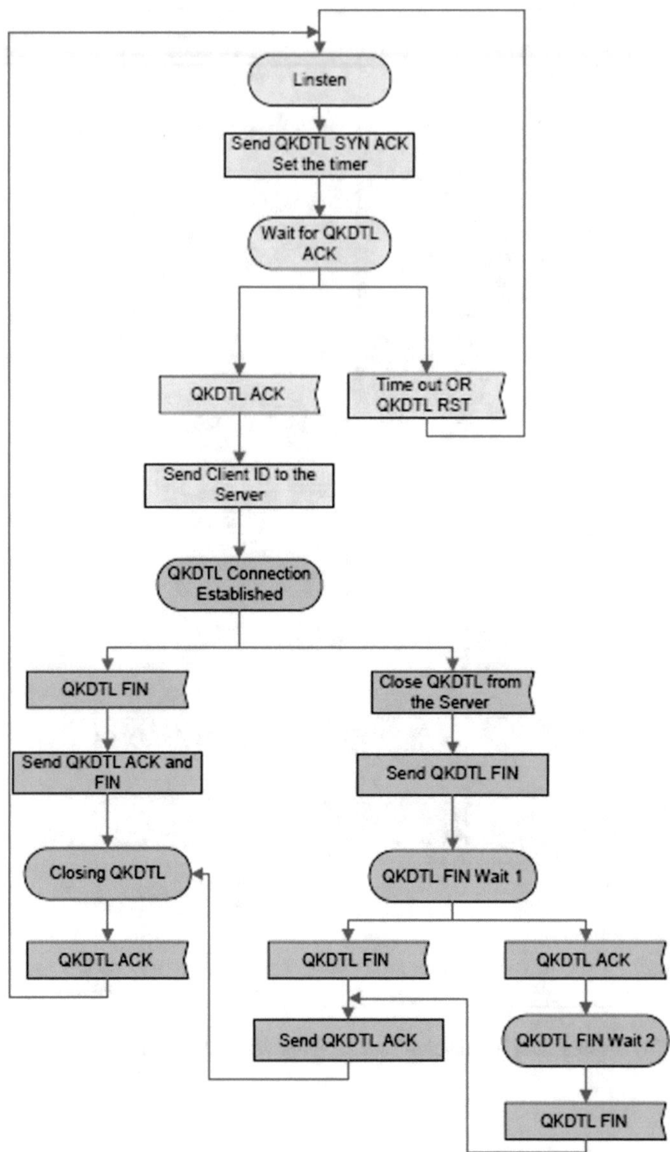

Fig. 17 *QKDTL* connection termination state diagram when server initiates the process

References

1. Bennett, C., Brassard, G.: Quantum cryptography: public key distribution and coin tossing. Theoret. Comput. Sci. **560**, 7–11 (2014). doi:10.1016/j.tcs.2014.05.025
2. Gisin, N., Ribordy, G., Tittel, W., Zbinden, H.: Quantum cryptography. Rev. Mod. Phys. **74** (1), 145 (2002)
3. Chuang, J.C.I., Sirbu, M.A.: pricing multicast communication: a cost-based approach. Telecommun. Syst. **17**(3), 281–297 (2001)
4. Phillips, G., Shenker, S., Tangmunarunkit, H.: Scaling of multicast trees: comments on the Chuang-Sirbu Scaling Law. ACM SIGCOMM Comput. Commun. Rev. **29**(4), 41–51). (1999) (ACM)
5. Kumar, Y., Munjal, R., Sharma, H.: Comparison of symmetric and asymmetric cryptography with existing vulnerabilities and countermeasures. Int. J. Comput. Sci. Manage. Stud. **11**(03) (2011)
6. Pathak, A.: Elements of Quantum Computation and Quantum Communication. Taylor & Francis (2013)
7. Forouzan, A.B.: Data Communications and Networking (sie). Tata McGraw-Hill Education (2006)
8. Friend, G.: Understanding Data Communications. Texas Instruments, Dallas, Tx. (1984)
9. Hughes, L.: Data Communications. McGraw-Hill, New York (1992)
10. Stallings, W.: Data and Computer Communications. Pearson/Prentice Hall (2007)
11. Ferguson, N., Schneier, B.: Practical Cryptography. Indianapolis, IN [etc.]. Wiley (2003)
12. Van Lint, J.H.: Introduction to Coding Theory, vol. 86. Springer Science & Business Media (1999)
13. Metwaly, A., Rashad, M.Z., Omara, F.A., Megahed, A.A.: Architecture of Point to Multipoint QKD Communication Systems (QKDP2MP). In 8th International Conference on Informatics and Systems (INFOS), Cairo, pp. NW 25–31. IEEE (2012)
14. Mink, A., Frankel, S., Perlner, R.: Quantum Key Distribution (QKD) and Commodity Security Protocols: Introduction and Integration (2010). arXiv preprint arXiv:1004.0605
15. Hughes, R.J., Luther, G.G., Morgan, G.L., Peterson, C.G., Simmons, C.: Quantum cryptography over underground optical fibers. In: Advances in Cryptology—CRYPTO'96, pp. 329–342. Springer, Heidelberg, January 1996
16. Stucki, D., Gisin, N., Guinnard, O., Ribordy, G., Zbinden, H.: Quantum key distribution over 67 km with a plug & play system. New J. Phys. **4**(1), 41 (2002)
17. Bethune, D.S., Risk, W.P.: Autocompensating quantum cryptography. New J. Phys. **4**(1), 42 (2002)
18. Rarity, J.G., Tapster, P.R., Gorman, P.M., Knight, P.: Ground to satellite secure key exchange using quantum cryptography. New J. Phys. **4**, 82 (2002)
19. Bienfang, J., Gross, A., Mink, A., Hershman, B., Nakassis, A., Tang, X., Wen, J.: Quantum key distribution with 1.25 Gbps clock synchronization. Opt. Express **12**(9), 2011–2016 (2004)
20. Tang, X., Ma, L., Mink, A., Nakassis, A., Xu, H., Hershman, B., Williams, C.: Quantum key distribution system operating at sifted-key rate over 4 Mbit/s. In: Defense and Security Symposium, pp. 62440P–62440P. International Society for Optics and Photonics, May 2006
21. Xu, H., Ma, L., Mink, A., Hershman, B., Tang, X.: 1310-nm quantum key distribution system with up-conversion pump wavelength at 1550 nm. Opt. Express **15**(12), 7247–7260 (2007)
22. Jackson, D.J., Giliam, D.P., Dowling, J.P.: Quantum Network Protocols (2001)
23. Curcic, T., Filipkowski, M.E., Chtchelkanova, A., D'Ambrosio, P.A., Wolf, S.A., Foster, M., Cochran, D.: Quantum networks: from quantum cryptography to quantum architecture. ACM SIGCOMM Comput. Commun. Rev. **34**(5), 3–8 (2004)
24. Dierks, T.: The Transport Layer Security (TLS) Protocol Version 1.2 (2008)
25. Gisin, N., Thew, R.: Quantum communication. Nat. Photonics **1**(3), 165–171 (2007)

26. Nguyen, T.M.T., Sfaxi, M.A., Ghernaouti-Hélie, S.: 802.11 i encryption key distribution using quantum cryptography. J. Netw. **1**(5), 9–20 (2006)
27. Ghernaouti-Hélie, S., Sfaxi, M.A.: Upgrading PPP security by quantum key distribution. In: Network Control and Engineering for QoS, Security and Mobility, vol. IV, pp. 45–59. Springer US (2007)
28. Sfaxi, M.A., Ghernaouti-Hélie, S., Ribordy, G., Gay, O.: Using quantum key distribution within IPSEC to secure MAN communications. In: Proceedings of Metropolitan Area Networks (MAN2005) (2005)
29. Peng, C.Z., Yang, T., Bao, X.H., Zhang, J., Jin, X.M., Feng, F.Y., Pan, J.W.: Experimental free-space distribution of entangled photon pairs over 13 km: towards satellite-based global quantum communication. Phys. Rev. Lett. **94**(15), 150501 (2005)
30. Hughes, R.J., Nordholt, J.E., Derkacs, D., Peterson, C.G.: Practical free-space quantum key distribution over 10 km in daylight and at night. New J. Phys. **4**(1), 43 (2002)
31. Metwaly, A.F., Rashad, M.Z., Omara, F.A., Megahed, A.A.: Architecture of multicast centralized key management scheme using quantum key distribution and classical symmetric encryption. Eur. Phys. J. Spec. Top. **223**(8), 1711–1728 (2014)
32. Deering, S.: Host Extension for IP Multicasting. RFC 1112 (1989)
33. Deering, S.E.: Multicast Routing in a Datagram Internetwork (No. STAN-CS-92–1415). Stanford University CA Department of Computer Science (1991)
34. Poppe, A., Peev, M., Maurhart, O.: Outline of the SECOQC quantum-key-distribution network in Vienna. Int. J. Quantum Inf. **6**(02), 209–218 (2008)
35. Peev, M., Pacher, C., Alléaume, R., Barreiro, C., Bouda, J., Boxleitner, W., Tualle-Brouri, R.: The SECOQC quantum key distribution network in Vienna. New J. Phys. **11**(7), 075001 (2009)
36. Elliott, C., Colvin, A., Pearson, D., Pikalo, O., Schlafer, J., Yeh, H.: Current status of the DARPA quantum network. In: Defense and Security, pp. 138–149. International Society for Optics and Photonics May 2005
37. Alleaume, R., Riguidel, M., Weinfurter, H., Gisin, N., Grangier, P., Dianati, M., Godfrey, M., et al.: SECOQC White Paper on Quantum Key Distribution and Cryptography. No. quant-ph/0701168 (2007)
38. Khan, M.M., Hyder, S., Pathan, M.K., Sheikh, K.H.: A Quantum key distribution network through single mode optical fiber. In: 2006 International Symposium on Collaborative Technologies and Systems, 2006. CTS 2006, pp. 386–391. IEEE May 2006
39. Le, Q.C., Bellot, P.: Enhancement of AGT telecommunication security using quantum cryptography. In: 2006 International Conference on Research, Innovation and Vision for the Future, pp. 7–16. IEEE, February 2006
40. Kimble, H.: The quantum internet. Nature **453**(7198), 1023–1030 (2008)
41. Dianati, M., Alléaume, R.: Architecture of the Secoqc Quantum Key Distribution nNetwork (2006). arXiv preprint quant-ph/0610202
42. Naseri, M., Heidari, S., Batle, J., Baghfalaki, M., Gheibi, R., Farouk, A., Habibi, A.: A new secure quantum watermarking scheme. Optik-Int. J. Light Electron Opt. **139**, 77–86 (2017)
43. Batle, J., Ciftja, O., Naseri, M., Ghoranneviss, M., Farouk, A., Elhoseny, M.: Equilibrium and uniform charge distribution of a classical two-dimensional system of point charges with hard-wall confinement. Phys. Scr. **92**(5), 055801 (2017)
44. Geurdes, H., Nagata, K., Nakamura, T., Farouk, A.: A Note on the Possibility of Incomplete Theory (2017). arXiv preprint arXiv:1704.00005
45. Batle, J., Farouk, A., Alkhambashi, M., Abdalla, S.: Multipartite correlation degradation in amplitude-damping quantum channels. J. Korean Phys. Soc. **70**(7), 666–672 (2017)
46. Batle, J., Naseri, M., Ghoranneviss, M., Farouk, A., Alkhambashi, M., Elhoseny, M.: Shareability of correlations in multiqubit states: optimization of nonlocal monogamy inequalities. Phys. Rev. A **95**(3), 032123 (2017)
47. Batle, J., Farouk, A., Alkhambashi, M., Abdalla, S.: Entanglement in the linear-chain Heisenberg antiferromagnet Cu (C 4 H 4 N 2) (NO 3) 2. Eur. Phys. J. B **90**, 1–5 (2017)

48. Batle, J., Alkhambashi, M., Farouk, A., Naseri, M., Ghoranneviss, M.: Multipartite non-locality and entanglement signatures of a field-induced quantum phase transition. Eur. Phys. J. B **90**(2), 31 (2017)
49. Nagata, K., Nakamura, T., Batle, J., Abdalla, S., Farouk, A.: Boolean approach to dichotomic quantum measurement theories. J. Korean Phys. Soc. **70**(3), 229–235 (2017)
50. Abdolmaleky, M., Naseri, M., Batle, J., Farouk, A., Gong, L.H.: Red-green-blue multi-channel quantum representation of digital images. Optik-Int. J. Light Electron Opt. **128**, 121–132 (2017)
51. Farouk, A., Elhoseny, M., Batle, J., Naseri, M., Hassanien, A.E.: A proposed architecture for key management schema in centralized quantum network. In: Handbook of Research on Machine Learning Innovations and Trends, pp. 997–1021. IGI Global (2017)
52. Zhou, N.R., Li, J.F., Yu, Z.B., Gong, L.H., Farouk, A.: New quantum dialogue protocol based on continuous-variable two-mode squeezed vacuum states. Quantum Inf. Process. **16**(1), 4 (2017)
53. Batle, J., Abutalib, M., Abdalla, S., Farouk, A.: Persistence of quantum correlations in a XY spin-chain environment. Eur. Phys. J. B **89**(11), 247 (2016)
54. Batle, J., Abutalib, M., Abdalla, S., Farouk, A.: Revival of Bell nonlocality across a quantum spin chain. Int. J. Quantum Inf. **14**(07), 1650037 (2016)
55. Batle, J., Ooi, C.R., Farouk, A., Abutalib, M., Abdalla, S.: Do multipartite correlations speed up adiabatic quantum computation or quantum annealing? Quantum Inf. Process. **15**(8), 3081–3099 (2016)
56. Batle, J., Bagdasaryan, A., Farouk, A., Abutalib, M., Abdalla, S.: Quantum correlations in two coupled superconducting charge qubits. Int. J. Mod. Phys. B **30**(19), 1650123 (2016)
57. Batle, J., Ooi, C.R., Abutalib, M., Farouk, A., Abdalla, S.: Quantum information approach to the azurite mineral frustrated quantum magnet. Quantum Inf. Process. **15**(7), 2839–2850 (2016)
58. Batle, J., Ooi, C.R., Farouk, A., Abdalla, S.: Nonlocality in pure and mixed n-qubit X states. Quantum Inf. Process. **15**(4), 1553–1567 (2016)
59. Metwaly, A.F., Mastorakis, N.E.: Architecture of decentralized multicast network using quantum key distribution and hybrid WDM-TDM. In: Advances in Information Science and Computer Engineering, pp. 504–518 (2015)
60. Metwaly, A.F., Rashad, M.Z., Omara, F.A., Megahed, A.A. (2015). Architecture of Multicast Network Based on Quantum Secret Sharing and Measurement
61. Metwaly, A.F., Rashad, M.Z., Omara, F.A., Megahed, A.A.: Architecture of multicast centralized key management scheme using quantum key distribution and classical symmetric encryption. Eur. Phys. J. Spec. Topics **223**(8), 1711–1728 (2014)

IPsec Multicast Architecture Based on Quantum Key Distribution, Quantum Secret Sharing and Measurement

Ahmed Farouk, O. Tarawneh, Mohamed Elhoseny, J. Batle, Mosayeb Naseri, Aboul Ella Hassanien and M. Abedl-Aty

Abstract In this chapter, securing the transmitted multicast information can be achieved through IPsec multicast architecture. The process of IPsec involves the sender and destinations to agree on IPsec keys. These keys are used for protection transmitted information among communicated peers over IPsec network. IPsec depends on classical algorithm for key generation and distribution. These algorithms proved their conditional security which means intruder can break the algorithm and intercept the communication process. A new IPsec multicast architecture is proposed. The proposed architecture is divided into five main processes. The most important process is key generation and distribution. The key generation and distribution through IPsec multicast network is achieved using quantum algorithms. Quantum keys proved their unconditional security according to their physical characteristics. Sender and receivers communicate through two channels; quantum and classical. Encryption and decryption processes depend on agreed quantum keys and classical cryptographic algorithms. IPsec depends on quantum key distribution

---

A. Farouk (✉) · M. Elhoseny
Faculty of Computer and Information Sciences, Mansoura University, Mansoura, Egypt
e-mail: dr.ahmedfarouk85@yahoo.com

A. Farouk · M. Abedl-Aty
University of Science and Technology, Zewail City of Science and Technology, Giza, Egypt

O. Tarawneh
Information Technology Department, Al-Zahra College for Women, P.O. Box 3365, Muscat, Oman

J. Batle
Departament de Física, Universitat de Les Illes Balears, 07122 Palma de Mallorca, Balearic Islands, Spain

M. Naseri
Department of Physics, Islamic Azad University, Kermanshah Branch, Kermanshah, Iran

A.E. Hassanien
Faculty of Computers and Information, Cairo University, Giza, Egypt

A.E. Hassanien et al. (eds.), *Quantum Computing: An Environment for Intelligent Large Scale Real Application*, Studies in Big Data 33,
https://doi.org/10.1007/978-3-319-63639-9_6

for creating keys for IPsec security associations. The confidentiality and authentication of the proposed architecture is analyzed.

Keywords Quantum key distribution • Quantum • *QKD* node-by-node routing • Quantum key generation and distribution

1 Introduction

Securing the transmitted multicast information can be achieved through IPsec multicast architecture. The process of IPsec involves the sender and destinations to agree on IPsec keys. These keys are used for protection transmitted information among communicated peers over IPsec network. IPsec depends on a classical algorithm for key generation and distribution. These algorithms proved their conditional security which mean intruder can break the algorithm and intercept the communication process. A new IPsec multicast architecture is proposed. The proposed architecture is divided into five main processes. The most important process is key generation and distribution. The key generation and distribution through IPsec multicast network is achieved using quantum algorithms. Quantum keys proved their unconditional security according to their physical characteristics. The sender and receivers communicate through two channels; quantum and classical. Encryption and decryption processes depend on agreed quantum keys and classical cryptographic algorithms. IPsec depends on quantum key distribution for creating keys for IPsec security associations. The confidentiality and authentication of the proposed architecture are analyzed.

IPsec guarantees security for IP communications at the network layer of the Open System Interconnections Model (*OSIModel*). Sensitive information sent over the internet can be encrypted via IPsec to maintain the secrecy of the information. IPsec encrypts data at layer 3 IP packet layer proposing exhaustively secured solution by providing data authentication, anti-replay protection, data confidentiality, and data integrity protection [1–3]. IPsec consists of set of protocols which each protocol concentrates on particular characteristics of the IPsec purposes to protect IP communications over untrusted networks. Internet key exchange is an example of IPsec protocols which focuses on message delivery authenticity, as well as, Encapsulating Security Payload which focuses on data confidentiality. IPsec uses encapsulated security payload (*ESP*) and authentication header (*AH*) to accomplish desired security objectives [1, 2, 4–8]. *AH* is constructed to enhance the security principles by protecting IP packet header. The protection objective is delivered by cryptographic authentication [9, 10]. The authentication service confirms that any interference with IP traffic will be identified. *ESP* provides protection and confidentiality for IP packet data. This protection objective is achieved by encoding the content of data packet using symmetric encryption algorithm as Data Encryption Standard (*DES*), Triple Data Encryption Standard (3*DES*), and Advanced Encryption Standard (*AES*) [3, 9–13].

Traditional multicast IPsec architecture is based on classical key generation and authentication. The Diffie–Hellman algorithm is mainly used for securing key transmission over unsafe media. Furthermore, it is used excessively in present key management for delivery keying information for IPsec [1, 2, 14]. Key generation and management based on Diffie–Hellman algorithm impaired from many weaknesses. Firstly, the frequency of changing the distributed keys between the communicated IPsec peers is limited. Secondly, the generated keys are conditional security which means intruder can spy on the communication channel and copy keys without any given warning for the communicated IPsec peers. Lastly, the whole architecture of traditional multicast IPsec deteriorates from man in the middle attack [3, 5, 6, 14].

In 1984, Bennett and Brassard [3, 15–17] achieved that the transfer of quantum states is very essential for securing communication system. Consequently, they developed the first quantum key distribution method which latterly is known as BB84 protocol [3, 15, 18]. BB84 protocol was applied practically in 1989. Quantum communication delivers an innovative technique for securing the confidentiality and authentication of modern communication systems. Unlike the classical communication, the quantum communication relies on physical characteristics of used quantum signals. Quantum communication security depends on the corresponding quantum physics laws, such as the well-known Heisenberg uncertainty principle and no-cloning theorem [3, 19–21]. By way of 2017, the improvement and growth of a real quantum computer is still in early stages but many poetical and theoretical experimentations were implemented by many research groups [25–46].

According to the work in this thesis, to secure IPsec multicast network, a proposed scheme has been introduced by replacing the classical key generation and distribution to be quantum one. According to the proposed scheme, the sender and multiple receipts have to go through quantum key distribution steps for generating and distributing keys. The sender and receivers communicate through two channels; quantum and classical. Quantum channel is used for generation and distributing quantum keys. Classical channel is used for negotiation for policies, security associations and security parameters. Encryption and decryption processes depend on agreed quantum keys and classical cryptographic algorithms. The entire process of quantum key generation and distribution inside the multicast IPsec architecture is managed by quantum key distribution. Based on agreed security parameters and policies, the sender and receipts start the negotiation of IPsec security associations. After negotiation is finished, IPsec relies on quantum key distribution for creating keys for IPsec security associations. The confidentiality and authentication of proposed architecture are analyzed [3, 22–24].

2 General Logical and Physical Architecture of the Decentralized Multicast *QKD–VPN*

In our proposed scheme as illustrated in Fig. 1, a secured multicast group is broken into numerous smaller groups called *sub-groups*. Each *sub-group* will be assigned to its perspective QM_{KC}. Each QM_{KC} requires two different channels; quantum and classical. A quantum channel is to transmit and delivery of encoded quantum signals to its *sub-group's* members as well as, *QKD* protocol for generation authentication and private keys. A classical channel is used to transmit the messages and distribute the raw private keys among *sub-group's* members as well as, transmit encrypted messages through *VPN*. The generated keys are used for encryption and decryption also play an important role for designing a decentralized secured multicast network from *QKD* protocols [3, 22–24].

3 The Architecture of Decentralized *QKD* Node by Node

Quantum network structure for transmitting quantum signals and moving photons from one *sub-group* to another *sub-group*, in other words transmitting quantum signals' between two different QM_{KC}, is done through *QKD* nodes routing. *QKD* nodes obviously act as mutualized resources for all QM_{KC} as shown in Fig. 2 [3, 22–24].

QKD nodes are also responsible for moving/routing messages and signals. Individual users send interesting traffic to its local QM_{KC}. Each QM_{KC} accumulates interesting traffic from its perspective members and sends it to the nearest *QKD* node. The interesting traffics then is transmitted and routed through nodes until is received by the desired QM_{KC} in other side [3, 22–24].

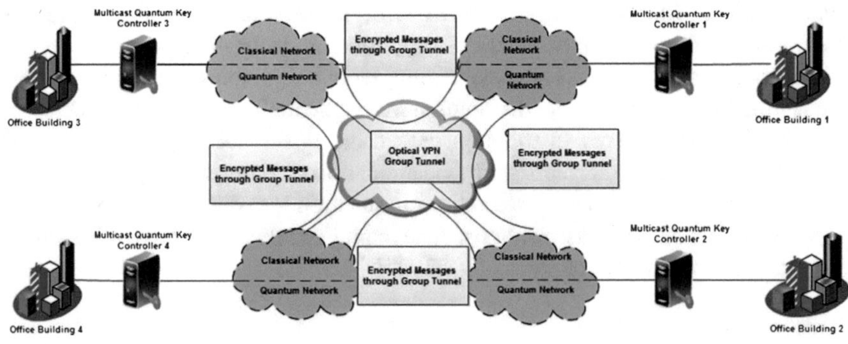

Fig. 1 General logical architecture of decentralized multicast *QKD–VPN*

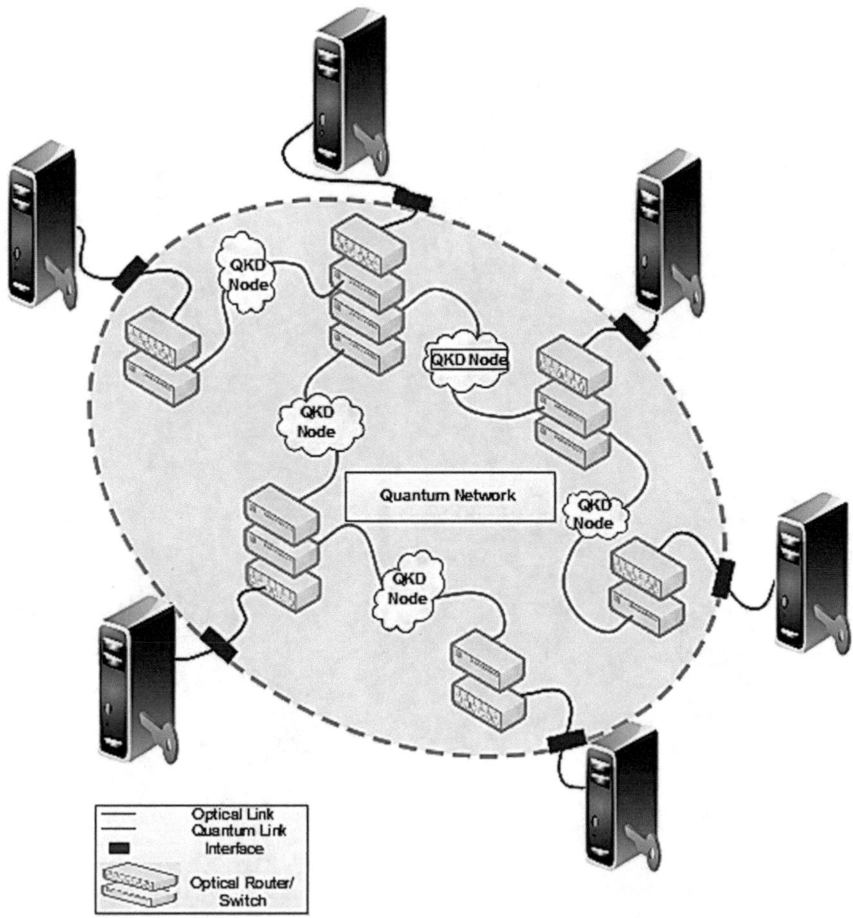

Fig. 2 Decentralized *QKD* node-by-node routing

4 The Architecture of Decentralized *VPN* Node by Node

In Fig. 3, site-to-site multicast group *VPN* tunnel is used to connect multiple *sub-group* securely over the classical channel, as well as, to ensure data integrity, confidentiality and protection of transmitted messages among different *sub-group* members. Tunneling mechanism means adding a security layer for all transmitted packets. In the tunnel mode, virtual channels have to build between *sub-groups*. These virtual channels are familiarized with virtual private network (*VPN*). A Multicast *VPN* provides *sub-groups* to clearly communicate its private network over the network backbone of a service provider, as well as, delivers a dynamically scalable high-speed information transmission for several sites concurrently. Multicast *VPN* protects segments transmission over an open network by using tunneling

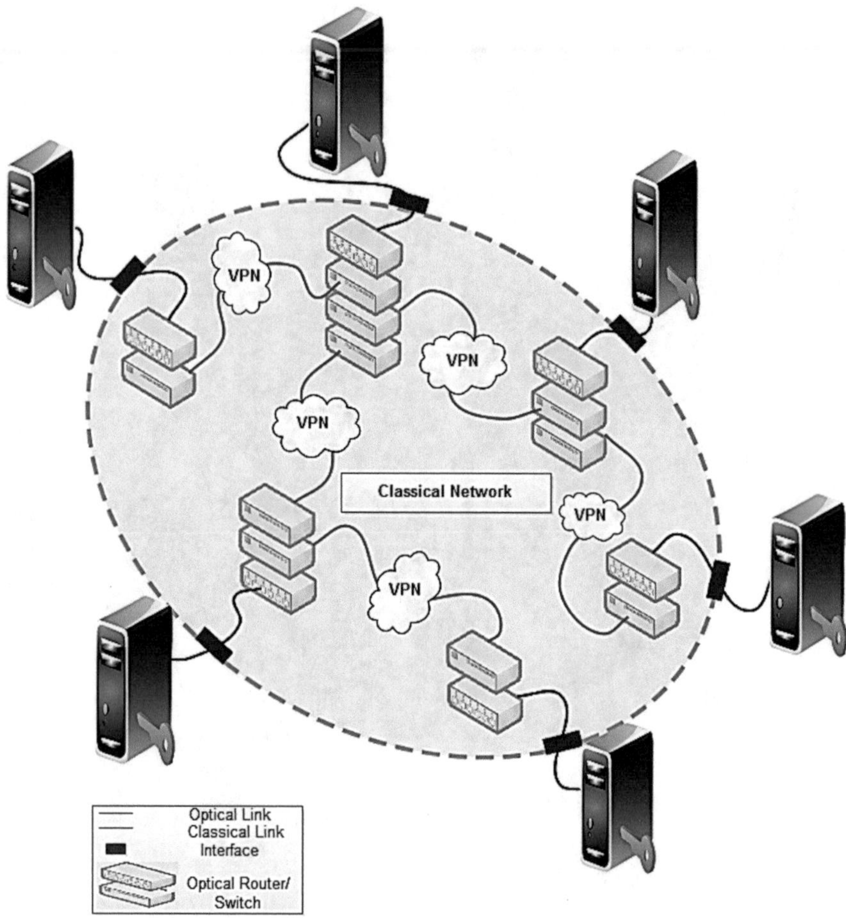

Fig. 3 Decentralized *VPN* node-by-node tunneling

mechanism. In the virtual private network, data encryption and decryption are performed at inbound and outbound tunnel interfaces respectively [3, 22–24].

5 The Physical Architecture of *WDM–TDM* Decentralized Network

In order to provide high utilization of the fiber bandwidth capacity in our proposed physical architecture, a hybrid *WDM–TDM* is used between QM_{KC} and *sub-groups*. Hybrid *WDM–TDM* combines advantages of both techniques. Advantages of *WDM* include increasing capability of delivered capacity as each *sub-group* has its own λ

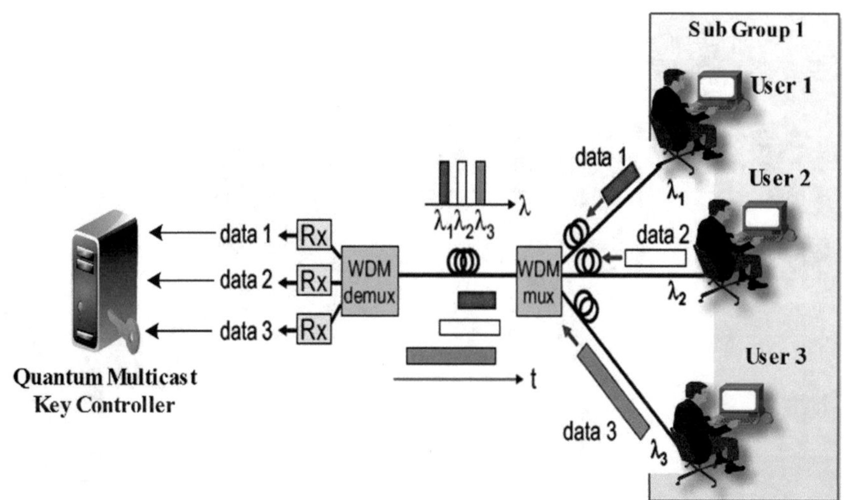

Fig. 4 Physical Architecture of *WDM–TDM* Decentralized network

channel and λ-routing, congestion is stopped between QM_{KC} and *sub-groups*, as well as, virtual communication is established between QM_{KC} and *sub-groups*. Advantages of *TDM* include sharing, power splitting and flexibility of delivered capacity between QM_{KC} and *sub-groups*, as illustrated in Fig. 4 [3, 22–24].

Transmitting the whole wavelength between QM_{KC} and *sub-groups* is achieved by *TDM*. The channels between QM_{KC} and *sub-groups* are spitted into time slots. Every *sub-groups* has provided a slot and the slots are turned amongst the *sub-groups*. After the last time slot for channel is handled, the cycle begins around once again with a new frame, starting with the second sample, byte or data block from the channel. *TDM* and *WDM* operation are illustrated in Figs. 5 and 6 respectively [3, 22–24].

QM_{KC} has two line cards; uplink and downlink cards, which are used for upstream and downstream between QM_{KC} and *sub-groups* respectively. Each *sub-group* has one uplink and one downlink card, for example *sub-group* 1 has uplink and downlink card 1 connected to QM_{KC}. The number of uplink and downlink cards depends on the number of the connected *sub-groups*. For *sub-groups*, QM_{KC} requires *N* uplink and *N* downlink cards. In the downstream path, the data transfer "*OLT* to the *ONUs*" is transmitted from QM_{KC} to *sub-groups*, while in the upstream path, the data transfer "*ONUs* to the *OLT*" is transmitted from *sub-groups* to QM_{KC} [3, 22–24].

The transformation between the electrical waves is managed by QM_{KC}'s equipment and the fiber optic signals used by *sub-groups* is achieved through an optical line termination (*OLT*), as well as, synchronize

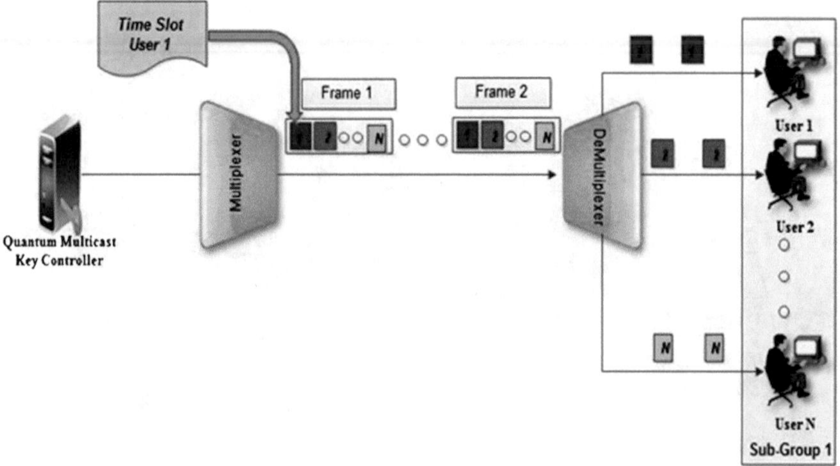

Fig. 5 *TDM* Operation for *sub-groups* 1

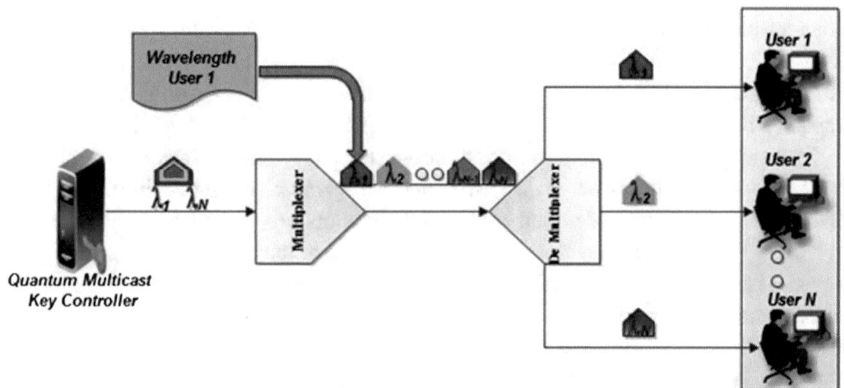

Fig. 6 *WDM* Operation for *sub-groups* 1

the multiplexing between the transformation tools of optical network unit (*ONU*). Altogether *ONUs* are linked to the *OLT* through a combiner/splitter, the transmission between *ONUs* is achieved only through the *OLT*. *ONU* provides access to the fiber distribution cable between QM_{KC} and *sub-groups*, as well as, deterrence of unauthorized Access. The performance and diagnostics monitoring between QM_{KC} and *sub-groups* are achieved using fault monitor. Fault monitor connected through Ethernet switch to *ONU*, is illustrated in Fig. 7 [3, 22–24].

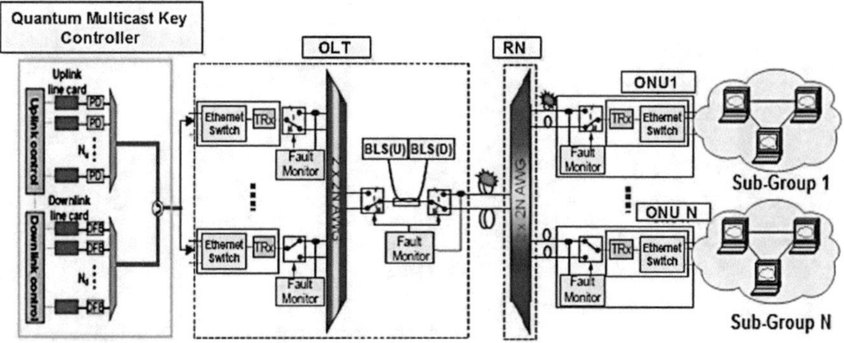

Fig. 7 Detailed physical architecture of *WDM–TDM* decentralized network

6 The Proposed Centralized IPsec Multicast Architecture

The sequence of procedures for communicating between a sender and multiple receipts in our proposed scheme are as follows:

(1) The sender wishes to transmit a packet to multiple receipts simultaneously. Since no security associations are created yet for protecting transmitted traffic, IPsec starts for creating security associations.
(2) The sender's IPsec process starts the conferring with each receiver's IPsec process. This conferring involves policy negotiation, transmission of quantum keys for securing established IPsec session and identity verification of communicated participants. This step is known as phase one.
(3) Based on the agreed security parameters and policies from phase one. The sender and receipts start the negotiation of IPsec security associations. After negotiation is finished, IPsec relies on quantum key distribution for creating keys for IPsec security associations.
(4) Now, the sender and receipts can exchange packets securely over IPsec tunnel.
(5) After transmission is over, the established session will be terminated [3, 22–24].

6.1 Phase One

The goal of the first phase in our proposed scheme is to exchange the policy, distribute and manage security keys, check the identity of the communicated participants and establish a secured media among the sender and his perspective multiple receipts. With the purpose of finalizing first phase, the communicated participants must agree for security factors as verification process, encryption method and key generation process as specified in Fig. 8 [3, 22–24].

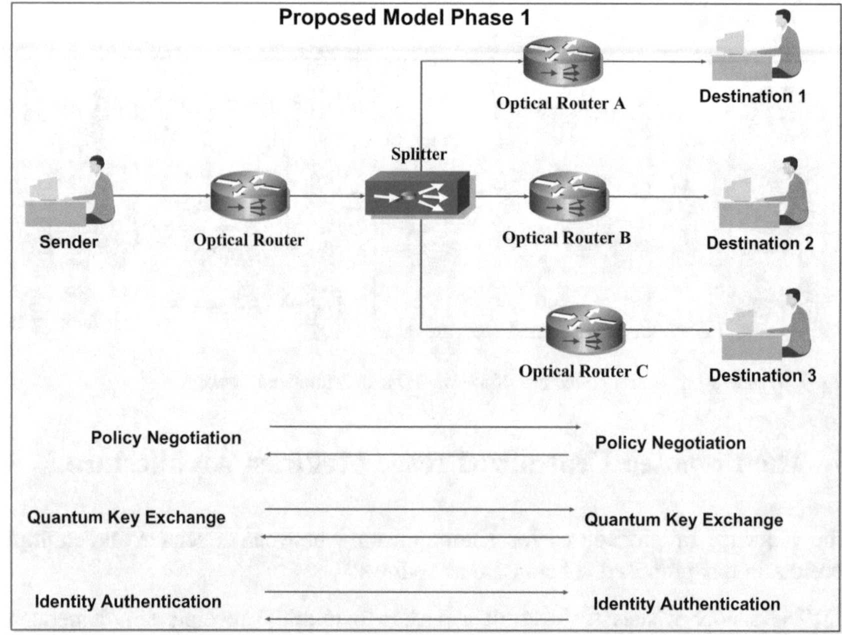

Fig. 8 The proposed model phase 1 process

The required steps for establishing phase one is demonstrated in Fig. 9. According to Fig. 9, the process of symmetrical key agreement between sender and destination 1 is as follows:

(1) Authentication between a sender and multiple destinations in a multicast network can be achieved through using a pre-shared quantum keys.
(2) Quantum key distribution generates a group of random quantum bits. Each user derives a private and public keys after negotiation with quantum key distribution through quantum key generation process including key distillation, sifted keys and raw key exchange. Furthermore, public keys are exchanged between the sender and multicast destination users.
(3) Each user produces a shared secret key from their private key and the other's public key. Shared secret key is the generated agreed quantum key.
(4) *QKD* is used to exchange key information, as well as, the corresponding methods which are used for cryptography among the users.
(5) Based on *QKD* and agreement key material, each user produces an independent symmetrical key [3, 22–24].

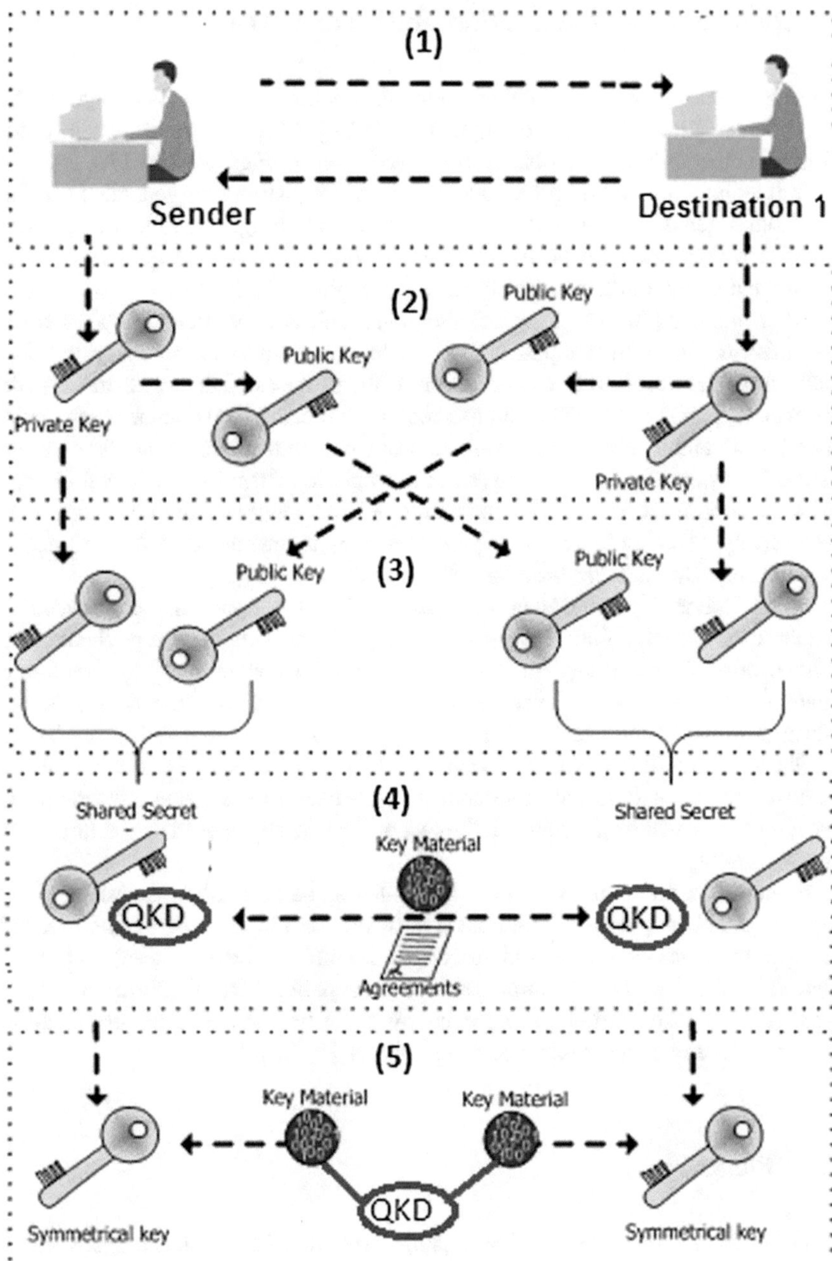

Fig. 9 Phase 1 steps

7 Quantum Key Generation and Distribution

The key generation and distribution among the communicated peers will go through four phases. The four phases are Quantum coding, Quantum transmission, Eavesdropper detection and Key distillation as described in Figs. 10 and 11.

In the Quantum coding phase, the quantum key distribution generates quantum bits from a quantum source to encode a random-bit string. Each random-bit string encodes with a probabilistic distribution function. The resulted quantum bits are transmitted to the multicast user in the second phase [3, 22–24].

After finishing the first phase, the communicators participate the quantum transmission phase. In this phase, the encoded quantum bits are transmitted physically from the quantum key distribution to the multicast user over a transmission channel. Apparently, a *QKD* technique requires a transmission channel such that the encoding quantum bits are transmitted from one transmitter to another through quantum carriers. Two common types of transmission channels are optical fiber and open air often used for telecommunication networks and satellite communications respectively. The multicast user generates measurements on received encoded quantum bits by selecting basis on his realized [3, 22–24].

In the Eavesdropper Detection phase, during the transmission between the sender and a destination; eavesdropper might listen on quantum channel and recover possible secret key bits. Using quantum laws, eavesdropper operation on quantum channel can be detected. Eavesdropping is discovered as follows; an arbitrary subset of the raw key is agreed by the communicators, and those bits are evaluated openly. If whichever two agreeing bits vary, this specifies the existence of an eavesdropper and so the communicators go back to stage one. Otherwise, the exchanged bits will be aborted and the rest of the raw key is used as the final secret key [3, 22–24].

In the key distillation phase, both the sender and a destination use several bases for measurements purpose. The objective of this step is to recognize and exclude those bit positions where the communicators use different bases. These positions are then discarded by both communicators over a public channel. Furthermore, the security of the key string is improved by correcting the resulted errors during transmission and eavesdropper detection phases [3, 22–24].

7.1 Phase Two

In phase two, IPsec starts for protecting the transmitting packets between the communicated participants by negotiation and establishing security associations of IPsec. This is achieved by the protection of established IPsec policies and appropriate keying information exchanged using Quantum Key Distribution system as shown in Fig. 12 [3, 22–24].

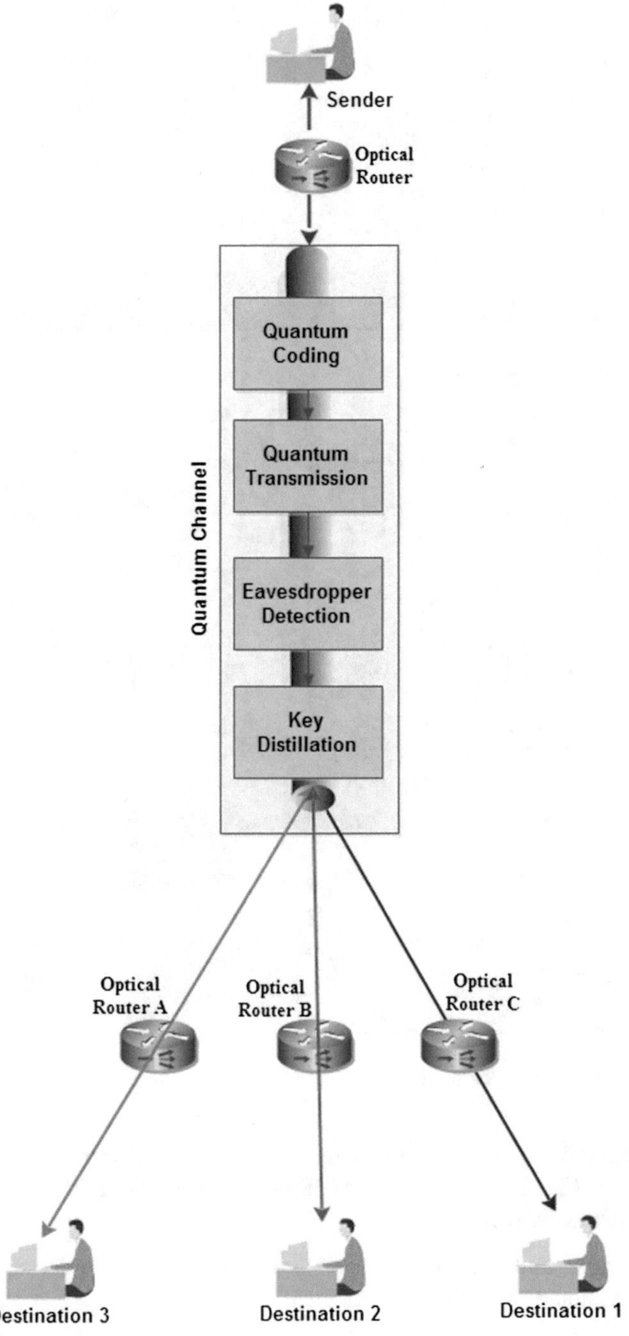

Fig. 10 Quantum key distribution process

Sender's bit	0	1	1	0	1	0	0	1
Sender's Basis	+	+	X	+	X	X	X	+
Sender's Polarization	↑	→	↖	↑	↖	↗	↗	→
Destination's Basis	+	X	X	X	+	X	+	+
Destination's Measurement	↑	↗	↖	↗	→	↗	→	→
Public discussion								
Shared Secret key	0		1			0		1

Fig. 11 Quantum shared secret key

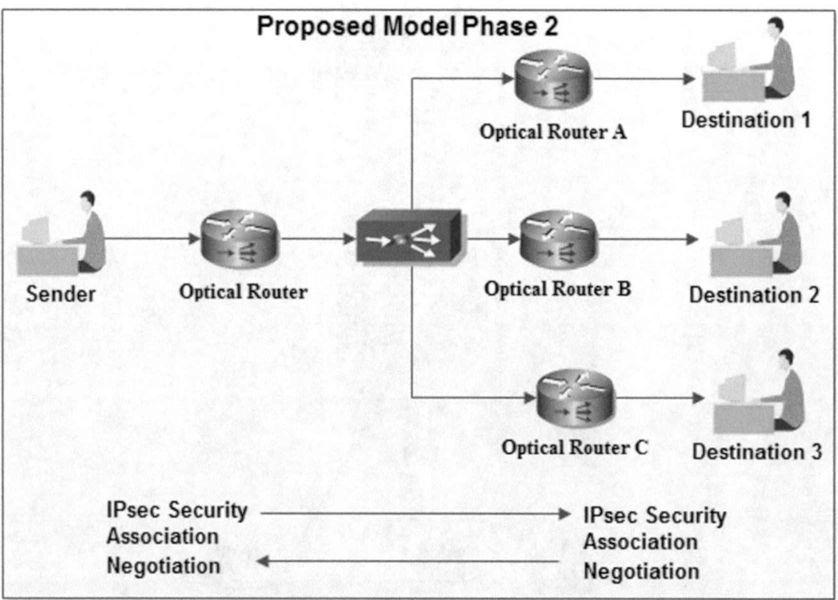

Fig. 12 Proposed model phase 2 process

The required steps for establishing phase two is demonstrated in Fig. 13. According to Fig. 13, the process of IPsec key agreement between sender and destination is as follows:

(1) The sender and destinations exchange more key material and agree on encryption and integrity methods for IPsec.

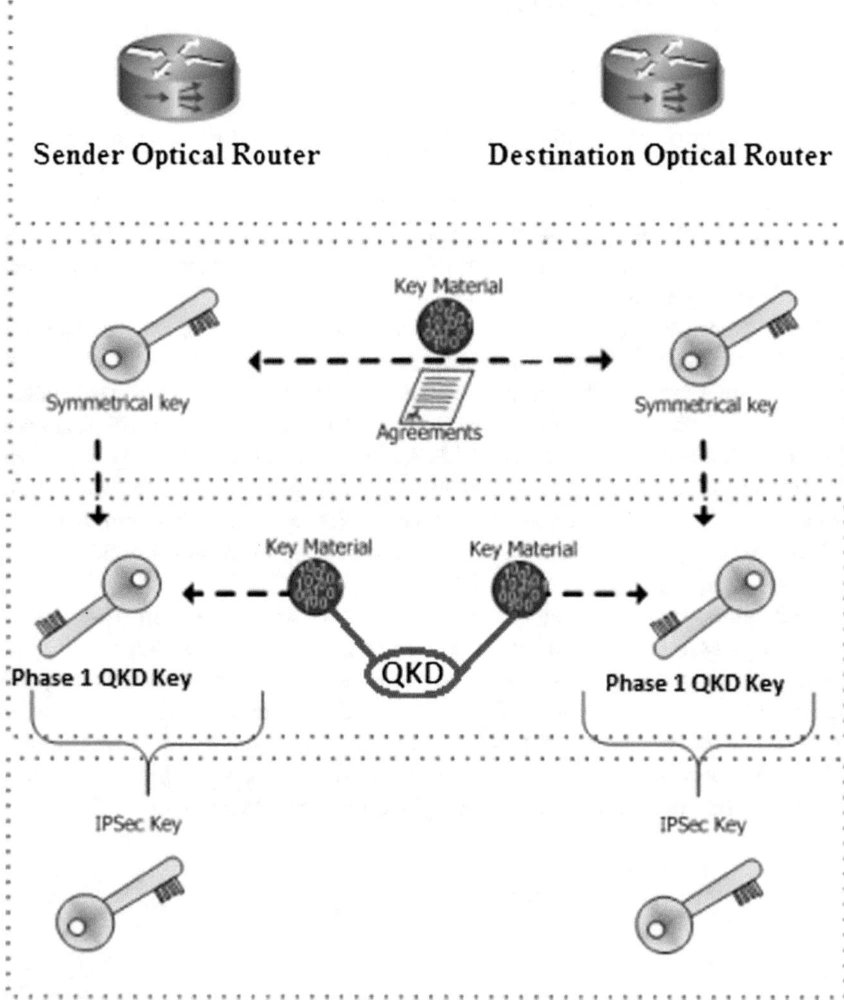

Fig. 13 Phase 2 Steps

(2) Quantum keys are combined with the key material to generate the symmetrical IPsec keys.
(3) The symmetrical IPsec keys are used for protecting transmitted data between the sender and multiple receipts simultaneously [3, 22–24].

8 Encryption and Decryption

Before starting a secured communication process between the sender and multiple receipts, they must agree upon an IPsec key and keep it secret between themselves. For distributing and generating IPsec keys, the sender and destinations are communicated through two channels; quantum and classical channels (see Fig. 14). The quantum channel is responsible for generating agreed IPsec keys. IPsec keys have to go through quantum key distribution steps to make sure the confidentiality of the generated keys. Classical channel is responsible for encryption, decryption and authentication processes [3, 22–24].

These processes depend on the generated quantum keys. The sender wishes to transmit the original message that can be read and understood without any special measures. Encryption of the original message is performed at the sender side and is achieved by combined original message, sender IPsec key and cryptographic algorithm. A cryptographic algorithm works in combination with a key to encrypt the original message. Encrypting the original message results in unreadable form called cipher message. The sender uses encryption to ensure that information is hidden from anyone for whom it is not intended, even those who can see the encrypted data (see Fig. 15) [3, 22–24].

Decryption of received cipher message is performed at receipt side and is achieved by combined cipher message, receipt IPsec key and cryptographic algorithm. Each receipt uses its own IPsec key and cryptographic algorithm. Decryption cipher message results the original message which sent by the sender as shown in Fig. 16. The security parameters are agreed between IPsec peers before transmitting information between the communicated peers, as well as, after the connection is established the transmitted traffic is encrypted through IPsec tunnel [22–24].

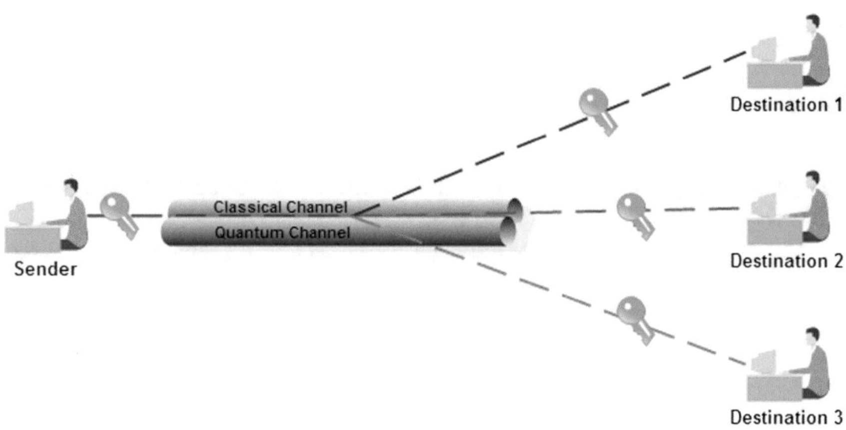

Fig. 14 IPsec key distribution through quantum and classical channels

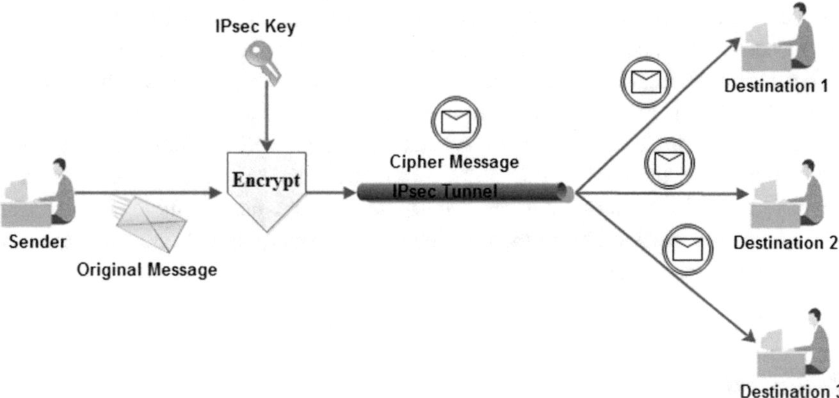

Fig. 15 Encryption process

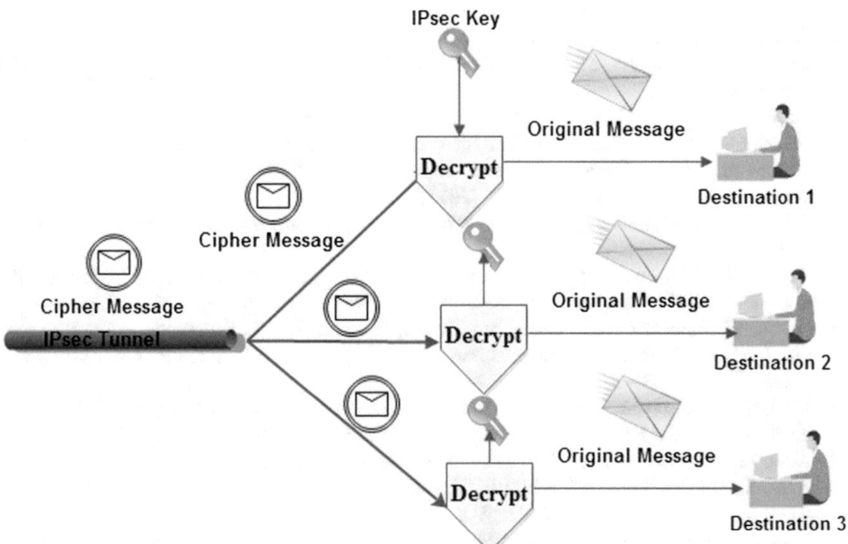

Fig. 16 Decryption process

9 Confidentiality and Authentication

The confidentiality of transmitted messages between the sender and multiple receipts over a communication channel is reached based on quantum no-cloning and Heisenberg uncertainty principle. So, the eavesdropper or even the intruder cannot retrieve any useable information or be familiar with the contents of the transmitted messages. According to Heisenberg uncertainty principle when the eavesdropper try to spy the communication channel, the eavesdropper operation intercepts the quantum channel and creates a result with erroneous probability 50%. So, eavesdropper has no any information about the transmitted messages [22–24].

By Diffie–Hellman key exchange, the communicated peers don't authenticate each other. So, it vulnerable to man-in-the-middle attack. Based on our proposed IPsec multicast architecture, each communicated peer will be authenticated before the beginning of transmitted messages. Authentication is done on phase one based on generated shared keys. Authentication mechanism uses combined technique. Which means keys used for authentication comes from quantum and authentication algorithm is a classical one [22–24].

References

1. Bellovin, S.M.: Problem areas for the IP security protocols. In: Proceedings of the 6th conference on USENIX Security Symposium, Focusing on Applications of Cryptography, Vol. 6, p. 21. USENIX Association (1996)
2. Paterson, K.G., Yau, A.K.: Cryptography in theory and practice: the case of encryption in IPsec. In: Advances in Cryptology-EUROCRYPT 2006, pp. 12–29. Springer Berlin, Heidelberg (2006)
3. Farouk, A., Omara, F., Zakria, M., Megahed, A.: Secured IPsec multicast architecture based on quantum key distribution. In: The International Conference on Electrical and Bio-medical Engineering, Clean Energy and Green Computing. The Society of Digital Information and Wireless Communication, pp. 38–47 (2015)
4. Schneier, B.: Applied Cryptography: Protocols, Algorithms, and Source Code in C. Wiley (2007)
5. Degabriele, J.P., Paterson, K.G.: Attacking the IPsec standards in encryption-only configurations. In IEEE Symposium on Security and Privacy, Vol. 161, pp. 335–349)
6. Aiello, W., Bellovin, S.M., Blaze, M., Canetti, R., Ioannidis, J., Keromytis, A.D., Reingold, O.: Just fast keying: key agreement in a hostile internet. ACM Trans. Inf. Syst. Secur. (TISSEC) **7**(2), 242–273 (2004)
7. Kent, S., Atkinson, R.: RFC 2401: Security architecture for the Internet protocol (1998)
8. Kent, S., Seo, K.: RFC 4301: Security architecture for the Internet protocol (2005)
9. Atkinson, R., Header, I.A.: RFC 1826. Naval Research Laboratory (1995)
10. Kent, S., Header, I.A.: RFC 4302: IETF, December (2005)
11. Kent, S., Atkinson, R.: RFC 2402: IP authentication header (1998)
12. Kent, S., Atkinson, R.: RFC 2401: Security architecture for the internet protocol (1998)
13. Kent, S., Seo, K.: RFC 4301: Security architecture for the internet protocol (2005)
14. Diffie, W., Hellman, M.E.: New directions in cryptography. IEEE Trans. Inf. Theory **22**(6), 644–654 (1976)
15. Bennett, C.H., Bessette, F., Brassard, G., Salvail, L., Smolin, J.: Experimental quantum cryptography. J. Cryptol. **5**(1), 3–28 (1992)
16. Bennett, C.H., Brassard, G., Breidbart, S., Wiesner, S.: Quantum cryptography, or unforgeable subway tokens. In: Advances in Cryptology, pp. 267–275. Springer, US (1983)
17. Bennett, C.H., Brassard, G. An update on quantum cryptography. In: Advances in Cryptology, pp. 475–480. Springer Berlin, Heidelberg (1985)
18. Bennett, C., Brassard, G.: Quantum cryptography: public key distribution and coin tossing. Theor. Comput. Sci. **560**, 7–11 (2014). doi:10.1016/j.tcs.2014.05.025
19. Wootters, W., Zurek, W.: A single quantum cannot be cloned. Nature **299**(5886), 802–803 (1982). doi:10.1038/299802a0
20. Zeng, G.H.: Quantum Cryptology. Science Press (2006)

21. Zeng, G.: Quantum Private Communication. Higher Education Press, Beijing (2010)
22. Metwaly, A.F., Mastorakis, N.E.: Architecture of decentralized multicast network using quantum key distribution and hybrid WDM-TDM. In: Proceedings of the 9th International Conference on Computer Engineering and Applications (CEA'15). Advances in Information Science and Computer Engineering, pp. 504–518 (2015)
23. Metwaly, A.F., Rashad, M.Z., Omara, F.A., Megahed, A.A.: Architecture of multicast centralized key management scheme using quantum key distribution and classical symmetric encryption. Euro. Phys. J. Spec. Top. **223**(8), 1711–1728 (2014)
24. Metwaly, A., Rashad, M.Z., Omara, F.A., Megahed, A.A.: Architecture of point to multipoint QKD communication systems (QKDP2MP). In: 8th International Conference on Informatics and Systems (INFOS), Cairo, pp. NW 25–31. IEEE (2012)
25. Farouk, A., Omara, F., Zakria, M., Megahed, A.: Secured IPsec multicast architecture based on quantum key distribution. In: The International Conference on Electrical and Bio-medical Engineering, Clean Energy and Green Computing. The Society of Digital Information and Wireless Communication, pp. 38–47 (2015)
26. Farouk, A., Zakaria, M., Megahed, A., Omara, F.A.: A generalized architecture of quantum secure direct communication for N disjointed users with authentication. Sci. Rep. **5**, 16080 (2014)
27. Wang, M.M., Wang, W., Chen, J.G., Farouk, A.: Secret sharing of a known arbitrary quantum state with noisy environment. Quantum Inf. Process **14**(11), 4211–4224 (2015)
28. Naseri, M., Heidari, S., Batle, J., Baghfalaki, M., Gheibi, R., Farouk, A., Habibi, A.: A new secure quantum watermarking scheme. Opt-Int. J. Light Electron Opt. **139**, 77–86 (2017)
29. Batle, J., Ciftja, O., Naseri, M., Ghoranneviss, M., Farouk, A., Elhoseny, M.: Equilibrium and uniform charge distribution of a classical two-dimensional system of point charges with hard-wall confinement. Phys. Scr. **92**(5), 055801 (2017)
30. Geurdes, H., Nagata, K., Nakamura, T., Farouk, A.: A note on the possibility of incomplete theory (2017). arXiv preprint arXiv:1704.00005
31. Batle, J., Farouk, A., Alkhambashi, M., Abdalla, S.: Multipartite correlation degradation in amplitude-damping quantum channels. J. Korean Phys. Soc. **70**(7), 666–672 (2017)
32. Batle, J., Naseri, M., Ghoranneviss, M., Farouk, A., Alkhambashi, M., Elhoseny, M.: Shareability of correlations in multiqubit states: optimization of nonlocal monogamy inequalities. Phys. Rev. A **95**(3), 032123 (2017)
33. Batle, J., Farouk, A., Alkhambashi, M., Abdalla, S.: Entanglement in the linear-chain Heisenberg antiferromagnet $Cu(C_4H_4N\ _2)(NO_3)\ _2$. Eur. Phys. J. B **90**, 1–5 (2017)
34. Batle, J., Alkhambashi, M., Farouk, A., Naseri, M., Ghoranneviss, M.: Multipartite non-locality and entanglement signatures of a field-induced quantum phase transition. Eur. Phys. J. B **90**(2), 31 (2017)
35. Nagata, K., Nakamura, T., Batle, J., Abdalla, S., Farouk, A.: Boolean approach to dichotomic quantum measurement theories. J. Korean Phys. Soc. **70**(3), 229–235 (2017)
36. Abdolmaleky, M., Naseri, M., Batle, J., Farouk, A., Gong, L.H.: Red-Green-Blue multi-channel quantum representation of digital images. Opt-Int. J. Light Electron Opt. **128**, 121–132 (2017)
37. Farouk, A., Elhoseny, M., Batle, J., Naseri, M., Hassanien, A.E.: A proposed architecture for key management schema in centralized quantum network. In: Handbook of Research on Machine Learning Innovations and Trends, pp. 997–1021. IGI Global
38. Zhou, N.R., Li, J.F., Yu, Z.B., Gong, L.H., Farouk, A.: New quantum dialogue protocol based on continuous-variable two-mode squeezed vacuum states. Quantum Inf. Process. **16**(1), 4 (2017)
39. Batle, J., Abutalib, M., Abdalla, S., Farouk, A.: Persistence of quantum correlations in a XY spin-chain environment. Euro. Phys. J. B **89**(11), 247 (2016)
40. Batle, J., Abutalib, M., Abdalla, S., Farouk, A.: Revival of Bell nonlocality across a quantum spin chain. Int. J. Quantum Inf. **14**(07), 1650037 (2016)

41. Batle, J., Ooi, C.R., Farouk, A., Abutalib, M., Abdalla, S.: Do multipartite correlations speed up adiabatic quantum computation or quantum annealing? Quantum Inf. Process **15**(8), 3081–3099 (2016)
42. Batle, J., Bagdasaryan, A., Farouk, A., Abutalib, M., Abdalla, S.: Quantum correlations in two coupled superconducting charge qubits. Int. J. Mod. Phys. B **30**(19), 1650123 (2016)
43. Batle, J., Ooi, C.R., Abutalib, M., Farouk, A., Abdalla, S.: Quantum information approach to the azurite mineral frustrated quantum magnet. Quantum Inf. Process **15**(7), 2839–2850 (2016)
44. Batle, J., Ooi, C.R., Farouk, A., Abdalla, S.: Nonlocality in pure and mixed n-qubit X states. Quantum Inf. Process **15**(4), 1553–1567 (2016)
45. Batle, J., Ooi, C.R., Farouk, A., Abutalib, M., Abdalla, S.: Do multipartite correlations speed up adiabatic quantum computation or quantum annealing? Quantum Inf. Process **15**(8), 3081–3099 (2016)
46. Metwaly, A.F., Rashad, M.Z., Omara, F.A., Megahed, A.A.: Architecture of Multicast Network Based on Quantum Secret Sharing and Measurement (2015)

Multi-parties Quantum Secure Direct Communication with Authentication

Ahmed Farouk, O. Tarawneh, Mohamed Elhoseny, J. Batle, Mosayeb Naseri, Aboul Ella Hassanien and M. Abedl-Aty

Abstract In this chapter, a generalized architecture of quantum secure direct communication for N disjoint users with partial and full cooperation of quantum server is proposed. So, $N-1$ disjointed users $u_1, u_2, \ldots, u_{N-1}$ can transmit a secret message of classical bits to a remote user u_N by utilizing the property of dense coding and Pauli unitary transformations. The authentication process between the quantum server and users validated by *EPR* entangled pair and *CNOT* gate. Afterward, the remaining *EPR* will be intended for generating shared *GHZ* states which used for directly transmitting the secret message. The partial cooperation process involved that $N-1$ users can transmit a secret message directly to a remote user u_N through quantum channel. Furthermore, $N-1$ users and a remote user u_N can communicate without an established quantum channel among them by full cooperation process. The security analysis of authentication and communication processes against many types of attacks proved that the attacker can't gain any information during intercepting either authentication or communication processes. Hence, the security of transmitted message among *N* users is ensured as the attacker introducing an error probability irrespective of the sequence of measurement.

A. Farouk (✉) · M. Elhoseny
Faculty of Computer and Information Sciences, Mansoura University, Mansoura, Egypt
e-mail: dr.ahmedfarouk85@yahoo.com

A. Farouk · M. Abedl-Aty
University of Science and Technology, Zewail City of Science and Technology, Giza, Egypt

O. Tarawneh
Information Technology Department, Al-Zahra College for Women, P.O.Box 3365, Muscat, Oman

J. Batle
Departament de Física, Universitat de Les Illes Balears, 07122 Palma de Mallorca, Balearic Islands, Spain

M. Naseri
Department of Physics, Kermanshah Branch, Islamic Azad University, Kermanshah, Iran

A.E. Hassanien
Faculty of Computers and Information, Cairo University, Giza, Egypt

A.E. Hassanien et al. (eds.), *Quantum Computing: An Environment for Intelligent Large Scale Real Application*, Studies in Big Data 33,
https://doi.org/10.1007/978-3-319-63639-9_7

Keywords Quantum key distribution • Quantum identity authentication • Quantum communication • Entanglement

1 Introduction

In this chapter, a generalized architecture of quantum secure direct communication for N disjoint users with partial and full cooperation of quantum server is proposed. So, $N-1$ disjointed users u_1, u_2, …, u_{N-1} can transmit a secret message of classical bits to a remote user u_N by utilizing the property of dense coding and Pauli unitary transformations. The authentication process between the quantum server and users validated by *EPR* entangled pair and *CNOT* gate. Afterward, the remaining *EPR* will be intended for generating shared *GHZ* states which are used for directly transmitting the secret message. The partial cooperation process involved that $N-1$ users can transmit a secret message directly to a remote user u_N through the quantum channel. Furthermore $N-1$ users and a remote user u_N can communicate without an established quantum channel among them by a full cooperation process. The security analysis of authentication and communication processes against many types of attacks proved that the attacker can't gain any information during intercepting either authentication or communication processes. Hence, the security of the transmitted message among *N* users is ensured as the attacker introducing an error probability irrespective of the sequence of measurement.

Bennett and Brassard [1] have introduced been the quantum cryptography. Quantum cryptography is one of the most significant prospects associated with laws of quantum mechanics in order to ensure unconditional security [2–5]. The quantum cryptography proves unconditional security characteristics through no cloning theory where the transmitted quantum bit can't be replicated or copied but its state can be teleported [6]. The most used quantum principles are quantum teleportation and dense coding. In quantum teleportation, the quantum information can be transmitted between distant parties based on both classical communication and maximally shared quantum entanglement among the distant parties [7–10]. In Dense coding, the classical information can be encoded and transmitted between distant parties based on both one quantum bit and maximally shared quantum entanglement among the distant parties as each quantum bit can transmit two classical bits [11, 12]. There are number of approaches and prototypes for the exploitation of quantum principles to secure the communication between two parties and multi-parties [13–18]. While these approaches used different techniques for achieving a private communication among authorized users, but still most of them depend on the generation of secret random keys [19].

Recently, quantum secure direct communication concept is introduced for transmitting the secured messages between the communicated participants without establishing secret keys to encode them [20–37]. In [20] a ping pong protocol is introduced for directly decrypted the transmitted encoded bits between the communicated participants in every corresponding transmission without the need of *QKD*. The authors [38] enhance the capability of ping pong protocol by adding two more unitary operations. In [22], a two-step quantum secure direct communication is proposed for transferring of quantum information by utilizing *EPR* pair blocks to secure the transmission. In [21] the authentication and communication process is performed using *GHZ* states. Firstly, *GHZ* states are used for authentication purpose then the remaining *GHZ* will be used for directly transmitting the secret message. In [18], architecture of centralized multicast scheme is proposed based on a hybrid model of quantum key distribution and classical symmetric encryption. The proposed scheme solved the key generation and management problem using a single entity called centralized Quantum Multicast Key Distribution Centre. In [39], a novel multiparty concurrent quantum secure direct communication based on *GHZ* states and dense coding is introduced. In [40], a managed quantum secure direct communication protocol has been introduced based on quantum encoding and incompletely entangled states. Different quantum authentication approaches have been developed for preventing various types of attacks and especially man in the middle attack [41–45]. By way of 2017, the improvement and growth of a real quantum computer is still in early stages but many poetical and theoretical experimentations were implemented by many research groups [46–67].

However, these quantum secure direct communication approaches still prone against low effectiveness and inadequate security assurance. Here, we propose a convenient and efficient scheme for transmitting a serial of classical message between two, three and generalized to N users. So, $N-1$ disjointed users $u_1, u_2, \ldots, u_{N-1}$ can transmit a secret message of classical bits to a remote user u_N with partial and full cooperation of the quantum server by utilizing the property of dense coding and Pauli unitary transformations. Firstly, with the objective of protection versus man-in-the-middle, Masquerade as Dishonest and Exchange Fake attacks, the quantum server has to verify and authenticate the identity of communicated disjoint users, so they can transmit quantum messages in a secured manner. Authentication between the quantum server and users is achieved by the generated entangled shared key and $Controlled-NOT$ gate. After completion of authentication, the remaining generated entangled shared key is used for generating shared *GHZ* states which designed for directly transmitting the secret message. By using our partial cooperation, there is a quantum channel among the users and $N-1$ disjointed users generate a random sequence bit strings of transmitted plain message. Next, each user applies appropriate unitary transformation according to his plain message bit string value and transmit the transformed message to u_N. u_N retrieves the original sent secret message by applying $N-GHZ$ measurement on his particle and $u_1, u_2, \ldots, u_{N-1}$ particles. Afterwards, the quantum server calculates the status of his particle according to x basis and announces his measurement results. u_N uses his measurements and the quantum server's publication for retrieving the original sent secret bits by $u_1, u_2, \ldots, u_{N-1}$. If no quantum channel among the users, they can use our full cooperation, but in this case the transformed message will be sent to the quantum server instead of u_N. The security analysis of authentication and communication processes of our proposed scheme against many types of attacks is proved that it's unconditionally secured and the attacker will not reveal any information about the key or the transmitted message in case of directly calculating the transferred particles over the communicated channel from the quantum server to the disjoint user and vice versa. In our scheme, the four Bell states (*EPR* pairs) are used during both authentication and communication processes and are defined by (Eq. (1)). Bell measurement is used in the communication process between two disjoint users for determining which unitary operation is used to transform the original classical message, so the receiver can retrieve it.

$$|\Phi^{\pm}\rangle = \frac{1}{\sqrt{2}}(|00\rangle \pm |11\rangle), |\psi^{\pm}\rangle = \frac{1}{\sqrt{2}}(|01\rangle \pm |10\rangle) \tag{1}$$

The unitary transformation operations are defined by (Eq. (2)). For simplicity we used X, Y, Z instead of σ_x, $i\sigma_y$, σ_z respectively. They are used to transform *GHZ* state at sender(s) side into unreadable form according to the generated original classical message before transmitting it to the receiver.

$$\begin{aligned} I &= |0\rangle\langle 0| + |1\rangle\langle 1| \\ X &= |0\rangle\langle 1| + |1\rangle\langle 0| \\ Y &= |0\rangle\langle 1| - |1\rangle\langle 0| \\ Z &= |0\rangle\langle 0| - |1\rangle\langle 1| \end{aligned} \tag{2}$$

The *GHZ* state is defined by (Eq. (3))

$$|GHZ\rangle = \frac{|0\rangle^{\otimes \text{qubits}} + |1\rangle^{\otimes \text{qubits}}}{\sqrt{2}} \tag{3}$$

According to the proposed scheme, when the quantum server receives a request from user(s) for a communication with another user. Consequently, the quantum server distributes *GHZ* entanglement states among involved participants' users in the communication process. The distribution will be established after successfully completion the authentication process and prior starting the communication process. The quantum server distributes all generated particles but holds one for himself. As a consequence, the quantum server and participated users are entangled because they keep only one particle for every distributed *GHZ* state. Additionally, *GHZ* measurement is used by the receiver or quantum server according to which type of cooperation is used during the communication among the users. Consistent with *GHZ* measurement result, the receiver can conclude which unitary operations are used by the senders for transforming *GHZ* state according to the generated original classical message before transmitting it to the receiver. By obtaining the applied unitary operations, the receiver can retrieve the original message. The Eight *GHZ* States are defined by (Eq. (4))

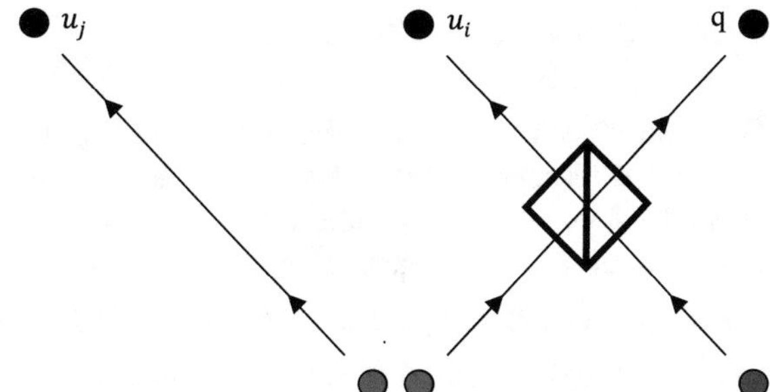

Fig. 1 Generation of *GHZ* states based on Einstein Podolsky-Rosen (*EPR*)

$$
\begin{aligned}
|\Psi^{\pm}\rangle &= \frac{1}{\sqrt{2}}(|000\rangle \pm |111\rangle), |\psi^{\pm}\rangle = \frac{1}{\sqrt{2}}(|011\rangle \pm |100\rangle) \\
|\phi^{\pm}\rangle &= \frac{1}{\sqrt{2}}(|010\rangle \pm |101\rangle), |\varphi^{\pm}\rangle = \frac{1}{\sqrt{2}}(|001\rangle \pm |110\rangle)
\end{aligned} \tag{4}
$$

In [42], an experimental entanglement of six photons *GHZ* state and cluster state in graph states has been proposed. The generation of six-photon *GHZ* states and cluster state is achieved by Einstein Podolsky-Rosen (*EPR*) entangled photon pairs. In our scheme, we will use the same concept for generating shared *GHZ* states. After successfully completion of the authentication process between the quantum server and specified user, the remaining *EPR* will be used for generating shared *GHZ* states intended for a directly transmitting the secret message among the communicated users. Figure 1 illustrates how *GHZ* states among u_i, quantum server and u_j are generated according to the remaining *EPR*. Suppose that the generated *EPR* for authentication process between the quantum server and u_i is given by (Eq. (5)). But, the quantum server particle is a part from another generated *EPR* to authenticate u_j (see Eq. (6)). So, the result will be shared *GHZ* state among u_i, quantum server and u_j (see Eq. (7)). By the same way $|\Psi^{+}_{iqjl}\rangle$, $|\Psi^{+}_{iqjlm}\rangle$, … and so on can be generated.

$$
|\Phi^{+}_{iq}\rangle = \frac{1}{\sqrt{2}}(|0_i 0_q\rangle + |1_i 1_q\rangle \tag{5}
$$

$$
|\Phi^{+}_{qj}\rangle(|0\rangle_i + |1\rangle_i) = \frac{1}{\sqrt{2}}(|0_q 0_j\rangle + |1_q 1_j\rangle)(|0\rangle_i > + |1\rangle_i) \tag{6}
$$

$$
|\Psi^{+}_{iqj}\rangle = \frac{1}{\sqrt{2}}(|0_i 0_q 0_j\rangle + |1_i 1_q 1_j\rangle) \tag{7}
$$

2 Authentication Process

With the objective for protection versus man-in-the-middle, Masquerade as Dishonest and Exchange Fake attacks, the quantum server has to verify and authenticate the identity of communicated disjoint users, so they can transmit quantum messages in a secured manner. Authentication between the quantum server and users is achieved by the generated entangled shared key $\left|\Phi_{qu}^{+}\right\rangle$ and $Controlled-NOT$ gate as illustrated by (Eqs. (8–14)) (see Fig. 2).

1. At the time of registration of disjoint users with the quantum server, the quantum server and disjoint user have a shared binary authentication key A_K as shown by (Eq. (8)).

$$A_K = \{A_1, A_2, \ldots, A_{2N}\} \tag{8}$$

2. The quantum server and disjoint user sent one of the entangled particles making up an *EPR* pair $\left|\Phi_{qu}^{+}\right\rangle$ as illustrated in Fig. 1. $\left|\Phi_{qu}^{+}\right\rangle$ is corresponding to two Particles q and u which are associated with quantum server and corresponding disjoint user respectively. The quantum server preserves q at his location and transmits u particle to the intended disjoint user as shown by (Eq. (9)).

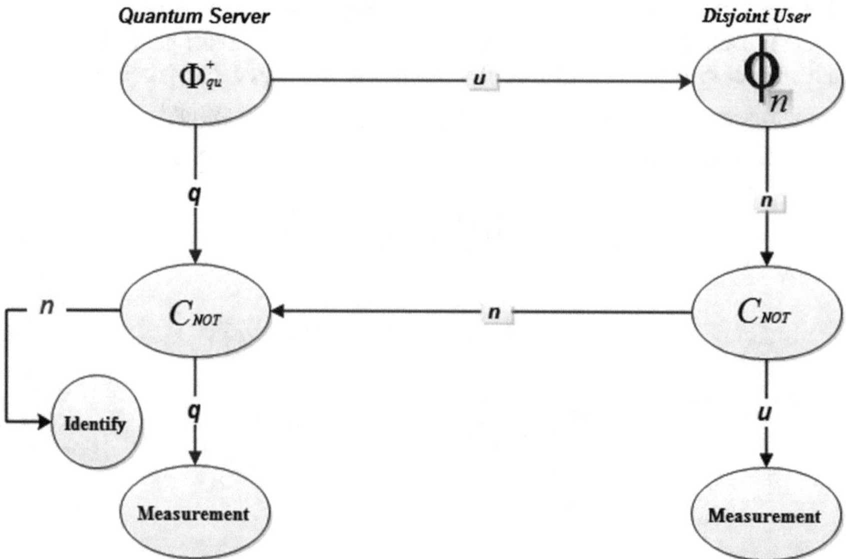

Fig. 2 Authentication process between quantum server and disjoint user

$$\left|\Phi_{qu}^{+}\right\rangle = \frac{1}{\sqrt{2}}\left(\left|0_q 0_u\right\rangle + \left|1_q 1_u\right\rangle\right) \tag{9}$$

3. Once the disjoint user obtains his u particle, he prepares a new state n see (Eq. (10)) by encoding the shared authentication information according to the specified operation.

$$|\phi_n\rangle = |A_{2i-1} \otimes A_{2i}\rangle \tag{10}$$

where $1 \leq i \leq N$ and $\otimes$ denotes the specified user operation.

4. By performing the quantum *CNOT* gate $₵_{OP}$ on both the transmitted particle and n (new state particle). The resulted particle r will be a three entanglement particles state see (Eq. (11)).

$$|\Phi_r\rangle = ₵_{OP}(|\phi_n\rangle \otimes |\Phi_{qu}^{+}\rangle) \tag{11}$$

Where $₵_{OP} = ₵_0$ at $A_{2i-1} = 0$ and $₵_{OP} = ₵_1$ at $A_{2i-1} = 1$. $₵_0$ and $₵_1$ are described by (Eq. (12))

$$\begin{aligned} ₵_0 &= |0\rangle\langle 0| \otimes I + |1\rangle\langle 1| \otimes X, \\ ₵_1 &= |+\rangle\langle +| \otimes I + |-\rangle\langle -| \otimes X \end{aligned} \tag{12}$$

5. After applying the requested operation, the disjoint user keeps particle u at his side and sends the resulted particle $|\Phi_r\rangle$ to the quantum server.
6. Once the quantum server receives the resulted particle $|\Phi_r\rangle$, he decodes it by applying a quantum *CNOT* gate $₵_{OP}$ on both the local particle q and n (new state particle) see (Eqs. (13, 14)).

$$|\Phi'_r\rangle = ₵_{OP}(|\Phi_r\rangle) \tag{13}$$

$$|\Phi'_r\rangle = ₵_{OP}(|\phi_n\rangle \otimes |\Phi_{qu}^{+}\rangle) \tag{14}$$

7. Quantum server starts to verify the identity of a disjoint user by measuring $|\phi_n\rangle$ in the basis of σ_z. The resulted state have to be in either 0 or 1. If the measurement result is equal to $|A_{2i-1}, A_{2i}\rangle$, then the disjoint user is authenticated. However, if the resulted measurement is erroneous which is greater than the agreed threshold then the authentication process will be terminated.
8. Afterward, the key is increased to authenticate the next disjoint user and quantum server recursively back to step one until all disjoint users are authenticated.

3 Communication Process Between Two Disjoint Users with Partial and Full Cooperation of Quantum Server

3.1 Partial Cooperation Process

Here, we will use the property of dense coding for encoding and transmitting a serial of classical messages between two disjoint users with partial support of the quantum server. In other words according to quantum server's publication and received user's measurement. In this scenario, u_i informs the quantum server about his request to transmit a message to distant u_j. The quantum server creates a *GHZ* three particle states $|\Psi_{iqj}\rangle$ where i represents particle state for the transmitted user u_i, j represents particle state for the received user u_j and q represents particle state for quantum server. u_i prepares his plain message which consists of classical bits. According to the value of transmitted message classical bits, u_i performs one of the unitary transformations on his qubit of the entangled *GHZ* three particle states $|\Psi_{iqj}\rangle$. After that, the *GHZ* states will be converted according to the transmitted bits and u_i transformation. Thenceforth, u_i transmits the transformed message to the received disjoint user u_j. u_j retrieves the original sent secret message by applying his bell measurements and the quantum server's publication on the received encoded message. Figure 3 illustrates the flow processes to communicate disjoint user 1 and user 2 with partial support of the quantum server. The required steps are listed below with supportive equations see (Eqs. (15–17)).

The quantum server creates a GHZ three particle states secret key sequence $|\Psi_i\rangle$

$$|\Psi_i\rangle = \{\Psi_1, \Psi_2, \Psi_3, \ldots, \Psi_N\} \tag{15}$$

Suppose that

$$\Psi_i = \frac{1}{\sqrt{2}}(|000_{iqj}\rangle + |111_{iqj}\rangle) \quad \text{where} \quad 1 \le i \le N \tag{16}$$

$$\Psi_{iqj} = \frac{1}{\sqrt{2}}\left(|\Phi^+\rangle_{ij}|+\rangle_q + |\Phi^-\rangle_{ij}|-\rangle_q\right),$$

As

$$|+\rangle = \frac{1}{\sqrt{2}}(|0\rangle + |1\rangle),\ |-\rangle = \frac{1}{\sqrt{2}}(|0\rangle - |1\rangle) \tag{17}$$

1. The disjoint user u_i chooses a randomly subset of *GHZ* particle sequences Ψ and keeps it confident.
2. u_i generates a random sequence bits string of transmitted plain message. According to each two transmitted bits which $(00, 01, 10, 11)$, The disjoint user

Table 1 Correlation between received classical value and its corresponding unitary transformations

First bit	Second bit	u_i transformation
0	0	I_{u_i}
0	1	X_{u_i}
1	0	Y_{u_i}
1	1	Z_{u_i}

Table 2 Correlation between received classical value and its corresponding *GHZ* transformations

First bit	Second bit	u_i transformation	GHZ transformation
0	0	I_{u_i}	$\frac{1}{\sqrt{2}}\left(\|\Phi^+\rangle_{ij}\|+\rangle_q+\|\Phi^-\rangle_{ij}\|-\rangle_q\right)$
0	1	X_{u_i}	$\frac{1}{\sqrt{2}}\left(\|\psi^+\rangle_{ij}\|+\rangle_q-\|\psi^-\rangle_{ij}\|-\rangle_q\right)$
1	0	Y_{u_i}	$\frac{1}{\sqrt{2}}\left(\|\psi^-\rangle_{ij}\|+\rangle_q-\|\psi^+\rangle_{ij}\|-\rangle_q\right)$
1	1	Z_{u_i}	$\frac{1}{\sqrt{2}}\left(\|\Phi^-\rangle_{ij}\|+\rangle_q+\|\Phi^+\rangle_{ij}\|-\rangle_q\right)$

u_i applies one of the unitary transformation operations $\breve{\mathrm{U}}=\left\{\breve{\mathrm{U}}_1,\breve{\mathrm{U}}_2,\breve{\mathrm{U}}_3,\breve{\mathrm{U}}_4\right\}$ which corresponds to four Pauli operations $\{I,X,Y,Z\}$ respectively as shown in Table 1.

3. Afterward, the *GHZ* states will be converted according to transmitted bits and u_i transformation as illustrated in Table 2 and by (Eqs. (18)–(21)).

- When the transmitted two bits = 00, u_i applies I operation on his bit

$$
\begin{aligned}
I_{u_i}|\Psi\rangle &= \frac{1}{\sqrt{2}}\left(|000\rangle_{iqj}+|111\rangle_{iqj}\right)\\
&=1/2\left\{\left(|\Phi^+\rangle_{ij}+|\Phi^-\rangle_{ij}\right)|0\rangle_q+\left(|\Phi^+\rangle_{ij}-|\Phi^-\rangle_{ij}\right)|1\rangle_q\right\}\\
&=\frac{1}{\sqrt{2}}\left(|\Phi^+\rangle_{ij}|+\rangle_q+|\Phi^-\rangle_{ij}|-\rangle_q\right)
\end{aligned}
\tag{18}
$$

- When the transmitted two bits = 01, u_i applies X operation on his bit

$$
\begin{aligned}
X_{u_i}|\Psi\rangle &= \frac{1}{\sqrt{2}}\left(|100\rangle_{iqj}+|011\rangle_{iqj}\right)\\
&=1/2\left\{\left(|\psi^+\rangle_{ij}-|\psi^-\rangle_{ij}\right)|0\rangle_q+\left(|\psi^+\rangle_{ij}+|\psi^-\rangle_{ij}\right)|1\rangle_q\right\}\\
&=\frac{1}{\sqrt{2}}\left(|\psi^+\rangle_{ij}|+\rangle_q-|\psi^-\rangle_{ij}|-\rangle_q\right)
\end{aligned}
\tag{19}
$$

- When the transmitted two bits = 10, u_i applies Y operation on his bit

$$
\begin{aligned}
Y_{u_i}|\Psi\rangle &= \frac{1}{\sqrt{2}}\left(|011\rangle_{iqj} - |100\rangle_{iqj}\right) \\
&= 1/2\left\{\left(|\psi^+\rangle_{ij} + |\psi^-\rangle_{ij}\right)|1\rangle_q - \left(|\psi^+\rangle_{ij} - |\psi^-\rangle_{ij}\right)|0\rangle_q\right\} \\
&= \frac{1}{\sqrt{2}}\left(|\psi^-\rangle_{ij}|+\rangle_q - |\psi^+\rangle_{ij}|-\rangle_q\right)
\end{aligned}
\tag{20}
$$

- When the transmitted two bits = 11, u_i applies Z operation on his bit

$$
\begin{aligned}
Z_{u_i}|\Psi\rangle &= \frac{1}{\sqrt{2}}\left(|000\rangle_{iqj} - |111\rangle_{iqj}\right) \\
&= 1/2\left\{\left(|\Phi^+\rangle_{ij} + |\Phi^-\rangle_{ij}\right)|0\rangle_q - \left(|\Phi^+\rangle_{ij} - |\Phi^-\rangle_{ij}\right)|1\rangle_q\right\} \\
&= \frac{1}{\sqrt{2}}\left(|\Phi^-\rangle_{ij}|+\rangle_q + |\Phi^+\rangle_{ij}|-\rangle_q\right)
\end{aligned}
\tag{21}
$$

4. After applying the proper *GHZ* transformation, $\mathrm{u_i}$ transmits the encoded message to the received disjoint user $\mathrm{u_j}$.
5. u_j performs a Bell measurement on his particle and u_i particle. Also, the quantum server calculates the status of his particle according to x basis $\{+, -\}$ and announces his measurement results.
6. u_j uses his measurements and the quantum server's publication for retrieving the original secret bits by u_i as shown in Table 3. For example, when u_j measurement is equivalent to $|\psi^-\rangle$ and Quantum Server's Publication is $|-\rangle$, so u_j concludes that u_i applied X operation and the sent bits are 01. Table 4 shows an illustrative example for transmitting a message 100111 from u_i to u_j with partial support of the quantum server (Fig. 3).

Table 3 Correlation between quantum server's publication, u_j measurement, u_i operation and sent bits

Quantum server's publication	u_j measurement	u_i operation	Sent bits
$\|+\rangle_q$	$\|\Phi^+\rangle_{ij}$	I	00
$\|+\rangle_q$	$\|\psi^+\rangle_{ij}$	X	01
$\|+\rangle_q$	$\|\psi^-\rangle_{ij}$	Y	10
$\|+\rangle_q$	$\|\Phi^-\rangle_{ij}$	Z	11
$\|-\rangle_q$	$\|\Phi^-\rangle_{ij}$	I	00
$\|-\rangle_q$	$\|\psi^-\rangle_{ij}$	X	01
$\|-\rangle_q$	$\|\psi^+\rangle_{ij}$	Y	10
$\|-\rangle_q$	$\|\Phi^+\rangle_{ij}$	Z	11

Table 4 Transmitting a message 100111 from u_i to u_j with partial support of quantum server

u_i plain message	10	01	11
u_i operation	Y	X	Z
GHZ transformation	$\frac{1}{\sqrt{2}}\left(\vert\psi^-\rangle_{ij}\vert+\rangle_q - \vert\psi^+\rangle_{ij}\vert-\rangle_q\right)$	$\frac{1}{\sqrt{2}}\left(\vert\psi^+\rangle_{ij}\vert+\rangle_q - \vert\psi^-\rangle_{ij}\vert-\rangle_q\right)$	$\frac{1}{\sqrt{2}}\left(\vert\Phi^-\rangle_{ij}\vert+\rangle_q + \vert\Phi^+\rangle_{ij}\vert-\rangle_q\right)$
u_j bell measurement	ψ^-	ψ^+	Φ^+
Quantum server's publication	+	+	−
u_j retrieved message	10	01	11

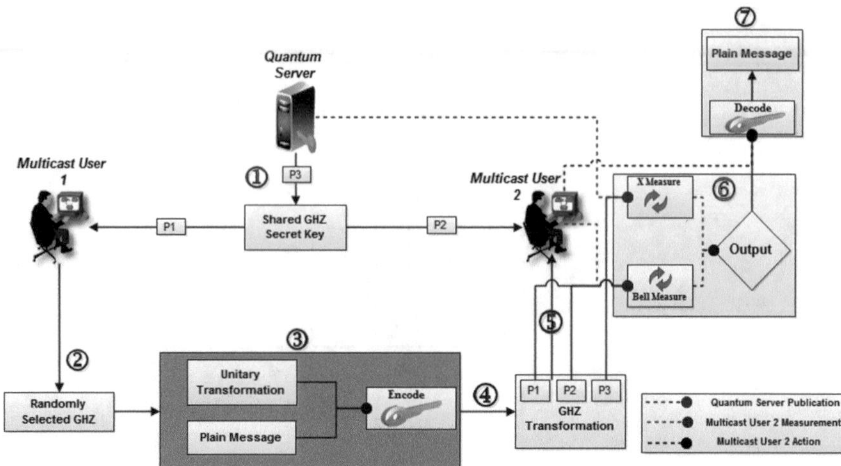

Fig. 3 Communication process between two disjoint users with partial support of quantum server

Table 5 Correlation between received classical value and its corresponding *GHZ* transformations

First bit	Second bit	u_i transformation	GHZ transformation
0	0	I_{u_i}	$\frac{1}{\sqrt{2}}\left(\lvert\Phi^+\rangle_{iq}\lvert+\rangle_j + \lvert\Phi^-\rangle_{iq}\lvert-\rangle_j\right)$
0	1	X_{u_i}	$\frac{1}{\sqrt{2}}\left(\lvert\psi^+\rangle_{iq}\lvert+\rangle_j - \lvert\psi^-\rangle_{iq}\lvert-\rangle_j\right)$
1	0	Y_{u_i}	$\frac{1}{\sqrt{2}}\left(\lvert\psi^-\rangle_{iq}\lvert+\rangle_j - \lvert\psi^+\rangle_{iq}\lvert-\rangle_j\right)$
1	1	Z_{u_i}	$\frac{1}{\sqrt{2}}\left(\lvert\Phi^-\rangle_{iq}\lvert+\rangle_j + \lvert\Phi^+\rangle_{iq}\lvert-\rangle_j\right)$

3.2 Full Cooperation Process

This process consists of the required steps for transmitting the classical message between two disjoint users with full support of the quantum server. In other words, the quantum server is functioning as message passing center between the communicated disjoint users. To accomplish this function, u_i transmits the transformed message to the quantum server instead of u_j as illustrated in Table 5. Afterwards, u_j retrieves the original sent secret message according to his publication and quantum server's bell measurement on the received transformed message. Figure 4 illustrates the flow processes to communicate disjoint user 1 and user 2 with full support of the quantum server. The required steps which are listed below as the first three steps are same partial mechanism but the process is just changed from step five as indicated in Fig. 4 green box.

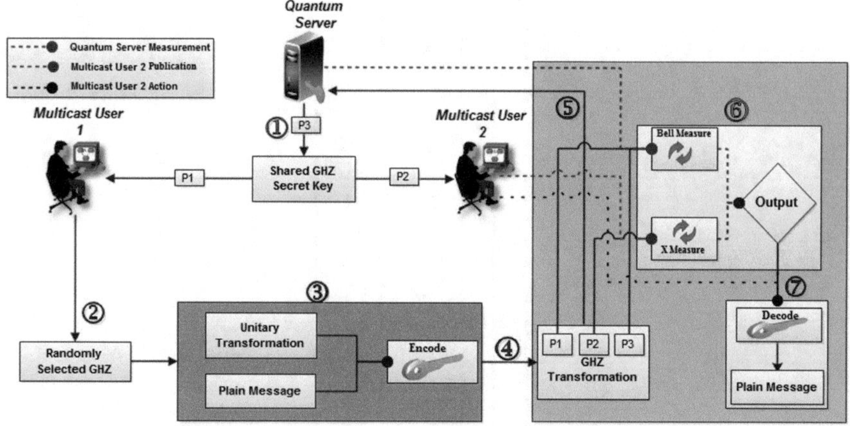

Fig. 4 Communication process between two disjoint users with full support of quantum server: *Green Box* indicates different steps from Fig. 3

Table 6 Correlation between quantum server's measurement, u_j publication, u_i operation and sent bits

Quantum server's measurement	u_j publication	u_i operation	Sent bits
$\|\Phi^+\rangle_{iq}$	$\|+\rangle_j$	I	00
$\|\Phi^+\rangle_{iq}$	$\|-\rangle_j$	Z	11
$\|\psi^+\rangle_{iq}$	$\|+\rangle_j$	X	01
$\|\psi^+\rangle_{iq}$	$\|-\rangle_j$	Y	10
$\|\Phi^-\rangle_{iq}$	$\|+\rangle_j$	Z	11
$\|\Phi^-\rangle_{iq}$	$\|-\rangle_j$	I	00
$\|\psi^-\rangle_{iq}$	$\|+\rangle_j$	X	10
$\|\psi^+\rangle_{iq}$	$\|-\rangle_j$	Y	01

5. After applying the proper GHZ transformation, u_i transmits the encoded message to the quantum server.
6. The quantum server performs a Bell measurement his particle and u_i particle. u_j calculates the status of his particle according to x basis $\{+, -\}$ and announces his measurement results.
7. u_j employs quantum server's measurement and his publication for retrieving the original sent secret bits by u_i as shown in Table 6. For example, when u_j publication is equivalent to $|+\rangle$ and Quantum Server's measurement is $|\Phi^-\rangle$, so u_j can conclude that u_i applied Z operation and the sent bits are 11. Table 7 shows an illustrative example for transmitting a message 100111 from u_i to u_j with the full support of quantum server.

Table 7 Transmitting a message 100111 from u_i to u_j with full support of quantum server

u_i plain message	10	01	11
u_i operation	Y	X	Z
GHZ transformation	$\frac{1}{\sqrt{2}}\left(\lvert\psi^-\rangle_{iq}\lvert+\rangle_j - \lvert\psi^+\rangle_{iq}\lvert-\rangle_j\right)$	$\frac{1}{\sqrt{2}}\left(\lvert\psi^+\rangle_{iq}\lvert+\rangle_j - \lvert\psi^-\rangle_{iq}\lvert-\rangle_j\right)$	$\frac{1}{\sqrt{2}}\left(\lvert\Phi^-\rangle_{iq}\lvert+\rangle_j + \lvert\Phi^+\rangle_{iq}\lvert-\rangle_j\right)$
Quantum server's bell measurement	ψ^-	ψ^+	Φ^+
u_j publication	$+$	$+$	$-$
u_j retrieved message	10	01	11

4 Communication Process Between Three Disjoint Users with Partial and Full Cooperation of Quantum Server

4.1 Partial Cooperation Process

This process consists of the required steps when two disjointed disjoint users u_i and u_j would like to transmit a secret message to a remote user u_l with the partial support of the quantum server. In this scenario, the quantum server creates three particles of four particles *GHZ* state $|\Psi_{ijql}\rangle$ and transmits it to u_i, u_j and u_l. Both u_i and u_j prepare their plain message which consists of classical bits. According to the value of transmitted message classical bits, both u_i and u_j apply one of their specified unitary transformation, each on his corresponding qubit of the entangled GHZ four particle states $|\Psi_{ijql}\rangle$ as shown in Tables (8, 9). After that, the *GHZ* states will be converted according to transmitted bits, u_i and u_j transformations. Thenceforth, u_i and u_j sent the transformed message to a remote user u_l. Afterwards, u_l retrieves the original sent secret message by applying his *GHZ* measurements and the quantum server's publication on the received transformed message. Figure 5 illustrates the flow processes to communicate disjoint user 1, user 2 and user 3 with partial support of the quantum server. The required steps listed below with supportive equations from Eqs. (22) to (24).

1. The quantum server creates four particles *GHZ* state secret key $|\Psi_i\rangle$

$$|\Psi_i\rangle = \{\Psi_1, \Psi_2, \Psi_3, \ldots, \Psi_N\} \tag{22}$$

   Suppose that

$$\Psi_i = \frac{1}{\sqrt{2}}\left(|0000_{ijql}\rangle + |1111_{ijql}\rangle\right)\text{where} \quad 1 \le i \le N \tag{23}$$

$$\Psi_i = \frac{1}{\sqrt{2}}\left(|\Psi^+\rangle_{ijl}|+\rangle_q + |\Psi^-\rangle_{ijl}|-\rangle_q\right)$$

   As

$$|+\rangle = \frac{1}{\sqrt{2}}(|0\rangle + |1\rangle), |-\rangle = \frac{1}{\sqrt{2}}(|0\rangle - |1\rangle) \tag{24}$$

Table 8 Correlation between received classical value and its corresponding unitary transformations of u_j

Transmitted bit	u_j transformation
0	I_{u_i}
1	X_{u_i}

Table 9 Correlation between received classical value and its corresponding unitary transformations of u_i

First bit	Second bit	u_i transformation
0	0	I_{u_i}
0	1	X_{u_i}
1	0	Y_{u_i}
1	1	Z_{u_i}

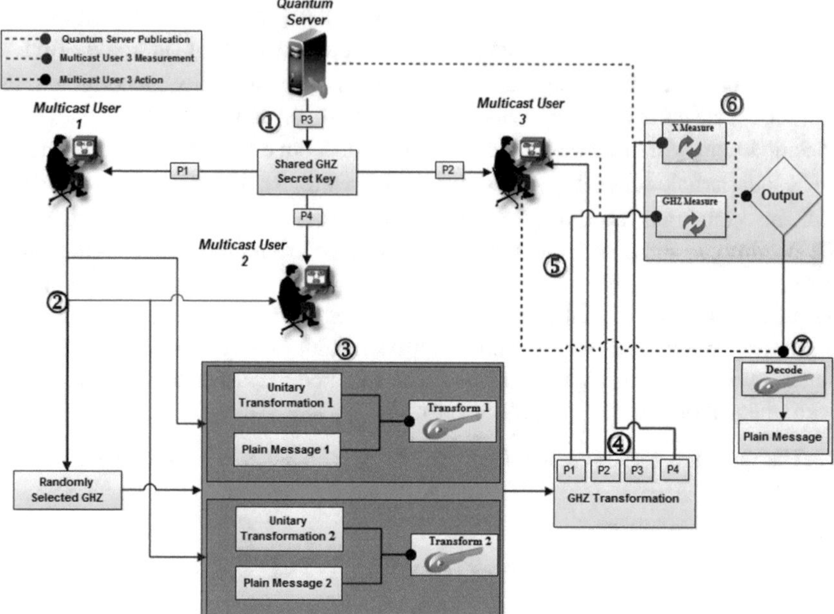

Fig. 5 Communication process between three disjoint users with partial support of quantum server

2. Both u_i and u_j choose a randomly subset of *GHZ* particle sequences Ψ and keep it confident.
3. u_i generates a random sequence of bits string of the transmitted plain message. According to each two transmitted bits which $(00, 01, 10, 11)$, The disjoint user u_i applies one of the unitary transformation operations $\breve{\mathrm{U}} = \left\{\breve{\mathrm{U}}_1, \breve{\mathrm{U}}_2, \breve{\mathrm{U}}_3, \breve{\mathrm{U}}_4\right\}$ which correspond to four Pauli operations $\{I, X, Y, Z\}$ respectively as shown in Table 8. u_j generates a random sequences of bits string of the transmitted plain message. u_j applying $I_{u_j}|\Psi\rangle$ or $X_{u_j}|\Psi\rangle$ according to the value of particle 0 or 1 respectively. The full operation is shown in Table 9.
4. Afterward, the *GHZ* states will be converted according to the transmitted bits, u_i and u_j transformations as shown in Table 10 and equations from Eq. (25) to (32).

Table 10 Correlation between received classical values and their corresponding *GHZ* transformations

Value	u_i and u_j transformation	GHZ transformation
0	$(I_{u_i} \otimes I_{u_j})$	$\frac{1}{\sqrt{2}}\left(\|\Psi^+\rangle_{ijl}\|+\rangle_q + \|\Psi^-\rangle_{ijl}\|-\rangle_q\right)$
1	$(I_{u_i} \otimes X_{u_j})$	$\frac{1}{\sqrt{2}}\left(\|\phi^+\rangle_{ijl}\|+\rangle_q + \|\phi^-\rangle_{ijl}\|-\rangle_q\right)$
2	$(X_{u_i} \otimes I_{u_j})$	$\frac{1}{\sqrt{2}}\left(\|\psi^+\rangle_{ijl}\|+\rangle_q - \|\psi^-\rangle_{ijl}\|-\rangle_q\right)$
3	$(X_{u_i} \otimes X_{u_j})$	$\frac{1}{\sqrt{2}}\left(\|\varphi^+\rangle_{ijl}\|+\rangle_q - \|\varphi^-\rangle_{ijl}\|-\rangle_q\right)$
4	$(Y_{u_i} \otimes I_{u_j})$	$\frac{1}{\sqrt{2}}\left(\|\psi^-\rangle_{ijl}\|+\rangle_q - \|\psi^+\rangle_{ijl}\|-\rangle_q\right)$
5	$(Y_{u_i} \otimes X_{u_j})$	$\frac{1}{\sqrt{2}}\left(\|\varphi^-\rangle_{ijl}\|+\rangle_q - \|\varphi^+\rangle_{ijl}\|-\rangle_q\right)$
6	$(Z_{u_i} \otimes I_{u_j})$	$\frac{1}{\sqrt{2}}\left(\|\Psi^-\rangle_{ijl}\|+\rangle_q + \|\Psi^+\rangle_{ijl}\|-\rangle_q\right)$
7	$(Z_{u_i} \otimes X_{u_j})$	$\frac{1}{\sqrt{2}}\left(\|\phi^-\rangle_{ijl}\|+\rangle_q + \|\phi^+\rangle_{ijl}\|-\rangle_q\right)$

- When the transmitted bits = 000, both u_i and u_j apply I operation on their bits

$$\begin{aligned}(I_{u_i} \otimes I_{u_j})|\Psi\rangle &= \frac{1}{\sqrt{2}}\left(|0000\rangle_{ijql} + |1111\rangle_{ijql}\right) \\ &= 1/2\left\{\left(|\Psi^+\rangle_{ijl} + |\Psi^-\rangle_{ijl}\right)|0\rangle_q + \left(|\Psi^+\rangle_{ijl} - |\Psi^-\rangle_{ijl}\right)|1\rangle_q\right\} \\ &= \frac{1}{\sqrt{2}}\left(|\Psi^+\rangle_{ijl}|+\rangle_q + |\Psi^-\rangle_{ijl}|-\rangle_q\right)\end{aligned} \tag{25}$$

- When the transmitted bits = 001, u_i applies I and u_j applies X operations respectively on their bits

$$\begin{aligned}(I_{u_i} \otimes X_{u_j})|\Psi\rangle &= \frac{1}{\sqrt{2}}\left(|0100\rangle_{ijql} + |1011\rangle_{ijql}\right) \\ &= 1/2\left\{\left(|\phi^+\rangle_{ijl} + |\phi^-\rangle_{ijl}\right)|0\rangle_q + \left(|\phi^+\rangle_{ijl} - |\phi^-\rangle_{ijl}\right)|1\rangle_q\right\} \\ &= \frac{1}{\sqrt{2}}\left(|\phi^+\rangle_{ijl}|+\rangle_q + |\phi^-\rangle_{ijl}|-\rangle_q\right)\end{aligned} \tag{26}$$

- When the transmitted bits = 010, u_i applies X and u_j applies I operations respectively on their bits

$$
\begin{aligned}
(X_{u_i} \otimes I_{u_j})|\Psi\rangle &= \frac{1}{\sqrt{2}}\left(|1000\rangle_{ijql} + |0111\rangle_{ijql}\right) \\
&= 1/2\left\{\left(|\psi^+\rangle_{ijl} - |\psi^-\rangle_{ijl}\right)|0\rangle_q + \left(|\psi^+\rangle_{ijl} + |\psi^-\rangle_{ijl}\right)|1\rangle_q\right\} \\
&= \frac{1}{\sqrt{2}}\left(|\psi^+\rangle_{ijl}|+\rangle_q - |\psi^-\rangle_{ijl}|-\rangle_q\right)
\end{aligned} \tag{27}
$$

- When the transmitted bits = 011, both u_i and u_j apply X operation on their bits

$$
\begin{aligned}
(X_{u_i} \otimes X_{u_j})|\Psi\rangle &= \frac{1}{\sqrt{2}}\left(|1100\rangle_{ijql} + |0011\rangle_{ijql}\right) \\
&= 1/2\left\{\left(|\varphi^+\rangle_{ijl} - |\varphi^-\rangle_{ijl}\right)|0\rangle_q + \left(|\varphi^+\rangle_{ijl} + |\varphi^-\rangle_{ijl}\right)|1\rangle_q\right\} \\
&= \frac{1}{\sqrt{2}}\left(|\varphi^+\rangle_{ijl}|+\rangle_q - |\varphi^-\rangle_{ijl}|-\rangle_q\right)
\end{aligned} \tag{28}
$$

- When the transmitted bits = 100, u_i applies Y and u_j applies I operations respectively on their bits

$$
\begin{aligned}
(Y_{u_i} \otimes I_{u_j})|\Psi\rangle &= \frac{1}{\sqrt{2}}\left(|0111\rangle_{ijql} - |1000\rangle_{ijql}\right) \\
&= 1/2\left\{\left(|\psi^-\rangle_{ijl} - |\psi^+\rangle_{ijl}\right)|0\rangle_q + \left(|\psi^+\rangle_{ijl} + |\psi^-\rangle_{ijl}\right)|1\rangle_q\right\} \\
&= \frac{1}{\sqrt{2}}\left(|\psi^-\rangle_{ijl}|+\rangle_q - |\psi^+\rangle_{ijl}|-\rangle_q\right)
\end{aligned} \tag{29}
$$

- When the transmitted bits = 101, u_i applies Y and u_j applies X operations respectively on their bits

$$
\begin{aligned}
(Y_{u_i} \otimes X_{u_j})|\Psi\rangle &= \frac{1}{\sqrt{2}}\left(|0011\rangle_{ijql} - |1100\rangle_{ijql}\right) \\
&= 1/2\left\{\left(|\varphi^-\rangle_{ijl} - |\varphi^+\rangle_{ijl}\right)|0\rangle_q + \left(|\varphi^+\rangle_{ijl} + |\varphi^-\rangle_{ijl}\right)|1\rangle_q\right\} \\
&= \frac{1}{\sqrt{2}}\left(|\varphi^-\rangle_{ijl}|+\rangle_q - |\varphi^+\rangle_{ijl}|-\rangle_q\right)
\end{aligned} \tag{30}
$$

- When the transmitted bits = 110, u_i applies Z and u_j applies I operations respectively on their bits

$$\begin{aligned}(Z_{u_i}\otimes I_{u_j})|\Psi\rangle &= \frac{1}{\sqrt{2}}\left(|0000\rangle_{ijql} - |1111\rangle_{ijql}\right)\\ &= 1/2\left\{\left(|\Psi^+\rangle_{ijl} + |\Psi^+\rangle_{ijl}\right)|0\rangle_q + \left(|\Psi^-\rangle_{ijl} + |\Psi^+\rangle_{ijl}\right)|1\rangle_q\right\} \\ &= \frac{1}{\sqrt{2}}\left(|\Psi^-\rangle_{ijl}|+\rangle_q + |\Psi^+\rangle_{ijl}|-\rangle_q\right)\end{aligned} \tag{31}$$

- When the transmitted bits = 111, u_i applies Z and u_j applies X operations respectively on their bits

$$\begin{aligned}(Z_{u_i}\otimes X_{u_j})|\Psi\rangle &= \frac{1}{\sqrt{2}}\left(|0100\rangle_{ijql} - |1011\rangle_{ijql}\right)\\ &= 1/2\left\{\left(|\phi^+\rangle_{ijl} + |\phi^-\rangle_{ijl}\right)|0\rangle_q + \left(|\phi^-\rangle_{ijl} - |\phi^+\rangle_{ijl}\right)|1\rangle_q\right\} \\ &= \frac{1}{\sqrt{2}}\left(|\phi^-\rangle_{ijl}|+\rangle_q + |\phi^+\rangle_{ijl}|-\rangle_q\right)\end{aligned} \tag{32}$$

5. Afterward, both u_i and u_j transmit the transformed message to the received disjoint user u_l, u_l performs a *GHZ* measurement on his particle, u_i and u_j particles.
6. The quantum server calculates the status of his particle according to x basis $\{+, -\}$ and announces his measurement results.
7. u_l uses his measurements and the quantum server's publication for retrieving the original sent secret bits by both u_i and u_j as shown in Table 11. For example when u_l measurement is equivalent to $|\varphi^+\rangle$ and Quantum Server's Publication is $|-\rangle$, so u_l can conclude that u_i and u_j applied Y and X operations respectively and the sent bits are 101.

Table 11 Correlation between quantum server's publications, u_l measurement. u_i, u_j operations and sent bits

Quantum server's publication	u_l measurement	u_i operation	Sent bits	u_j operation	Sent bits	Message sent
$\|+\rangle_q$	$\|\Psi^+\rangle_{ijl}$	I	00	I	0	000
$\|+\rangle_q$	$\|\phi^+\rangle_{ijl}$	I	00	X	1	001
$\|+\rangle_q$	$\|\psi^+\rangle_{ijl}$	X	01	I	0	010
$\|+\rangle_q$	$\|\varphi^+\rangle_{ijl}$	X	01	X	1	011
$\|+\rangle_q$	$\|\psi^-\rangle$	Y	10	I	0	100
$\|+\rangle_q$	$\|\varphi^-\rangle_{ijl}$	Y	10	X	1	101

(continued)

Table 11 (continued)

Quantum server's publication	u_l measurement	u_i operation	Sent bits	u_j operation	Sent bits	Message sent
$\|+\rangle_q$	$\|\Psi^-\rangle_{ijl}$	Z	11	I	0	110
$\|+\rangle_q$	$\|\phi^-\rangle_{ijl}$	Z	11	X	1	111
$\|-\rangle_q$	$\|\Psi^-\rangle_{ijl}$	I	00	I	0	000
$\|-\rangle_q$	$\|\phi^-\rangle_{ijl}$	I	00	X	1	001
$\|-\rangle_q$	$\|\psi^-\rangle_{ijl}$	X	01	I	0	010
$\|-\rangle_q$	$\|\varphi^-\rangle_{ijl}$	X	01	X	1	011
$\|-\rangle_q$	$\|\psi^+\rangle_{ijl}$	Y	10	I	0	100
$\|-\rangle_q$	$\|\varphi^+\rangle_{ijl}$	Y	10	X	1	101
$\|-\rangle_q$	$\|\Psi^+\rangle_{ijl}$	Z	11	I	0	110
$\|-\rangle_q$	$\|\phi^+\rangle_{ijl}$	Z	11	X	1	111

Table 12 Correlation between received classical value and its corresponding *GHZ* transformations

Value	Transformation	Encoded/new state
0	$(I_{u_i} \otimes I_{u_j})$	$\frac{1}{\sqrt{2}}\left(\|\Psi^-\rangle_{ijq}\|+\rangle_l - \|\Psi^-\rangle_{ijq}\|-\rangle_l\right)$
1	$(I_{u_i} \otimes X_{u_j})$	$\frac{1}{\sqrt{2}}\left(\|\phi^+\rangle_{ijq}\|+\rangle_l + \|\phi^-\rangle_{ijq}\|-\rangle_l\right)$
2	$(X_{u_i} \otimes I_{u_j})$	$\frac{1}{\sqrt{2}}\left(\|\psi^+\rangle_{ijq}\|+\rangle_l - \|\psi^-\rangle_{ijq}\|-\rangle_l\right)$
3	$(X_{u_i} \otimes X_{u_j})$	$\frac{1}{\sqrt{2}}\left(\|\varphi^+\rangle_{ijq}\|+\rangle_l - \|\varphi^+\rangle_{ijq}\|-\rangle_l\right)$
4	$(Y_{u_i} \otimes I_{u_j})$	$\frac{1}{\sqrt{2}}\left(\|\psi^-\rangle_{ijq}\|+\rangle_l - \|\psi^+\rangle_{ijq}\|-\rangle_l\right)$
5	$(Y_{u_i} \otimes X_{u_j})$	$\frac{1}{\sqrt{2}}\left(\|\varphi^-\rangle_{ijq}\|+\rangle_l - \|\varphi^+\rangle_{ijq}\|-\rangle_l\right)$
6	$(Z_{u_i} \otimes I_{u_j})$	$\frac{1}{\sqrt{2}}\left(\|\Psi^-\rangle_{ijq}\|+\rangle_l + \|\Psi^+\rangle_{ijq}\|-\rangle_l\right)$
7	$(Z_{u_i} \otimes X_{u_j})$	$\frac{1}{\sqrt{2}}\left(\|\phi^-\rangle_{ijq}\|+\rangle_l + \|\phi^+\rangle_{ijq}\|-\rangle_l\right)$

4.2 Full Cooperation Process

This process consists of the required steps when two disjointed disjoint users u_i and u_j would like to transmit a secret message to a remote user u_l with full support of the quantum server. In other words, the quantum server is functioning as a message passing center between the communicated disjoint users. To achieve this function, both u_i and u_j transmit the transformed message to the quantum server instead of u_l as illustrated in Table 12. Afterwards, u_l retrieves the original sent secret message according to his publication and quantum server's *GHZ* measurement on the received transformed message as illustrated in Table 13. Figure 6 illustrates the flow processes to communicate disjoint user 1 and user 2 with a full support of the quantum server. The required steps listed below as the first three steps are the same partial mechanism but the process is just changed from step four as indicated in Fig. 6 the green box.

Table 13 Correlation between quantum server's measurements, u_l publication. u_i, u_j operations and sent bits

Quantum server's measurement	u_l publication	u_i operation	Sent bits	u_j operation	Sent bits	Message sent
$\vert\Psi^+\rangle_{ijq}$	$\vert+\rangle_l$	I	00	I	0	000
$\vert\Psi^+\rangle_{ijq}$	$\vert-\rangle_l$	Z	11	I	0	110
$\vert\psi^+\rangle_{ijq}$	$\vert+\rangle_l$	X	01	I	0	010
$\vert\psi^+\rangle_{ijq}$	$\vert-\rangle_l$	Y	10	I	0	100
$\vert\phi^+\rangle_{ijq}$	$\vert+\rangle_l$	I	00	X	1	001
$\vert\phi^+\rangle_{ijq}$	$\vert-\rangle_l$	Z	11	X	1	111
$\vert\varphi^+\rangle_{ijq}$	$\vert+\rangle_l$	X	01	X	1	011
$\vert\varphi^+\rangle_{ijq}$	$\vert-\rangle_l$	Y	10	X	1	101
$\vert\Psi^-\rangle_{ijq}$	$\vert+\rangle_l$	Z	11	I	0	110
$\vert\Psi^-\rangle_{ijq}$	$\vert-\rangle_l$	I	00	I	0	000
$\vert\psi^-\rangle_{ijq}$	$\vert+\rangle_l$	Y	10	I	0	100
$\vert\psi^-\rangle_{ijq}$	$\vert-\rangle_l$	X	01	I	0	010
$\vert\phi^-\rangle_{ijq}$	$\vert+\rangle_l$	Z	11	X	1	111
$\vert\phi^-\rangle_{ijq}$	$\vert-\rangle_l$	I	00	X	1	001
$\vert\varphi^-\rangle_{ijq}$	$\vert+\rangle_l$	Y	10	X	1	101
$\vert\varphi^-\rangle_{ijq}$	$\vert-\rangle_l$	X	01	X	1	011

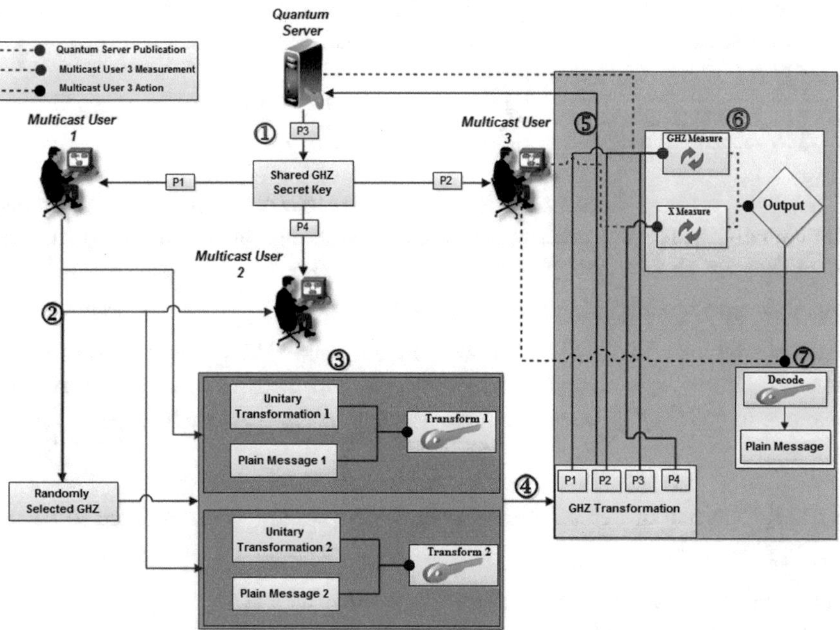

Fig. 6 Communication process between three disjoint users with full support of quantum server: *Green Box* indicates different steps from Fig. 5

1. Afterward, both u_i and u_j transmit the transformed message to the quantum server, quantum server performs a *GHZ* measurement on his particle, u_i and u_j particles.
2. u_l calculates the status of his particle according to x basis $\{+, -\}$ and announces his measurement results.
3. u_l uses his publication and the quantum server's measurement for retrieving the original sent secret bits which sent by both u_i and u_j as shown in Table 13. For example, when the quantum server measurement is equivalent to $|\phi^+\rangle$ and u_l Publication is $|+\rangle$, so u_l can conclude that u_i and u_j applied I and X operations respectively and the sent bits are 001.

5 Generalization of Communication Process Between N Disjoint Users with Partial and Full Cooperation of Quantum Server

5.1 *Partial Cooperation Process*

Now, our approach will be generalized for Communication Process between N Disjoint Users with Partial Cooperation of Quantum Server. So, $N-1$ disjointed disjoint users $u_1, u_2, \ldots, u_{N-1}$ can transmit a secret message of classical bits to a remote user u_N with partial cooperation of the quantum server. Firstly, the quantum server distributes N particles of $N+1$ particles *GHZ* state $|GHZ\rangle_{1 \ldots N} =$

$$\frac{1}{\sqrt{2}}\left(\left|\underbrace{00}_{N-1}\cdots\cdots\underbrace{0}_{q}\underbrace{0}_{N}\right\rangle + \left|\underbrace{11}_{N-1}\cdots\cdots\underbrace{1}_{q}\underbrace{1}_{N}\right\rangle\right) \text{ to } u_1,\ u_2,\ \ldots,\ u_N.$$

$u_1, u_2, \ldots, u_{N-1}$ choose a randomly subset of $|GHZ\rangle_{1 \ldots N}$ and keep it confident additionally, generate a random sequence of bit strings of transmitted plain message. Next, each user is applying appropriate unitary transformation according to his plain message bit string value $\left(\breve{U}_1 \otimes \breve{U}_2 \ldots \otimes \breve{U}_{N-1}\right)$ $\breve{U}$ corresponds to four Pauli operations $\{I, X, Y, Z\}$. Afterward, the selected $|GHZ\rangle_{1 \ldots N}$ will be transformed according to $u_1, u_2, \ldots, u_{N-1}$ plain messages and their applied unitary transformations to $|GHZ\rangle_{1' \ldots N'} = \frac{1}{\sqrt{2}}\big(\underbrace{|GHZ\rangle_{N'}}_{\substack{u_1, \ldots u_N \\ Users}} \underbrace{|\pm\rangle_q}_{\substack{N+1 \\ quantum \\ server}} \pm \underbrace{|GHZ\rangle_{N''}}_{\substack{u_1, \ldots u_N \\ Users}} \underbrace{|\pm\rangle_q}_{\substack{N+1 \\ quantum \\ server}}\big)$ where $\underbrace{|GHZ\rangle_{N'}}_{\substack{u_1, \ldots u_N \\ Users}}$ and $\underbrace{|GHZ\rangle_{N''}}_{\substack{u_1, \ldots u_N \\ Users}}$ are one of defined *GHZ* states. Next, $|GHZ\rangle_{1' \ldots N'}$ transmitted to u_N, u_N performs $N-GHZ$ measurement on his particle and $u_1, u_2, \ldots, u_{N-1}$ particles. Afterwards, the quantum server calculates the status of his particle according to x

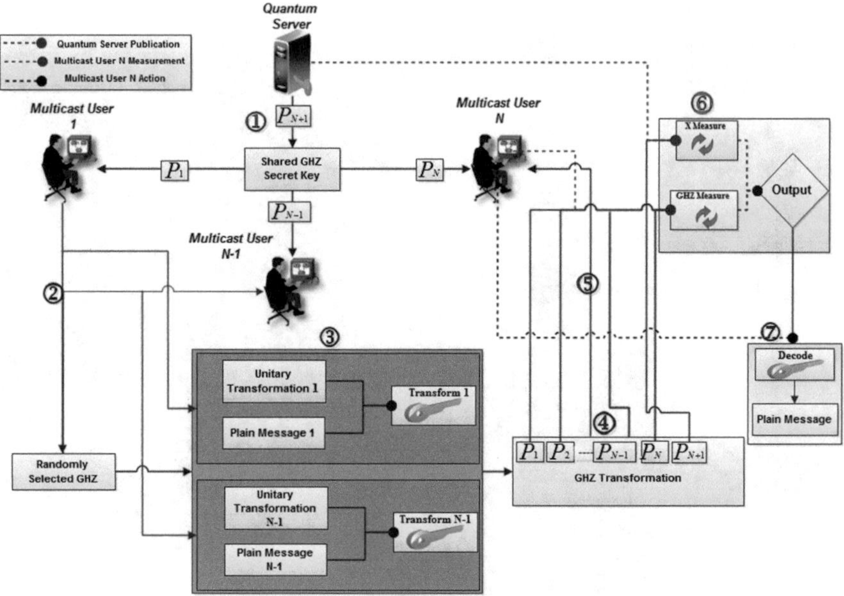

Fig. 7 Generalization of communication process between N disjoint users with partial cooperation of quantum server

basis $\{+, -\}$ and announces his measurement results. u_N uses his measurements and the quantum server's publication for retrieving the original sent secret bits by $u_1, u_2, \ldots, u_{N-1}$ see (Fig. 7).

5.2 *Full Cooperation Process*

Our approach will be generalized for Communication Process between N Disjoint Users with a full Cooperation of Quantum Server. The sequence of steps Similar to partial one except that the selected $|GHZ\rangle_{1\ldots N}$ will be transformed to $|GHZ\rangle_{1'\ldots N'} = \frac{1}{\sqrt{2}}(\underbrace{|GHZ\rangle_{N'}}_{\substack{u_1, \ldots, q, u_{N-1} \\ Senders + \\ quantumserver}} \underbrace{|\pm\rangle_N}_{\substack{u_N \\ Reciever}} \pm \underbrace{|GHZ\rangle_{N''}}_{\substack{u_1, \ldots, q, u_{N-1} \\ Senders + \\ quantumserver}} \underbrace{|\pm\rangle_N}_{\substack{u_N \\ Reciever}})$ where $\underbrace{|GHZ\rangle_{N'}}_{\substack{u_1, \ldots, q, u_{N-1} \\ Senders + \\ quantumserver}}$ and $\underbrace{|GHZ\rangle_{N''}}_{\substack{u_1, \ldots, q, u_{N-1} \\ Senders + \\ quantumserver}}$ are one of defined GHZ states. Next, $|GHZ\rangle_{1'\ldots\ldots\ldots N'}$ transmitted to the quantum server, quantum server performs $N-GHZ$ measurement on his particle and $u_1, u_2, \ldots, u_{N-1}$ particles. Afterwards, u_N calculates the status of his particle according to x basis .. and announces his measurement results. u_N uses his

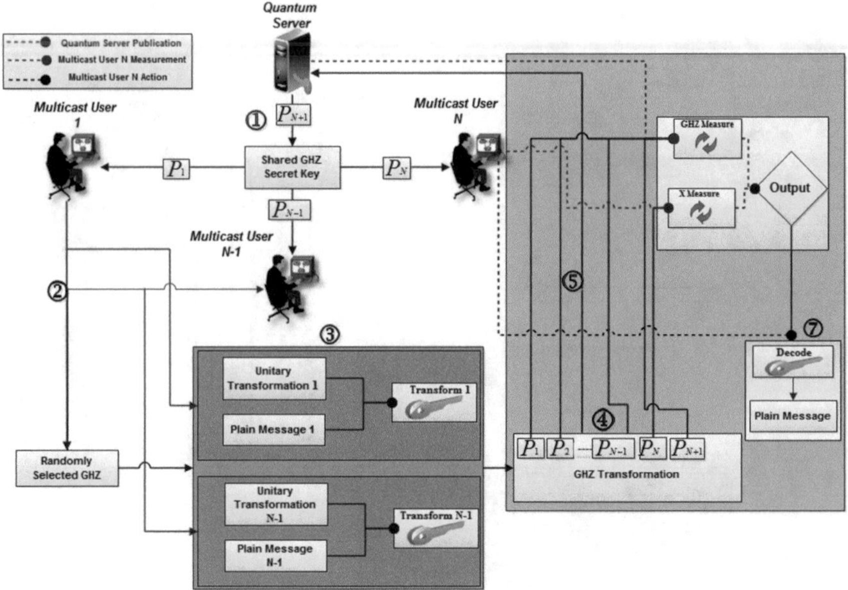

Fig. 8 Generalization of communication process between N disjoint users with full cooperation of quantum server: *Green Box* indicates different steps from Fig. 7

publication' and the quantum server's measurement for retrieving the original sent secret bits by $u_1, u_2, \ldots, u_{N-1}$ see (Fig. 8).

6 Authentication Security Analysis

6.1 Masquerade as Dishonest Disjoint User

If an attacker would like to masquerade as dishonest disjoint user, then the attacker will work on the transmitting particle u from the quantum server to the disjoint user. By assuming that the attacker applies a universal operation $\mathfrak{R}$ on u see (Eqs. (33, 34)).

$$|0_u\mathfrak{R}\rangle \rightarrow \alpha_0|0_u0_a\rangle + \beta_0|0_u1_a\rangle + \gamma_0|1_u0_a\rangle + \delta_0|1_u1_a\rangle \tag{33}$$

$$|1_u\mathfrak{R}\rangle \rightarrow \alpha_1|0_u0_a\rangle + \beta_1|0_u1_a\rangle + \gamma_1|1_u0_a\rangle + \delta_1|1_u1_a\rangle \tag{34}$$

Where $|\mathfrak{R}\rangle$ represents an additional state which is created by the attacker, a represents the attacker particle and,

$$\left|\alpha_0^2\right| + \left|\beta_0^2\right| + \left|\gamma_0^2\right| + \left|\delta_0^2\right| = \left|\alpha_1^2\right| + \left|\beta_1^2\right| + \left|\gamma_1^2\right| + \left|\delta_1^2\right| = 1 \tag{35}$$

When the attacker applies its operation, a new shared key state will be created see (Eqs. (36, 37)).

$$\left|\Phi_{qu}^{+}\right\rangle \rightarrow \left|\Phi_{qu}^{+'}\right\rangle \tag{36}$$

$$\begin{aligned} \left|\Phi_{qu}^{+'}\right\rangle = \frac{1}{\sqrt{2}}(&\alpha_0\left|0_q 0_u 0_a\right\rangle + \beta_0\left|0_q 0_u 1_a\right\rangle + \gamma_0\left|0_q 1_u 0_a\right\rangle + \delta_0\left|0_q 1_u 1_a\right\rangle \\ &+ \alpha_1\left|1_q 0_u 0_a\right\rangle + \beta_1\left|1_q 0_u 1_a\right\rangle + \gamma_1\left|1_q 1_u 0_a\right\rangle + \delta_1\left|1_q 1_u 1_a\right\rangle) \end{aligned} \tag{37}$$

The attacker transmits the new formulated state $\left|\Phi_{qu}^{+'}\right\rangle$ to the quantum server. Afterwards, the quantum server performs $₵_{OP}$ on the received state as result one of four states $\left|\Phi_{qu}^{00}\right\rangle, \left|\psi_{qu}^{01}\right\rangle, \left|\psi_{qu}^{10}\right\rangle, \left|\Phi_{qu}^{11}\right\rangle$ which are equivalent to the dual bits 00, 01, 10 and 11. By assuming when the dual bits are equivalent to 00 then the quantum server operation is equivalent (see Eq. (38)).

$$\left|\Phi_{qu}^{00}\right\rangle = ₵_0 \left|\Phi_{qu}^{+'}\right\rangle \tag{38}$$

By performing $₵_0$ the result is obtained by (Eq. (39))

$$\begin{aligned} \left|\Phi_{qu}^{00}\right\rangle = \frac{1}{\sqrt{2}}(&\alpha_0\left|0_q 0_u 0_a\right\rangle + \beta_0\left|0_q 0_u 1_a\right\rangle + \gamma_0\left|0_q 1_u 0_a\right\rangle + \delta_0\left|0_q 1_u 1_a\right\rangle \\ &+ \alpha_1\left|1_q 0_u 1_a\right\rangle + \beta_1\left|1_q 0_u 0_a\right\rangle + \gamma_1\left|1_q 1_u 1_a\right\rangle + \delta_1\left|1_q 1_u 0_a\right\rangle) \end{aligned} \tag{39}$$

So, the chance for discovering the attacker for $\left|\Phi_{qu}^{00}\right\rangle$ is $\acute{P}_{00}$ can be computed from (Eq. (39)) as by (Eq. (40))

$$\acute{P}_{00} = \frac{1}{2}\left(\left|\alpha_1^2\right| + \left|\gamma_1^2\right| + \left|\beta_0^2\right| + \left|\delta_0^2\right|\right) \tag{40}$$

Also, when the dual bits are equivalent to 01, so the chance for discovering the attacker for $\left|\psi_{qu}^{01}\right\rangle$ is $\acute{P}_{01}$ can be computed from Eq. (37) as shown by (Eq. (41))

$$\acute{P}_{01} = \frac{1}{2}\left(\left|\alpha_0^2\right| + \left|\gamma_0^2\right| + \left|\beta_1^2\right| + \left|\delta_1^2\right|\right) \tag{41}$$

When the dual bits are equivalent to 10, so the chance for discovering the attacker for $\left|\psi_{qu}^{10}\right\rangle$ is $\acute{P}_{10}$ and equivalent to $\acute{P}_{01}$. Furthermore, the chance for discovering the attacker for $\left|\Phi_{qu}^{11}\right\rangle$ is $\acute{P}_{11}$ and equivalent to $\acute{P}_{00}$. Accordingly, the total discovering probability of attacker $\acute{P}_{Total}$ for each disjoint user is equivalent ½ see (Eqs. (42, 43))

$$\acute{P}_{Total} = \frac{1}{4} (\acute{P}_{00} + \acute{P}_{01} + \acute{P}_{10} + \acute{P}_{11}) \quad (42)$$

$$\acute{P}_{Total} = \frac{1}{2} \quad (43)$$

As stated by Simmons theory [68, 69], the result of the previous equation proved that the proposed scheme is unconditionally secured under this type of attacks.

6.2 Substitution Fraudulent Attack

In this type of attacks, the eavesdropper should get the authentication key, so there are two possibilities. One is to extract information about the authentication key by directly calculating the transferred particles over the communicated channel from the quantum server to the disjoint user. The other possibility is achieved by eavesdropping the transmitted particles through two-way communicated channel between the quantum server and the disjoint user.

6.2.1 One-way Channel Substitution Fraudulent Attack

As the transmitted particle from the quantum server to the disjoint user doesn't contain any fact about the authentication key, so in this type of attacks only the restored *n* (new state particle) from the disjoint user to the quantum server has to be measured. The maximum reachable information which the attacker may obtain over the communicated channel between quantum server and a disjoint user can be computed by Holevo theory see (Eq. (44)) [70].

$$\chi(\rho) = S(\rho) - \sum_i P_i S(\rho_i) \quad (44)$$

As $S(\rho)$ is equivalent to Von Neumann entropy $-Tr(\rho \log_2 \rho)$, ρ_i is a component in the hybrid status and P_i is the possibility of ρ_i in the universe ρ. So, the eavesdropper just has information about the authentication key by directly calculating the *n* (new state particle), so the resulted $\chi(\rho)$ relies on the reduced density matrix of *n*, by substitution of (Eq. (44)) as shown by (Eq. (45))

$$\chi(\rho_n) = S(\rho_n) - \sum_i P_i S(\rho_{ni}) \quad (45)$$

As both Þ_n and Þ_{ni} require reduced density matrix for Þ and Þ_i respectively. For any authentication key, the reduced density matrix of n can be represented in the form of in (Eq. (46))

$$\text{Þ}_n = Tr_{qu}\left(|\Phi_r\rangle\langle\Phi_r|\right) = \frac{1}{2}I \tag{46}$$

In addition to, Þ_{ni} is equivalent to the subsequent equations in (Eqs. (47)–(51))

$$|\Phi_r^{00}\rangle = \frac{1}{\sqrt{2}}\left(|0_q 0_u 0_n\rangle + |1_q 1_u 1_n\rangle\right) \tag{47}$$

$$|\Phi_r^{01}\rangle = \frac{1}{\sqrt{2}}\left(|0_q 0_u 1_n\rangle + |1_q 1_u 0_n\rangle\right) \tag{48}$$

$$|\Phi_r^{10}\rangle = \frac{1}{\sqrt{2}}\left(|+_q +_u 1_n\rangle + |-_q -_u 0_n\rangle\right) \tag{49}$$

$$|\Phi_r^{11}\rangle = \frac{1}{\sqrt{2}}\left(|+_q +_u 0_n\rangle + |-_q -_u 1_n\rangle\right) \tag{50}$$

Therefore;

$$\text{Þ}_{ni} = Tr_{qu}\left(|\Phi_r^i\rangle\langle\Phi_r^i|\right) = \frac{1}{2}I \tag{51}$$

By replacing values for both Þ_n and Þ_{ni} in Eq. (45), Ӽ () = 0. So, the eavesdropper will not reveal any information about the key in case of directly calculating the transferred particles over the communicated channel from the quantum server to the disjoint user.

6.2.2 Two-Way Channel Substitution Fraudulent Attack

Firstly, the eavesdropper listens to the communication channel and intercept the transmitting particle from the quantum server to the disjoint user. The attacker applies an operation Θ_1 at his side on the transmitted particle u and supportive particle ε. Afterward, the attacker transmits the resulted particle to the disjoint user. When the disjoint user receives the transmitted particle, he doesn't realize there is an attacker and he did an operation. The disjoint user applies his normal operation and transmits the resulted particle to the quantum server. The attacker intercepts the information particle which is sent by the disjoint user. The attacker applies an operation Θ_2 at his side on the information particle and supportive particle η. Afterward, the attacker transmits the resulted particle to the quantum server. The attacker attempts to retrieve certain amount of information about the key by employing two supportive particles ε and η as illustrated in Fig. 9.

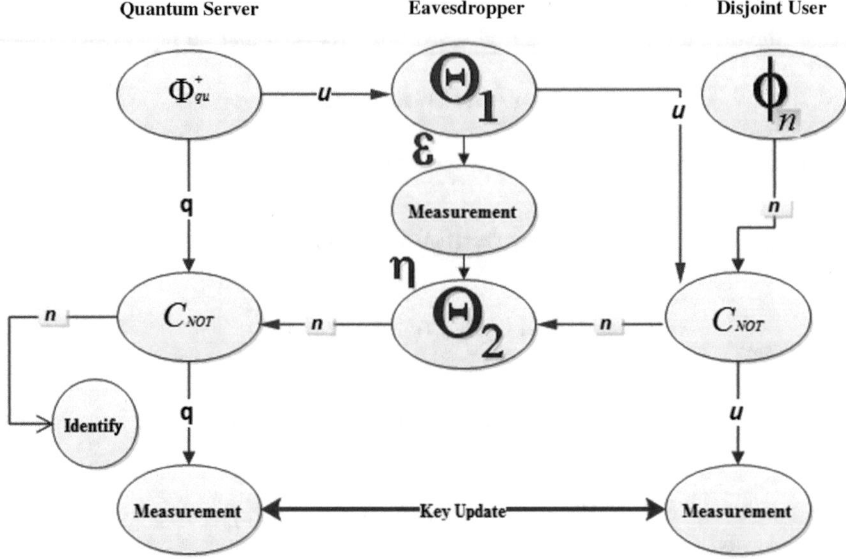

Fig. 9 Two-way channel substitution fraudulent attack with attacker's operation $\boldsymbol{\Theta}_1$ and $\boldsymbol{\Theta}_2$

The operation of the attacker Θ_1 and ε on the transmitted particle u is given by (Eqs. (52)–(55))

$$||0_u\varepsilon\rangle \rightarrow \alpha_\varepsilon|0_u\varepsilon_{00}\rangle + \beta_\varepsilon|1_u\varepsilon_{01}\rangle \tag{52}$$

$$||1_u\varepsilon\rangle \rightarrow \beta_\varepsilon|0_u\varepsilon_{10}\rangle + \alpha_\varepsilon|1_u\varepsilon_{11}\rangle \tag{53}$$

$$\begin{aligned}|+_u\varepsilon\rangle \rightarrow &\frac{1}{2}|+_u\rangle(\alpha_\varepsilon|\varepsilon_{00}\rangle + \alpha_\varepsilon|\varepsilon_{11}\rangle + \beta_\varepsilon|\varepsilon_{01}\rangle + \beta_\varepsilon|\varepsilon_{10}\rangle)\\ &+\frac{1}{2}|-_u\rangle(\alpha_\varepsilon|\varepsilon_{00}\rangle - \alpha_\varepsilon|\varepsilon_{11}\rangle - \beta_\varepsilon|\varepsilon_{01}\rangle + \beta_\varepsilon|\varepsilon_{10}\rangle)\end{aligned} \tag{54}$$

$$\begin{aligned}|-_u\varepsilon\rangle \rightarrow &\frac{1}{2}|+_u\rangle(\alpha_\varepsilon|\varepsilon_{00}\rangle - \alpha_\varepsilon|\varepsilon_{11}\rangle + \beta_\varepsilon|\varepsilon_{01}\rangle - \beta_\varepsilon|\varepsilon_{10}\rangle)\\ &+\frac{1}{2}|-_u\varepsilon\rangle(\alpha_\varepsilon|\varepsilon_{00}\rangle + \alpha_\varepsilon|\varepsilon_{11}\rangle - \beta_\varepsilon|\varepsilon_{01}\rangle - \beta_\varepsilon|\varepsilon_{10}\rangle)\end{aligned} \tag{55}$$

Correspondingly, the operation of the attacker Θ_2 and η on the transmitted information particle n is given by (Eqs. (56)–(59))

$$|0_n\eta\rangle \rightarrow \alpha_\eta|0_n\eta_{00}\rangle + \beta_\eta|1_n\eta_{01}\rangle \tag{56}$$

$$|1_n\eta\rangle \rightarrow \beta_\eta|0_n\eta_{10}\rangle + \alpha_\eta|1_n\eta_{11}\rangle \tag{57}$$

$$
\begin{aligned}
|+_{\mathrm{n}}\eta\rangle \rightarrow & \frac{1}{2}|+_{\mathrm{n}}\rangle(\alpha_\eta|\eta_{00}\rangle+\alpha_\eta|\eta_{11}\rangle+\beta_\eta|\eta_{01}\rangle+\beta_\eta|\eta_{10}\rangle) \\
& +\frac{1}{2}|-_{\mathrm{n}}\rangle(\alpha_\eta|\eta_{00}\rangle-\alpha_\eta|\eta_{11}\rangle-\beta_\eta|\eta_{01}\rangle+\beta_\eta|\eta_{10}\rangle)
\end{aligned} \tag{58}
$$

$$
\begin{aligned}
|-_{\mathrm{n}}\eta\rangle \rightarrow & \frac{1}{2}|+_{\mathrm{n}}\rangle(\alpha_\eta|\eta_{00}\rangle-\alpha_\eta|\eta_{11}\rangle+\beta_\eta|\eta_{01}\rangle-\beta_\eta|\eta_{10}\rangle) \\
& +\frac{1}{2}|-_{\mathrm{n}}\rangle(\alpha_\eta|\eta_{00}\rangle+\alpha_\eta|\eta_{11}\rangle-\beta_\eta|\eta_{01}\rangle-\beta_\eta|\eta_{10}\rangle)
\end{aligned} \tag{59}
$$

Correspondingly, applying the unitary operation requires the following conditions see (Eqs. (60)–(62))

$$|\alpha_\varepsilon^2|+|\beta_\varepsilon^2|=1,\ |\alpha_\eta^2|+|\beta_\eta^2|=1 \tag{60}$$

$$\langle\varepsilon_{00}|\varepsilon_{10}\rangle+\langle\varepsilon_{01}|\varepsilon_{11}\rangle=0 \tag{61}$$

$$\langle\eta_{00}|\eta_{10}\rangle+\langle\eta_{01}|\eta_{11}\rangle=0 \tag{62}$$

As well for the shorten discussion, supposes that equations for orthogonal conditions and equations for non-orthogonal conditions are given by (Eqs. (63, 64)) and (Eqs. (65, 66)) respectively.

$$\langle\varepsilon_{00}|\varepsilon_{01}\rangle = \langle\varepsilon_{10}|\varepsilon_{11}\rangle = \langle\varepsilon_{00}|\varepsilon_{10}\rangle = \langle\varepsilon_{01}|\varepsilon_{11}\rangle = 0 \tag{63}$$

$$\langle\eta_{00}|\eta_{01}\rangle = \langle\eta_{10}|\eta_{11}\rangle = \langle\eta_{00}|\eta_{10}\rangle = \langle\eta_{01}|\eta_{11}\rangle = 0 \tag{64}$$

$$\langle\varepsilon_{00}|\varepsilon_{11}\rangle=\cos\theta_\varepsilon,\ \langle\varepsilon_{01}|\varepsilon_{10}\rangle=\cos\varphi_\varepsilon \tag{65}$$

$$\langle\eta_{00}|\eta_{11}\rangle=\cos\theta_\eta,\ \langle\eta_{01}|\eta_{10}\rangle=\cos\varphi_\eta \tag{66}$$

When the two-bit key $A_i\ A_{i+1} = 00$, so the resulting decoding state by the quantum server is given by (Eqs. (67, 68))

$$|\Phi_{qu}^{00}\rangle = \not{C}_0\,\Theta_2\,\{\not{C}_0\,[\Theta_1\,(|\Phi_{qu}^{+}\rangle|\varepsilon\rangle)\,|\,|\phi_{\mathrm{n}}\rangle]\,|\eta\rangle\} \tag{67}$$

$$
\begin{aligned}
|\Phi_{qu}^{00}\rangle = \frac{1}{\sqrt{2}}(&\alpha_\varepsilon\,\alpha_\eta\,|0_{\mathrm{q}}0_{\mathrm{u}}0_{\mathrm{n}}\,\varepsilon_{00}\,\eta_{00}\rangle+\alpha_\varepsilon\,\beta_\eta|0_{\mathrm{q}}0_{\mathrm{u}}1_{\mathrm{n}}\,\varepsilon_{00}\,\eta_{01}\rangle \\
&+\beta_\varepsilon\,\beta_\eta|0_{\mathrm{q}}1_{\mathrm{u}}0_{\mathrm{n}}\,\varepsilon_{01}\,\eta_{10}\rangle+\beta_\varepsilon\,\alpha_\eta|0_{\mathrm{q}}1_{\mathrm{u}}1_{\mathrm{n}}\,\varepsilon_{01}\,\eta_{11}\rangle \\
&+\beta_\varepsilon\,\alpha_\eta|1_{\mathrm{q}}0_{\mathrm{u}}1_{\mathrm{n}}\,\varepsilon_{10}\,\eta_{00}\rangle+\beta_\varepsilon\,\beta_\eta|1_{\mathrm{q}}0_{\mathrm{u}}0_{\mathrm{n}}\,\varepsilon_{01}\,\eta_{01}\rangle \\
&+\alpha_\varepsilon\,\alpha_\eta|1_{\mathrm{q}}1_{\mathrm{u}}1_{\mathrm{n}}\,\varepsilon_{11}\,\eta_{10}\rangle+\alpha_\varepsilon\,\alpha_\eta|1_{\mathrm{q}}1_{\mathrm{u}}0_{\mathrm{n}}\,\varepsilon_{11}\,\eta_{11}\rangle)
\end{aligned} \tag{68}
$$

The attacker will be discovered if the transmitted particle state is not $|0_n\rangle$, in such situation, the possibility of distinguishing the attacker when $A_i A_{i+1} = 00$ is given by (Eq. (69))

$$\dot{\mathrm{P}}_{\mathrm{Total}}(\mathrm{A_i\, A_{i+1}} = 00) = \left(\alpha_\varepsilon\, \beta_\eta\right)^2 + \left(\beta_\varepsilon\, \alpha_\eta\right)^2 \tag{69}$$

By applying the similar sequences from (Eq. (67) to Eq. (69)) for $A_i A_{i+1} = 01$ showing that the possibility of distinguishing the attacker is equivalent to $\dot{\mathrm{P}}_{Total}(A_i A_{i+1} = 00)$. The possibility of distinguishing the attacker when $A_i A_{i+1} = 10$ is given by (Eq. (70))

$$\begin{aligned}\dot{\mathrm{P}}_{\mathrm{Total}}(\mathrm{A_i\, A_{i+1}} = 10) = \frac{1}{2}[&\left(\alpha_\varepsilon\, \beta_\eta\right)^2(1 + \cos\,\theta_\varepsilon) + \left(\beta_\varepsilon\, \beta_\eta\right)^2(1 + \cos\,\varphi_\varepsilon) \\ &+ \left(\alpha_\varepsilon\, \alpha_\eta\right)^2(1 - \cos\,\theta_\varepsilon) + \left(\beta_\varepsilon\, \alpha_\eta\right)^2(1 - \cos\,\varphi_\varepsilon)]\end{aligned} \tag{70}$$

By applying the similar sequences from (Eqs. (67)–(69)) for $A_i A_{i+1} = 11$ showing that the possibility of distinguishing the attacker is equivalent to $\dot{\mathrm{P}}_{Total}(A_i A_{i+1} = 10)$. From the above equations, we can conclude that when $A_i = 0$ the possibility of distinguishing the attacker is equivalent to $\left(\alpha_\varepsilon \beta_\eta\right)^2 + \left(\beta_\varepsilon \alpha_\eta\right)^2$ $\left(\alpha_\varepsilon \beta_\eta\right)^2 + \left(\beta_\varepsilon \alpha_\eta\right)^2$ and when $A_i = 1$ is equivalent to $\frac{1}{2}[\left(\alpha_\varepsilon \beta_\eta\right)^2(1 + \cos\,\theta_\varepsilon) + \left(\beta_\varepsilon \beta_\eta\right)^2$ $(1 + \cos\,\varphi_\varepsilon) + \left(\alpha_\varepsilon \alpha_\eta\right)^2(1 - \cos\,\theta_\varepsilon) + \left(\beta_\varepsilon \alpha_\eta\right)^2(1 - \cos\,\varphi_\varepsilon)]$ as illustrated in Fig. 10.

By combining Eqs. (69) and (70), we can calculate the total possibility for discovering the attacker in the authentication process which is given by (Eq. (71))

$$\dot{\mathrm{P}}_{Total} = 1/2[\dot{\mathrm{P}}_{Total}(A_i = 0) + \dot{\mathrm{P}}_{Total}(A_i = 1)] \tag{71}$$

If the attacker would like to minimize his detection probability then the attacker has to adjust $\dot{\mathrm{P}}_{Total}$ as minimum discovering probability see (Eq. (72)). Equation (72) is calculated under the condition of $\alpha_\varepsilon = \alpha_\eta = 1$

$$Total = Min\left(\dot{\mathrm{P}}_{Total}\right) = \frac{1}{4}(1 - cos\;\theta_\varepsilon) \tag{72}$$

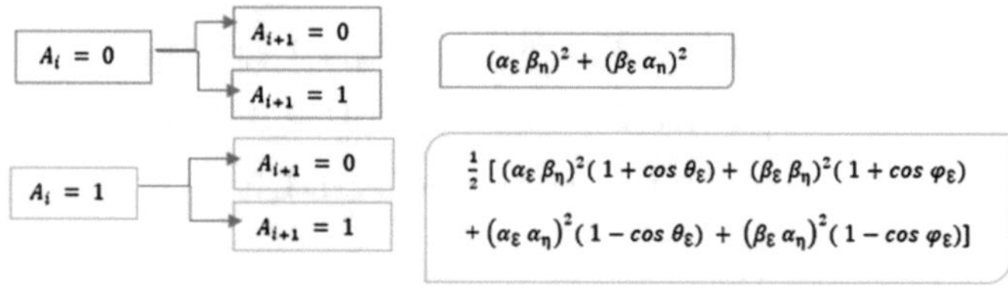

Fig. 10 Relation between two-bit key $A_i\, A_{i+1}$ and $\dot{\mathrm{P}}_{Total}$

From Eq. (72) it is shown that $Min(\dot{P}_{Total})$ is depends on $cos\,\theta_\varepsilon$ and unrelated to θ_η. So, the attacker's total information amount on the transmitted key bits between the quantum server and the disjoint user can be estimated by (Eq. (73)).

$$\mathfrak{T}(A_K, \Theta_{Total}) = \sum_{x,y} {}'\mathrm{P}(A_K, \Theta_{Total}) \log_2 \frac{{}'\mathrm{P}(A_K, \Theta_{Total})}{{}'\mathrm{P}(A_K)\,{}'\mathrm{P}(\Theta_{Total})} \tag{73}$$

Where Θ_{Total} represents the total operation performed by the attacker Θ_1 and Θ_2, x represents the key values (00, 01, 10, 11) with probability ${}'\mathrm{P}(x) = \frac{1}{4}$, A_K indicates the selected random values from variable x, $y = \varepsilon_{ij}\eta_{\mu\tau}$ with $i, j, \mu, \tau \in \{0, 1\}$ which represents 16 possibilities of the mutual measurement output of the attacker at positions Θ_1 and Θ_2. For retrieving the value of attacker's total amount of information from Eq. (73). We should only attain the ${}'\mathrm{P}(A_K)$ and ${}'\mathrm{P}(\Theta_{Total}|A_K)$ by (Eq. (74))

$${}'\mathrm{P}(A_K, \Theta_{Total}) = {}'\mathrm{P}(A_K)\,{}'\mathrm{P}(\Theta_{Total}|A_K) \tag{74}$$

Assuming a special case ${}'\mathrm{P}(\varepsilon_{00}\eta_{00}|00)$ by substituting in Eq. (68), the minimum detection probability of the attacker's total operation Θ_{Total} is either $\varepsilon_{00}\eta_{00}$ or ${}'\mathrm{P}_{11}\eta_{11}$ with equal possibility of ½ when $A_{2i-1}A_{2i} = 00$ see (Eq. (75))

$$\left|\Phi_{Total_{qu}^{00}}\right\rangle = \frac{1}{\sqrt{2}}\left(\left|0_q 0_u 0_n\, \varepsilon_{00}\, \eta_{00}\right\rangle + \left|1_q 1_u 0_n\, \varepsilon_{11}\, \eta_{11}\right\rangle\right) \tag{75}$$

Since $< \varepsilon_{00}|\varepsilon_{11} > = cos\,\theta_\varepsilon$ from Eq. (65) so,

$${}'\mathrm{P}(\varepsilon_{00}\,\eta_{00}|00) = (1 + sin\,\theta_\varepsilon)/2 \tag{76}$$

Since ${}'\mathrm{P}(x) = \frac{1}{4}$, so ${}'\mathrm{P}(00) = \frac{1}{4}$ and from (Eq. (76)) ${}'\mathrm{P}(\varepsilon_{00}\,\eta_{00}|00) = (1 + sin\,\theta_\varepsilon)/2$. Therefore, by substitution in Eq. (74), the value of attacker's total amount of information on the transmitted key values $A_{2i-1}A_{2i} = 00$ is given by (Eq. (77))

$${}'\mathrm{P}(00, \varepsilon_{00}\,\eta_{00}) = {}'\mathrm{P}(00)\,{}'\mathrm{P}(\varepsilon_{00}\,\eta_{00}|00) = \frac{1 + sin\,\theta_\varepsilon}{8} \tag{77}$$

By the same way when $A_{2i-1}A_{2i} = 01$ the attacker's measurement output can be $\varepsilon_{00}\,\eta_{11}$ $or\,\varepsilon_{11}\,\eta_{00}$, as well when $A_{2i-1}A_{2i} = 10|11$ one of four probable outputs $\{\varepsilon_{00}\,\eta_{00}, \varepsilon_{00}\,\eta_{11}, \varepsilon_{11}\,\eta_{00}, \varepsilon_{11}\,\eta_{11}\}$. Consequently, the joint gained information by attacker's total operation Θ_{Total} is given by (Eq. (78))

$$\mathfrak{T} = \frac{1}{4}[(1 + sin\,\theta_\varepsilon)\log_2(1 + sin\,\theta_\varepsilon) + (1 - sin\,\theta_\varepsilon)\log_2(1 - sin\,\theta_\varepsilon)] \tag{78}$$

Table 14 Correlation between joint information $\mathfrak{T}$ and the minimum discovering probability *Total*

Total	5%	10%	15%	20%	25%
$\mathfrak{T}$ (bits)	0.139	0.2655	0.378795	0.459653	0.5

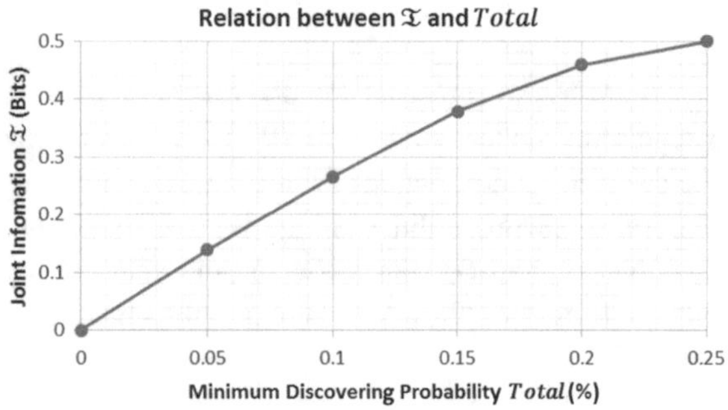

Fig. 11 Correlation between joint information $\mathfrak{T}$ (Bits) and the minimum discovering probability *Total* (%)

Since $\sin\theta_\varepsilon = \sqrt{8\times Total - 16\times Total^2}$, by substitution in Eq. (78)

$$\begin{aligned}\mathfrak{T} = \frac{1}{4}[(1+\sqrt{8\times Total - 16\times Total^2})\log_2(1+\sqrt{8\times Total - 16\times Total^2}) \\ +\left(1-\sqrt{8\times Total - 16\times Total^2}\right)\log_2(1-\sqrt{8\times Total - 16\times Total^2})]\end{aligned} \tag{79}$$

The correlation between the joint information $\mathfrak{T}$ and the minimum discovering probability *Total* for the attacker is shown in Table 14 and Fig. 11. It is shown Fig. 11 that the possibility for discovering the attacker while attempting for retrieving any information about the key bits is equal to non-zero. For example if $Total = 25\%$ means the attacker can gain maximal joint information $\mathfrak{T} = 0.5\ bits$ on the transmitted keys between the quantum server and disjoint user.

In the case of the attacker positively receiving the transmission key A_K between the quantum server and disjoint user, so for each transmitted key, the attacker has to determine which two-bits are used. As per the attacker's measurement output $y = \varepsilon_{ij}\eta_{\mu\tau}$ with $i, j, \mu, \tau \in \{0, 1\}$ and key values are $(00, 01, 10, 11)$, he can estimate the possibility of the key bits. For instance, if $y = \varepsilon_{00}\eta_{11}$ then the attacker can estimate that the transmitted key bits are either 00, 10 or 11 with occurrence possibility 0.5, 0.25 and 0.25 correspondingly. By assuming that the attacker chooses the possibility for detecting transmitted key values 00 or 01 is $'\mathrm{P}$ and for detecting 10 and 11 is $1 - {'\mathrm{P}}$ and take into account the inconclusive measurement output. Therefore, the total estimation probability $'\mathrm{P}_e$ of A_K is given by (Eq. (80))

$$'\mathrm{P}_e = \frac{(1+\sin\theta_\varepsilon)}{2}\left[\frac{1}{2}{'\mathrm{P}} + \frac{1}{4}\left(1-{'\mathrm{P}}\right)\right] + \frac{(1-\sin\theta_\varepsilon)}{2}\left[\frac{1}{4}\left(1-{'\mathrm{P}}\right)\right] \tag{80}$$

Table 15 Correlation between maximized total estimation probability $'P_e^m$ and the minimum discovering probability *Total*

Total	0%	5%	10%	12.5%	25%	50%
$'P_e^m$	0.25	0.40	0.45	0.19	0.50	0.25

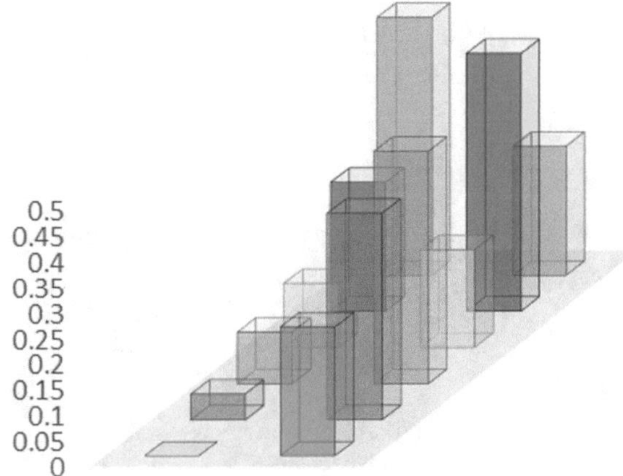

Fig. 12 Correlation between maximized total estimation probability $'P_e^m$ and the minimum discovering probability *Total*

By simplification of Eq. (80), $'P_e$ of A_K is given by (Eq. (81)),

$$'P_e = \frac{1}{8}\left[\left(\sin\theta_\varepsilon(3\times{}'P - 1\right) + 2\right] \tag{81}$$

If $'P = 1$ indicates that the total estimation probability $'P_e$ is maximized see (Eq. (82))

$$'P_e^m = \frac{1}{4}\left(\sqrt{8\times Total - 16\times Total^2} + 1\right) \tag{82}$$

Table 15 and Fig. 12 show the Correlation between Maximized Total Estimation Probability $'P_e^m$ and the minimum discovering probability *Total*. From both, we can conclude that when the total estimation probability is maximized which meant reaching to one, the attacker can positively maximum retrieving 0.5 bits of the transmitted key A_K while the maximum total estimation probability $'P_e^m$ equals to 25%.

Therefore, the probability of the attacker for successfully retrieving the transmitted keys $'\mathrm{P}_e^r$ *for* $A_k = \{A_1, A_2, A_3 \ldots A_{2N}\}$ are expressed as in (Eq. (83)).

$$'\mathrm{P}_e^r = \left['\mathrm{P}_e^m (1 - Total) \right]^{N/2} \tag{83}$$

By substituting (Eq. (82)) in equation (Eq. (83)), so

$$'\mathrm{P}_e^r = \left[\frac{1}{4} \left(\sqrt{8 \times Total - 16 \times Total^2} + 1 \right) (1 - Total) \right]^{N/2} \tag{84}$$

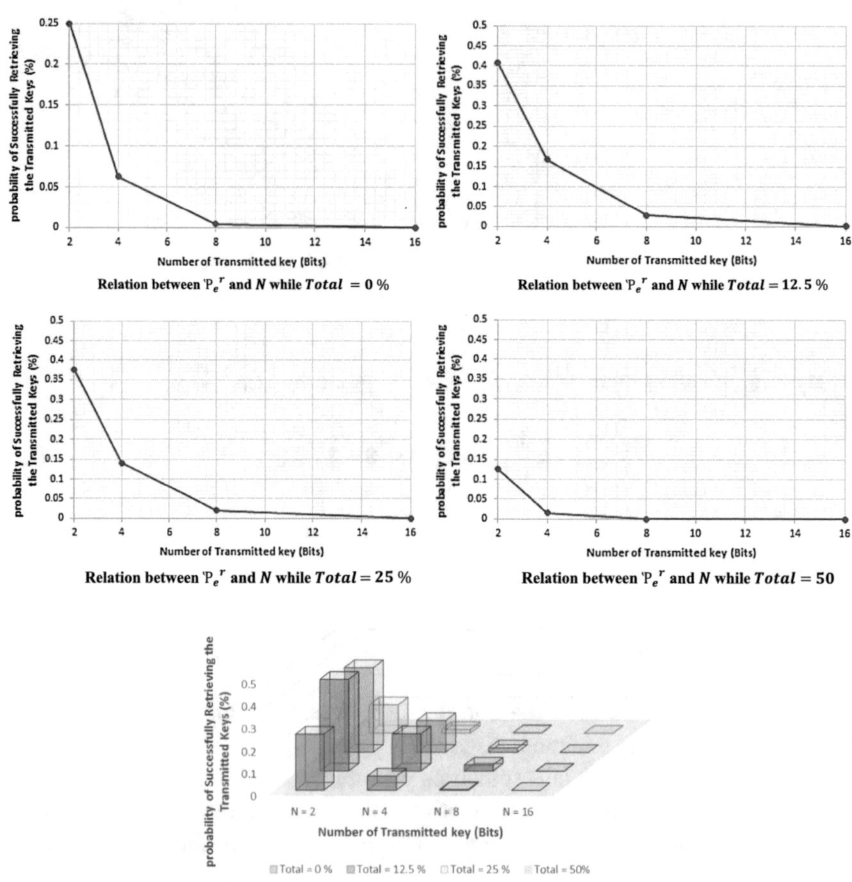

Fig. 13 Relation between $'\mathrm{P}_e^r$, $N = [2, 4, 8, 16]$ and $Total = [0, 12.5, 25, 50]\%$

Figure 13 shows the relation between $'P_e^r$, $N = [2, 4, 8, 16]$ and $Total = [0, 12.5, 25, 50]\%$. The relation illustrates that if the minimum discovery probability is equal to $[0, 12.5, 25, 50]\%$ and $= 2$, so the maximum values for successfully retrieving the information of the transmitted keys A_k by the attacker are $[0.25,\ 4.08 \times 10^{-1},\ 4.08 \times 10^{-1},\ 1.25 \times 10^{-1}]$ respectively. Also, for $= 16$, the maximum successfully of retrieving information of the transmitted keys A_k by the attacker is $\left[1.53 \times 10^{-5}, 7.7 \times 10^{-4}, 3.91 \times 10^{-4}, 5.96 \times 10^{-8}\right]$ respectively. So,

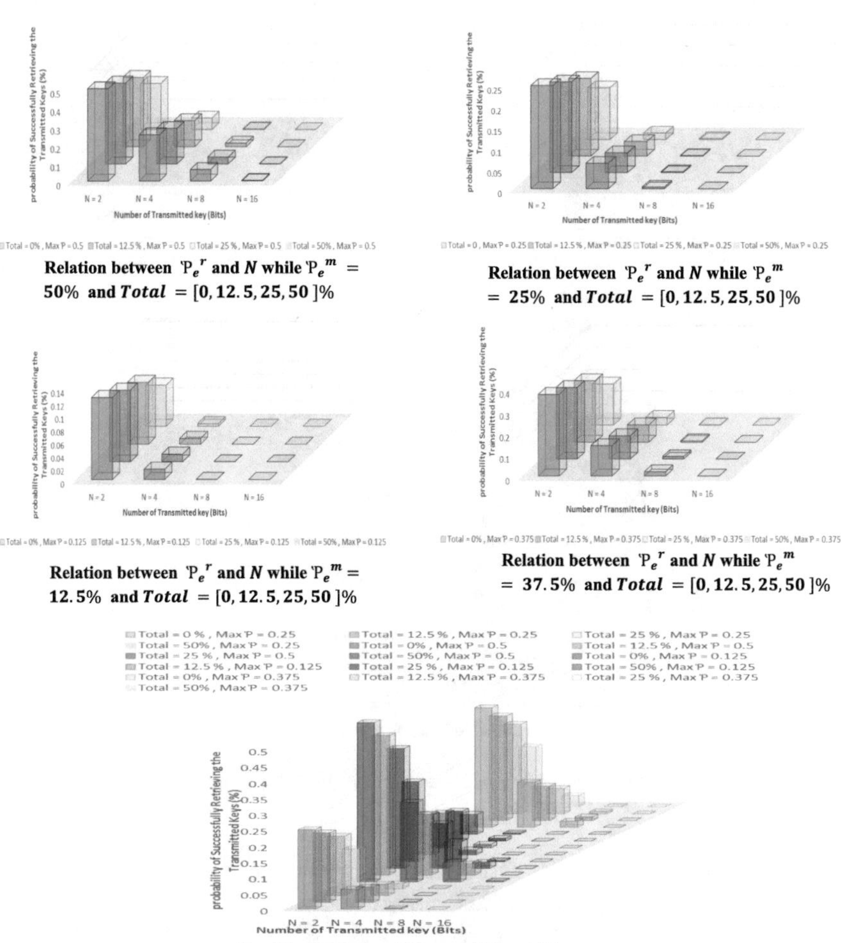

Fig. 14 Relation between $'P_e^r$ and N while $'P_e^m = [12.5, 25, 37.5, 50]\%$ and $Total = [0, 12.5, 25, 50]\%$

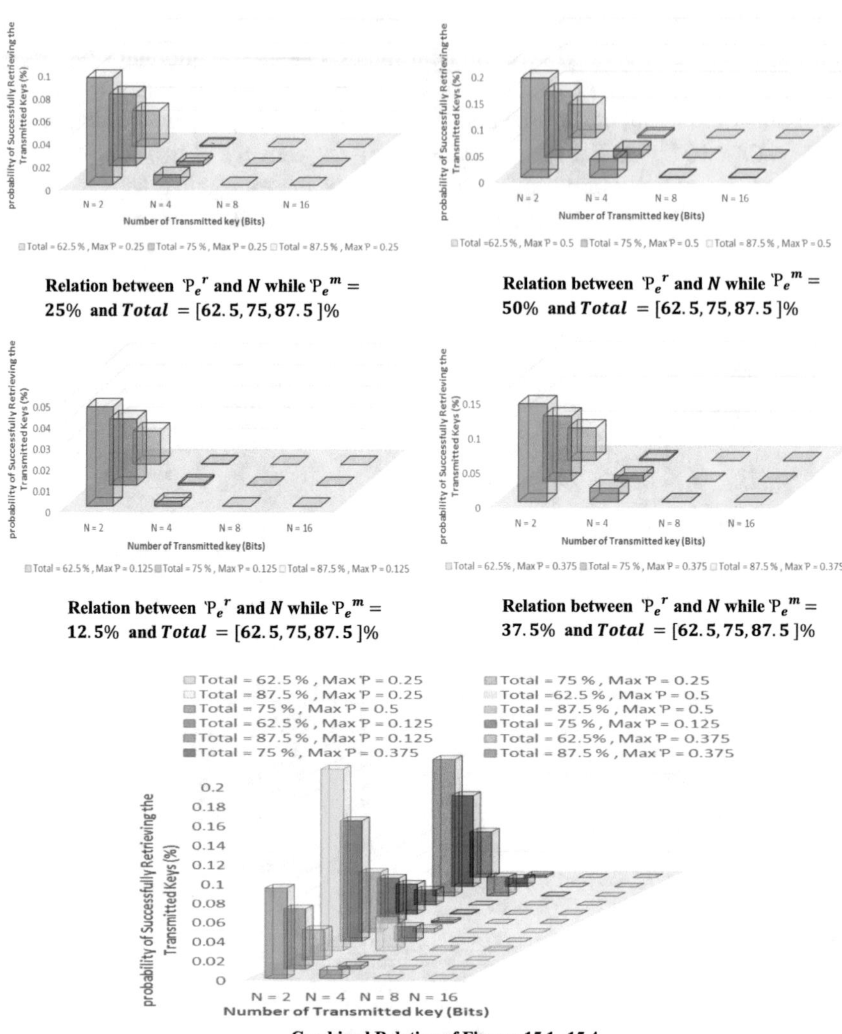

Relation between $'P_e^r$ and N while $'P_e^m$ = 25% and $Total$ = [62.5, 75, 87.5]%

Relation between $'P_e^r$ and N while $'P_e^m$ = 50% and $Total$ = [62.5, 75, 87.5]%

Relation between $'P_e^r$ and N while $'P_e^m$ = 12.5% and $Total$ = [62.5, 75, 87.5]%

Relation between $'P_e^r$ and N while $'P_e^m$ = 37.5% and $Total$ = [62.5, 75, 87.5]%

Combined Relation of Figures 15.1 –15.4

Fig. 15 Relation between $'P_e^r$ and N while $'P_e^m = [12.5, 25, 37.5, 50]\%$ and $Total = [62.5, 75, 87.5]\%$

while the number of transmitted bits increasing, the possibility for successfully retrieving the transmitted information is decreased.

Figure 14 shows the relation between $'P_e^r$ and N while $'P_e^m = [12.5, 25, 37.5, 50]\%$ and $Total = [0, 12.5, 25, 50]\%$. The relation illustrates the maximum and minimum values for successfully retrieving the information of the transmitted keys A_k by the attacker while $N = [2, 4, 8, 16]$ is $[0.5, 0.25, 0.0625, 3.91 \times 10^{-3}]$ and $[0.0625, 3.91 \times 10^{-3}, 1.53 \times 10^{-5}, 2.32 \times 10^{-10}]$ respectively. We can conclude that all maximum values are corresponding when $'\mathrm{P}_e^m$ and *Total* equal to $[50, 0]\%$. So, the attacker can gain maximum information about the transmitted keys A_k at this situation. Furthermore, the minimum value for the attacker to gain maximum information about the transmitted keys A_k $[2.32 \times 10^{-10}]$ is corresponding when the value of $'\mathrm{P}_e^m$ and *Total* equal to $[12.5, 50]\%$.

Figure 15 shows the relation between $'\mathrm{P}_e^r$ and N while $'\mathrm{P}_e^m = [12.5, 25, 37.5, 50]\%$ and $Total = [62.5, 75, 87.5]\%$. The relation illustrates the maximum and minimum values for successfully retrieving the information of the transmitted keys A_k by the attacker while $N = [2, 4, 8, 16]$ is $[1.87 \times 10^{-1}, 3.51 \times 10^{-2}, 1.23 \times 10^{-3}, 1.53 \times 10^{-6}]$ and $[1.56 \times 10^{-2}, 2.5 \times 10^{-4}, 5.96 \times 10^{-8}, 3.55 \times 10^{-15}]$ respectively. From Fig. 15, we can conclude that all maximum and minimum values are corresponding when $'\mathrm{P}_e^m$ and *Total* equal to $[50, 62.5]\%$ and $[12.5, 87.5]\%$ respectively. So, the attacker can gain maximum information about the transmitted keys A_k when $'\mathrm{P}_e^m$ and *Total* equal to $[50, 62.5]\%$ and minimum when $'\mathrm{P}_e^m$ and *Total* equal to $[12.5, 87.5]\%$.

From Figs. 16 and 17, we can conclude that while the number of transmitted key bits N becomes larger, the possibility for successfully retrieving A_k becomes smaller and reach zero. So, the attacker will not reveal an enormous amount of information which can be ignored and avoided by updating key between the disjoint user and the quantum server periodically. In this case, the information of the attacker on the old key will be useless.

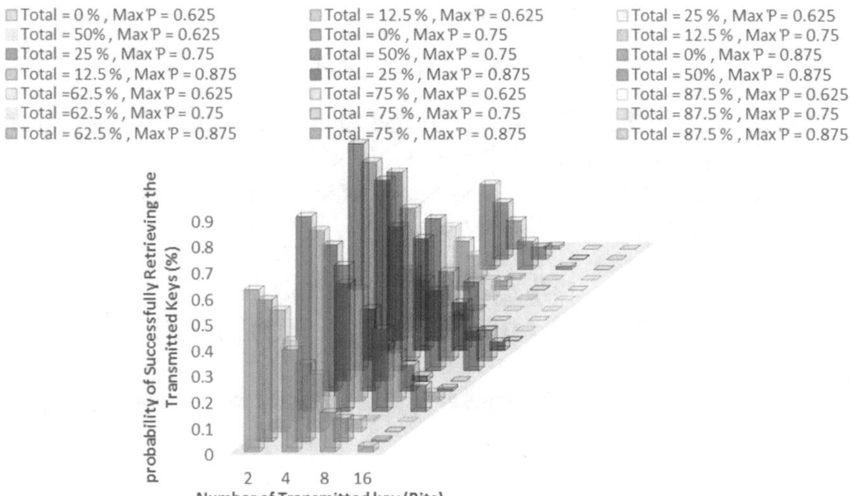

Fig. 16 Relation between $'\mathrm{P}_e^r$ and N while $'\mathrm{P}_e^m = [62.5, 75, 87.5]\%$ and $Total = [0, 12.5, 25, 37.5, 50, 62.5, 75, 87.5]\%$

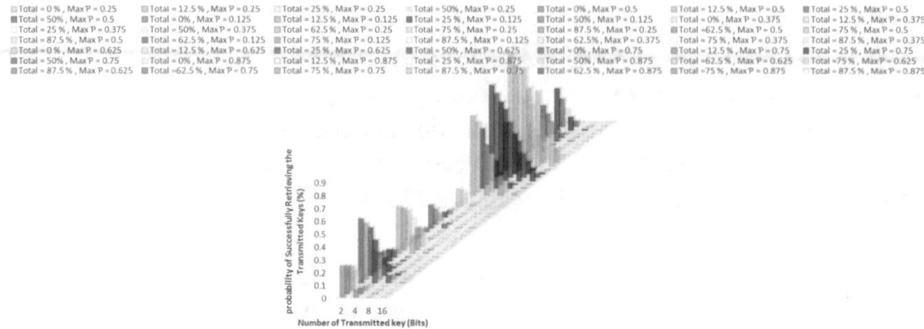

Fig. 17 Combined relation of Figs. 14, 15 and 16

7 Communication Process Security Analysis

After u_j retrieves the original sent secret bits which sent by u_i, u_i informs u_j about the positions of the transmitted particles and the selected unitary transformation applied on them. Afterward, u_j verifies the selected particles by u_i and obtains an approximation of error percentage in the communication process. If the error percentage under the specified threshold, both u_i and u_j can continue the transmission of secret messages. Otherwise, the communication process will be terminated. If an attacker tries to spy on the transmitted *GHZ* particles, the attacker at most can obtain one particle. So, the attacker couldn't decide which operation is applied by u_i consequently couldn't retrieve any transmitted secret bits. Suppose that the attacker applies an operation Θ_{u_iA} on u_i and his qubit $|A\rangle$ see (Eq. (85))

$$\begin{array}{l} |0_A\varepsilon\rangle_{u_iA} \rightarrow \alpha_\varepsilon|0\rangle_{u_i}|\varepsilon_{00}\rangle_A + \beta_\varepsilon|1\rangle_{u_i}|\varepsilon_{01}\rangle_A \\ |1_A\varepsilon\rangle_{u_iA} \rightarrow \beta'_\varepsilon|0\rangle_{u_i}|\varepsilon_{10}\rangle_A + \alpha'_\varepsilon|1\rangle_{u_i}|\varepsilon_{11}\rangle_A \end{array} \tag{85}$$

Correspondingly, applying the unitary operation requires the following conditions $\left|\alpha_\varepsilon^2\right| + \left|\beta_\varepsilon^2\right| = 1$ and $\left|\alpha_\varepsilon'^2\right| + \left|\beta_\varepsilon'^2\right| = 1$

So, the state of the protocol is transformed as follow, Firstly, the states after u_i made a unitary operation see (Eq. (86))

$$\begin{array}{c} |\Psi_1\rangle_{iqjA} = \Theta_{u_i}\left|\Psi_{iqj}\right\rangle \otimes |A\rangle_A \\ = 1/2\Big(|000\rangle_{iqj} \mp |100\rangle_{iqj} + |011\rangle_{iqj} \pm |111\rangle_{iqj}\Big) \otimes |A\rangle_A \end{array} \tag{86}$$

Secondly, the states after the attacker applies a unitary transformation on u_i and his qubit $|A\rangle$ see (Eqs. (87, 88))

$$|\Psi_2\rangle_{iqjA} = |\Theta_{u_iA}\Psi_1\rangle_{iqjA} = 1/2\left[\left(\begin{array}{c} |000\rangle_{iqj}\left(\alpha_\varepsilon|\varepsilon_{00}\rangle \pm \beta_\varepsilon'|\varepsilon_{01}\rangle\right)_A + |100\rangle_{iqj}\left(\beta_\varepsilon|\varepsilon_{01}\rangle \pm \alpha_\varepsilon'|\varepsilon_{11}\rangle\right)_A \\ + |011\rangle_{iqj}\left(\alpha_\varepsilon|\varepsilon_{00}\rangle \mp \beta_\varepsilon'|\varepsilon_{01}\rangle\right)_A + |111\rangle_{iqj}\left(\beta_\varepsilon|\varepsilon_{01}\rangle \mp \alpha_\varepsilon'|\varepsilon_{11}\rangle\right)_A \end{array}\right)\right] \tag{87}$$

$$= \frac{1}{2\sqrt{2}}\left[\begin{array}{l} |\phi_{ij}^{+}\rangle\left\{\begin{array}{c} |+\rangle_q\left(\alpha_\varepsilon|\varepsilon_{00}\rangle \pm \beta_\varepsilon'|\varepsilon_{01}\rangle + \beta_\varepsilon|\varepsilon_{01}\rangle \mp \alpha_\varepsilon'|\varepsilon_{11}\rangle\right)_A \\ + \\ |-\rangle_q\left(\alpha_\varepsilon|\varepsilon_{00}\rangle \pm \beta_\varepsilon'|\varepsilon_{01}\rangle - \beta_\varepsilon|\varepsilon_{01}\rangle \pm \alpha_\varepsilon'|\varepsilon_{11}\rangle\right)_A \end{array}\right\} + \\ |\phi_{ij}^{-}\rangle\left\{\begin{array}{c} |+\rangle_q\left(\alpha_\varepsilon|\varepsilon_{00}\rangle \pm \beta_\varepsilon'|\varepsilon_{01}\rangle - \beta_\varepsilon|\varepsilon_{01}\rangle \pm \alpha_\varepsilon'|\varepsilon_{11}\rangle\right)_A \\ + \\ |-\rangle_q\left(\alpha_\varepsilon|\varepsilon_{00}\rangle \pm \beta_\varepsilon'|\varepsilon_{01}\rangle + \beta_\varepsilon|\varepsilon_{01}\rangle \mp \alpha_\varepsilon'|\varepsilon_{11}\rangle\right)_A \end{array}\right\} + \\ |\Psi_{ij}^{+}\rangle\left\{\begin{array}{c} |\rangle_q\left(\alpha_\varepsilon|\varepsilon_{00}\rangle \mp \beta_\varepsilon'|\varepsilon_{01}\rangle + \beta_\varepsilon|\varepsilon_{01}\rangle \pm \alpha_\varepsilon'|\varepsilon_{11}\rangle\right)_A \\ - \\ |-\rangle_q\left(\alpha_\varepsilon|\varepsilon_{00}\rangle \mp \beta_\varepsilon'|\varepsilon_{01}\rangle - \beta_\varepsilon|\varepsilon_{01}\rangle \mp \alpha_\varepsilon'|\varepsilon_{11}\rangle\right)_A \end{array}\right\} + \\ |\Psi_{ij}^{-}\rangle\left\{\begin{array}{c} |+\rangle_q\left(\alpha_\varepsilon|\varepsilon_{00}\rangle \mp \beta_\varepsilon'|\varepsilon_{01}\rangle - \beta_\varepsilon|\varepsilon_{01}\rangle \mp \alpha_\varepsilon'|\varepsilon_{11}\rangle\right)_A \\ - \\ |-\rangle_q\left(\alpha_\varepsilon|\varepsilon_{00}\rangle \mp \beta_\varepsilon'|\varepsilon_{01}\rangle + \beta_\varepsilon|\varepsilon_{01}\rangle \pm \alpha_\varepsilon'|\varepsilon_{11}\rangle\right)_A \end{array}\right\} \end{array}\right] \tag{88}$$

As shown from (Eqs. (87, 88)), the attacker can't gain any information during intercepting the communication process. Also, the attacker introduces an error probability of ½ irrespective of the sequence of measurement. For example, suppose that the attacker measurement is the same as u_i, it means the applied unitary transformation may be I or Z. If the measurements are different then the possible applied unitary transformation may be X or Y.

References

1. Bennett, C., Brassard, G.: Quantum cryptography: public key distribution and coin tossing. Theoret. Comput. Sci. **560**, 7–11 (2014). doi:10.1016/j.tcs.2014.05.025
2. Nielsen, M., Chuang, I.: Quantum Computation and Quantum Information. Cambridge University Press, Cambridge (2000)
3. Zeng, G.H.: Quantum Cryptology. Science Press (2006)
4. Van Assche, G.: Quantum Cryptography and Secret-key Distillation. Cambridge University Press, Cambridge (2006)
5. Sharbaf, M.S.: Quantum cryptography: a new generation of information technology security system. In: Sixth International Conference on Information Technology: New Generations, ITNG'09, pp. 1644–1648. IEEE (2009)
6. Wootters, W., Zurek, W.: A single quantum cannot be cloned. Nature **299**(5886), 802–803 (1982). doi:10.1038/299802a0
7. Bennett, C., Brassard, G., Crépeau, C., Jozsa, R., Peres, A., Wootters, W.: Teleporting an unknown quantum state via dual classical and Einstein-Podolsky-Rosen channels. Phys. Rev. Lett. **70**(13), 1895–1899 (1993). doi:10.1103/physrevlett.70.1895
8. Stinson, D.: Cryptography. CRC Press, Boca Raton (1995)

9. Chakrabarty, I.: Teleportation via a mixture of a two qubit subsystem of a N-qubit W and GHZ state. Eur. Phys. J. D **57**(2), 265–269 (2010). doi:10.1140/epjd/e2010-00017-8
10. Liang, H., Liu, J., Feng, S., Chen, J.: Quantum teleportation with partially entangled states via noisy channels. Quantum Inf. Process. **12**(8), 2671–2687 (2013). doi:10.1007/s11128-013-0555-3
11. Bennett, C., Wiesner, S.: Communication via one- and two-particle operators on Einstein-Podolsky-Rosen states. Phys. Rev. Lett. **69**(20), 2881–2884 (1992). doi:10.1103/physrevlett.69.2881
12. Kaye, P., Laflamme, R.: An Introduction to Quantum Computing. Oxford University Press (2007)
13. Poppe, A., Peev, M., Maurhart, O.: Outline of the SECOQC quantum-key-distribution network in Vienna. Int. J. Quantum Inf. **6**(02), 209–218 (2008)
14. Peev, M., Pacher, C., Alléaume, R., Barreiro, C., Bouda, J., Boxleitner, W., Tualle-Brouri, R.: The SECOQC quantum key distribution network in Vienna. New J. Phys. **11**(7), 075001 (2009)
15. Elliott, C.: Building the quantum network. New J. Phys. **4**(1), 46 (2002)
16. Elliott, C., Colvin, A., Pearson, D., Pikalo, O., Schlafer, J., Yeh, H.: Current Status of the DARPA Quantum Network. In Defense and Security, pp. 138–149. International Society for Optics and Photonics (2005)
17. Metwaly, A.F., Mastorakis, N.E.: Architecture of decentralized multicast network using quantum key distribution and hybrid WDM-TDM. In: Proceedings of the 9th International Conference on Computer Engineering and Applications (CEA'15). Advances in Information Science And Computer Engineering, pp. 504–518 (2015)
18. Metwaly, A.F., Rashad, M.Z., Omara, F.A., Megahed, A.A.: Architecture of multicast centralized key management scheme using quantum key distribution and classical symmetric encryption. Eur. Phys. J. Spec. Top. **223**(8), 1711–1728 (2014)
19. Gisin, N., Ribordy, G., Tittel, W., Zbinden, H.: Quantum cryptography. Rev. Mod. Phys. **74** (1), 145 (2002)
20. Boström, K., Felbinger, T.: Deterministic secure direct communication using entanglement. Phys. Rev. Lett. **89**(18), 187902 (2002)
21. Deng, F.G., Long, G.L.: Secure direct communication with a quantum one-time pad. Phys. Rev. A **69**(5), 052319 (2004)
22. Deng, F.G., Long, G.L., Liu, X.S.: Two-step quantum direct communication protocol using the Einstein-Podolsky-Rosen pair block. Phys. Rev. A **68**(4), 042317 (2003)
23. Lucamarini, M., Mancini, S.: Secure deterministic communication without entanglement. Phys. Rev. Lett. **94**(14), 140501 (2005)
24. Yan, F.L., Zhang, X.Q.: A scheme for secure direct communication using EPR pairs and teleportation. Eur. Phys. J. B-Condens. Matter Complex Syst. **41**(1), 75–78 (2004)
25. Cai, Q.Y.: The Ping-Pong protocol can be attacked without eavesdropping (2004). arXiv: quant-ph/0402052
26. Zhu, A.D., Xia, Y., Fan, Q.B., Zhang, S.: Secure direct communication based on secret transmitting order of particles. Phys. Rev. A **73**(2), 022338 (2006)
27. Xue, P., Han, C., Yu, B., Lin, X.M., Guo, G.C.: Entanglement preparation and quantum communication with atoms in optical cavities. Phys. Rev. A **69**(5), 052318 (2004)
28. Lee, H., Lim, J., Yang, H.: Quantum direct communication with authentication. Phys. Rev. A **73**(4), 042305 (2006)
29. Zhang, Z.J., Liu, J., Wang, D., Shi, S.H.: Comment on "Quantum direct communication with authentication". Phys. Rev. A **75**(2), 026301 (2007)
30. Wang, C., Deng, F.G., Li, Y.S., Liu, X.S., Long, G.L.: Quantum secure direct communication with high-dimension quantum superdense coding. Phys. Rev. A **71**(4), 044305 (2005)
31. Gao, T., Yan, F.L., Wang, Z.X.: Quantum secure direct communication by EPR pairs and entanglement swapping. Nuovo Cimento B Serie **119**, 313 (2004)
32. Ting, G., Feng-Li, Y., Zhi-Xi, W.: A simultaneous quantum secure direct communication scheme between the central party and other M parties. Chin. Phys. Lett. **22**(10), 2473 (2005)

33. Wang, C., Deng, F.G., Long, G.L.: Multi-step quantum secure direct communication using multi-particle Green–Horne–Zeilinger state. Optics communications **253**(1), 15–20 (2005)
34. Wang, J., Zhang, Q., Tang, C.J.: Quantum secure direct communication based on order rearrangement of single photons. Phys. Lett. A **358**(4), 256–258 (2006)
35. Qing-Yu, C., Bai-Wen, L.: Deterministic secure communication without using entanglement. Chin. Phys. Lett. **21**(4), 601 (2004)
36. Cai, Q.Y.: Eavesdropping on the two-way quantum communication protocols with invisible photons. Phys. Lett. A **351**(1), 23–25 (2006)
37. Long, G.L., Deng, F.G., Wang, C., Li, X.H., Wen, K., Wang, W.Y.: Quantum secure direct communication and deterministic secure quantum communication. Front. Phys. China **2**(3), 251–272 (2007)
38. Cai, Q.Y., Li, B.W.: Improving the capacity of the Boström-Felbinger protocol. Phys. Rev. A **69**(5), 054301 (2004)
39. Ying, S., Qiao-Yan, W., Fu-Chen, Z.: Multiparty quantum chatting scheme. Chin. Phys. Lett. **25**(3), 828 (2008)
40. Jin, X.M., Ren, J.G., Yang, B., Yi, Z.H., Zhou, F., Xu, X.F., Pan, J.W.: Experimental free-space quantum teleportation. Nat. Photonics **4**(6), 376–381 (2010)
41. Curty, M., Santos, D.J.: Quantum authentication of classical messages. Phys. Rev. A **64**(6), 062309 (2001)
42. Dušek, M., Haderka, O., Hendrych, M., Myška, R.: Quantum identification system. Phys. Rev. A **60**(1), 149 (1999)
43. Zeng, G., Zhang, W.: Identity verification in quantum key distribution. Phys. Rev. A **61**(2), 022303 (2000)
44. Ljunggren, D., Bourennane, M., Karlsson, A.: Authority-based user authentication in quantum key distribution. Phys. Rev. A **62**(2), 022305 (2000)
45. Biham, E., Huttner, B., Mor, T.: Quantum cryptographic network based on quantum memories. Phys. Rev. A **54**(4), 2651 (1996)
46. Metwaly, A., Rashad, M.Z., Omara, F.A., Megahed, A.A.: Architecture of point to multipoint QKD communication systems (QKDP2MP). In: 8th International Conference on Informatics and Systems (INFOS), Cairo, pp. NW 25–31. IEEE (2012)
47. Farouk, A., Omara, F., Zakria, M., Megahed, A.: Secured IPsec multicast architecture based on quantum key distribution. In: The International Conference on Electrical and Bio-medical Engineering, Clean Energy and Green Computing, pp. 38–47. The Society of Digital Information and Wireless Communication (2015)
48. Farouk, A., Zakaria, M., Megahed, A., Omara, F.A.: A generalized architecture of quantum secure direct communication for N disjointed users with authentication. Sci. Rep. **5**, 16080 (2014)
49. Wang, M.M., Wang, W., Chen, J.G., Farouk, A.: Secret sharing of a known arbitrary quantum state with noisy environment. Quantum Inf. Process. **14**(11), 4211–4224 (2015)
50. Naseri, M., Heidari, S., Batle, J., Baghfalaki, M., Gheibi, R., Farouk, A., Habibi, A.: A new secure quantum watermarking scheme. Optik-Int. J. Light Electron Opt. **139**, 77–86 (2017)
51. Batle, J., Ciftja, O., Naseri, M., Ghoranneviss, M., Farouk, A., Elhoseny, M.: Equilibrium and uniform charge distribution of a classical two-dimensional system of point charges with hard-wall confinement. Phys. Scr. **92**(5), 055801 (2017)
52. Geurdes, H., Nagata, K., Nakamura, T., Farouk, A.: A note on the possibility of incomplete theory (2017). arXiv:1704.00005
53. Batle, J., Farouk, A., Alkhambashi, M., Abdalla, S.: Multipartite correlation degradation in amplitude-damping quantum channels. J. Korean Phys. Soc. **70**(7), 666–672 (2017)
54. Batle, J., Naseri, M., Ghoranneviss, M., Farouk, A., Alkhambashi, M., Elhoseny, M.: Shareability of correlations in multiqubit states: optimization of nonlocal monogamy inequalities. Phys. Rev. A **95**(3), 032123 (2017)
55. Batle, J., Farouk, A., Alkhambashi, M., Abdalla, S.: Entanglement in the linear-chain Heisenberg antiferromagnet Cu $(C_4H_4N_2)$ $(NO_3)_2$. Eur. Phys. J. B **90**, 1–5 (2017)

56. Batle, J., Alkhambashi, M., Farouk, A., Naseri, M., Ghoranneviss, M.: Multipartite non-locality and entanglement signatures of a field-induced quantum phase transition. Eur. Phys. J. B **90**(2), 31 (2017)
57. Nagata, K., Nakamura, T., Batle, J., Abdalla, S., Farouk, A.: Boolean approach to dichotomic quantum measurement theories. J. Korean Phys. Soc. **70**(3), 229–235 (2017)
58. Abdolmaleky, M., Naseri, M., Batle, J., Farouk, A., Gong, L.H.: Red-Green-Blue multi-channel quantum representation of digital images. Optik-Int. J. Light Electron Opt. **128**, 121–132 (2017)
59. Farouk, A., Elhoseny, M., Batle, J., Naseri, M., Hassanien, A.E.: A proposed architecture for key management schema in centralized quantum network. In: Handbook of Research on Machine Learning Innovations and Trends, pp. 997–1021. IGI Global (2017)
60. Zhou, N.R., Li, J.F., Yu, Z.B., Gong, L.H., Farouk, A.: New quantum dialogue protocol based on continuous-variable two-mode squeezed vacuum states. Quantum Inf. Process. **16**(1), 4 (2017)
61. Batle, J., Abutalib, M., Abdalla, S., Farouk, A.: Persistence of quantum correlations in a XY spin-chain environment. Eur. Phys. J. B **89**(11), 247 (2016)
62. Batle, J., Abutalib, M., Abdalla, S., Farouk, A.: Revival of Bell nonlocality across a quantum spin chain. Int. J. Quantum Inf. **14**(07), 1650037 (2016)
63. Batle, J., Ooi, C.R., Farouk, A., Abutalib, M., Abdalla, S.: Do multipartite correlations speed up adiabatic quantum computation or quantum annealing? Quantum Inf. Process. **15**(8), 3081–3099 (2016)
64. Batle, J., Bagdasaryan, A., Farouk, A., Abutalib, M., Abdalla, S.: Quantum correlations in two coupled superconducting charge qubits. Int. J. Mod. Phys. B **30**(19), 1650123 (2016)
65. Batle, J., Ooi, C.R., Abutalib, M., Farouk, A., Abdalla, S.: Quantum information approach to the azurite mineral frustrated quantum magnet. Quantum Inf. Process. **15**(7), 2839–2850 (2016)
66. Batle, J., Ooi, C.R., Farouk, A., Abdalla, S.: Nonlocality in pure and mixed n-qubit X states. Quantum Inf. Process. **15**(4), 1553–1567 (2016)
67. Metwaly, A.F., Rashad, M.Z., Omara, F.A., Megahed, A.A.: Architecture of Multicast Network Based on Quantum Secret Sharing and Measurement (2015)
68. Simmons, G.J.: How to insure that data acquired to verify treaty compliance are trustworthy. Proc. IEEE **76**(5), 621–627 (1988)
69. Simmons, G.J.: Authentication theory/coding theory. In: Advances in Cryptology, pp. 411–431. Springer, Berlin, Heidelberg (1985)
70. Holevo, A.S.: Statistical problems in quantum physics. In: Proceedings of the Second Japan-USSR Symposium on Probability Theory, pp. 104–119. Springer, Berlin, Heidelberg (1973)

Quantum Cryptography, Quantum Communication, and Quantum Computing in a Noisy Environment

Koji Nagata, Tadao Nakamura and Ahmed Farouk

Abstract First, we study several information theories based on quantum computing in a desirable noiseless situation. (1) We present quantum key distribution based on Deutsch's algorithm using an entangled state. (2) We discuss the fact that the Bernstein-Vazirani algorithm can be used for quantum communication including an error correction. Finally, we discuss the main results. We study the Bernstein-Vazirani algorithm in a noisy environment. The original algorithm determines a noiseless function. Here we consider the case that the function has an environmental noise. We introduce a noise term into the function $f(x)$. So we have another noisy function $g(x)$. The relation between them is $g(x) = f(x) \pm O(\epsilon)$. Here $O(\epsilon) \ll 1$ is the noise term. The goal is to determine the noisy function $g(x)$ with a success probability. The algorithm overcomes classical counterpart by a factor of N in a noisy environment.

PACS numbers 03.67.Lx (Quantum computation architectures and implementations) · 03.67.Ac (Quantum algorithms, protocols, and simulations) · 03.67.Dd (Quantum cryptography) · 03.67.Hk (Quantum communication)

1 Introduction

Quantum mechanics (cf. [1–6]) gives approximate but frequently remarkably accurate numerical predictions. Much experimental data approximately fit to the

K. Nagata (✉)
Department of Physics, Korea Advanced Institute of Science and Technology, Daejeon, Korea
e-mail: ko_mi_na@yahoo.co.jp

T. Nakamura
Department of Information and Computer Science, Keio University, 3-14-1 Hiyoshi, Kohoku-ku, Yokohama, Japan

A. Farouk
Computer Sciences Department, Faculty of Computers and Information, Mansoura University, Mansoura, Egypt

A.E. Hassanien et al. (eds.), *Quantum Computing: An Environment for Intelligent Large Scale Real Application*, Studies in Big Data 33,
https://doi.org/10.1007/978-3-319-63639-9_8

quantum predictions for the past some 100 years. We do not doubt the correctness of the quantum theory. The quantum theory also says new science with respect to information theory. The science is called the quantum information theory [6]. Therefore, the quantum theory gives us very useful another theory in order to create new information science and to explain the handling of raw experimental data in our physical world.

As for the foundations of the quantum theory, Leggett-type non-local variables theory [7] is experimentally investigated [8–10]. The experiments report that the quantum theory does not accept Leggett-type non-local variables interpretation. However there are debates for the conclusions of the experiments. See Refs. [11–13].

As for the applications of the quantum theory, implementation of a quantum algorithm to solve Deutsch's problem [14] on a nuclear magnetic resonance quantum computer is reported firstly [15]. Implementation of the Deutsch-Jozsa algorithm on an ion-trap quantum computer is also reported [16]. There are several attempts to use single-photon two-qubit states for quantum computing. Oliveira et al. implement Deutsch's algorithm with polarization and transverse spatial modes of the electromagnetic field as qubits [17]. Single-photon Bell states are prepared and measured [18]. Also the decoherence-free implementation of Deutsch's algorithm is reported by using such single-photon and by using two logical qubits [19]. More recently, a one-way based experimental implementation of Deutsch's algorithm is reported [20]. In 1993, the Bernstein-Vazirani algorithm was reported [21]. It can be considered as an extended Deutsch-Jozsa algorithm.

In 1994, Simon's algorithm was reported [22]. Implementation of a quantum algorithm to solve the Bernstein-Vazirani parity problem without entanglement on an ensemble quantum computer is reported [23]. Fiber-optics implementation of the Deutsch-Jozsa and Bernstein-Vazirani quantum algorithms with three qubits is discussed [24]. Quantum learning robust against noise is studied [25]. A quantum algorithm for approximating the influences of Boolean functions and its applications is recently reported [26]. It is discussed that the Deutsch-Jozsa algorithm can be used for quantum key distribution [27]. Transport implementation of the Bernstein-Vazirani algorithm with ion qubits is more recently reported [28].

Quantum communication is the art of transferring a quantum state from one place to another. Traditionally, the sender is named Alice and the receiver Bob. The basic motivation is that quantum states code quantum information—called qubits in the case of 2-dimensional Hilbert spaces and that quantum information allows one to perform tasks that could only be achieved far less efficiently, if at all, using classical information.

On the other hand, the earliest quantum algorithm, the Deutsch-Jozsa algorithm, is representative to show that quantum computation is faster than classical counterpart with a magnitude that grows exponentially with the number of qubits. In 2015, it is discussed that the Deutsch-Jozsa algorithm can be used for quantum key distribution [27].

There are many researches concerning quantum computing. In a real experiment, we cannot avoid an environmental noise. We address this problem by providing more concrete way rather than [25].

In this paper, first, we study several information theories based on quantum computing in a noiseless environment. We present secure quantum key distribution based on Deutsch's algorithm. The security of the protocol is based on it of Ekert 91 protocol [29].

Next, we study quantum communication including an error correction based on the Bernstein-Vazirani algorithm. The original algorithms determine a bit-strings. Here we discuss the fact that the Bernstein-Vazirani algorithm can be used for quantum communication including an error correction. Let us explain the situation. Alice has a bit-strings $b = (b_1, b_2, \dots, b_N)$. Bob has another bit-strings $c = (c_1, c_2, \dots, c_N)$. The goal is to correct errors of them. We have discussed the fact that the quantum communication overcomes classical counterpart by a factor of N in the protocol.

Finally, we study the Bernstein-Vazirani algorithm in a noisy environment. The original algorithm determines a noiseless function. Here we consider the case that the function has an environmental noise. Let us explain the situation. We introduce a noise term into the function $f(x)$. So we have another noisy function $g(x)$. The relation between them is $g(x) = f(x) \pm O(\epsilon)$. Here $O(\epsilon) \ll 1$ is the noise term. The goal is to determine the noisy function $g(x)$ with a success probability. We discuss the fact that the quantum algorithm overcomes classical counterpart by a factor of N.

This paper is organized as follows:

In Sect. 2, we review Deutsch's algorithm along with Ref. [6].

In Sect. 3, we study Deutsch's algorithm by using another input state. In this case, we cannot perform Deutsch's algorithm.

In Sect. 4, we study Deutsch's algorithm by using the Bell state.

In Sect. 5, we discuss the fact that Deutsch's algorithm can be used for quantum key distribution by using an entangled state.

In Sect. 6, we review the Bernstein-Vazirani algorithm.

In Sect. 7, we study quantum communication based on the Bernstein-Vazirani algorithm.

In Sect. 8, we present an error correction based on the Bernstein-Vazirani algorithm.

In Sect. 9, we present the Bernstein-Vazirani algorithm in a noisy environment.

Section 10 concludes this paper.

2 A Review of Deutsch's Algorithm

In this section, we review Deutsch's algorithm along with Ref. [6].

Quantum parallelism is a fundamental feature of many quantum algorithms. It allows quantum computers to evaluate the values of a function $f(x)$ for many different values of x simultaneously. Suppose

$$f : \{0, 1\} \rightarrow \{0, 1\} \tag{1}$$

is a function with a one-bit domain and range. A convenient way of computing this function on a quantum computer is to consider a two-qubit quantum computer which starts in the state

$$|x, y\rangle. \tag{2}$$

With an appropriate sequence of logic gates it is possible to transform this state into

$$|x, y \oplus f(x)\rangle, \tag{3}$$

where $\oplus$ indicates addition modulo 2. We give the transformation defined by the map

$$|x, y\rangle \rightarrow |x, y \oplus f(x)\rangle \tag{4}$$

a name, U_f.

Deutsch's algorithm combines quantum parallelism with a property of quantum mechanics known as interference. Let us use the Hadamard gate to prepare the first qubit

$$|0\rangle \tag{5}$$

as the superposition

$$(|0\rangle + |1\rangle)/\sqrt{2}, \tag{6}$$

but let us prepare the second qubit as the superposition

$$(|0\rangle - |1\rangle)/\sqrt{2}, \tag{7}$$

using the Hadamard gate applied to the state

$$|1\rangle. \tag{8}$$

The Hadamard gate is as

$$H = \frac{1}{\sqrt{2}}(|0\rangle\langle 1| + |1\rangle\langle 0| + |0\rangle\langle 0| - |1\rangle\langle 1|). \tag{9}$$

Let us follow the states along to see what happens in this circuit. The input state

$$|\psi_0\rangle = |01\rangle \tag{10}$$

is sent through two Hadamard gates to give

$$|\psi_1\rangle = \left[\frac{|0\rangle + |1\rangle}{\sqrt{2}}\right]\left[\frac{|0\rangle - |1\rangle}{\sqrt{2}}\right]. \tag{11}$$

A little thought shows that if we apply U_f to the state

$$|x\rangle(|0\rangle - |1\rangle)/\sqrt{2} \tag{12}$$

then we obtain the state

$$(-1)^{f(x)}|x\rangle(|0\rangle - |1\rangle)/\sqrt{2}. \tag{13}$$

Applying U_f to $|\psi_1\rangle$ therefore leaves us with one of the two possibilities:

$$|\psi_2\rangle = \begin{cases} \pm\left[\frac{|0\rangle + |1\rangle}{\sqrt{2}}\right]\left[\frac{|0\rangle - |1\rangle}{\sqrt{2}}\right] & \text{if } f(0) = f(1) \\ \pm\left[\frac{|0\rangle - |1\rangle}{\sqrt{2}}\right]\left[\frac{|0\rangle - |1\rangle}{\sqrt{2}}\right] & \text{if } f(0) \neq f(1). \end{cases} \tag{14}$$

The final Hadamard gate on the qubits thus gives us

$$|\psi_3\rangle = \begin{cases} \pm|0\rangle|1\rangle & \text{if } f(0) = f(1) \\ \pm|1\rangle|1\rangle & \text{if } f(0) \neq f(1). \end{cases} \tag{15}$$

so by measuring the first qubit we may determine $f(0) \oplus f(1)$. This is very interesting indeed: the quantum circuit gives us the ability to determine a global property of $f(x)$, namely $f(0) \oplus f(1)$, using only one evaluation of $f(x)$! This is faster than is possible with a classical apparatus, which would require at least two evaluations.

3 Failing Deutsch's Algorithm

In this section, we study Deutsch's algorithm by using another input state. In this case, we cannot perform Deutsch's algorithm as shown below.

The input state

$$|\psi_0\rangle = |10\rangle \tag{16}$$

is sent through two Hadamard gates to give

$$|\psi_1\rangle = \left[\frac{|0\rangle - |1\rangle}{\sqrt{2}}\right]\left[\frac{|0\rangle + |1\rangle}{\sqrt{2}}\right]. \tag{17}$$

We apply U_f to the following state

$$\frac{|0\rangle - |1\rangle}{\sqrt{2}}|x\rangle. \tag{18}$$

If $x = 1$

$$\frac{|0\rangle|1\rangle - |1\rangle|1\rangle}{\sqrt{2}} \tag{19}$$

we have

$$\frac{|0\rangle|\overline{f(0)}\rangle - |1\rangle|\overline{f(1)}\rangle}{\sqrt{2}} \tag{20}$$

and if $x = 0$

$$\frac{|0\rangle|0\rangle - |1\rangle|0\rangle}{\sqrt{2}} \tag{21}$$

we have

$$\frac{|0\rangle|f(0)\rangle - |1\rangle|f(1)\rangle}{\sqrt{2}}. \tag{22}$$

Thus,

$$\frac{|0\rangle(|f(0)\rangle + |\overline{f(0)}\rangle) - |1\rangle(|f(1)\rangle + |\overline{f(1)}\rangle)}{\sqrt{2}}. \tag{23}$$

Applying U_f to $|\psi_1\rangle$ therefore leaves us with one of the two possibilities:

$$|\psi_2\rangle = \begin{cases} \pm\left[\frac{|0\rangle - |1\rangle}{\sqrt{2}}\right]\left[\frac{|0\rangle + |1\rangle}{\sqrt{2}}\right] & \text{if } f(0) = f(1) \\ \pm\left[\frac{|0\rangle - |1\rangle}{\sqrt{2}}\right]\left[\frac{|0\rangle + |1\rangle}{\sqrt{2}}\right] & \text{if } f(0) \neq f(1). \end{cases} \tag{24}$$

The final Hadamard gate on the qubits thus gives us

$$|\psi_3\rangle = \begin{cases} \pm|1\rangle|0\rangle & \text{if } f(0) = f(1) \\ \pm|1\rangle|0\rangle & \text{if } f(0) \neq f(1). \end{cases} \tag{25}$$

In this case we fail to perform Deutsch's algorithm.

4 Deutsch's Algorithm Using the Bell State

In this section, we study Deutsch's algorithm by using the Bell state.

The input state

$$|\psi_0\rangle = \frac{|10\rangle + |01\rangle}{\sqrt{2}} \tag{26}$$

is sent through two Hadamard gates to give

$$|\psi_1\rangle = \frac{1}{\sqrt{2}}(\left[\frac{|0\rangle - |1\rangle}{\sqrt{2}}\right]\left[\frac{|0\rangle + |1\rangle}{\sqrt{2}}\right] + \left[\frac{|0\rangle + |1\rangle}{\sqrt{2}}\right]\left[\frac{|0\rangle - |1\rangle}{\sqrt{2}}\right]). \tag{27}$$

Applying U_f to $|\psi_1\rangle$ therefore leaves us with one of the two possibilities:

$$|\psi_2\rangle = \pm\frac{1}{\sqrt{2}}(\left[\frac{|0\rangle - |1\rangle}{\sqrt{2}}\right]\left[\frac{|0\rangle + |1\rangle}{\sqrt{2}}\right] \pm \left[\frac{|0\rangle + |1\rangle}{\sqrt{2}}\right]\left[\frac{|0\rangle - |1\rangle}{\sqrt{2}}\right]) \tag{28}$$

if $f(0) = f(1)$, or

$$|\psi_2\rangle = \pm\frac{1}{\sqrt{2}}(\left[\frac{|0\rangle - |1\rangle}{\sqrt{2}}\right]\left[\frac{|0\rangle + |1\rangle}{\sqrt{2}}\right] \pm \left[\frac{|0\rangle - |1\rangle}{\sqrt{2}}\right]\left[\frac{|0\rangle - |1\rangle}{\sqrt{2}}\right]). \tag{29}$$

if $f(0) \neq f(1)$. The final Hadamard gate on the qubits thus gives us

$$|\psi_3\rangle = \begin{cases} \pm\dfrac{|1\rangle|0\rangle \pm |0\rangle|1\rangle}{\sqrt{2}} & \text{if } f(0) = f(1) \text{ entanglement} \\ \\ \pm\dfrac{|1\rangle|0\rangle \pm |1\rangle|1\rangle}{\sqrt{2}} & \text{if } f(0) \neq f(1) \text{ separable.} \end{cases} \tag{30}$$

So by measuring the qubits (by means of the Bell measurement) we may determine $f(0) \oplus f(1)$. The Bell measurement is explained as follows: Alice and Bob prepare the Bell basis

$$\begin{aligned} |\Psi_+\rangle &= \frac{|1\rangle|0\rangle + |0\rangle|1\rangle}{\sqrt{2}} \\ |\Psi_-\rangle &= \frac{|1\rangle|0\rangle - |0\rangle|1\rangle}{\sqrt{2}} \\ |\Phi_+\rangle &= \frac{|1\rangle|1\rangle + |0\rangle|0\rangle}{\sqrt{2}} \\ |\Phi_-\rangle &= \frac{|1\rangle|1\rangle - |0\rangle|0\rangle}{\sqrt{2}} \end{aligned} \tag{31}$$

If the state $|\psi_3\rangle$ is an entangled state, we have

$$\begin{aligned} &|\langle\psi_3|\Psi_+\rangle|^2 = 1 \text{ or } |\langle\psi_3|\Psi_-\rangle|^2 = 1 \text{ or} \\ &|\langle\psi_3|\Phi_+\rangle|^2 = 1 \text{ or } |\langle\psi_3|\Phi_-\rangle|^2 = 1. \end{aligned} \tag{32}$$

Therefore the measurement outcome should be 1 if the function is constant. If the state $|\psi_3\rangle$ is a separable state, we have

$$\begin{aligned} &|\langle\psi_3|\Psi_+\rangle|^2 = 1/2 \text{ or } |\langle\psi_3|\Psi_-\rangle|^2 = 1/2 \text{ or} \\ &|\langle\psi_3|\Phi_+\rangle|^2 = 1/2 \text{ or } |\langle\psi_3|\Phi_-\rangle|^2 = 1/2. \end{aligned} \tag{33}$$

Therefore the measurement outcome should be not 1 if the function is balanced.

5 Quantum Key Distribution Based on Deutsch's Algorithm

We discuss the fact that Deutsch's algorithm can be used for quantum key distribution by using an entangled state.

Alice and Bob have promised to use a function f which is of one of the two kinds; either the value of f is constant or balanced. To Eve, it is secret. Alice's and Bob's goal is to determine with certainty whether they have chosen a constant or a balanced function without information of the function to Eve. If the function is constant the output qubits are entangled, otherwise separable. Alice and Bob perform the Bell measurement. Alice and Bob share one secret bit if they determine the function f by getting a suitable measurement outcome. The existence of Eve destroys entanglement. The security of our protocol is based on it of Ekert 91 protocol [29].

- First Alice prepares the entangled qubits, applies the Hadamard transformation to the state, and sends the output state described in the Bell state to Bob.
- Next, Bob randomly picks a function "f" that is either balanced or constant and Bob applies U_f. He then sends the one qubit to Alice.
- Finally, Alice and Bob perform the Bell measurement. She learns whether f was balanced or constant. If the final qubits are entangled, then the function is constant. If the final qubits are not entangled, then the function is balanced—Alice and Bob now share a secret bit of information (the "type" of $f(x)$).
- The result of the Bell measurement is 1 if the function is constant.
- Alice and Bob compare a subset of all the results of the Bell measurements when the function is constant; all of them should be 1.
- The existence of Eve must destroy entanglement (Ekert 91).
- Eve is detected in the following case; The result of the Bell measurement is not 1 and the function is constant.

In conclusion, we have shown that Deutsch's algorithm can be used for secure quantum key distribution. The security is based on it of Ekert 91 protocol.

6 A Review of the Bernstein-Vazirani Algorithm

In this section, we review the Bernstein-Vazirani algorithm. Suppose

$$f : \{0, 1\}^N \to \{0, 1\} \tag{34}$$

is a function with a N-bit domain and a 1-bit range. We assume the following case

$$\begin{aligned} f(x) &= a \cdot x = \sum_{i=1}^{N} a_i x_i (\text{mod} 2) \\ &= a_1 x_1 \oplus a_2 x_2 \oplus a_3 x_3 \oplus \cdots \oplus a_N x_N, \\ & a \in \{0, 1\}^N \end{aligned} \tag{35}$$

The goal is to determine $f(x)$. Let us follow the quantum states through the Bernstein-Vazirani algorithm. The input state is

$$|\psi_0\rangle = |0\rangle^{\otimes N}|1\rangle. \tag{36}$$

After the Hadamard transformation on the state we have

$$|\psi_1\rangle = \sum_{x \in \{0,1\}^N} \frac{|x\rangle}{\sqrt{2^N}} \left[\frac{|0\rangle - |1\rangle}{\sqrt{2}} \right]. \tag{37}$$

Next, the function f is evaluated (by Bob) using

$$U_f : |x, y\rangle \rightarrow |x, y \oplus f(x)\rangle, \tag{38}$$

giving

$$|\psi_2\rangle = \pm \sum_x \frac{(-1)^{f(x)}|x\rangle}{\sqrt{2^N}} \left[\frac{|0\rangle - |1\rangle}{\sqrt{2}} \right]. \tag{39}$$

Here

$$y \oplus f(x) \tag{40}$$

is the bitwise XOR (exclusive OR) of y and $f(x)$. To determine the result of the Hadamard transformation it helps to first calculate the effect of the Hadamard transformation on a state

$$|x\rangle. \tag{41}$$

By checking the cases $x = 0$ and $x = 1$ separately we see that for a single qubit

$$H|x\rangle = \sum_z (-1)^{xz}|z\rangle/\sqrt{2}. \tag{42}$$

Thus

$$H^{\otimes N}|x_1, \ldots, x_N\rangle = \frac{\sum_{z_1,\ldots,z_N}(-1)^{x_1 z_1+\cdots+x_N z_N}|z_1, \ldots, z_N\rangle}{\sqrt{2^N}}. \tag{43}$$

This can be summarized more succinctly in the very useful equation

$$H^{\otimes N}|x\rangle = \frac{\sum_z(-1)^{x\cdot z}|z\rangle}{\sqrt{2^N}}, \tag{44}$$

where

$$x \cdot z \tag{45}$$

is the bitwise inner product of x and z, modulo 2. Using this equation and (39) we can now evaluate $|\psi_3\rangle$,

$$|\psi_3\rangle = \pm\sum_z\sum_x \frac{(-1)^{x\cdot z+f(x)}|z\rangle}{2^N}\left[\frac{|0\rangle - |1\rangle}{\sqrt{2}}\right]. \tag{46}$$

Thus,

$$|\psi_3\rangle = \pm\sum_z\sum_x \frac{(-1)^{x\cdot z+a\cdot x}|z\rangle}{2^N}\left[\frac{|0\rangle - |1\rangle}{\sqrt{2}}\right]. \tag{47}$$

We notice

$$\sum_x(-1)^{x\cdot z+a\cdot x} = 2^N\delta_{a,z}. \tag{48}$$

Thus,

$$\begin{aligned}|\psi_3\rangle &= \pm\sum_z\sum_x \frac{(-1)^{x\cdot z+a\cdot x}|z\rangle}{2^N}\left[\frac{|0\rangle - |1\rangle}{\sqrt{2}}\right] \\ &= \pm\sum_z \frac{2^N\delta_{a,z}|z\rangle}{2^N}\left[\frac{|0\rangle - |1\rangle}{\sqrt{2}}\right] \\ &= \pm|a\rangle\left[\frac{|0\rangle - |1\rangle}{\sqrt{2}}\right] \\ &= \pm|a_1a_2a_3\cdots a_N\rangle\left[\frac{|0\rangle - |1\rangle}{\sqrt{2}}\right].\end{aligned} \tag{49}$$

Alice now observes

$$|a_1 a_2 a_3 \cdots a_N\rangle. \tag{50}$$

Summarizing, if Alice measures $|a_1 a_2 a_3 \cdots a_N\rangle$ the function is

$$\begin{aligned} & f(x_1, x_2, ..., x_N) \\ & = a_1 x_1 \oplus a_2 x_2 \oplus a_3 x_3 \oplus \cdots \oplus a_N x_N. \end{aligned} \tag{51}$$

7 Quantum Communication Based on the Bernstein-Vazirani Algorithm

We study quantum communication based on the Bernstein-Vazirani algorithm.

Alice and Bob have promised to select a function $f(x_1, x_2, ..., x_N) = a_1 x_1 \oplus a_2 x_2 \oplus a_3 x_3 \oplus \cdots \oplus a_N x_N$. Alice does not know $a_1, a_2, ..., a_N$. Bob knows $a_1, a_2, ..., a_N$. Alice's goal is to determine with certainty what $a_1, a_2, ..., a_N$ Bob has chosen. In the classical theory, Alice has to ask Bob N questions. In the quantum theory, Alice has to ask Bob "one" question! Alice prepares a suitable $N + 1$ partite uncorrelated state, performs the Hadamard transformation to the state, and sends to the output state to Bob. And Bob performs the Bernstein-Vazirani algorithm and inputs the information of a into the finall state. Alice asks him what state is. Alice measures the finall state and she knows the a. If the a is learned by Alice, Alice and Bob share N bits of information, by one communication with each other. The speed to share N bits improves by a factor of N by comparing the classical case. This shows quantum communication overcomes classical communication by a factor of N.

- First Alice prepares the qubits in (37) and sends the $N + 1$ qubits to Bob.
- Next, Bob picks N bits "a"and Bob applies U_f Eq. (38) evolving the $N + 1$ qubits to Eq. (39). He then sends the N qubit to Alice.
- Finally, Alice applies the Hadamard transformation to each of the qubits and measures. She learns $f(x) = a \cdot x = \sum_{i=1}^{N} a_i x_i (\text{mod}2) = a_1 x_1 \oplus a_2 x_2 \oplus a_3 x_3 \oplus \cdots \oplus a_N x_N$. Alice and Bob now share N bits of information (the "type" of $f(x)$).
- In the classical case (without this quantum computing), Alice needs at least N-communication with Bob to share N bits of information.

In conclusion, we have shown quantum communication overcomes classical communication by a factor of N in the Bernstein-Vazirani algorithm case. However there may be an error between Alice's bits-strings and Bob's one. In the next section, we discuss an error correction based on the Bernstein-Vazirani algorithm.

8 An Error Correction Based on the Bernstein-Vazirani Algorithm

In this section, we present an error correction based on the Bernstein-Vazirani algorithm. Suppose

$$f : \{0,1\}^N \rightarrow \{0,1\} \tag{52}$$

is a function with a N-bit domain and a 1-bit range. We introduce two functions $g(x)$ and $h(x)$. The relation with the function $f(x)$ is as follows:

$$f(x) = g(x) \oplus h(x). \tag{53}$$

We assume the following case

$$\begin{aligned}
g(x) &= b \cdot x = \sum_{i=1}^{N} b_i x_i (\text{mod}2) \\
&= b_1x_1 \oplus b_2x_2 \oplus b_3x_3 \oplus \cdots \oplus b_Nx_N, \\
h(x) &= c \cdot x = \sum_{i=1}^{N} c_i x_i (\text{mod}2) \\
&= c_1x_1 \oplus c_2x_2 \oplus c_3x_3 \oplus \cdots \oplus c_Nx_N, \\
f(x) &= \sum_{i=1}^{N} (b_i \oplus c_i) x_i (\text{mod}2) \\
&= (b_1 \oplus c_1)x_1 \oplus (b_2 \oplus c_2)x_2 \oplus (b_3 \oplus c_3)x_3 \oplus \cdots \\
&\oplus (b_N \oplus c_N)x_N, \\
b_j, c_j &= 0, 1, \ x_j = 0, 1.
\end{aligned} \tag{54}$$

Alice has a bit-strings $b = (b_1, b_2, \ldots, b_N)$. Bob has another bit-strings $c = (c_1, c_2, \ldots, c_N)$. We want to correct errors of them.

Let us follow the quantum states through the algorithm. The input state is

$$|\psi_0\rangle = |0\rangle^{\otimes N}|1\rangle. \tag{55}$$

After the Hadamard transformation on the state we have

$$|\psi_1\rangle = \sum_{x \in \{0,1\}^N} \frac{|x\rangle}{\sqrt{2^N}} \left[\frac{|0\rangle - |1\rangle}{\sqrt{2}} \right]. \tag{56}$$

Next, the function f is evaluated using

$$U_f : |x, y\rangle \rightarrow |x, y \oplus f(x)\rangle, \tag{57}$$

giving

$$|\psi_2\rangle = \pm \sum_x \frac{(-1)^{f(x)}|x\rangle}{\sqrt{2^N}} \left[\frac{|0\rangle - |1\rangle}{\sqrt{2}} \right]. \tag{58}$$

After the Hadamard transformation, by using (58) we can now evaluate $|\psi_3\rangle$,

$$|\psi_3\rangle = \pm \sum_z \sum_x \frac{(-1)^{x\cdot z + f(x)}|z\rangle}{2^N} \left[\frac{|0\rangle - |1\rangle}{\sqrt{2}} \right]. \tag{59}$$

We have

$$\begin{aligned} |\psi_3\rangle &= \pm \sum_z \sum_x \frac{(-1)^{x\cdot z + g(x)\oplus h(x)}|z\rangle}{2^N} \left[\frac{|0\rangle - |1\rangle}{\sqrt{2}} \right] \\ &= \pm \sum_z \sum_x \frac{(-1)^{x\cdot z + x\cdot b \oplus x\cdot c}|z\rangle}{2^N} \left[\frac{|0\rangle - |1\rangle}{\sqrt{2}} \right] \\ &= \pm \sum_z \sum_x \frac{(-1)^{x\cdot z + x\cdot (b+c)}|z\rangle}{2^N} \left[\frac{|0\rangle - |1\rangle}{\sqrt{2}} \right], \end{aligned} \tag{60}$$

where

$$b + c = (b_1 \oplus c_1, b_2 \oplus c_2, \ldots, b_N \oplus c_N). \tag{61}$$

We notice

$$\sum_x (-1)^{x\cdot z + x\cdot (b+c)} = 2^N \delta_{(b+c),z}. \tag{62}$$

Thus,

$$\begin{aligned} |\psi_3\rangle &= \pm \sum_z \sum_x \frac{(-1)^{x\cdot z + x\cdot (b+c)}|z\rangle}{2^N} \left[\frac{|0\rangle - |1\rangle}{\sqrt{2}} \right] \\ &= \pm \sum_z \frac{2^N \delta_{(b+c),z}|z\rangle}{2^N} \left[\frac{|0\rangle - |1\rangle}{\sqrt{2}} \right] \\ &= \pm |b + c\rangle \left[\frac{|0\rangle - |1\rangle}{\sqrt{2}} \right] \\ &= \pm |b_1 \oplus c_1, b_2 \oplus c_2, b_3 \oplus c_3, \ldots, b_N \oplus c_N\rangle \\ &\times \left[\frac{|0\rangle - |1\rangle}{\sqrt{2}} \right]. \end{aligned} \tag{63}$$

Alice now observes

$$|b_1 \oplus c_1, b_2 \oplus c_2, b_3 \oplus c_3, \dots, b_N \oplus c_N\rangle. \tag{64}$$

Summarizing, if Alice measures $|100 \cdots 0\rangle$ the relation is

$$b_1 \oplus c_1 = 1, b_2 \oplus c_2 = 0, \dots, b_N \oplus c_N = 0. \tag{65}$$

Thus there is an errors for the first bit:

$$b_1 \neq c_1, b_2 = c_2, \dots, b_N = c_N. \tag{66}$$

Hence Alice detects the error. In general, Alice can know where such errors are.

If Alice measures $|000 \cdots 0\rangle$ the relation is

$$b_1 \oplus c_1 = 0, b_2 \oplus c_2 = 0, \dots, b_N \oplus c_N = 0. \tag{67}$$

Thus Alice and Bob share N-bits of information.

$$b_1 = c_1, b_2 = c_2, \dots, b_N = c_N. \tag{68}$$

We discuss the fact that the quantum error correction overcomes classical counterpart by a factor of N in this case.

9 The Bernstein-Vazirani Algorithm in a Noisy Environment

In this section, we present the Bernstein-Vazirani algorithm in a noisy environment. Suppose

$$f : \{0,1\}^N \to \{0,1\} \tag{69}$$

is a noiseless function with a N-bit domain and a 1-bit range. We introduce a noisy function g by using the function $f(x)$

$$g(x) = f(x) \pm O(\epsilon). \tag{70}$$

Here $O(\epsilon) \ll 1$ is the noise term.

The noise is explained as follows. Suppose two qubits are described by a superposition state and the value of a function ($f(1)$) has an error. Then there must be two error states. (For example, when we treat 100 bits and there are two errors, the error probability is $2/100 = 1/50$).

Let us explain by using a quantum state:

$$|\psi\rangle = \frac{|1\rangle_1 + |0\rangle_1}{\sqrt{2}} \frac{|1\rangle_2 + |0\rangle_2}{\sqrt{2}}$$
$$= \frac{|1\rangle_1|1\rangle_2 + |1\rangle_1|0\rangle_2 + |0\rangle_1|1\rangle_2 + |0\rangle_1|0\rangle_2}{2} \tag{71}$$

is the superposition state. The function f is evaluated using

$$U_f : |x, y\rangle \rightarrow |x, y \oplus f(x)\rangle. \tag{72}$$

Thus,

$$U_f|\psi\rangle = U_f \frac{|1,1\rangle + |1,0\rangle + |0,1\rangle + |0,0\rangle}{2}$$
$$= (1/2)(|1, 1 \oplus f(1)\rangle + |1, 0 \oplus f(1)\rangle$$
$$+|0, 1 \oplus f(0)\rangle + |0, 0 \oplus f(0)\rangle). \tag{73}$$

Therefore, there are two $f(1)$s in the output state. If there is an error for $f(1)$, then the following two states

$$|1, 1 \oplus f(1)\rangle, |1, 0 \oplus f(1)\rangle \tag{74}$$

have an error, simultaneously. Thus the number of errors is even. Here we globally treat such errors in a statistical model.

We assume the following case

$$g(x) = a \cdot x = \sum_{i=1}^{N} a'_i x_i (\mathrm{mod} 2) \pm O(\epsilon)$$
$$= a'_1 x_1 \oplus a'_2 x_2 \oplus a'_3 x_3 \oplus \cdots \oplus a'_N x_N$$
$$\pm \epsilon (x_1 + x_2 \cdots + x_N) = f(x) \pm O(\epsilon),$$
$$a_j = a'_j \pm \epsilon,\ a'_j = 0, 1,\ x_j = 0, 1. \tag{75}$$

We want to determine $a_1, a_2, ..., a_N$ with a success probability simultaneously so that we determine the noisy function $g(x)$ with the success probability. It is the Bernstein-Vazirani algorithm in a noisy environment.

Let us follow the quantum states through the algorithm. The input state is

$$|\psi_0\rangle = |0\rangle^{\otimes N}|1\rangle. \tag{76}$$

After the Hadamard transformation on the state we have

$$|\psi_1\rangle = \sum_{x\in\{0,1\}^N} \frac{|x\rangle}{\sqrt{2^N}} \left[\frac{|0\rangle - |1\rangle}{\sqrt{2}}\right]. \tag{77}$$

Next, the function g is approximately evaluated using

$$U_g : |x, y\rangle \to |x, y \oplus [g(x)]\rangle. \tag{78}$$

On a real line, $[g(x)]$ is the nearest natural number from $g(x)$. Here we see $[g(x)] = 0, 1$ and

$$[g(x)] = f(x). \tag{79}$$

We have

$$|\psi_2\rangle = \pm \sum_x \frac{(-1)^{[g(x)]}|x\rangle}{\sqrt{2^N}} \left[\frac{|0\rangle - |1\rangle}{\sqrt{2}}\right]. \tag{80}$$

After the Hadamard transformation, by using (80) we can now evaluate $|\psi_3\rangle$,

$$|\psi_3\rangle = \pm \sum_z \sum_x \frac{(-1)^{x\cdot z+[g(x)]}|z\rangle}{2^N} \left[\frac{|0\rangle - |1\rangle}{\sqrt{2}}\right]. \tag{81}$$

So we have

$$|\psi_3\rangle = \pm \sum_z \sum_x \frac{(-1)^{x\cdot z+g(x)\pm O(\epsilon)}|z\rangle}{2^N} \left[\frac{|0\rangle - |1\rangle}{\sqrt{2}}\right]. \tag{82}$$

We notice

$$(-1)^{\pm O(\epsilon)}|z\rangle = (e^{\pm i\pi O(\epsilon)})|z\rangle \simeq |z\rangle \tag{83}$$

because $(e^{\pm i\pi O(\epsilon)}) \simeq 1$.

Thus we have

$$|\psi_3\rangle \simeq \pm \sum_z \sum_x \frac{(-1)^{x\cdot z+a\cdot x}|z\rangle}{2^N} \left[\frac{|0\rangle - |1\rangle}{\sqrt{2}}\right]. \tag{84}$$

In what follows, we evaluate $\sum_x (-1)^{x\cdot z+a\cdot x}$. We notice

$$\sum_x (e^{i\pi})^{x_1 z_1 + \cdots + x_N z_N} (e^{i\pi})^{x_1 a_1 + \cdots + x_N a_N}$$
$$= \sum_x (e^{i\pi})^{x_1(z_1+a_1)+\cdots+x_N(z_N+a_N)}$$
$$= \sum_{x_1} (e^{i\pi})^{x_1(z_1+a_1)} \ldots \sum_{x_N} (e^{i\pi})^{x_N(z_N+a_N)}. \tag{85}$$

We have the following:

$$\sum_{x_1} (e^{i\pi})^{x_1(z_1+a_1)} = (1 + (e^{i\pi z_1})(e^{i\pi a_1})). \tag{86}$$

By checking the cases $z_1 = 0$ and $z_1 = 1$ separately we see that

$$\begin{aligned}
& (1 + (e^{i\pi a_1}))|0\rangle_1 + (1 - (e^{i\pi a_1}))|1\rangle_1 \\
= \; & 2e^{i\pi a_1/2} \frac{(e^{-i\pi a_1/2} + (e^{i\pi a_1/2})}{2} |0\rangle_1 \\
& + 2ie^{i\pi a_1/2} \frac{(e^{-i\pi a_1/2} - (e^{i\pi a_1/2})}{2i} |1\rangle_1 \\
& = 2(i)^{a_1} \cos(a_1\pi/2)|0\rangle_1 - i(i)^{a_1} 2\sin(a_1\pi/2)|1\rangle_1.
\end{aligned} \tag{87}$$

Thus we have

$$\begin{aligned}
|\psi_3\rangle & \simeq \pm \sum_z \sum_x \frac{(-1)^{x\cdot z + a\cdot x}|z\rangle}{2^N} \left[\frac{|0\rangle - |1\rangle}{\sqrt{2}}\right] \\
& = \pm \sum_z \frac{(1 + (-1)^{z_1} e^{i\pi a_1})) \cdots (1 + (-1)^{z_N} e^{i\pi a_N}))|z\rangle}{2^N} \\
& \times \left[\frac{|0\rangle - |1\rangle}{\sqrt{2}}\right] \\
& = \pm \prod_{j=1\ldots N} \left[(i)^{a_j} \cos(a_j\pi/2)|0\rangle_j - i(i)^{a_j} \sin(a_j\pi/2)|1\rangle_j\right] \\
& \times \left[\frac{|0\rangle - |1\rangle}{\sqrt{2}}\right].
\end{aligned} \tag{88}$$

We now observe a quantum state $|100\cdots 1\rangle$ with high probability if

$$
\begin{aligned}
(a_1 = 1 \pm \epsilon), |\cos(a_1\pi/2)|^2 &\ll |\sin(a_1\pi/2)|^2, \\
(a_2 = \pm\epsilon), |\cos(a_2\pi/2)|^2 &\gg |\sin(a_2\pi/2)|^2, \\
(a_3 = \pm\epsilon), |\cos(a_3\pi/2)|^2 &\gg |\sin(a_3\pi/2)|^2, ..., \\
(a_N = 1 \pm \epsilon), |\cos(a_N\pi/2)|^2 &\ll |\sin(a_N\pi/2)|^2.
\end{aligned} \tag{89}
$$

Therefore, we present the Bernstein-Vazirani algorithm in a noisy environment.

We introduce a success probability of finding a_1: It is the probability of detecting $|1\rangle_1$ if $a_1 = 1 \pm \epsilon$. On the other hand, an error probability of finding a_1 is as follows: It is the probability of detecting $|0\rangle$ if $a_1 = 1 \pm \epsilon$. In what follows, we evaluate the success probability of the algorithm. It is the probability that we detect the desirable quantum states for all $a_1, a_2, \ldots, a_N$.

The error probability for a_1 is

$$
|\cos(a_1\pi/2)|^2 = E_1. \tag{90}
$$

The error probability for a_2 is

$$
|\sin(a_2\pi/2)|^2 = E_2, \tag{91}
$$

and so on. The success probability for a_1 is

$$
|\sin(a_1\pi/2)|^2 = 1 - E_1. \tag{92}
$$

The success probability for a_2 is

$$
|\cos(a_2\pi/2)|^2 = 1 - E_2, \tag{93}
$$

and so on. The success probability S for the algorithm is

$$
S = (1 - E_1)(1 - E_2) \cdots (1 - E_N). \tag{94}
$$

The algorithm we discussed determines $a_1, a_2, ..., a_N$ simultaneously with the success probability S. So we can know the noisy function $g(x)$ with the success probability S.

We discuss the fact that the quantum algorithm overcomes classical counterpart by a factor of N in the algorithm over an environmental noise.

10 Conclusions

In conclusion, first, we have presented quantum key distribution based on Deutsch's algorithm by using an entangled state. The idea of the security of the protocol has been based on it of Ekert 91 protocol. The existence of eavesdroppers must has destroyed entanglement.

Next, we have studied quantum communication including an error correction. It has been based on the Bernstein-Vazirani algorithm. The original algorithm has determined a bit-strings. Here we have discussed the fact that the Bernstein-Vazirani algorithm can be used for quantum communication including an error correction. Let us explain the situation. Alice has had a bit-strings $b = (b_1, b_2, \ldots, b_N)$. Bob has had another bit-strings $c = (c_1, c_2, \ldots, c_N)$. The goal has been to correct errors of them. We have discussed the fact that the quantum communication overcomes classical counterpart by a factor of N in the Bernstein-Vazirani algorithm.

Finally, we have studied the Bernstein-Vazirani algorithm having an environmental noise. The original algorithm has determined a noiseless function. Here we have considered the case that the function has an environmental noise. Let us explain the situation this. We have introduced a noise term into the original function $f(x)$. So we have had another noisy function $g(x)$. The relation between them has been $g(x) = f(x) \pm O(\epsilon)$. Here $O(\epsilon) \ll 1$ has been the noise term. The goal has been to determine the noisy function $g(x)$ with a success probability. We have discussed the fact that the quantum algorithm overcomes classical counterpart by a factor of N in the algorithm including the noise function case.

References

1. von Neumann, J.: Mathematical Foundations of Quantum Mechanics. Princeton University Press, Princeton, New Jersey (1955)
2. Feynman, R.P., Leighton, R.B., Sands, M.: Lectures on Physics, Volume III, Quantum mechanics. Addison-Wesley Publishing Company (1965)
3. Redhead, M.: Incompleteness, Nonlocality, and Realism. 2nd ed. Clarendon Press, Oxford (1989)
4. Peres, A.: Quantum Theory: Concepts and Methods. Kluwer Academic, Dordrecht, The Netherlands (1993)
5. Sakurai, J.J.: Modern Quantum Mechanics. Revised ed. Addison-Wesley Publishing Company (1995)
6. Nielsen, M.A., Chuang, I.L.: Quantum Computation and Quantum Information. Cambridge University Press (2000)
7. Leggett, A.J.: Found. Phys. **33**, 1469 (2003)
8. Gröblacher, S., Paterek, T., Kaltenbaek, R., Brukner, Č., Żukowski, M., Aspelmeyer, M., Zeilinger, A.: Nature (London) **446**, 871 (2007)
9. Paterek, T., Fedrizzi, A., Gröblacher, S., Jennewein, T., Żukowski, M., Aspelmeyer, M., Zeilinger, A.: Phys. Rev. Lett. **99**, 210406 (2007)
10. Branciard, C., Ling, A., Gisin, N., Kurtsiefer, C., Lamas-Linares, A., Scarani, V.: Phys. Rev. Lett. **99**, 210407 (2007)
11. Suarez, A.: Found. Phys. **38**, 583 (2008)
12. Żukowski, M.: Found. Phys. **38**, 1070 (2008)
13. Suarez, A.: Found. Phys. **39**, 156 (2009)
14. Deutsch, D.: Proc. Roy. Soc. London Ser. A **400**, 97 (1985)
15. Jones, J.A., Mosca, M.: J. Chem. Phys. **109**, 1648 (1998)
16. Gulde, S., Riebe, M., Lancaster, G.P.T., Becher, C., Eschner, J., Häffner, H., Schmidt-Kaler, F., Chuang, I.L., Blatt, R.: Nature (London) **421**, 48 (2003)
17. de Oliveira, A.N., Walborn, S.P., Monken, C.H.: J. Opt. B: Quantum Semiclass. Opt. **7**, 288–292 (2005)

18. Kim, Y.-H.: Phys. Rev. A **67**, 40301(R) (2003)
19. Mohseni, M., Lundeen, J.S., Resch, K.J., Steinberg, A.M.: Phys. Rev. Lett. **91**, 187903 (2003)
20. Tame, M.S., Prevedel, R., Paternostro, M., Böhi, P., Kim, M.S., Zeilinger, A.: Phys. Rev. Lett. **98**, 140501 (2007)
21. Bernstein, E., Vazirani, U.: Proceedings of the Twenty-Fifth Annual ACM Symposium on Theory of Computing (STOC '93), pp. 11–20 (1993). doi:10.1145/167088.167097; SIAM J. Comput. 26–5, pp. 1411–1473 (1997)
22. Simon, D.R.: Foundations of Computer Science. In: Proceedings of the 35th Annual Symposium on: 116-123, retrieved 2011-06-06 (1994)
23. Du, J., Shi, M., Zhou, X., Fan, Y., Ye, B.J., Han, R., Wu, J.: Phys. Rev. A **64**, 42306 (2001)
24. Brainis, E., Lamoureux, L.-P., Cerf, N.J., Emplit, P., Haelterman, M., Massar, S.: Phys. Rev. Lett. **90**, 157902 (2003)
25. Cross, A.W., Smith, G., Smolin, J.A.: Phys. Rev. A **92**, 12327 (2015)
26. Li, H., Yang, L.: Quantum Inf. Process. **14**, 1787 (2015)
27. Nagata, K., Nakamura, T.: Open Access Library Journal **2**, e1798 (2015). doi:10.4236/oalib.1101798
28. Fallek, S.D., Herold, C.D., McMahon, B.J., Maller, K.M., Brown, K.R., Amini, J.M.: New J. Phys. **18**, 083030 (2016)
29. Ekert, A.K.: Phys. Rev. Lett. **67**, 661 (1991)

An Efficient Scheme for Video Delivery in Wireless Networks

Abdulaziz Shehab, Mohamed Elhoseny and Aboul Ella Hassanien

Abstract This chapter presents a theoretical background for the history of wireless networks. It gives a comprehensive overview and performance evaluation for IEEE 802.11, 802.15 and 802.16 standards focusing on different standards and coverage area. Then, this chapter proposes an efficient scheme for P2P VoD system based on a smart recommender taking into account the analysis of the user's behavior. The chapter describes the proposed mobility scheme that describes network entry process and channel scanning process. The proposed models are examined using different video resolutions (low and high). Then, a mobility model is presented to study the influence of different scanning schemes (light and dense) on some performance metrics like throughput, data dropped, and particularly handover latency. Finally, simulation results are documented and analyzed. The simulation results show that the proposed scheme can efficiently improve both server's load and the initial playout latency.

A. Shehab (✉) · M. Elhoseny
Faculty of Computers and Information, Mansoura University, Mansoura, Egypt
e-mail: Abdulaziz_shehab@mans.edu.eg

M. Elhoseny
e-mail: mohamed_elhoseny@mans.edu.eg
URL: http://www.egyptscience.net

A.E. Hassanien
Faculty of Computers and Information, Cairo University, Cairo, Egypt
e-mail: ewees@du.edu.eg; aboitcairo@gmail.com

M. Elhoseny · A.E. Hassanien
Scientific Research Group in Egypt (SRGE), Cairo, Egypt

A.E. Hassanien et al. (eds.), *Quantum Computing: An Environment for Intelligent Large Scale Real Application*, Studies in Big Data 33,
https://doi.org/10.1007/978-3-319-63639-9_9

1 Introduction

1.1 Video Streaming Over Wireless Networks

The demand for streaming video over wireless networks has been gradually increased over the years. Wireless Communication Technology is changing rapidly. Wireless broadband technologies promise to make all kind of information available anywhere, anytime, at a low cost to a large scale of the population.

Typical example of wireless networks is office wireless local area networks (WLANs) where wireless AP serves all wireless devices within a specific radius. Example of IEEE 802.11 standards [1] described as a group of nodes within a particular zone, wirelessly connected to each other with the help of limited battery powered devices, well interfacing and efficient routing protocol that helps them to maintain quality of communication, while they are changing their position rapidly. Therefore, routing in wireless networks plays a significant role for data forwarding, where each MS can continuously access the Internet from any location in any time. MS can change its point of connecting from one station to another while still being reachable.

Wireless networks can be broadly categorized into two categories: infrastructure based wireless networks and infrastructure-less networks (Ad hoc wireless networks). WLAN technologies are infrastructure based networks. They provide free wireless connectivity to the end users, offering an easy and viable access to a network and its services [2–4].

- The infrastructure mode: it uses fixed, network APs over which MSs can communicate. These APs are usually connected to landlines to widen the LAN's capability by linking wireless nodes to other wired nodes. If service areas of APs overlap, mobile nodes handed over between them.
- The infrastructure-less mode: in this case, a set of mobile devices are brought together to form a pilot network, there is no fixed structure to the network, there are no fixed points, and usually every node is able to communicate with every other node in its communication range with no APs. WLAN/WiFi network is based on the IEEE 802.11 standard. WiMAX, which based on IEEE 802.16, is a standard with similar principles. The existing WiFi and WiMAX wireless networks offer flexibility to support real-time applications such as audio and video streaming. WiFi technology shows great success as an inexpensive wireless Internet access. Whereas WiMAX provides large coverage area (approximately 50 km) and high data rates (up to 75 Mbps) using radio links [5, 6].

WiMAX and WiFi are frequently used for wireless Internet access. WiMAX network operators usually provide a WiMAX subscriber unit that connects the user to the metropolitan WiMAX network and provide WiFi within homes or business for connecting local devices, such as laptops, WiFi handsets, and smartphones.

WiFi has two types of components: a wireless client station and an AP as shown at Fig. 1. Wireless client station is any user device such as computer or laptop that has

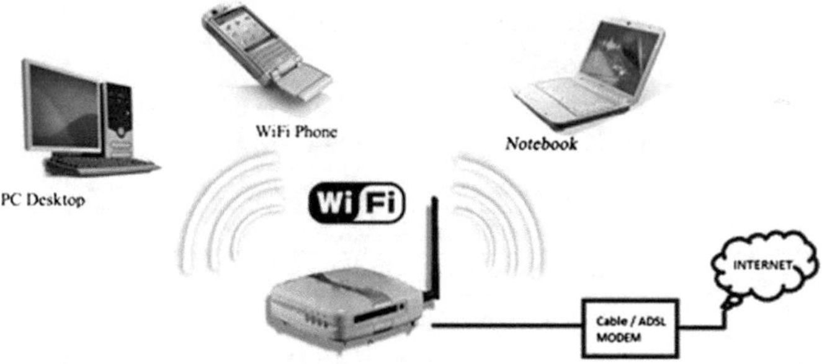

Fig. 1 Different WiFi client/station connections

a wireless network card [7]. AP acts as a bridge between devices and wireless networks. It connects to the cable modem or ADSL modem, provides Internet services for wireless and wired Ethernet clients, and organizes and grants access from multiple MSs to the fixed network. Users can now connect smartphones to the wireless router for WiFi services to access the Internet over mobile phone [8].

1.2 WLAN/WiFi Networks

WLANs based on the IEEE 802.11 (a/b/g, WiFi technology) specification family [9] have gained popularity as being low-cost solutions that are easy to install and provide broadband connectivity, and such networks are being widely deployed in private spaces (e.g. homes and workplaces and as hotspots in public spaces, waiting areas and hotel lobbies).

In wireless network communication, nodes communicate with others using wireless channels. Two important issues are employed in the wireless networks the spectrum frequency ranges and different data rates. For example IEEE 802.11a/g use 54 Mbit/s where IEEE 802.11b used 11 Mbit/s [9, 10]. The signal strength in a wireless medium decreased when the signal travels further beyond a certain distance, the strength reduced to the point where reception is not possible. The topology of the wireless network can be dynamic with time because of the mobility feature where the host or the subnet may move from one place to another [11]. Traditional networks require re-configuration of IP address used by these host or subnet at the new location. An interface enables with Mobile IP allows these hosts or subnet to move without any manual address re-configuration. In an attempt to conclude the different IEEE 802.11 standards, Table 1 provides a summary of IEEE 802.11 standards descriptions [12].

Table 1 Summary of IEEE 802.11 standards description

Standard	Description
IEEE 802.11a	Supports rates of up to 54 Mbps in the 5 GHz ISM band
IEEE 802.11b	Supports rates of up to 11 Mbps in the 2.4 GHz ISM band
IEEE 802.11c	Supports wireless AP bridge operations
IEEE 802.11d	Supports internationalization
IEEE 802.11e	Supports QoS enhancement mechanisms
IEEE 802.11f	Addresses interoperability of Apps from different vendors
IEEE 802.11g	Supports rates of up to 54 Mbps in the 2.4 GHz ISM band
IEEE 802.11h	Supports power control for 5 GHz range requirements
IEEE 802.11i	Deals with security issues
IEEE 802.11n	Supports high data rates up to100 Mbps

1.3 WiMAX Networks

Over the last few years, the demand for high speed mobile broadband access to multimedia services and Internet applications has increased. This demand created new interest to explore new technologies like WiMAX to offer such services at low cost. WiMAX is a standard-based wireless technology that support high throughput and long distance connections. There are two main versions of WiMAX: The first one is IEEE 802.16-2004 WiMAX, which is based on the IEEE 802.16 standard and on ETSI HiperMAN [10]. It utilizes OFDM (Orthogonal Frequency Division Multiplexing) and supports access in Line of Sight (LOS) and Non-line-of-sight (NLOS). The other one is IEEE 802.16e WiMAX version based on the 802.16e amendment, which enables handover and roaming. This release uses a multi-carrier modulation scheme called SOFDMA (Scalable Orthogonal Frequency Division Multiplexing Access) that employs sub-channelization [2, 10]. WiMAX has a group of IEEE 802.16 standards (PHY and MAC layers) suitable for fixed and mobile networks. In 2004, IEEE formalized a specification for fixed wireless networks [13–15]. WiMAX uses 10–66 GHz band with LOS communications using a Single Carrier (SC) air interface. The IEEE 802.16a standard portrayed NLOS communications in the 2–11 GHz band depending on one of the three air interfaces: SC, OFDM, and OFDMA. OFDM and OFDMA enable carriers to increase their data transmission and information capacity. This expanded proficiency is attained by dividing subcarriers nearly together without impedance because subcarriers are orthogonal to one another [16].

With OFDM, number of subcarriers increases as a given channel data transmission increase, subcarriers are allotted as: pilot subcarriers, data subcarriers, null subcarriers, and DC subcarriers. Routine modulation methods with various inward code rates are used to modulate those subcarriers [17]: Binary Phase Shift Keying (BPSK), Quadrature Phase Shift Keying (QPSK), and Quadrature Amplitude Modulation (QAM). WiMAX can accomplish QoS by utilizing transmission capacity requirements [14, 16, 18].

WiMAX networks strengthen connection to the metropolitan WiMAX network, while Wi-Fi networks are utilized for connecting nearby local devices inside homes or organizations. As they are both wireless technology, much of the individuals treat WiMAX as a robust of Wi-Fi. The essential purpose of enthusiasm of Wimax over Wi-Fi is that it spans larger geographic areas and has higher data rates. Because of standardization issues and relatively high price, WiMAX does not acting well performance in market position. While the initial versions of 802.16/a/d focused on fixed applications, the latest versions of 802.16e amendment were formed in 2005 [19], which have included many new features and functionalities that support mobility issues.

1.4 Mobile WiMAX

Video communication can be greatly facilitated by providing adequate QoS. The IEEE 802.11e standard specifies a number of QoS enhancement mechanisms at the MAC level. These IEEE 802.11e improvements can be exploited to increase the WLAN throughput and decrease packet latencies. IEEE 802.16e-2005 Air Interface is the basis of Mobile WiMAX, which makes mobility features and attributes are applicable [20]. Mobile WiMAX supports cell phone applications, for example, it supports video streaming to all subscribers from a vehicle at 70 MPH. Mobile WiMAX becomes precious for promising applications, such as gaming and mobile TV.

IEEE 802.16m-2011 Advanced Air Interface with data rates of 100 Mbps mobile and 1 Gbps fixed enable new 4G mobile radio communication systems [21, 22]. Also known as Mobile WiMAX Release 2 or Wireless MAN-Advanced. It offers significantly higher data rates than and supports different handover mechanisms and QoS features.

To guarantee functionality for BSs that are totally interoperable, there are a number of frameworks that offer scalability and network architecture. Therefore, a lot of adaptabilities and flexibilities will be available. Mobile WiMAX features are summarized as follow [23]:

- High Data Rates: Mobile WiMAX support downlink data rates up to 63 Mbps and uplink data rates up to 28 Mbps.
- Mobility: Mobile WiMAX supports the handover latency less than 50 ms
- Scalability: Mobile WiMAX works in different bandwidths from 1.25 to 20 MHz to conform to various demands.
- Security: Enhanced safety in IEEE 802.16e-2005 is applicable.

There are two critical issues for mobile applications: battery life (power) and handover. Mobile WiMAX supports both idle mode and sleep mode to enable power-efficient MS operation. Mobile WiMAX likewise supports seamless handover to empower the MS to switch starting with one BS then to the next at vehicular rates

Table 2 Summary of various wireless communication technologies

Technology	Standard approved (Year)	Maximum data rate (Mbps)	Modulation scheme	Channel frequency band (GHZ)	Coverage area (m)
IEEE 802.11a (WiFi)	1999	54	OFDM	5	50–100
IEEE 802.11b (WiFi)	1999	11	DSSS, CCK	2.4	50–100
IEEE 802.11g (WiFi)	2003	54	OFDM, DSSS, CCK	2.4	50–100
IEEE 802.11n (WiFi)	2009	60	OFDM, DSSS, CCK	2.5	100–200
IEEE 802.15 (Bluetooth)	2003	2	FHSS	2.45	20–30
IEEE 802.16a (WiMAX)	2003	75	QPSK, 16–64 QAM	2–11	50,000
IEEE 802.16e (Mobile WiMAX)	2005	15	QPSK, 16–64 QAM	2.3–7	30,000
IEEE 802.16m (Release 2)	2011	100	QPSK, 16–64 QAM	2.3–7	40,000

without interrupting the connection [23]. Mobility can be configured through either pre-defined trajectories or using Random Waypoint mobility model. Trajectories and Orbits specify deterministic paths for MS (more details about trajectories will be presented in Chap. 4). Table 2 presents a summary of various wireless communication technologies through focusing on the date approved, data rate, modulation scheme, frequency band, and coverage area [5].

In a summarized manner, The IEEE 802.11 (Wi-Fi) technology achieve incredible success and low-cost wireless Internet access while the IEEE 802.16 (WiMAX) covers large geographic area up to approximately 50 km and high data rates up to 75 Mbps [10]. IEEE 802.16e standard supports wide bandwidth and high-speed mobility. IEEE 802.16e (Mobile WiMAX.) standard is considered a WiMAX solution for mobile applications. It is proposed to enable mobility up to 60 mile/h with the data transmission rate up to 2 Mbps [21].

In fact, the most salient differences between WLAN and WiMAX are: WLAN is meant for short range applications while WiMAX is meant for extended range applications.

- WLAN can deliver much faster speeds compared to WiMAX.
- WiMAX provides a much better method of bandwidth distribution compared to WLAN.
- Both technologies are still possible to overloading.

2 Related Work

There are a number of performance studies focusing on wireless networks. The reader can get some interesting related information and publications in the literature. For example, Etemad [19] presented a high-level survey of a roadmap of the mobile WiMAX from both the radio and network viewpoints. He provided a high-level overview of standardization roadmap and timelines in IEEE 802.16 and the WiMAX forum. Peh et al. [8] presents in achieving seamless handover of VoIP sessions within a hybrid WLAN and WiMAX network. Durantini and Petracca [24] could make 1–1.8 Mbps data rate with WiMAX at a distance between the BS and the MS up to a number of miles. De Bruyne et al. [25] noticed that there are few relationships between the distance and the performance, but in contrary, there are many relationships between the performance and the Carrier to Interference Noise Ratio (CINR). Kim et al. [26] showed that the delay could be vast when using TCP over WiMAX. However, they could setup seamless communications with a moving car in a city. The developed system manages the QoS per user but not per flow.

This issue considered an important factor that affects their performance results. Daan et al. [10] have explained the diverse activities that happened inside the three paramount associations: The 802.16 working group of the IEEE, the WiMAX, and the ITU (International Telecommunication Union). They displayed a complete review about the advancement of WiMAX in terms of standardization and certification. At last, WiMAX trend analysis has additionally presented. Caiyong [23] proposed a velocity adaptive scheme to minimize handover latency and upgrade the system resource usage. This scheme has merits since it balances the delay and channel resources wastage. Ouni et al. [27] examined the energy, latency, and throughput tradeoff in OFDMA wireless channel. They assumed omni-directional antennas on BS and fixed amount of traffic flow in the downlink direction. Ray et al. [28] presented an efficient and simple technique for MS-controlled handover mechanism. They proposed a measure for the distance between MS and any neighboring BSs and referred to it as "Received Signal Strength (RSS)". The MS screens the RSS that serves particular BS periodically. The MS-controlled methodology simulation outcomes demonstrate a recognizable decrease to handover latency accompanied by increment in network scalability.

3 The Proposed Schema for Video Delivery in Wireless Networks

3.1 The Mobility Scheme

IEEE 802.16e standard supports the needs of wide bandwidth and high-speed mobility. The standard is proposed to support the mobility up to 70 mile/h with the data transmission rate up to 2 Mbps as mentioned earlier. Therefore, handover has become one of the elements that affect the performance of IEEE 802.16e system. While a MS is moving, it continuously scans for neighboring BSs and data transferred between MS and BS as long as the connection is established. Generally, in wireless broadband networks, handovers typically consist of three phases: (1) neighbor discovery via scanning, (2) a handover decision, and (3) link switching and association to the new AP [29]. In order to locate the target BS that best fits the mobility path and quality of service (QoS) requirements of the user, a wireless MS needs to scan multiple channels [30–34]. Once the target network (horizontal or vertical) and BS have been determined, the MS perform a handover process and then switches the current link to the new BS and authentication and association procedures are followed. During a handover, a certain level of service disruption characterized by delays and packet losses is inevitable. Specially, scanning and link switching are the main contributors to the disruptions of running services. As channel scanning can be a relatively time-consuming procedure, QoS degradation during a scan is a critical issue. Therefore, to support user QoS demands and seamless service, service disruptions during the scanning procedure should be controlled and scheduled efficiently [22, 35]. Using the previous channel selection history or information regarding the neighboring network topology, different algorithms aim at scanning fewer channels (i.e., selective scan) in an effort to reduce the handover latency. One common approach is that an individual MS determines the channels that may have working APs and scans only those channels. Another technique involves the current Base Station (BS) reporting its neighboring information to the MS. In some wireless Media access control (MAC) protocols such as WLAN and WiMAX, the current BS provides a specified level of information about the same network type neighbor APs to the MSs. With WLAN, the Neighbor Report frame from the current AP includes a list of the neighbor APs, and with WiMAX, a neighbor advertisement message (MOB NBR-ADV) is sent by the current BS at specified intervals to identify the neighbor network systems and to define the characteristics of neighboring BSs. Existing scanning algorithms can be classified into two categories. While the first category aims to reduce the total required channel scanning time by reducing the number of channels to scan, the goal of the second category is to minimize the QoS degradation during the scanning period [29]. The IEEE 802.11specifies two scanning methods. For a passive scan, the MS switches to a new channel and waits for beacon frames from neighboring APs. Therefore, a large amount of time may be required to discover all neighboring APs. With an active scan, the wireless station broadcasts a probe request frame on a selected new channel and waits for a probe response frame from neigh-boring APs operating on that same channel. In IEEE 802.16e WiMAX networks, when the serving BS has obtained downlink/uplink channel descriptor (DCD/UCD) information via the wired backbone net-works, it broadcasts this information periodically using MOBNBR-ADV messages. Scanning is an inevitable procedure for seeking the neighbor BS that is more suitable to be a potential target BS. The purpose of these scans is to decide if it possible to acquire a connection with a more appropriate BS that may have a better wireless signal SNR, a lower traffic, etc. When a MS first communicates with a BS, it is responding to an advertising message that is sent out periodically by the BS informing clients with various conditions of the BS.

The scanning response requires bandwidth usage of the BS along with three basic parameters: scan duration, interleaving interval, and the number of iterations. The scan duration is a period of N frames during which the MS scans neighboring BSs and acquires information about them. The interleaving interval is a period of P frames during which the MS handles regular data transmission between itself and currently connected BS. It repeats pairs of N scan frames and P interleaving interval frames T times. At end, it must decide either still connected to current BS or reconnect to a new one.

3.2 Network Entry Process

The scanning procedure in the IEEE 802.16e protocol is designed as a periodical scanning process. Typically scanning procedures are conducted within the overlapping area of two BSs. The 802.16e extension defines several mechanisms related to BS communication and channel scanning to facilitate neighbor discovery and handovers. Regarding BS interface, the assumption in IEEE 802.16e is that neighboring BSs exchange downlink and uplink channel descriptors (DCD and UCD messages) over the backbone. The information is then embedded in messages sent periodically by the serving BS to the MSs. As shown at Fig. 2, four steps starting with downlink and uplink synchronization, followed by initial ranging, and ending with registration are performed to reach normal operation state. These actions allow MS to acquire channel information prior to any scanning.

3.3 Channel Scanning Process

When the fading SNR reaches the scanning threshold value, the MS begins the scanning process by sending the $MOB - SCN_{R}EQ$ message as illustrated in Fig. 3.

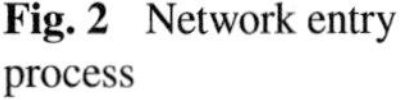

Fig. 2 Network entry process

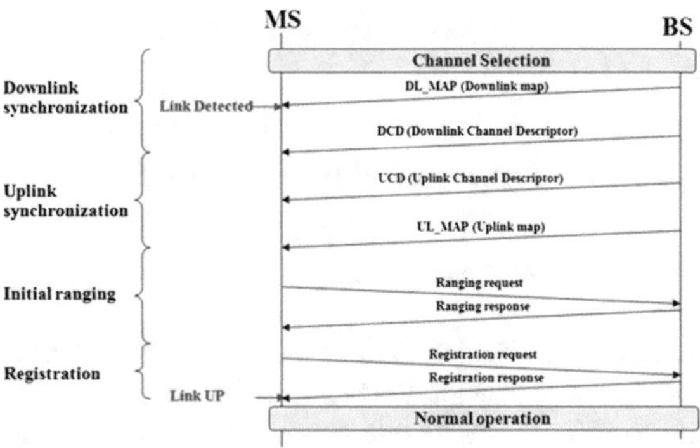

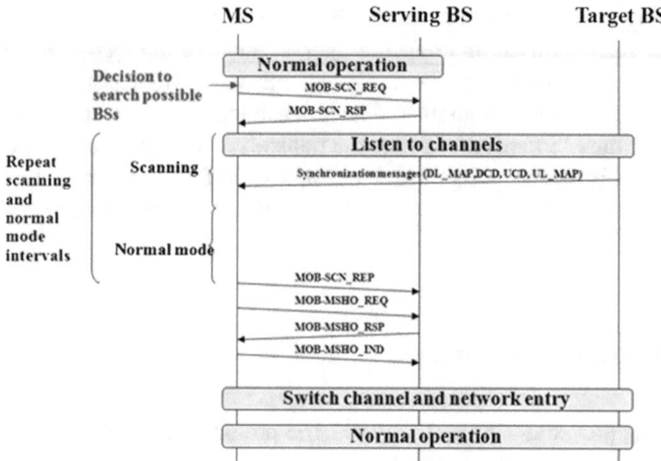

Fig. 3 Channel scanning and handover process

The serving BS reply with the $MOB - SCN_RSP$ message that permits the scanning. The message contains the parameters for scan duration N, interleaving interval P, and the start frame M. After receiving the $MOB - SCN_RSP$ management-message from serving BS, the MS starts the scanning after M received frames (start frame). The MS changes after M frames to the next channel and stays there for N frames period (scanning interval/duration) to detect a BS and to assess its SNR. After a scanning interval, the MS returns to the base channel of the serving BS. These scenarios are meaningful in order to keep the service interruption as short as possible since no payload transmissions are possible during the scanning process. If no preferable BS could be detected, it reinitiates the scanning mode after P frames period (interleaving interval) in a trial to find a new BS. The total number of allowed repetitions of the scanning process is given with the parameter T. At the end of the scanning, the MS reports the scan status to serving BS in the form of a MOB-SCN REP message. The MS sends an $MOB - MSHO_REQ$ message to serving BS indicating the potential target BSs for handover. The serving BS sends an $MOB - MSHO_RSP$ message to the MS to acknowledge the handover request.

3.4 The Proposed Mobility Model

3.4.1 Model Description

The proposed model highlights the mobility effects created by MS. The server was configured to provide VoD telecommunication technology that allows video streaming to MS. The MS node is a mobile IPv4 enabled, with a home agent set to an

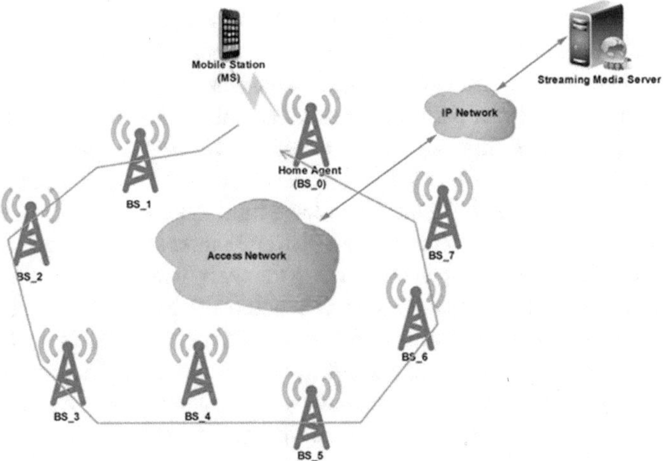

Fig. 4 The mobile WiMAX model structure

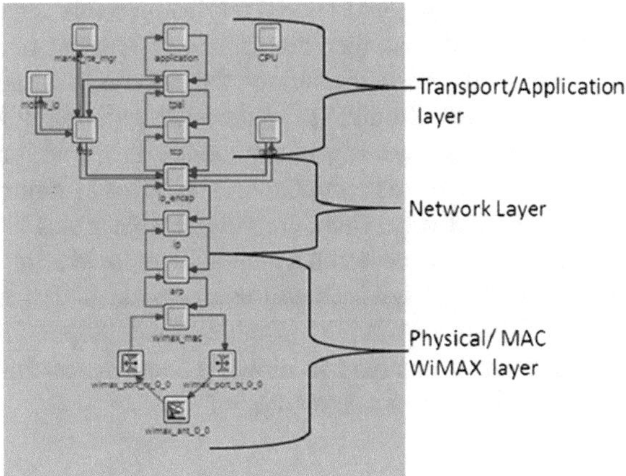

Fig. 5 MS node model architecture

initial BS. The MS node moves away from the home agent and visits seven foreign BS nodes, before returning to its start location as depicted in Fig. 4. Figure 5 shows a representation of MS node model. The Physical and MAC layers are those of the WiMAX 802.16e standard. The node model can be divided into three main layers—Physical/MAC WiMAX layer, Network Layer, and Transport/Application Layer. The Physical/MAC WiMAX layer models the WiMAX physical layer and MAC layer. This layer responsible for data forwarding the classifying higher-layer traffic to service flows, encapsulation/de-capsulation of higher-layer packets in MAC frames, bandwidth request/grant mechanism, and packet delivery. It also handles admission control initiation, activation of service flows, handover initiation, starting the scanning neighbor BSs, ranging (initial and periodic) and selecting the serving BS. The network layer models the IP routing functions, fragmentation, and reassembly.

The Transport layer uses the UDP protocol. RTCP is implemented at the application layer to allow feedback from the receiver to sender. The trajectory in OPNET Modeler has two types: segment-based and vector-based trajectories. Segment-based trajectories draw movement path with a set of segments determined by pre-defined points [1]. Vector-based trajectories set movement via bearing, ground speed, and ascent rate attributes of a MS. A segment-based trajectory is composed of a set of path segments. Each path segment can have individual features. Segment-based trajectory has two types: fixed-interval and variable-interval. For fixed-interval trajectory, a MS takes the same period to navigate every path segment. For variable-interval trajectory, each station has its own specified altitude, wait time, segment traversal time, and orientation. The wait time required for each MS. It makes stations pause at each segment point before navigating the next segment. Creating a segment-based trajectory object with a fixed interval is similar to creating a segment-based trajectory with variable interval. The main difference is that when defining each path segment, the "Segment Information" dialog will show up to ask you to specify parameters for this specified segment, since each segment can have individual parameters in variable-interval trajectory. The OPNET Modeler enables the path specification for a MS using trajectories. OPNET stores this segment-based trajectory data set in .trj files which can then be associated with one or more MSs. Table 3 details the segment-based trajectory derived for this model. Distance and altitude are reported in meters where Ground speed rates are reported in km/h. The MS is then configured to use this trajectory in the model. When MS moves, the mobility will affect SNR as a systematic increment or decrement as it moves closer or further way from the BS, respectively. There are comfortable zones that correspond to areas around each BS. Outside these zones, the SNR of an MS transmission becomes lower than the scanning threshold. Once the smoothed SNR becomes less than the scanning threshold, scanning activity is started in the hope of identifying other BSs as target for handover. When the MS moves towards a new BS, scanning continues until the MS receiver enters a stable zone.

3.4.2 Parameters Setup

This section describes the necessary steps needed to implement WiMAX functionality in a network model. The points listed below are sufficient for running a simulation with default WiMAX parameters.

- **Define Service Classes**: A service class groups the QoS requirements of service flows. The service classes must be defined in the MAC Service Class Definitions attribute of the WiMAX configuration object. Each scenario that includes WiMAX must have a WiMAX configuration object. Service classes are associated with service flows on subscriber nodes.
- **Configure Efficiency Mode**: Depending on the goals of WiMAX simulation study, one can configure the model to use a limited set of features.

Table 3 Details of segment-based trajectory

	X Pos (m)	Y Pos (m)	Distance (m)	Altitude (m)	Traverse time	Ground speed	Wait time	Accum time
1	0	0	n/a	0	n/a	n/a	1 m 40.00 s	1m 40.00 s
2	−281.04065	0	281.040643	0	09.67 s	65.012411	10.00 s	1 m 59.67 s
3	−667.47155	0	386.430901	0	13.30 s	64.994083	10.00 s	2 m 22.97 s
4	−1,065.61	11.710027	398.311789	0	13.71 s	64.98892	10.00 s	2 m 46.68 s
5	−1,475.46	35.130082	410.515759	0	14.13 s	64.989214	10.00 s	3 m 10.81 s
6	−1,592.56	632.341473	608.561088	0	20.96 s	64.94811	10.00 s	3 m 41.77 s
7	−1,592.56	1,018.77	386.430901	0	13.30 s	64.994083	10.00 s	4 m 05.07 s
8	−1,569.14	1,452.04	433.908236	0	14.93 s	65.011727	10.00 s	4 m 30.00 s
9	−1,487.17	1,826.76	383.597179	0	13.19 s	65.055531	10.00 s	4 m 53.19 s
10	−667.47155	1,744.79	823.779039	0	28.36 s	64.976771	10.00 s	5 m 31.55 s
11	−222.49051	2,014.12	520.154497	0	17.86 s	65.148515	10.00 s	5 m 59.41 s
12	304.460709	1,791.63	571.997258	0	19.72 s	64.884454	10.00 s	6 m 29.13 s
13	925.092156	1,955.57	641.906461	0	22.07 s	65.061344	10.00 s	7 m 01.20 s
14	1,323.23	2,014.12	402.414765	0	13.84 s	65.041632	10.00 s	7 m 25.04 s
15	1,838.47	1,850.18	540.725378	0	18.63 s	64.92583	10.00 s	7 m 53.67 s
16	2,306.88	1,791.63	472.061617	0	16.25 s	64.982878	10.00 s	8 m 19.92 s
17	2,342.01	1,393.49	399.698143	0	13.76 s	64.978146	10.00 s	8 m 43.68 s
18	2,342.01	1,007.06	386.430924	0	13.30 s	64.994087	10.00 s	9 m 06.98 s
19	2,166.36	526.951228	511.186209	0	17.57 s	65.082014	10.00 s	9 m 34.55 s
20	2,049.25	11.710027	528.349863	0	18.16 s	65.081772	10.00 s	10 m 02.71 s
21	1,768.21	23.420055	281.287329	0	09.68 s	65.002256	10.00 s	10 m 22.39 s
22	1,323.23	105.390246	452.483802	0	15.58 s	64.966459	10.00 s	10 m 47.97 s
23	889.962074	93.680218	433.427584	0	14.91 s	65.02682	10.00 s	11 m 12.88 s
24	538.661255	210.780491	370.313702	0	12.76 s	64.919135	10.00 s	11 m 35.64 s

- **Associate Subscriber Stations with BSs**: By default, this attribute is set to Distance Based, which means that the subscriber station connects to the closest BS. This association is permanent when mobility is disabled.
- **Configure Mobility**: To model mobility in WiMAX networks, a set of parameters should be taken into consideration to configure the network like WiMAX efficiency mode (the WiMAX efficiency mode must be set to Mobility and Ranging Enabled), handover and scanning (Attributes that control handovers and scanning are configured), ASN (Access Service Network) Gateway IP address, and Mobile IPv4 (facilitate handoff between BS interfaces).

The mobility parameters in the MS node must contain the necessary information to access and query any neighbor BS. To achieve seamless mobility, MobileIPv4 had to be configured. This is a critical detail in mobile communications systems since the mobile subscriber needs to retain the same virtual address across different subnets. Consequently, each BS had to be configured with a specific unique IP subnet on the WiMAX interface. Additionally, the MobileIPv4 specific configuration on each BS was set to use this interface along with the specification of whether the site was the home or foreign agent. Only one site will act as a home agent and all other sites will act as a foreign agent. The primary task of the simulations is to determine the parameters that have the most significant influence on the handover latency while video streaming session. The chosen parameters to the proposed model are presented at Tables 4, 5, 6, and 7. The WiMAX parameters are given in Table 4. The BS parameters, for BS_0, for example, are shown in table Table 5. The parameters applied for MS that represent the mobility attributes are indicated in table Table 6. Video traffic parameters are given in table Table 7. In our simulation model, we try to study the ideal scanning duration to use in a way that allows the MS to have maximum throughput from the current BS it is connected to, while also being able to scan for new and better BSs to connect to. For a stationary node, long scanning duration

Table 4 The WIMAX parameters

Efficiency mode	Mobility and ranging enabled
OFDM PHY profile	Wireless OFDMA 5MHZ
Frame duration (ms)	5
Number of subcarriers	512

Table 5 The BS_0 parameters

Antenna gain (dBi)	15 dBi
PHY profile	Wireless OFDMA 5MHZ
Maximum transmission power (w)	2.0
MAC address	0
Neighbor advertisement interval (Frames)	10

Table 6 The mobility attributes for MS

	Light scanning	Dense scanning
Scanning threshold	27	54
Scan duration	4	20
Interleaving interval	240	140
Scan iterations	10	10

Table 7 Video traffic parameters

Frame interarrival time	10 frames/s
Frame size information (bytes)	128 × 120 pixels
Type of service	Streaming multimedia (4)

would be somewhat useless. For a very mobile node, such as a vehicle, the scanning duration could be crucial in maintaining an active connection to delay and bandwidth sensitive services such as voice-over-IP or video conferencing.

3.4.3 Performance Evaluation

In this section, we study the influence of different scanning schemes (light and dense) on the performance of video streaming through focusing on some performance metrics like throughput, data dropped, and handover delay. **Handover Delay (s)** Handover delay is computed for the instance that MS transmits a message that indicates starting handover process until initial ranging with the new BS is successfully completed. Figure 6 shows the average handover delay for both light and

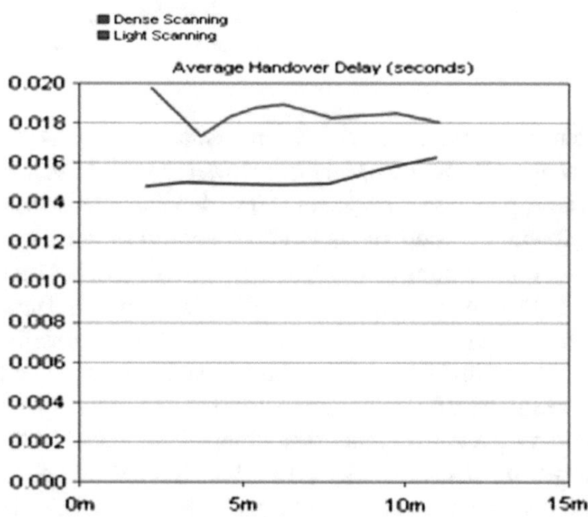

Fig. 6 The average handover delay

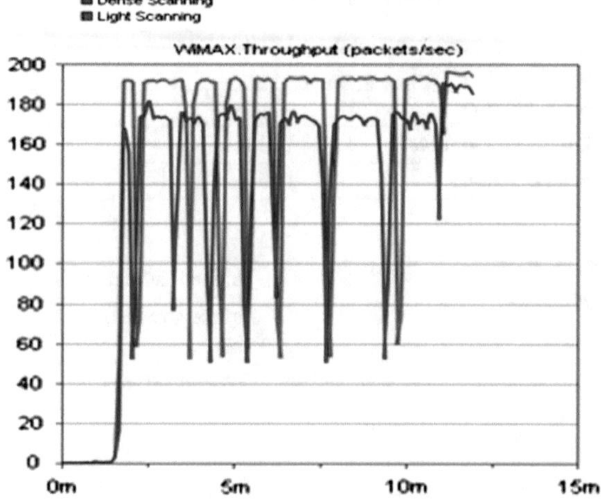

Fig. 7 The WiMAX throughput (packets/s)

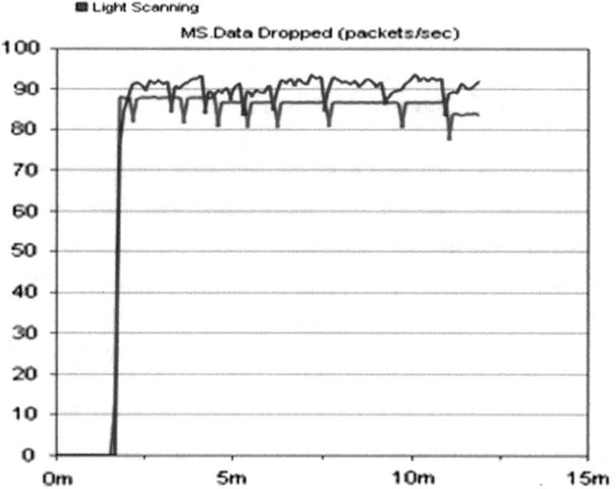

Fig. 8 The data dropped at MS's side (packets/s)

dense scanning. The results demonstrate that the handover latency in dense scanning is approximately 0.015 s that is lower than light scanning. **Throughput (Packets/s)** The application throughput shows that more dense scanning reduces the application throughput (approximately 1% reduction) as illustrated at Fig. 7. This was expected in dense scanning since the WiMAX MAC will be "blocked" more packets forwarded to higher layers while it is scanning neighbor BSs. **Data Dropped During Handover (Packets/s)** Data dropped records the uplink packets dropped (Packets/s) due to physical layer impairments. At the MS side, this statistic represents the bit drops measured at the BS for all packets arriving from particular MS. Figure 8 shows that dense scanning records higher data dropped compared to light scanning. **Video Application End-to-End Delay (s)** End-to-End statistic delay represents the time interval between the transmission of a packet by a video called side and its reception by video

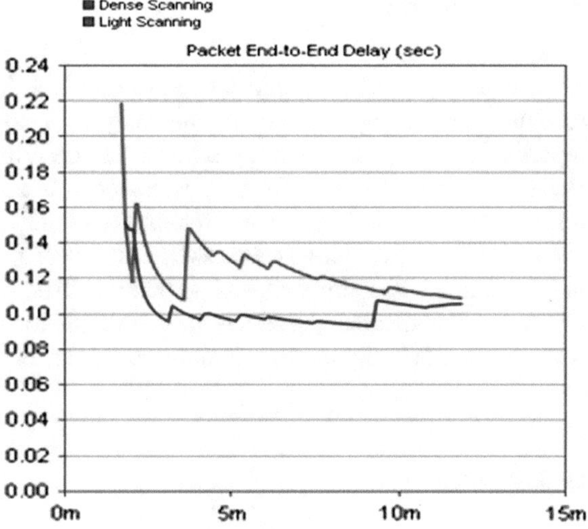

Fig. 9 The packet End-to-End delay (s)

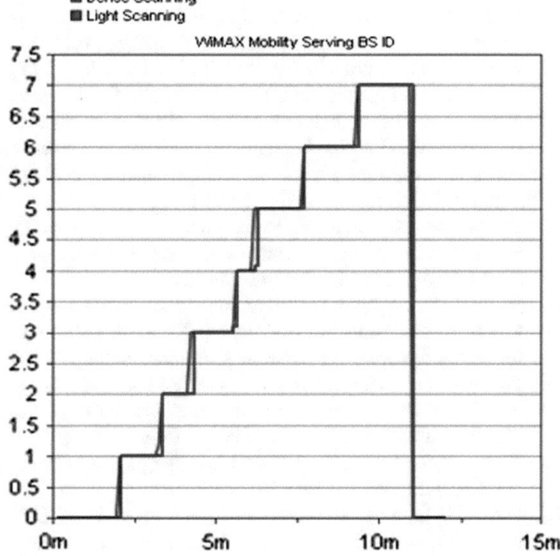

Fig. 10 The service provides BS with ID during the simulation time

calling side. As illustrated at Fig. 9, the dense scanning has lower end-to-end delay. Dense scanning speeds-up handoff process reaching target BS in small attempts. **Mobility service providing BS** For the sake of easy understanding of the service providing BS statistics, each BS has been designated a MAC address corresponding to its name, e.g. MAC 0 for BS_0, MAC 1 for BS 1 ... etc. Figure 10 shows the service providing BS ID during the simulation time. In both scenarios, the serving BS is approximately identical.

4 Conclusion

This chapter proposes an efficient scheme for P2P VoD system based on a smart recommender taking into account the analysis of the user's behavior. The chapter described the proposed mobility scheme that describes network entry process and channel scanning process. The proposed models are examined using different video resolutions (low and high). Then, a mobility model is presented to study the influence of different scanning schemes (light and dense) on some performance metrics like throughput, data dropped, and particularly handover latency. Finally, simulation results are documented and analyzed. The simulation results show that the proposed scheme can efficiently improve both server's load and the initial playout latency.

References

1. Walke, B., Mangold, S., Berlemann, L.: IEEE 802 Wireless Systems: Protocols, Multi-Hop Mesh/Relaying, Performance and Spectrum Coexistence, Wiley, Nov 2006
2. Niyato, D., Hossain, E.: Wireless broadband access: wimax and beyond—integration of wimax and wifi: optimal pricing for bandwidth sharing. IEEE Commun. Mag. **45**, 140–146 (2007)
3. Elhoseny, M., Yuan, X., ElMinir, H., Riad, A.: Extending self-organizing network availability using genetic algorithm. In: International Conference on Computing Communication and Networking Technologies (ICCCNT), IEEE, July 2014
4. Elhoseny, M., Elleithy, K., Elminir, H., Yuan, X., Riad, A.: Dynamic clustering of heterogeneous wireless sensor networks using a genetic algorithm, towards balancing energy exhaustion. Int. J. Sci. Eng. Res. **6**(8) (2015)
5. Gracias, M., Knezevic, V., Esmailpour, A.: Interoperability between wimax and wifi in a testbed environment. In: 2011 24th Canadian Conference on Electrical and Computer Engineering (CCECE), pp. 001144–001148, May 2011
6. Hrudey, W., Trajkovi, L.: Streaming video content over IEEE 802.16/wimax broadband access
7. Elhoseny, M., Yuan, X., El-Minir, H.K., Riad, A.M.: An energy efficient encryption method for secure dynamic WSN. Secur. Commun. Netw. **9**, 2024–2031 (2016)
8. Peh, E., Seah, W.-G., Chew, Y., Ge, Y.: Experimental study of voice over ip services over broadband wireless networks. In: 22nd International Conference on Advanced Information Networking and Applications, 2008. AINA 2008, pp. 834–839, Mar 2008
9. Kuran, M.S., Tugcu, T.: A survey on emerging broadband wireless access technologies. Comput. Netw. **51**, 3013–3046 (2007)
10. Pareit, D., Lannoo, B., Moerman, I., Demeester, P.: The history of wimax: a complete survey of the evolution in certification and standardization for ieee 802.16 and wimax. IEEE Commun. Surv. Tutor. **14**, 1183–1211 (Fourth 2012)
11. Andrews, N., Kondareddy, Y., Agrawal, P.: Prioritized resource sharing in wimax and wifi integrated networks. In: 2010 IEEE Wireless Communications and Networking Conference (WCNC), pp. 1–6, Apr 2010
12. Geier, J.: Designing and Deploying 802.11N Wireless Networks. 1st edn. (2010)
13. Yuan, X., Elhoseny, M., ElMinir, H., Riad, A.: A genetic algorithm-based, dynamic clustering method towards improved wsn longevity. J. Netw. Syst. Manag. **1–26**, 2016 (2016)
14. Retnasothie, F., Ozdemir, M., Yucek, T., Celebi, H., Zhang, J., Muththaiah, R.: Wireless IPTV over WiMAX: challenges and applications. In: Wireless and Microwave Technology Conference, 2006. WAMICON '06. IEEE Annual, pp. 1–5, Dec 2006
15. Elhoseny, M., Elminir, H., Riad, A., Yuan, X.: Recent advances of secure clustering protocols in wireless sensor networks. Int. J. Comput. Netw. Commun. Secur. **2**(11), 400–413 (2014)

16. Chatterjee, M., Sengupta, S., Ganguly, S.: Feedback-based real-time streaming over wimax. IEEE Wirel. Commun. **14**, 64–71 (2007)
17. Elayoubi, S.-E., Fourestie, B.: Performance evaluation of admission control and adaptive modulation in ofdma wimax systems. IEEE/ACM Trans. Netw. **16**, 1200–1211 (2008)
18. Heegard, C., Coffey, J., Gummadi, S., Murphy, P., Provencio, R., Rossin, E., Schrum, S., Shoemake, M.: High performance wireless ethernet. IEEE Commun. Mag. **39**, 64–73 (2001)
19. Etemad, K.: Overview of mobile wimax technology and evolution. IEEE Commun. Mag. **46**, 31–40 (2008)
20. Ieee draft amendment standard for local and metropolitan area networks—part 16: air interface for fixed and mobile broadband wireless access systems—advanced air interface. IEEE P802.16m/D6, pp. 1–932 May, July 2010
21. Zreikat, A.I.: A new wimax/wi-fi interoperability model and its performance evaluation. Wirel. Pers. Commun. **72**, 1229–1257 (2013)
22. Zekri, M., Jouaber, B., Zeghlache, D.: Review: a review on mobility management and vertical handover solutions over heterogeneous wireless networks. Comput. Commun. **35**, 2055–2068 (2012)
23. HAO, C.: A velocity-adaptive handover scheme for mobile wimax. http://www.SciRP.org/journal/ijcns/ (2014)
24. Durantini, A., Petracca, M.: Test of 2.5 GHZ WiMAX performances for business and soho in a multi-service environment. In: IEEE 19th International Symposium on Personal, Indoor and Mobile Radio Communications, 2008. PIMRC 2008, pp. 1–5, Sept 2008
25. De Bruyne, J., Joseph, W., Verloock, L., Martens, L.: Measurements and evaluation of the network performance of a fixed wimax system in a suburban environment. In: IEEE International Symposium on Wireless Communication Systems. 2008. ISWCS '08, pp. 98–102, Oct 2008
26. Kim, D., Cai, H., Na, M., Choi, S.: Performance measurement over mobile WiMAX/IEEE 802.16e network. In: 2008 International Symposium on a World of Wireless, Mobile and Multimedia Networks, 2008. WoWMoM 2008, pp. 1–8, June 2008
27. Ouni, A., Rivano, H., Valois, F.: Wireless mesh networks: energy—capacity tradeoff and physical layer parameters. In: 2011 IEEE 22nd International Symposium on Personal Indoor and Mobile Radio Communications (PIMRC), pp. 1845–1849, Sept 2011
28. Ray, S., Sirisena, H., Deka, D.: Fast and reliable target base station selection scheme for mobile WiMAX handover. In: Telecommunication Networks and Applications Conference (ATNAC), 2012 Australasian, pp. 1–6, Nov 2012
29. Yoo, S.-J., Golmie, N.: Policy-based scanning with qos support for seamless handovers in wireless networks. Wirel. Commun. Mob. Comput. **10**, 405–425 (2010)
30. Elhoseny, M., Yuan, X., Yu, Z., Mao, C., El-Minir, H.K., Riad, A.M.: Balancing energy consumption in heterogeneous wireless sensor networks using genetic algorithm. IEEE Commun. Lett. **19**(12), 2194–2197 (2015)
31. Metawa, N., Elhoseny, M., Kabir Hassan, M., Hassanien, A.: Loan portfolio optimization using genetic algorithm: a case of credit constraints. In: 12th International Computer Engineering Conference (ICENCO), IEEE, pp. 59–64. doi:10.1109/ICENCO.2016.7856446
32. Metawa, N., Hassan, M.K., Elhoseny, M.: Genetic algorithm based model for optimizing bank lending decisions. Expert Syst. Appl. **80**, 75–82 (2017). ISSN 0957-4174. doi:10.1016/j.eswa.2017.03.021
33. Elhoseny, M., Farouk, A., Zhou, N., Wang, M.M., Abdalla, S., Batle, J.: Dynamic multi hop clustering in a wireless sensor network: performance improvement. Wirel. Pers. Commun., 1–21 (2017)
34. Elhoseny, M., Elminir, H., Riad, A., Yuan, X.: A secure data routing schema for WSN using elliptic curve cryptography and homomorphic encryption. J. King Saud Univ.—Comput. Inf. Sci. (2015)
35. Riad, A.M., El-minir, H.K., Elhoseny, M.: Secure routing in wireless sensor network: a state of the art. Int. J. Comput. Appl. **67**(7) (2013)

Part II
Quantum in Physics

QFT + NP = P Quantum Field Theory (QFT): A Possible Way of Solving NP-Complete Problems in Polynomial Time

Vladik Kreinovich, Luc Longpré and Adriana Beltran

Abstract It has been recently theoretically shown that the dependency of some (potential observable) quantities in quantum field theory (QFT) on the parameters of this theory is discontinuous. This discovery leads to the *theoretical* possibility of checking whether the value of a given physical quantity is equal to 0 or different from 0 (here, *theoretical* means that this checking requires very precise measurements and because of that, this conclusion has not yet been verified by a direct experiment). This result from QFT enables us to do what we previously could not: check whether two computable real numbers are equal or not. In this paper, we show that we can use this ability to solve NP-complete ("computationally intractable") problems in polynomial ("reasonable") time. Specifically, we will introduce a new model of computation. This new model is based on solid mainstream physics (namely, on quantum field theory). It is capable of solving NP-complete problems in polynomial time.

1 Pre-Introduction: Feasible and Intractable

Feasible. Some algorithms require lots of time to run. For example, some algorithms require the running time of $\geq 2^n$ computational steps on an input of size n. For reasonable sizes $n \approx 300$, the resulting running time exceeds the lifetime of the Universe and is, therefore, for all practical purposes, non-feasible.

In order to find out which algorithms are feasible and which are not, we must formalize what "feasible" means. This formalization problem has been studied in theoretical computer science; no completely satisfactory definition has yet been proposed.

The best known formalization is: an algorithm $\mathscr{U}$ is *feasible* iff it is *polynomial time*, i.e., iff there exists a polynomial P such that for every input x, the running

V. Kreinovich (✉) · L. Longpré · A. Beltran
Department of Computer Science, University of Texas at El Paso,
500 W. University, El Paso, TX 79968, USA
e-mail: vladik@utep.edu

A.E. Hassanien et al. (eds.), *Quantum Computing: An Environment for Intelligent Large Scale Real Application*, Studies in Big Data 33,
https://doi.org/10.1007/978-3-319-63639-9_10

time $t_{\mathscr{U}}(x)$ of the algorithm $\mathscr{U}$ on the input x is bounded by $P(|x|)$ (here, $|x|$ denotes the length of the input x).

This definition is not perfect, because there are algorithms that are polynomial time but that require billions of years to compute, and there are algorithms that require in a few cases exponential time but that are, in general, very practical. However, this is the best definition we have so far.

Intractable (NP-hard). For many mathematical problems, it is not yet known (2017) whether these problems can be solved in polynomial time or not. However, it is known that some combinatorial problems are as tough as possible, in the sense that:

- if we can solve any of these problems in polynomial time,
- then, crudely speaking, we can solve many practically important combinatorial problems in polynomial time.

The corresponding set of important combinatorial problems is usually denoted by NP, and problems whose fast solution leads to a fast solution of all problems from the class NP are called *NP-hard*.

The majority of computer scientists believe that NP-hard problems are not feasible. For that reason, NP-hard problems are also called *intractable*. For formal definitions and detailed descriptions, see, e.g., [19].

2 Introduction: Equality of Computable Real Numbers Is Algorithmically Undecidable

Computable numbers is what theory of computing started with. In 1936, Alan Turing wrote his classical paper [35], one of the papers that started what we now know as *theory of computing*. This paper was motivated by the necessity to distinguish between *computable* and non-computable *real numbers*. To give a precise definition of what a computable real number is, Turing:

- invented the Turing machine and
- proved several positive and negative results about the Turing machines in general and computable real numbers in particular.

Since then, the definition of a computable real number has slightly changed, but one negative results still stands: it is algorithmically impossible to check whether a real number is equal to 0. Turing defined a computable real number as a number for which an algorithm can compute the digits of its *decimal* expansion. This definition turned out to be not perfect, because some numbers computable in this sense stop being computable when we switch to *binary* expansions.

In view of this difficulty, nowadays, a slightly modified definition of a computable real is used:

a real number x is computable if there exist an algorithm that for every integer k returns a rational number r_k for which

$$|x - r_k| \leq 2^{-k}.$$

There is an area of mathematics called *constructive mathematics* which, crudely speaking, analyzes which problems with computable real numbers are algorithmically decidable and which are not (see, e.g., [1, 4, 10–12, 27]).

Equality of real numbers is not decidable: a seemingly counter-intuitive but in reality, a rather clear result. In every exposition of constructive mathematics, one of the first negative results is a theorem that no algorithm can tell whether a given computable real number is equal to 0 or not.

From the viewpoint of a computer science student who learns about this result in his first (and often last) Theory of Computation course, this result may seem counter-intuitive. Indeed, checking equality of real numbers is something that programmers do on a daily basis, without giving it much thought. To a practical-oriented programmer, the fact that this simple problem is, in theory of computing, proven to be algorithmically undecidable, is an indication that theory of computing may be (somewhat) far away from the actual computing.

At first glance, this result may seem to be counter-intuitive, but on second glance, it is very natural.

- If we have an algorithmic binary sequence $f(n)$, then the real number

  $$x = \sum_{n=1}^{\infty} f(n) 2^{-n}$$

  is computable, because we can easily, given k, compute x's 2^{-k}–rational approximation

  $$r_k = \sum_{n=1}^{k} f(n) 2^{-n}.$$

- If we were able to tell whether $x = 0$ or not, we would thus be able to tell whether the sequence $f(n)$ is all zeros or not.
- Thus, taking $f(n) = 0$ if a given algorithm U does not stop by time n, we would get an algorithm that solves the halting problem for algorithms, which is known to be impossible.

3 Introduction Continued: A Heuristic Use of the Result that Equality of Computable Real Numbers Is Undecidable

It is possible to put a positive twist on this negative result. What this negative result basically says is that:

- if we were able to check whether two given computable numbers are equal or not,
- then we would be able to solve many problems that are now considered algorithmically undecidable.

In this form, this result sounds purely negative, because until recently, there was no known physical way of checking whether two real numbers (i.e., values of two physical quantities) are equal or not. To be able to check, say, equality of x and 0, we must have a physical device that:

- produces 1 ("true") when $x = 0$ and
- produces 0 ("false") when $x \neq 0$.

In other words, we need a physical process for which the dependence of some observable quantity on an input parameter is discontinuous. So far, all the known physical dependencies between observables are continuous. (The dependencies that correspond to phase transitions have "abrupt" changes, but these changes are still not discontinuous: they are continuous but exceptionally fast.)

With this physical impossibility to check whether two real numbers are equal or not, the result that we are discussing seems purely *negative*. However, it is possible to put the following *positive* spin on this result: The computer-based comparison of two real numbers can be viewed as a *heuristic* method of checking whether two real numbers are equal or not. Thus, we can try to design a heuristic method of solving difficult-to-solve problems as follows:

- first, we formulate an "algorithm" for solving these problems that includes checking equality of computable real numbers as an intermediate step;
- second, we transform this "algorithm" into an actual algorithm by checking easy-to-check "computer equality" (i.e., approximate equality) of computed real numbers instead of the algorithmically undecidable actual equality.

This idea has led to reasonably efficient heuristic algorithms in computational chemistry. This idea is actively being used in *computational chemistry*. One of the main problems of chemistry is identification of chemical substances. In non-organic and organic chemistry, there exist experimental techniques that enable us to describe a *graph* structure of the unknown substance, i.e., to describe which atoms it consists of, and which of these atoms are connected by chemical bounds. In order to identify this substance, we must compare it with graphs that describe known substances. In mathematical terms, we need to check whether an (experimentally obtained) graph is isomorphic to one of the graphs that describe known substances.

Unfortunately, graph isomorphism problem is known to be hard to solve. For some substances, different nodes correspond to different types of atoms; in this case, it is relatively easy to check whether a given molecule coincides with this substance, because we can simply identify each atom with a similar atom in the standard substance and then check whether all connections are as in the standard model. For many other substances, however, atoms of the same type occur in different places of the structure in different roles; examples of such substances are organic substances and fullerenes. For these substances, we have to solve the difficult graph isomorphism problem.

One way of solving this problem is based on the idea described above:

- It is known that to every graph, we can assign a polynomial or several polynomials that uniquely determine this graph (i.e., the two tuples of polynomials coincide iff the graphs are isomorphic).
- Thus, to check whether the two graphs are isomorphic, we can compare the coefficients of the corresponding polynomials.

These methods are widely used in structural chemistry; see, e.g., [2, 5, 6, 24, 25, 30, 31].

We can further compress these polynomials into *numbers* (called *indices*) that also give complete information about the graph [32], and compare only these numbers.

The resulting method is, of course, only a *heuristic* method, because sometimes, due to computer inaccuracy, non-isomorphic substances get erroneously identified. However, the numerical experiments show that these errors are extremely rare (and that, therefore, this method works really well) [7]: among all the generated graphs, only 10^{-5} of them got mis-identified.

4 A New Physical Phenomenon

We have already mentioned that until recently, all known physical phenomena were *continuous*, and thus, the above idea could be applied only heuristically.

Recently, a *discontinuous* phenomenon has been discovered in quantum field theory [18, 20–23]. Namely, it was shown that the discontinuous dependence on the parameter λ usually takes place in the theories with the so-called *spontaneous symmetry breaking* (see explanations below). Specifically, the authors of the above papers considered theories of a complex-valued scalar field φ in which a non-linear term

$$\lambda \cdot \varphi^4$$

is added to the standard Lagrangian L_0 of the wave equations (i.e., to

$$\varphi_{,a}\varphi^{,a} + m_0^2 \cdot \varphi^2,$$

where $\varphi_{,a}$ denotes the partial derivative $\dfrac{\partial \varphi}{\partial x_a}$, and the repetition of the index a means summation; for an introduction to classical and quantum field theory, see, e.g., [3, 28]).

This theory can be equivalently described as adding a non-linear term

$$2\lambda \cdot \varphi^3$$

to the right-hand side of the wave equation

$$\Box\varphi + m_0^2 \cdot \varphi = 0;$$

here, $\Box$ indicates the d'Alembertian operator

$$\Box \stackrel{\text{def}}{=} \frac{\partial^2}{\partial x_0^2} - \frac{\partial^2}{\partial x_1^2} - \frac{\partial^2}{\partial x_2^2} - \frac{\partial^2}{\partial x_3^2}.$$

For the resulting non-linear theory, the characteristics of the vacuum state, such as the energy density ρ and pressure p of quantum fluctuations, have a discontinuity at $\lambda = 0$, i.e., these (potentially observable) physical quantities have different values

- for $\lambda = 0$ and
- for $\lambda \neq 0$.

This result leads to a potential possibility of checking whether a given physical quantity is equal to 0 or not. In this paper, we will show that this possibility can help us solve NP-complete ("computationally intractable") problems in polynomial time. Before we describe how, we would like to make this physical result slightly less mysterious by describing in plain words why the dependence of ρ on λ is discontinuous.

Why discontinuous? In classical (pre-quantum) physics, vacuum means absence of matter. In quantum field theory, due to Heisenberg's uncertainty principle, we cannot have a complete absence of matter: there always are so-called *quantum fluctuations*. As a result of these fluctuations, in quantum field theory, vacuum is a complicated quantum state. One way to analyze the quantum vacuum is to neglect quantum effects and to consider the (quasi) classical approximation to the vacuum state.

Crudely speaking, in this approximation, the vacuum is the state with the smallest energy. In the classical vacuum, the equations do not explicitly depend on the coordinates, and therefore, the solution should also be independent on the coordinates. Thus, the gradient $\varphi_{,a}$ is equal to 0, and the vacuum energy can be described as the integral

$$E = \int \rho\, dx_0\, dx_1\, dx_2\, dx_3,$$

where the energy density ρ is equal to

$$\rho = m_0 \cdot \varphi^2 + \lambda \cdot \varphi^4.$$

The minimum is attained at one of the stationary points, i.e., at one of the points where the derivative of the density w.r.t. φ is equal to 0.

- When $\lambda = 0$, the derivative is equal to $2m_0^2 \cdot \varphi$, and therefore, in the vacuum state, $\varphi = 0$. Hence, for $\lambda = 0$, the vacuum energy density $\rho(0)$ is also equal to 0.
- When $\lambda \neq 0$, the derivative is equal to

$$2m_0^2 \cdot \varphi + 4\lambda \cdot \varphi^3.$$

This derivative can be expressed as a product

$$\varphi \cdot (2m_0^2 + 4\lambda \cdot \varphi^2).$$

Thus, this derivative is equal to 0 if:

– either $\varphi = 0$;
– or $2m_0^2 + 4\lambda \cdot \varphi^2 = 0$, in which case $\varphi^2 = -m_0^2/2\lambda$, and $\varphi = \pm\dfrac{m_0}{\sqrt{-2\lambda}}$.

Thus, we have the three stationary points: $\varphi_1 = 0$ and $\varphi_{2,3} = \pm\dfrac{m_0}{\sqrt{-2\lambda}}$.

– For $\varphi_1 = 0$, the corresponding energy density ρ_1 is equal to 0.
– For two other stationary points, $\varphi_{2,3}^2 = \dfrac{m_0^2}{-2\lambda}$, and therefore, the corresponding vacuum energy density $\rho_2 = \rho_3$ is equal to the following expression:

$$\rho_2 = \rho_3 = m_0^2 \cdot \frac{m_0^2}{-2\lambda} + \lambda \cdot \frac{m_0^4}{4\lambda^2} = -\frac{m_0^4}{4\lambda}.$$

When $\lambda > 0$, $\rho_2 = \rho_3 < \rho_1$, and therefore, the true vacuum (i.e., the state with the smallest energy) corresponds to the one of the value $\varphi_{2,3}$. Thus, the energy density $\rho(\lambda)$ of the true vacuum is equal to

$$-\frac{m_0^4}{4\lambda}.$$

When $\lambda > 0$ and $\lambda \to 0$, we have $\rho(\lambda) \to -\infty \neq \rho(0) = 0$. Thus, the dependence of the vacuum energy density ρ on λ is indeed discontinuous.

Comment. In classical physics, there is usually a single vacuum state, and if the theory was invariant w.r.t. some symmetries, then this unique vacuum state is also invariant w.r.t. the same transformations. In our example, for $\lambda > 0$, we have *two* different vacuum states that correspond to the different value of α. The original theory is clearly invariant w.r.t. the transformation $\varphi \to -\varphi$, but neither of the two vacuum state φ_2 and φ_3 is invariant w.r.t. this transformation. Thus, we get a *symmetry breaking*. This symmetry breaking is called *spontaneous*, because it is not caused by any non-symmetric external fields.

Caution: what we described above was an oversimplified, qualitative explanation, *not* the proof. The actual proof requires much more complicated physical computations.

Discontinuity is not so surprizing after all. To some extent, quantum field theory can be viewed as a phenomenological theory, because it is based on the underlying notion of the non-quantized space-time. These discontuity results basically say that if we try to solve QFT equations within the natural assumption of a standard space-time in which all the fields are continuous, we inevitably run into discontinuities, which means that the underlying topology is no longer standard. Thus, to adequately describe quantuk fields, we need to quantize not only the fileds themselves, but also the underlying topology of space-time (see, e.g., [16, 17]).

When re-formulated in these terms, the discountuity stops being a strange mathematical phenomenon caused by a specific feature of the Lagrange function, but becomes a natural general effect of any sufficiently complicated quantum field theory.

5 How to Use the New Physical Phenomenon to Solve NP-Complete Problems in Polynomial Time: Definitions and the Main Result

5.1 Motivations for the New Model of Computation

Our goal is to describe the computational ability of a computer that can use the phenomena of Quantum Field Theory (QFT) in its computations; we will call such enriched computer a *QFT computer*. Before we introduce formal definitions, let us first describe informally what a computer can do if we equip it with the additional possibility stemming from the Quantum Field Theory.

General idea of QFT computer: analog operations in addition to discrete ones + ability to check whether two analog values are equal or not. To apply the physical phenomenon described above, we must have the real numbers represented as the actual values of physical quantities (i.e., we must have *analog* computations in addition to the usual ones). Thus:

- in addition to the standard memory, in which discrete values are stored, we need an additional (analog) memory for storing the actual values;
- in addition to the standard processing unit in which discrete values are processed, we need an additional (analog) processing unit for processing the analog values.

Comment. For simplicity, in this paper, we will add the operations with the analog values to the standard Turing machine. It is reasonably easy, however, to add these operations to RAM or to any other existing model of computation.

Operations with analog values. What operations can we perform with the analog quantities? First, we can *compare* these quantities. This possibility to check whether the two values are equal or not is a completely new phenomenon, and this is what make our computer more powerful that other known computation models.

In addition to this new operation of comparison, there are several physically meaningful operations that we can perform with the actual (analog) values:

- First of all, there are many physically reasonable ways to implement the *sum* of two given quantities. Let us give three examples:

- If we have two values v_1 and v_2 stored as masses, then we can simply add these masses together and get the mass

$$v = v_1 + v_2$$

  that is equal to the sum of the two masses.
- Similarly, if we have two currents v_1 and v_2, we can place the corresponding two wires (where these currents are) as inputs to the third wire, then this third wire will carry the current

$$v = v_1 + v_2$$

  that is equal to the sum of these two currents.
- If we have two values v_1 and v_2 represented as distances marked on straight-line rulers, then we can align these two rulers so that the second instance goes right after the first one, and as a result, get the distance

$$v = v_1 + v_2.$$

• If the two values v_1 and v_2 are represented as distances marked on the straight-line rulers, then we align the rulers in such a way that the second ruler goes in the different direction than the first one, then we get a physical implementation of the *difference*.
• In addition to physical processes that implement addition and difference, there are also known physical processes that implement, e.g.:

  - *multiplication*: e.g., Ohm's law, according to which the voltage V is a product of the current and the resistance:

$$V = I \cdot R;$$

  - *division*;
  - *absolute value* $|x|$;
  - *sine*: e.g., for a perfect resonator, the signal $x(t)$ is proportional to the sine of time t:

$$x(t) = A \cdot \sin(\omega \cdot t);$$

  - *arcsine*, etc.

The above described implementations are currently *approximate*, because so far, there was no way to check whether the given value is exactly correct. The new physical phenomenon enables us to check exactly this, and thus, will, hopefully, lead to a precise implementation of these analog operations.

5.2 Informal Description of the QFT Computer

Memory. In contrast to the standard (one-tape-one-head) Turing machine, this computer will have *two* different *tapes*:

- a *standard* tape (for storing symbols from a given finite alphabet), and
- a new *analog* tape, i.e., a potentially infinite sequence of storage devices in which the actual values are stored.

Initial contents of the memory. In the standard Turing machine, it is usually assumed that all the cells whose contents is not initially specified are empty.

Similarly, we will assume that on the *analog tape*, the initial stored values are all zeros.

Processing unit: states. The *head* (*processing unit*) must be able to store not only the symbol written on the standard tape, but also the value stored in the analog tape. It is convenient to make the processing unit able to store not only a single value, but several values. Thus, we want to equip the head with three *registers* in which the analog values can be stored;

- these registers will be denoted R_0, R_1, and R_2, and
- their contents will be denotes, correspondingly, by r_0, r_1, and r_2.

At any given moment of time, the state of the head is characterized:

- by the state of its discrete part (as in the standard Turing machine) +
- by the contents of the three registers.

Similarly to the tape, we assume that the initial values stored in these registers are zeros.

What influences the processing unit. In the standard Turing machine, the next action of the processing unit is determined by two things: by the state of the head and by the symbol that it is currently observing on the standard tape.

In the QFT computer, we can also make our decision based on whether the contents r_0 of the main register R_0 is equal to 0 or not.

Possible actions. At each moment of time, the standard Turing machine has the following options:

- First, it can change its state (in particular, it can get into the halting state, or into the states that correspond to "yes" and "no" answers).
- Second, it can move the cursor to the left, to the right, or it can *stay*, i.e., leave the cursor at the same place.
- Third, it can also write one of the possible symbols to the cell at which we are cur rently looking.

The new machine has two tapes and has, therefore, more options:

- In addition to moving the cursor of the standard tape, it can also move the cursor of the analog tape.

- We can also perform some operations with the register values. We can perform no operations at all; this case will be denoted by Ø. We can perform the simplest possible *copying* operations:
  - we can copy the contents of the main register R_0 into the corresponding cell on the analog tape (we will denote this operation by $C_{0\to}$);
  - we can copy the contents of the corresponding cell on the analog tape into the main register R_0 (we will denote this operation by $C_{\to 0}$); or
  - we can copy the contents of one of the registers into another one (copying from register R_i to the register R_j will be denoted by $C_{i\to j}$).

  In addition to copying, we can perform some *operations* with the analog values stored in the registers; e.g.:

  - the operation 1 will be performed as

$$r_0 := 1.0;$$

  - the operation + will be performed as

$$r_0 := r_1 + r_2;$$

  - the operation − will be performed as

$$r_0 := r_1 - r_2;$$

  - the operation sin will be performed as

$$r_0 := \sin(r_1),$$

  etc.

Now, we are ready for a formal definition. In this definition, we will extend the formal notion of a Turing machine as defined in [29].

5.3 *Definitions*

Definition 1 By a *Quantum Field Theory computer* (or *QFT computer*, for short), we will mean a quadruple $M = (K, \Sigma, \delta, s)$, where:

K is a finite set whose elements are called *states*;
s is an element of the set K;
Σ is a finite set of symbols (that will be called an *alphabet* of M); we assume that K and Σ are disjoint sets, and that the set Σ contains the special symbols $\sqcup$ and $\triangleright$ that are called the *blank* and the *first* symbol.

δ is a function (called *transition function*) from $K \times \Sigma \times \{T, F\}$ into

$$S \times \Sigma \times M_s \times M_a \times (\{\emptyset\} \cup C \cup O),$$

where:

- $S = K \cup \{h, \text{“yes”}, \text{“no”}\}$ is called the set of possible *resulting states*;
- $M_s = \{\leftarrow_s, \rightarrow_s, -_s\}$ is called *the set of possible cursor motions on the standard tape*;
- $M_a = \{\leftarrow_a, \rightarrow_a, -_a\}$ is called *the set of possible cursor motions on the analog tape*;
- $C = \{C_{0\rightarrow}, C_{\rightarrow 0}, C_{i\rightarrow j}\ (0 \leq i, j \leq 2)\}$ is called *the set of copying operations*;
- $O = \{1, +, -, \cdot, /, \sin, \arcsin\}$ is called the *set of analog operations*.

We assume that the following symbols are not in $K \cup \Sigma$:

- h (called the *halting state*);
- “yes” (called the *accepting* state);
- “not” (called the *rejecting* state);
- the *cursor directions*:
  - $\leftarrow_s, \leftarrow_a$ (called *left*);
  - $\rightarrow_s, \rightarrow_a$ (called *right*); and
  - $-_s, -_a$ (called *stay*), and
- the operation symbols for the analog tape.

Definition 2 By a *configuration* of a QFT computer, we mean a tuple

$$(q, w_s, u_s, w_a, u_a, r_0, r_1, r_2),$$

where:

q is a state (i.e., an element of the set K);

w_s is a finite sequence of symbols from Σ (it describes all the symbols to the left of the cursor, including the symbol scanned by the cursor);

u_s is a finite sequence of symbols from Σ (it describes all the symbols to the right of the cursor);

w_a is a finite sequence of real numbers (it describes all the real numbers on the analog tape that are to the left of the cursor, including the real number currently scanned by the cursor);

u_a is a finite sequence of real numbers (it describes all the real numbers on the analog tape that are to the right of the cursor; we can always add 0's to this sequence);

r_i are real numbers; r_i is called the *current contents of the register i*.

Comment. We will not go into the technicality of defining the initial configuration and a step of the newly defined machine.

Basically, we assume that in the initial configuration, only the standard tape contains the data, i.e., that in the initial configuration we have:

- $q = s$,
- $w_s = \triangleright$,
- $r_0 = r_1 = r_2 = 0$, and
- all zeros are written on the analog tape.

The transition to a new configuration is done according to the description of the operations given above.

The decision is based:

- on the current state ($q \in K$),
- on the symbol ($\sigma \in \Sigma$) currently scanned on the standard tape, and
- on whether the current contents r_0 of the main register R_0 is equal to 0 (this case is denoted by a symbol T), or not (this case is denoted by the symbol F).

Based on this information, the function δ determines what to do next. For each tuple (q, σ, t), where $t \in \{T, F\}$, this function δ returns a tuple

$$(q', \sigma', m_s, m_a, o),$$

where:

q' is a new state of the head;
σ' is a new symbol overwriting the old symbol on the standard tape's cell that is currently scanned by the cursor;
m_s is a motion of the cursor that scans the standard tape;
m_a is a motion of the cursor that scans the analog tape;
o is an operation that is performed with the registers.

(The knowledgeable reader can easily fill in the details.)

Comment. If we are not using the new tape and the new registers at all, then we have a standard Turing machine. Thus:

- every function computable on a standard Turing machine (i.e., every function computable in the usual sense of this word)
- is computable on a QFT computer as well.

5.4 Main Result and Its Proof

Theorem 1 *Every problem from the class NP can be solved on a QFT computer in polynomial time.*

Proof We will show that by using this new computer, we can solve the partition problem (known to be NP-complete) in polynomial time. Since partition problem is NP-complete, from the fact that we can solve it in polynomial time, it follows that we can solve any other problem from the class NP in polynomial time.

The partition problem is listed as Problem SP12 in [19]; its NP-completeness was proven in the pioneer paper by Karp [26]. This problem is as follows (we will slightly change notations):

GIVEN: an integer n, and n positive integers

$$s_1, \ldots, s_n.$$

QUESTION: do there exist values

$$\varepsilon_1, \ldots, \varepsilon_n \in \{-1, 1\}$$

for which

$$\varepsilon_1 \cdot s_1 + \ldots + \varepsilon_n \cdot s_n = 0?$$

We will show how to solve this problem on a QFT computer in polynomial time for $n \geq 3$. Let us first describe how we can solve several auxiliary problems in polynomial time:

1°. According to the definition of QFT computer, we can check *equality* of two real numbers in one step. Using this possibility, we can *check inequality* between real numbers by using the following simple property: for any real numbers a and b,

$$a \leq b \text{ if and only if } b - a = |b - a|.$$

Thus, we can check in three steps whether $a \leq b$ or not:

1. apply the subtraction operation to compute
$$b - a;$$
2. apply the absolute value to compute
$$|b - a|;$$
3. compare the real numbers obtained on Steps 1 and 2.

2°. Let us now show that, given a integer k, a QFT computer can compute the real number

$$P(k) = 2^{2^k}$$

in time bounded by a polynomial of k.

Indeed:

- We can easily generate a real number 1.0.
- Then, we can compute $P(0) = 2.0$ as $1.0 + 1.0$.
- From $P(j)$, we can compute $P(j + 1)$ by a single multiplication operation, as
$$P(j) \cdot P(j).$$

- So, we can compute $P(k)$ by computing sequentially

$$P(1), P(2), \dots, P(k).$$

This computation takes $k + 1$ steps.
3°. Let us now use this auxiliary result to show that, given an integer s, we can compute the real number 2^s in *polynomial time* (i.e., in time bounded by a polynomial of the length of the binary number s).

Indeed, if an integer is given in its binary form

$$s = d_k d_{k-1} \dots d_0,$$

this means that

$$s = d_k \cdot 2^k + d_{k-1} \cdot 2^{k-1} + \dots + d_0 \cdot 2^0,$$

and hence, 2^s is equal to the product of $k + 1$ factors $2^{d_j \cdot 2^j}$, $0 \le j \le k$. When $d_j = 0$, the corresponding factor is equal to 1, so it is sufficient to consider only the factors for which $d_j = 1$. Hence,

$$2^s = \prod_{j:\ d_j=1} 2^{2^j},$$

where the product is taken over all j for which $d_j = 1$.

According to Part 2 of this proof, the time for computing each factor is bounded by a linear function of k.

- There are at most k factors.
- Since the length of s is $k + 1$, we can thus compute 2^s in quadratic time.

4°. Similarly, for a given integer s, we can compute 2^{-s} in polynomial (quadratic) time, as $1/2^s$.

5°. To solve the partition problem, we will compute $\pi = 2 \cdot \arcsin(1.0)$,

$$\Pi = \prod_{i=1}^{n} (2^{2n \cdot s_i} + 2^{-2n \cdot s_i}),$$

and

$$\alpha = \sin(2\pi \cdot \Pi \cdot 2^{-2n}).$$

Then:

- If $\alpha \ge \sin(2\pi \cdot 2^{-2n})$, then we conclude that the desired partition exists, i.e., that there exist ε_i for which

$$\sum \varepsilon_i \cdot s_i = 0;$$

- otherwise, we conclude that the desired partition does not exist.

Due to Parts 3 and 4 of the proof, computing each term in the product Π takes polynomial (quadratic) time. Thus, the entire product can be actually computed in cubic time. The remaining parts of the algorithm are even faster, hence, the entire algorithm takes *cubic time* to run.

To complete the proof, we must show that this algorithm returns a correct answer. Indeed, the product Π of the sums can be represented as the sum of all possible products:

$$\Pi = \sum 2^{2n\cdot(\varepsilon_1\cdot s_1+\ldots+\varepsilon_n\cdot s_n)},$$

where the sum is taken over all 2^n possible combinations of signs

$$(\varepsilon_1, \ldots, \varepsilon_n).$$

This sum, in its turn, can be represented as

$$\Pi = \Sigma_+ + \Sigma_0 + \Sigma_-,$$

where:

- Σ_+ is the sum of all terms for which $\sum \varepsilon_i \cdot s_i > 0$;
- Σ_0 is the sum of all terms for which $\sum \varepsilon_i \cdot s_i = 0$;
- Σ_- is the sum of all terms for which $\sum \varepsilon_i \cdot s_i < 0$.

Hence,

$$\alpha = \sin(A),$$

where

$$A = 2\pi \cdot \Pi \cdot 2^{-2n} = 2\pi \cdot \Sigma_+ \cdot 2^{-2n} + 2\pi \cdot \Sigma_0 \cdot 2^{-2n} + 2\pi \cdot \Sigma_- \cdot 2^{-2n}.$$

Each term in Σ_+ has the form $2^{2n\cdot p}$ for a positive integer p and is, therefore, an integer multiple of 2^{2n}. Hence, the expression

$$2\pi \cdot \Sigma_+ \cdot 2^{-2n}$$

is an integer multiple of 2π and can, therefore, be omitted in the argument of sin. Thus,

$$\alpha = \sin(B),$$

where

$$B = 2\pi \cdot \Sigma_0 \cdot 2^{-2n} + 2\pi \cdot \Sigma_- \cdot 2^{-2n}.$$

- If a particular instance of the partition problem *has no solutions*, then $\Sigma_0 = 0$. Each terms in Σ_- is of the type $2^{-2n\cdot p}$ for a positive integer p and, therefore, it cannot exceed 2^{-2n}. There are, totally, no more than 2^n terms in the entire sum; therefore,

$$\Sigma_- \le 2^n \cdot 2^{-2n} = 2^{-n}.$$

Hence, the argument B of the expression $\alpha = \sin(B)$ is $\le 2\pi \cdot 2^{-3n}$. For $n > 0$, we have $2^{-3n} < 2^{-2n}$, and therefore,

$$B < 2\pi \cdot 2^{-2n}.$$

For all $n \ge 1$, we have

$$2^{-2n} \le 2^{-2} = 1/4,$$

hence,

$$2\pi \cdot 2^{-2n} \le 2\pi \cdot (1/4) = \pi/2.$$

Thence, B belongs to the interval $[0, \pi/2]$ on which the function $\sin(B)$ is strictly increasing. Hence, from

$$B < 2\pi \cdot 2^{-2n},$$

it follows that

$$\alpha = \sin(B) < \sin(2\pi \cdot 2^{-2n}).$$

- If a given instance of the partition problem *has a solution*, then the sum Σ_0 has at least one term. By definition of Σ_0, each of these terms is equal to 1, so, the fact that we have at least one term means that $\Sigma_0 \ge 1$ and hence, that

$$B \ge 2\pi \cdot \Sigma_0 \cdot 2^{-2n} \ge 2\pi \cdot 2^{-2n}.$$

Let us show that B is within the interval $[0, \pi/2]$ on which the sine function is strictly increasing. We have already proven that

$$\Sigma_- \le 2^{-n}.$$

In the total sum, there are 2^n terms, so

$$\Sigma_0 \le 2^n.$$

Hence,

$$B = 2\pi \cdot \Sigma_0 \cdot 2^{-2n} + 2\pi \cdot \Sigma_- \cdot 2^{-2n} \le$$

$$2\pi \cdot 2^n \cdot 2^{-2n} + 2\pi \cdot 2^{-n} \cdot 2^{-2n} = 2\pi \cdot (2^{-n} + 2^{-3n}).$$

From the condition $n \ge 3$, we can conclude that

$$B \le 2\pi \cdot (2^{-n} + 2^{-3n}) \le 2\pi \cdot (2^{-3} + 2^{-9}) < 2\pi \cdot (1/4) = \pi/2.$$

From $B \in [0, \pi/2]$ and

$$B \geq 2\pi \cdot 2^{-2n},$$

we can now conclude that

$$\alpha = \sin(B) > \sin(2\pi \cdot 2^{-2n}).$$

The theorem is proven.

5.5 *This Result Is Not so Surprising Since Our "computer" uses Quantum Effects*

The fact that our newly proposed "computer" can solve the problems faster than the traditional computers is not so surprising if we take into consideration that the proposed computation model uses specifically *quantum* effects (namely, quantum vacuum fluctuations).

The idea that the use of quantum effects can make computers faster and more efficient was first proposed by R. Feynman [15]. Shortly after a quantum computing was formally described by D. Deutsch [13], it was shown that *some* computational problems can indeed be drastically sped up by using such computers [14].

The most convincing example of such a speed-up was given by P. Shor [33], who showed that quantum computers can solve, in polynomial time, the problem of factoring integers into prime factors, the problem known to be very tough.

In view of this success of computers that use *quantum mechanics*, it is no wonder that computers that use the more powerful physical processes of *quantum field theory* can do even more: namely, solve NP-complete problems in polynomial time.

Comment. In the context of quantum computing, our result can be viewed as a partial answer to the question raised in [34]: can other models of quantum computing solve some problems that the existing models cannot yet solve. Our answer is: yes, the new QFT model can do what previous models could not: namely, it can solve all NP-problems in polynomial time.

6 Open Problems

6.1 *Engineering Problem: How Can We Implement These computations?*

The first important problem is of engineering nature: how can we implement these computations?

This is a very tough engineering problem, because, as we have already mentioned, even the discontinuity phenomenon itself, although it follows directly from the equations of mainstream physics, has not yet been confirmed by a direct experiment.

6.2 Theoretical Problem: What Else Can We Compute on a Quantum Field Theory Computer?

In this paper, we have proposed a new computation model, and we have shown that within this computation model, we can solve all problems from the class NP is polynomial time.

Similar to the class QP that was introduced in [9] as the class of all the problems that can be solved on a quantum computer in polynomial time, we can introduce a new class QFTP of all the problem that can be solved on the above-described Quantum Field Theory computer in polynomial time.

In terms of this denotation, our main result says that

$$\text{NP} \subseteq \text{QFTP}.$$

Since complement is easily handled by this computer, the class coNP also belongs to QFTP. So,

$$\text{NP} \cup \text{coNP} \subseteq \text{QFTP}.$$

The main open problem is:

What else is in QFTP?

Specifically:

- Does this new complexity class QFTP contain any higher-level classes in the polynomial hierarchy?
- For example, does this new complexity class contain the class Σ_2P?
- Does this complexity class contain all Turing computable functions?
- Can this new complexity class contain any function that is *not* Turing computable?
- What is the relation between this class QFTP and different structural complexity classes of *quantum computing* (e.g., the class QP and the classes introduced in [8])?

Acknowledgements This work was supported in part:
- by the National Science Foundation grants
  - HRD-0734825 and HRD-1242122 (Cyber-ShARE Center of Excellence) and
  - DUE-0926721, and
- by an award "UTEP and Prudential Actuarial Science Academy and Pipeline Initiative" from Prudential Foundation.

We are thankful:

• to Andrei A. Grib and Vladimir M. Mostepanenko who patiently explained their physical results to us,
• to Michael G. Gelfond and Yuri Gurevich for valuable discussions of constructive real numbers, and
• to James M. Salvador for valuable discussions of the corresponding chemical algorithms.

References

1. Aberth, O.: Precise Numerical Analysis. Wm. C. Brown Publishers, Dubuque, Iowa (1988)
2. Aihara, J., Hosoya, H.: Bull. Chem. Soc. Japan, 61, 2657–ff (1988)
3. Akhiezer, A.I., Berestetskii, V.B.: Quantum Electrodynamics. Pergamon Press, New York (1982)
4. Beeson, M.J.: Foundations of Constructive Mathematics. Springer-Verlag, New York (1985)
5. Beezer, R.A., Farrell, E.J.: The matching polynomial of a regular graph. Discrete Mathe. **137**(1), 7–18 (1995)
6. Beezer, R.A., Farrell, E.J., Riegsecker, J., Smith, B.: Graphs with a minimum number of pairs of independent edges I: Matching polynomials. Bulletin of the Institute of Combinatorics and Its Applications, 1996 (to appear)
7. Beltran, A., Salvador, J.M.: The Ulam index. Abstracts of the Second SC-COSMIC Conference in Computational Sciences, October 25–27, El Paso, TX, Rice University Center for Research on Parallel Computations and University of Texas at El Paso, p. 6 (1996)
8. Bernstein, E., Vazirani, U.: Quantum complexity theory. In: Proceedings of the 25th ACM Symposium on Theory of Computing, pp. 11–20 (1993)
9. Berthiaume, A., Brassard, G.: The quantum challenge to structural complexity theory. In: Proceedings of the 7th IEEE Conference on Structure in Complexity Theory, pp. 132–137 (1992)
10. Bishop, E.: Foundations of Constructive Analysis, McGraw-Hill (1967)
11. Bishop, E., Bridges, D.S.: Constructive Analysis. Springer, New York (1985)
12. Bridges, D.S.: Constructive Functional Analysis. Pitman, London (1979)
13. Deutsch, D.: Quantum theory, the Church-Turing principle, and the universal quantum computer. In: Proceedings of the Royal Society of London, Ser. A, Vol. 400, pp. 96–117 (1985)
14. Deutsch, D., Jozsa, R.: Rapid solution of problem by quantum computation. In: Proceedings of the Royal Society of London, Ser. A, Vol. 439, pp. 553–558 (1992)
15. Feynman, R.: Simulating physics with computers. Int. J. Theor. Phys. **21**, 467–488 (1982)
16. Finkelstein, D., Gibbs, J.M.: Quantum relativity. Int. J. Theor. Phys. **32**, 1801 (1993)
17. Finkelstein, D.: Quantum Relativity. Springer-Verlag, Berlin, Heidelberg (1996)
18. Frolov, V.M., Grib, A.A., Mostepanenko, V.M.: Conformal symmetry breaking and quantization in curved space-time. Phys. Lett. A **65**, 282–284 (1978)
19. Garey, M., Johnson, D.: Computers and Intractability: A Guide to the Theory of NP-Completeness. Freeman, San Francisco (1979)
20. Grib, A.A., Mamayev, S.G., Mostepanenko, V.M.: Vacuum Quantum Effects in Strong Fields. Friedmann Laboratory Publishing, St.Petersburg (1994). (Chapter 11)
21. Grib, A.A., Mostepanenko, V.M.: Spontaneous breaking of gauge symmetry in a homogeneous isotropic universe if the open type. Sov. Phys.-JETP Lett., **25**, 277–279 (1977)
22. Grib, A.A., Mostepanenko, V.M., Frolov, V.M.: Spontaneous breaking of gauge symmetry in a nonstationary isotropic metric. Theor. Math. Phys. **33**, 869–876 (1977)
23. Grib, A.A., Mostepanenko, V.M., Frolov, V.M.: Spontaneous breaking of CP symmetry in a nonstationary isotropic metric. Theor. Math. Phys. **37**(2), 975–983 (1978)
24. Hosoya, H.: Comp. Math. Appls. **12B**, 271–ff (1986)
25. Hosoya, H., Balasubramanian, K.: Computational algorithms for matching polynomials of graphs from the characteristic polynomials of edge-weighted graphs. J. Comput. Chem. **10**(5), 698–710 (1989)

26. Karp, R.M.: Reducibility among combinatorial problems. In: Miller, R.E., Thatcher, J.W. (eds.) Complexity of Computer Computations, pp. 85–103. Plenum Press, New York (1972)
27. Kushner, B.A.: Lectures on Constructive Mathematical Analysis, Translations of Mathematical Monographs, vol. 60. American Mathematical Society, Providence (1984)
28. Landau, L.D., Lifschitz, E.M.: Quantum Mechanics: Non-relativistic Theory. Pergamon Press, Oxford (1965)
29. Papadimitriou, C.H.: Computational Complexity. Addison Wesley, San Diego (1994)
30. Randic, M., Hosoya, H., Polansky, O.E.: On the construction of the matching polynomial for unbranched catacondensed benzenoids. J. Comput. Chemi. **10**(5), 683–697 (1989)
31. Rosenfeld, V.R., Gutman, I.: A novel approach to graph polynomials. Match, **24**, 191–ff (1989)
32. Salvador, J.M.: Topological indices and polynomials: the partial derivatives. In: Abstracts of the 5th International Conference on Mathematical and Computational Chemistry, May 17–21, Kansas City, Missouri, p. 154 (1993)
33. Shor, P.W.: Algorithms for quantum computations: discrete logarithms and factoring. In: Proceedings of the 35th Annual Symposium on Fundamentals of Computer Science (FOCS), pp. 124–134 (1994)
34. Simon, P.: On the power of quantum computation. In: Proceedings of the 35th Annual Symposium on Fundamentals of Computer Science (FOCS), pp. 116–123 (1994)
35. Turing, A.M.: On computable numbers, with an application to the em Entscheidungsproblem. In: Proceedings London Mathematical Society, vol. 42, pp. 230–265 (1936); see also Proceedings London Mathematics Society, vol. 43, pp. 544–546 (1937)

(Hypothetical) Negative Probabilities Can Speed Up Uncertainty Propagation Algorithms

Andrzej Pownuk and Vladik Kreinovich

Abstract One of the main features of quantum physics is that, as basic objects describing uncertainty, instead of (non-negative) probabilities and probability density functions, we have complex-valued probability amplitudes and wave functions. In particular, in quantum computing, negative amplitudes are actively used. In the current quantum theories, the actual probabilities are always non-negative. However, there have been some speculations about the possibility of actually negative probabilities. In this paper, we show that such hypothetical negative probabilities can lead to a drastic speed up of uncertainty propagation algorithms.

1 Introduction

From non-negative to more general description of uncertainty. In the traditional (non-quantum) physics, the main way to describe uncertainty—when we have several alternatives and we do not know which one is true—is by assigning probabilities p_i to different alternatives i.

The physical meaning of each probability p_i is that it represents the frequency with which the i-th alternative appears in similar situations. As a result of this physical meaning, probabilities are always non-negative.

In the continuous case, when the number of alternatives is infinite, each possible alternative has 0 probability. However, we can talk:

- about probabilities of values being in a certain interval and, correspondingly, item about the probability density $\rho(x)$—probability per unit length or per unit volume.

The corresponding probability density function is a limit of the ratio of two non-negative values:

A. Pownuk (✉) · V. Kreinovich
Computational Science Program, University of Texas at El Paso,
500 W. University, El Paso, Texas 79968, USA
e-mail: ampownuk@utep.edu

A.E. Hassanien et al. (eds.), *Quantum Computing: An Environment for Intelligent Large Scale Real Application*, Studies in Big Data 33,
https://doi.org/10.1007/978-3-319-63639-9_11

- probability and
- volume,

and is, thus, also always non-negative.

One of the main features of quantum physics is that in quantum physics, probabilities are no longer the basic objects for describing uncertainty; see, e.g., [4]. To describe a general uncertainty, we now need to describe the complex-valued *probability amplitudes* ψ_i corresponding to different alternatives i. In the continuous case:

- instead of a probability density function $\rho(x)$,
- we have a complex-valued *wave function* $\psi(x)$.

Non-positive and non-zero values of the probability amplitude and of the wave function are important: e.g., negative values of the amplitudes are actively used in many quantum computing algorithms; see, e.g., [9].

Can there be negative probabilities? In the current quantum theories, the actual probabilities are always non-negative. For example:

- the probability p_i of observing the i-th alternative is equal to a non-negative number

$$p_i = |\psi_i|^2,$$

and
- the probability density function is equal to a non-negative expression

$$\rho(x) = |\psi(x)|^2.$$

However, there have been some speculations about the possibility of actually negative probabilities, speculations actively explored by Nobel-rank physicists such as Dirac and Feynman; see, e.g., [2] and [3]. Because of the high caliber of these scientists, it makes sense to take these speculations very seriously.

What we do in this paper. In this paper, we show that such hypothetical negative probabilities can lead to a drastic speed up of uncertainty propagation algorithms.

2 Uncertainty Propagation: Reminder and Precise Formulation of the Problem

Need for data processing. In many practical situations, we are interested in the value of a physical quantity y which is difficult or even impossible to measure directly. For example, we may be interested:

- in tomorrow's temperature, or
- in a distance to a faraway star, or
- in the amount of oil in a given oil field.

Since we cannot measure the quantity y directly, a natural idea is:

- to measure easier-to-measure related quantities

$$x_1, \ldots, x_n,$$

  and then
- to use the known relation

$$y = f(x_1, \ldots, x_n)$$

  between these quantities to estimate y as

$$\widetilde{y} = f(\widetilde{x}_1, \ldots, \widetilde{x}_n),$$

  where $\widetilde{x}_i$ denotes the result of measuring the quantity x_i.

For example:

- To predict tomorrow's temperature y:
  - we measure temperature, humidity, and wind velocity at different locations, and
  - we use the known partial differential equations describing atmosphere to estimate y.
- To measure a distance to a faraway star:
  - we measure the direction to this star in two different seasons, when the Earth is on different sides of the Sun, and then
  - we use trigonometry to find y based on the difference between the two measured directions.

In all these cases, the algorithm f transforming our measurement results into the desired estimate $\widetilde{y}$ is an example of *data processing*.

Need for uncertainty propagation. Measurements are never absolutely accurate. The measurement result $\widetilde{x}_i$ is, in general, somewhat different from the actual (unknown) value of the corresponding quantity x_i. As a result, even when the relation

$$y = f(x_1, \ldots, x_n)$$

is exact, the result $\widetilde{y}$ of data processing is, in general, somewhat different from the the actual values $y = f(x_1, \ldots, x_n)$:

$$\widetilde{y} = f(\widetilde{x}_1, \ldots, \widetilde{x}_n) \neq y = f(x_1, \ldots, x_n).$$

It is therefore necessary to estimate

- how accurate is our estimation $\widetilde{y}$, i.e.,

- how big is the estimation error

$$\Delta y \stackrel{\text{def}}{=} \widetilde{y} - y.$$

The value of Δy depends on how accurate were the original measurements, i.e., how large were the corresponding measurement errors

$$\Delta x_i \stackrel{\text{def}}{=} \widetilde{x}_i - x_i.$$

Because of this, estimation of Δy is usually known as the *propagation* of uncertainty with which we know x_i through the data processing algorithm.

Uncertainty propagation: an equivalent formulation. By definition of the measurement error, we have

$$x_i = \widetilde{x}_i - \Delta x_i.$$

Thus, for the desired estimation error Δy, we get the following formula:

$$\Delta y = \widetilde{y} - y = f(\widetilde{x}_1, \ldots, \widetilde{x}_n) - f(\widetilde{x}_1 - \Delta x_1, \ldots, \widetilde{x}_n - \Delta x_n).$$

Our goal is to transform the available information about Δx_i into the information about the desired estimation error Δy.

What do we know about Δx_i: ideal case. Ideally, for each i, we should know:

- which values of Δx_i are possible, and
- how frequently can we expect each of these possible values.

In other words, in the ideal case, for every i, we should know the probability distribution of the corresponding measurement error.

Ideal case: how to estimate Δy? In some situations, we have analytical expressions for estimating Δy.

In other situations, since we know the exact probability distributions corresponding to all i, we can use *Monte-Carlo simulations* to estimate Δy. Namely, several times $\ell = 1, 2, \ldots, L$, we:

- simulate the values $\Delta x_i^{(\ell)}$ according to the known distribution of Δx_i, and
- estimate

$$\Delta y^{(\ell)} = \widetilde{y} - f(\widetilde{x}_1 - \Delta x_1^{(\ell)}, \ldots, \widetilde{x}_n - \Delta x_n^{(\ell)}).$$

Since the values $\Delta x_i^{(\ell)}$ have the exact same distribution as Δx_i, the computed values $\Delta y^{(\ell)}$ are a sample from the same distribution as Δy. Thus, from this sample

$$\Delta y^{(1)}, \ldots, \Delta y^{(L)},$$

we can find all necessary characteristics of the corresponding Δy-probability distribution.

What if we only have partial information about the probability distributions? In practice, we rarely full full information about the probabilities of different values of the measurement errors Δx_i, we only have partial information about these probabilities; see, e.g., [10]. In such situations, it is necessary to transform this partial information into the information about Δy.

What partial information do we have? What type of information can we know about Δx_i? To answer this question, let us take into account that the ultimate goal of all these estimations is to make a decision:

- when we estimate tomorrow's temperature, we make a decision of what to wear, or, in agriculture, a decision on whether to start planting the field;
- when we estimate the amount of oil, we make a decision whether to start drilling right now or to wait until the oil prices will go up since at present, the expected amount of oil is too large enough to justify the drilling expenses.

According to decision theory results (see, e.g., [5, 7, 8, 11]), a rational decision maker always selects an alternative that maximizes the expected value of some objective function $u(x)$—known as *utility*. From this viewpoint, it is desirable to select characteristics of the probability distribution that help us estimate this expected value – and thus, help us estimate the corresponding utility.

For each quantity x_i, depending on the measurement error Δx_i, we have different values of the utility $u(\Delta x_i)$. For example:

- If we overestimate the temperature and start planting the field too early, we may lose some crops and thus, lose potential profit.
- If we start drilling when the actual amount of oil is too low—or, vie versa, do not start drilling when there is actually enough of oil —we also potentially lose money.

The measurement errors Δx_i are usually reasonably small. So, we can expand the expression for the utility $u(\Delta x_i)$ in Taylor series and keep only the first few terms in this expansion:

$$u(\Delta x_i) \approx u(0) + u_1 \cdot \Delta x_i + u_2 \cdot (\Delta x_i)^2 + \ldots + u_k \cdot (\Delta x_i)^k,$$

where the coefficients u_i are uniquely determined by the corresponding utility function $u(\Delta x_i)$. By taking the expected value $E[\cdot]$ of both sides of the above equality, we conclude that

$$E[u(\Delta x_i)] \approx u(0) + u_1 \cdot E[\Delta x_i] + u_2 \cdot E[(\Delta x_i)^2] + \ldots + u_k \cdot E[(\Delta x_i)^k].$$

Thus, to compute the expected utility, it is sufficient to know the first few moments

$$E[\Delta x_i],\quad E[(\Delta x_i)^2], \ldots, E[(\Delta x_i)^k]$$

of the corresponding distribution.

From this viewpoint, a reasonable way to describe a probability distribution is via its first few moments. This is what we will consider in this paper.

From the computational viewpoint, it is convenient to use cumulants, not moments themselves. From the computational viewpoint, in computational statistics, it is often more convenient to use not the moments themselves but their combinations called *cumulants*; see, e.g., [13]. A general mathematical definition of the k-th order cumulant κ_{in} of a random variable Δx_i is that it is a coefficient in the Taylor expansion of the logarthm of the *characteristic function*

$$\chi_i(\omega) \stackrel{\text{def}}{=} E[\exp(\mathrm{i} \cdot \omega \cdot \Delta x_i)]$$

(where $\mathrm{i} \stackrel{\text{def}}{=} \sqrt{-1}$) in terms of ω:

$$\ln(E[\exp(\mathrm{i} \cdot \omega \cdot \Delta x_i)]) = \sum_{k=1}^{\infty} \kappa_{ik} \cdot \frac{(\mathrm{i} \cdot \omega)^k}{k!}.$$

It is known that the k-th order cumulant can be described in terms of the moments up to order k; for example:

- κ_{i1} is simply the expected value, i.e., the first moment;
- κ_{i2} is negative variance;
- κ_{i3} and κ_{i4} are related to skewness and excess, etc.

The convenient thing about cumulants (as opposed to moments) is that when we add two independent random variables, their cumulants also add:

- the expected value of the sum of two independence random variables is equal to the sum of their expected values (actually, for this case, we do not even need independence, in other cases we do);
- the variance of the sum of two independent random variables is equal to the sum of their variance, etc.

In addition to this important property, k-th order cumulants have many of the same properties of the k-th order moments. For example:

- if we multiply a random variable by a constant c,
- then both its k-th order moment and its k-th order cumulant will multiply by c^k.

Usually, we know the cumulants only approximately. Based on the above explanations, a convenient way to describe each measurement uncertainty Δx_i is by describing the corresponding cumulants κ_{ik}.

The value of these cumulants also come from measurements. As a result, we usually know them only approximately, i.e., have an approximate value $\widetilde{\kappa}_{ik}$ and the upper bound Δ_{ik} on the corresponding inaccuracy:

$$|\kappa_{ik} - \widetilde{\kappa}_{ik}| \le \Delta_{ik}.$$

In this case, the only information that we have about the actual (unknown) values κ_{ik} is that each of these values belongs to the corresponding interval

$$[\underline{\kappa}_{ik}, \overline{\kappa}_{ik}],$$

where

$$\underline{\kappa}_{ik} \stackrel{\text{def}}{=} \widetilde{\kappa}_{ik} - \Delta_{ik}$$

and

$$\overline{\kappa}_{ik} \stackrel{\text{def}}{=} \widetilde{\kappa}_{ik} + \Delta_{ik}.$$

Thus, we arrive at the following formulation of the uncertainty propagation problem.

Uncertainty propagation: formulation of the problem. We know:

- an algorithm

$$f(x_1, \ldots, x_n),$$

- the measurement results

$$\widetilde{x}_1, \ldots, \widetilde{x}_n,$$

  and
- for each i from 1 to n, we know intervals

$$[\underline{\kappa}_{ik}, \overline{\kappa}_{ik}] = [\widetilde{\kappa}_{ik} - \Delta_{ik}, \widetilde{\kappa}_{ik} + \Delta_{ik}]$$

  that contain the actual (unknown) cumulants κ_{ik} of the measurement errors

$$\Delta x_i = \widetilde{x}_i - x_i.$$

Based on this information, we need to compute the range

$$[\underline{\kappa}_k, \overline{\kappa}_k]$$

of possible values of the cumulants κ_k corresponding to

$$\Delta y = f(\widetilde{x}_1, \ldots, \widetilde{x}_n) - f(x_1, \ldots, x_n) = f(\widetilde{x}_1, \ldots, \widetilde{x}_n) - f(\widetilde{x}_1 - \Delta x_1, \ldots, \widetilde{x}_n - \Delta x_n).$$

3 Existing Algorithms for Uncertainty Propagation and Their Limitations

Usually, measurement errors are relatively small. As we have mentioned, in most practical cases, the measurement error is relatively small. So, we can safely ignore terms which are quadratic (or of higher order) in terms of the measurement errors. For example:

- if we measure something with 10% accuracy,
- then the quadratic terms are of order 1%, which is definitely much less than 1%.

Thus, to estimate Δy, we can expand the expression for Δy in Taylor series and keep only linear terms in this expansion. Here, by definition of the measurement error, we have

$$x_i = \widetilde{x}_i - \Delta x_i,$$

thus

$$\Delta y = f(\widetilde{x}_1, \ldots, \widetilde{x}_n) - f(\widetilde{x}_1 - \Delta x_1, \ldots, \widetilde{x}_n - \Delta x_n).$$

Expanding the right-hand side in Taylor series and keeping only linear terms in this expansion, we conclude that

$$\Delta y = \sum_{i=1}^{n} c_i \cdot \Delta x_i,$$

where c_i is the value of the i-th partial derivative $\dfrac{\partial f}{\partial x_i}$ at a point $(\widetilde{x}_1, \ldots, \widetilde{x}_n)$:

$$c_i \stackrel{\text{def}}{=} \frac{\partial f}{\partial x_i}(\widetilde{x}_1, \ldots, \widetilde{x}_n).$$

Let us derive explicit formulas for $\underline{\kappa}_k$ and $\overline{\kappa}_k$. Let us assume that we know the coefficients c_i.

Due to the above-mentioned properties of cumulants, if κ_{ik} is the k-th cumulant of Δx_i, then the k-th cumulant of the product $c_i \cdot \Delta x_i$ is equal to

$$(c_i)^k \cdot \kappa_{ik}.$$

In its turn, the k-th order cumulant κ_k for the sum Δy of these products is equal to the sum of the corresponding cumulants:

$$\kappa_k = \sum_{i=1}^{n} (c_i)^k \cdot \kappa_{ik}.$$

We can represent each (unknown) cumulant κ_{ik} as the difference

$$\kappa_{ik} = \widetilde{\kappa}_{ik} - \Delta\kappa_{ik},$$

where

$$\Delta\kappa_{ik} \stackrel{\text{def}}{=} \widetilde{\kappa}_{ik} - \kappa_{ik}$$

is bounded by the known value Δ_{ik}:

$$|\Delta\kappa_{ik}| \leq \Delta_{ik}.$$

Substituting the above expression for κ_{ik} into the formula for κ_k, we conclude that

$$\kappa_k = \widetilde{\kappa}_k - \Delta\kappa_k,$$

where we denoted

$$\widetilde{\kappa}_k \stackrel{\text{def}}{=} \sum_{i=1}^{n} (c_i)^k \cdot \widetilde{\kappa}_k$$

and

$$\Delta\kappa_k \stackrel{\text{def}}{=} \sum_{i=1}^{n} (c_i)^k \cdot \Delta\kappa_{ik}.$$

The value $\widetilde{\kappa}_k$ is well defined. The value $\Delta\kappa_k$ depends on the approximation errors $\Delta\kappa_{ik}$. To find the set of possible values κ_k, we thus need to find the range of possible values of $\Delta\kappa_k$.

This value is the sum of n independent terms, independent in the sense that each of them depends only on its own variable $\Delta\kappa_{ik}$. So, the sum attains its largest values when each of the terms

$$(c_i)^k \cdot \Delta\kappa_{ik}$$

is the largest.

- When $(c_i)^k > 0$, the expression $(c_i)^k \cdot \Delta\kappa_{ik}$ is an increasing function of $\Delta\kappa_{ik}$, so it attains its largest possible value when $\Delta\kappa_{ik}$ attains its largest possible value Δ_{ik}. The resulting largest value of this term is

$$(c_i)^k \cdot \Delta_{ik}.$$

- When $(c_i)^k < 0$, the expression $(c_i)^k \cdot \Delta\kappa_{ik}$ is a decreasing function of $\Delta\kappa_{ik}$, so it attains its largest possible value when $\Delta\kappa_{ik}$ attains its smallest possible value $-\Delta_{ik}$. The resulting largest value of this term is

$$-(c_i)^k \cdot \Delta_{ik}.$$

Both cases can be combined into a single expression $|(c_i)^k| \cdot \Delta_{ik}$ if we take into account that:

- when $(c_i)^k > 0$, then $|(c_i)^k| = (c_i)^k$, and
- when $(c_i)^k < 0$, then $|(c_i)^k| = -(c_i)^k$.

Thus, the largest possible value of $\Delta\kappa_k$ is equal to

$$\Delta_k \stackrel{\text{def}}{=} \sum_{i=1}^{n} |(c_i)^k| \cdot \Delta_{ik}.$$

Similarly, we can show that the smallest possible value of $\Delta\kappa_k$ is equal to $-\Delta_k$. Thus, we arrive at the following formulas for computing the desired range $[\underline{\kappa}_k, \overline{\kappa}_k]$.

Explicit formulas for $\underline{\kappa}_k$ and $\overline{\kappa}_k$. Here, $\underline{\kappa}_k = \widetilde{\kappa}_k - \Delta_k$ and $\overline{\kappa}_k = \widetilde{\kappa}_k + \Delta_k$, where

$$\widetilde{\kappa}_k = \sum_{i=1}^{n} (c_i)^k \cdot \widetilde{\kappa}_k$$

and

$$\Delta_k = \sum_{i=1}^{n} |(c_i)^k| \cdot \Delta_{ik}.$$

A resulting straightforward algorithm. The above formulas can be explicitly used to estimate the corresponding quantities. The only remaining question is how to estimate the corresponding values c_i of the partial derivatives.

- When $f(x_1, \ldots, x_n)$ is an explicit expression, we can simply differentiate the function f and get the values of the corresponding derivatives.
- In more complex cases, e.g., when the algorithm $f(x_1, \ldots, x_n)$ is given as a proprietary black box, we can compute all the values c_i by using numerical differentiation:
$$c_i \approx \frac{f(\widetilde{x}_1, \ldots, \widetilde{x}_{i-1}, \widetilde{x}_i + \varepsilon_i, \widetilde{x}_{i+1}, \ldots, \widetilde{x}_n) - \widetilde{y}}{\varepsilon_i}$$
for some small ε_i.

Main limitation of the straightforward algorithm: it takes too long. When $f(x_1, \ldots, x_n)$ is a simple expression, the above straightforward algorithm is very efficient.

However, in many cases—e.g., with weather prediction or oil exploration—the corresponding algorithm $f(x_1, \ldots, x_n)$ is very complex and time-consuming,

- requiring hours of computation on a high performance computer,
- while processing thousands of data values x_i.

In such situations, the above algorithm requires $n+1$ calls to the program that implements the algorithm $f(x_1, \ldots, x_n)$:

- one time to compute
$$\widetilde{y} = f(\widetilde{x}_1, \ldots, \widetilde{x}_n),$$
  and then
- n times to compute n values
$$f(\widetilde{x}_1, \ldots, \widetilde{x}_{i-1}, \widetilde{x}_i + \varepsilon_i, \widetilde{x}_{i+1}, \ldots, \widetilde{x}_n)$$
  needed to compute the corresponding partial derivatives c_i.

When each call to f takes hours, and we need to make thousands of such class, the resulting computation time is in years.

This makes the whole exercise mostly useless: when it takes hours to predict the weather, no one will wait more than a year to check how accurate is this prediction. It is therefore necessary to have faster methods for uncertainty propagation.

Much faster methods exist for moments (and cumulants) of even order k. For all k, the computation of the value

$$\kappa_k = \sum_{i=1}^{n} (c_i)^k \cdot \widetilde{\kappa}_{ik}$$

can be done much faster, by using the following Monte-Carlo simulations.

Several times $\ell = 1, 2, \ldots, L$, we:

- simulate the values $\Delta x_i^{(\ell)}$ according to some distribution of Δx_i with the given value $\widetilde{\kappa}_{ik}$, and
- estimate
$$\Delta y^{(\ell)} = \widetilde{y} - f(\widetilde{x}_1 - \Delta x_1^{(\ell)}, \ldots, \widetilde{x}_n - \Delta x_n^{(\ell)}).$$

One can show that in this case, the k-th cumulant of the resulting distribution for $\Delta y^{(\ell)}$ is equal to exactly the desired value

$$\kappa_k = \sum_{i=1}^{n} (c_i)^k \cdot \widetilde{\kappa}_{ik}.$$

Thus, by computing the sample moments of the sample

$$\Delta y^{(1)}, \ldots, \Delta y^{(L)},$$

we can find the desired k-th order cumulant.

For example, for $k = 2$, when the cumulant is the variance, we can simply use normal distributions with a given variance.

The main advantage of the Monte-Carlo method is that its accuracy depends only on the number of iterations: its uncertainty decreases with L as $1/\sqrt{L}$; see, e.g., [13]. Thus, for example:

- to get the moment with accuracy 20% (=1/5),
- it is sufficient to run approximately 25 simulations, i.e., approximately 25 calls to the algorithm f; this is much much faster than thousands of iterations needed to perform the straightforward algorithm.

For even k, the value $(c_i)^k$ is always non-negative, so

$$|(c_i)^k| = (c_i)^k,$$

and the formula for Δ_k get a simplified form

$$\Delta_k = \sum_{i=1}^{n} (c_i)^k \cdot \Delta_{ik}.$$

This is exactly the same form as for $\widetilde{\kappa}_k$, so we can use the same Monte-Carlo algorithm to estimate Δ_k—the only difference is that now, we need to use distributions of Δx_i with the k-th cumulant equal to Δ_{ik}.

Specifically, several times $\ell = 1, 2, \ldots, L$, we:

- simulate the values $\Delta x_i^{(\ell)}$ according to some distribution of Δx_i with the value Δ_{ik} of the k-th cumulant, and
- estimate
$$\Delta y^{(\ell)} = \widetilde{y} - f(\widetilde{x}_1 - \Delta x_1^{(\ell)}, \ldots, \widetilde{x}_n - \Delta x_n^{(\ell)}).$$

One can show that in this case, the k-th cumulant of the resulting distribution for $\Delta y^{(\ell)}$ is equal to exactly the desired value

$$\Delta_k = \sum_{i=1}^{n} (c_i)^k \cdot \Delta_{ik}.$$

Thus, by computing the sample moments of the sample

$$\Delta y^{(1)}, \ldots, \Delta y^{(L)},$$

we can find the desired bound Δ_k on the k-th order cumulant.

Odd order moments (such as skewness) remain a computational problem. For odd k, we can still use the same Monte-Carlo method to compute the value $\widetilde{\kappa}_k$.

However, we can no longer use this method to compute the bound Δ_k on the k-th cumulant, since for odd k, we no longer have the equality

$$|(c_i)^k| = (c_i)^k.$$

What we plan to do. We will show that the use of (hypothetical) negative probabilities enables us to attain the same speed up for the case of odd k as we discussed above for the case of even orders.

4 Analysis of the Problem and the Resulting Negative-Probability-Based Fast Algorithm for Uncertainty Quantification

Why the Monte-Carlo method works for variances? The possibility to use normal distributions to analyze the propagation of variances

$$V = \sigma^2$$

comes from the fact that if we have n independent random variables Δx_i with variances

$$V_i = \sigma_i^2,$$

then their linear combination

$$\Delta y = \sum_{i=1}^{n} c_i \cdot \Delta x_i$$

is also normally distributed, with variance

$$V = \sum_{i=1}^{n} (c_i)^2 \cdot V_i$$

– and this is exactly how we want to relate the variance (2-nd order cumulant) of Δy with the variances V_i of the inputs.

Suppose that we did not know that the normal distribution has this property. How would we then be able to find a distribution $\rho_1(x)$ that satisfies this property? Let us consider the simplest case of this property, when

$$V_1 = \ldots = V_n = 1.$$

In this case, the desired property has the following form:

- if n independent random variables $\Delta x_1, \ldots, \Delta x_n$ have exactly the same distribution, with variance 1,
- then their linear combination

$$\Delta y = \sum_{i=1}^{n} c_i \cdot \Delta x_i$$

has the same distribution, but re-scaled, with variance

$$V = \sum_{i=1}^{n} (c_i)^2.$$

Let $\rho_1(x)$ denote the desired probability distribution, and let

$$\chi_1(\omega) = E[\exp(\mathrm{i} \cdot \omega \cdot \Delta x_1)]$$

be the corresponding characteristic function. Then, for the product $c_i \cdot \Delta x_i$, the characteristic function has the form

$$E[\exp(\mathrm{i} \cdot \omega \cdot (c_i \cdot \Delta x_1)].$$

By re-arranging multiplications, we can represent this same expression as

$$E[\exp(\mathrm{i} \cdot (\omega \cdot c_i) \cdot \Delta x_1],$$

i.e., as $\chi_1(c_i \cdot \omega)$.

For the sum of several independent random variables, the characteristic function is equal to the product of characteristic functions (see, e.g., [13]); thus, the characteristic function of the sum

$$\sum_{i=1}^{n} c_i \cdot \Delta x_i$$

has the form

$$\chi_1(c_1 \cdot \omega) \cdot \ldots \cdot \chi_1(c_n \cdot \omega).$$

We require that this sum be distributed the same way as Δx_i, but with a larger variance. When we multiply a variable by c, its variable increases by a factor of c^2. Thus, to get the distribution with variance

$$V = \sum_{i=1}^{n} (c_i)^2,$$

we need to multiply the variable Δx_i by a factor of

$$c = \sqrt{\sum_{i=1}^{n} (c_i)^2}.$$

For a variable multiplied by this factor, the characteristic function has the form

$$\chi_1(c \cdot \omega).$$

By equating the two characteristic functions, we get the following functional equation:

$$\chi_1(c_1 \cdot \omega) \cdot \ldots \cdot \chi_1(c_n \cdot \omega) = \chi_1\left(\sqrt{\sum_{i=1}^{n}(c_i)^2} \cdot \omega\right).$$

In particular, for $n = 2$, we conclude that

$$\chi_1(c_1 \cdot \omega) \cdot \chi_1(c_2 \cdot \omega) = \chi_1\left(\sqrt{(c_1)^2 + (c_2)^2} \cdot \omega\right).$$

This expression can be somewhat simplified if we take logarithms of both sides. Then products turn to sums, and for the new function

$$\ell(\omega) \stackrel{\text{def}}{=} \ln(\chi_1(\omega)),$$

we get the equation

$$\ell(c_1 \cdot \omega) + \ell(c_2 \cdot \omega) = \ell\left(\sqrt{(c_1)^2 + (c_2)^2} \cdot \omega\right).$$

This equation can be further simplified if we consider an auxiliary function

$$F(\omega) \stackrel{\text{def}}{=} \ell(\sqrt{\omega}),$$

for which

$$\ell(x) = F(x^2).$$

Substituting the expression for $\ell(x)$ in terms of $F(x)$ into the above formula, we conclude that

$$F((c_1)^2 \cdot \omega^2) + F((c_2)^2 \cdot \omega^2) = F(((c_1)^2 + (c_2)^2) \cdot \omega^2).$$

One can easily check that for every two non-negative numbers a and b, we can take

$$\omega = 1,\ c_1 = \sqrt{a}, \text{ and } c_2 = \sqrt{b},$$

and thus turn the above formula into

$$F(a) + F(b) = F(a + b).$$

It is well known (see, e.g., [1]) that every measurable solution to this functional equation has the form

$$F(a) = K \cdot a$$

for some constant K. Thus,

$$\ell(\omega) = F(\omega^2) = K \cdot \omega^2.$$

Here,

$$\ell(\omega) = \ln(\chi_1(\omega)),$$

hence

$$\chi_1(\omega) = \exp(\ell(\omega)) = \exp(K \cdot \omega^2).$$

Based on the characteristic function, we can reconstruct the original probability density function $\rho_1(x)$. Indeed, from the purely mathematical viewpoint, the characteristic function

$$\chi(\omega) = E[\exp(\mathrm{i} \cdot \omega \cdot \Delta x_1)] = \int \exp(\mathrm{i} \cdot \omega \cdot \Delta x_1) \cdot \rho_1(\Delta x_1)\, d(\Delta x_1)$$

is nothing else but the Fourier transform of the probability density function $\rho_1(\Delta x_1)$. We can therefore always reconstruct the original probability density function by applying the inverse Fourier transform to the characteristic function.

For

$$\chi_1(\omega) = \exp(K \cdot \omega^2),$$

the inverse Fourier transform leads to the usual formula of the normal distribution, with

$$K = -\sigma^2.$$

Can we apply the same idea to odd k? Our idea us to use Monte-Carlo methods for odd k, to speed up the computation of the value

$$\Delta_k = \sum_{i=1}^{n} |(c_i)^k| \cdot \Delta_{ik}.$$

What probability distribution $\rho_1(x)$ can we use to do it?

Similar to the above, let us consider the simplest case when

$$\Delta_{1k} = \ldots = \Delta_{nk} = 1.$$

In this case, the desired property of the probability distribution takes the following form:

- if n independent random variables

$$\Delta x_1, \ldots, \Delta x_n$$

  have exactly the same distribution $\rho_1(x)$, with k-th cumulant equal to 1,
- then their linear combination

$$\Delta y = \sum_{i=1}^{n} c_i \cdot \Delta x_i$$

  has the same distribution, but re-scaled, with the k-th order cumulant equal to

$$\sum_{i=1}^{n} |c_i|^k.$$

Let $\rho_1(x)$ denote the desired probability distribution, and let

$$\chi_1(\omega) = E[\exp(\mathrm{i} \cdot \omega \cdot \Delta x_1)]$$

be the corresponding characteristic function. Then, as we have shown earlier, for the product $c_i \cdot \Delta x_i$, the characteristic function has the form $\chi_1(c_i \cdot \omega)$. For the sum

$$\sum_{i=1}^{n} c_i \cdot \Delta x_i,$$

the characteristic function has the form

$$\chi_1(c_1 \cdot \omega) \cdot \ldots \cdot \chi_1(c_n \cdot \omega).$$

We require that this sum be distributed the same way as Δx_i, but with a larger k-th order cumulant. As we have mentioned:

- when we multiply a variable by c,
- its k-th order cumulant increases by a factor of c^k.

Thus, to get the distribution with the value

$$\sum_{i=1}^{n} |c_i|^k,$$

we need to multiply the variable Δx_i by a factor of

$$c = \sqrt[k]{\sum_{i=1}^{n} |c_i|^k}.$$

For a variable multiplied by this factor, the characteristic function has the form

$$\chi_1(c \cdot \omega).$$

By equating the two characteristic functions, we get the following functional equation:

$$\chi_1(c_1 \cdot \omega) \cdot \ldots \cdot \chi_1(c_n \cdot \omega) = \chi_1\left(\sqrt[k]{\sum_{i=1}^{n} |c_i|^k} \cdot \omega\right).$$

In particular, for $n = 2$, we conclude that

$$\chi_1(c_1 \cdot \omega) \cdot \chi_1(c_2 \cdot \omega) = \chi_1\left(\sqrt[k]{|c_1|^k + |c_2|^k} \cdot \omega\right).$$

This expression can be somewhat simplified if we take logarithms of both sides. Then products turn to sums, and for the new function

$$\ell(\omega) \stackrel{\text{def}}{=} \ln(\chi_1(\omega)),$$

we get the equation

$$\ell(c_1 \cdot \omega) + \ell(c_2 \cdot \omega) = \ell\left(\sqrt[k]{(|c_1|^k + |c_2|^k} \cdot \omega\right).$$

This equation can be further simplified if we consider an auxiliary function

$$F(\omega) \stackrel{\text{def}}{=} \ell(\sqrt[k]{\omega}),$$

for which

$$\ell(x) = F(x^k).$$

Substituting the expression for $\ell(x)$ in terms of $F(x)$ into the above formula, we conclude that

$$F(|c_1|^k \cdot \omega^k) + F(|c_2|^k \cdot \omega^k) = F((|c_1|^k + |c_2|^k) \cdot \omega^k).$$

One can easily check that for every two non-negative numbers a and b, we can take

$$\omega = 1, c_1 = \sqrt[k]{a}, \text{ and } c_2 = \sqrt[k]{b}$$

and thus get

$$F(a) + F(b) = F(a + b).$$

As we have already shown, this leads to

$$F(a) = K \cdot a$$

for some constant K. Thus,

$$\ell(\omega) = F(\omega^k) = K \cdot \omega^k.$$

Here,

$$\ell(\omega) = \ln(\chi_1(\omega)),$$

hence

$$\chi_1(\omega) = \exp(\ell(\omega)) = \exp(K \cdot \omega^k).$$

Case of $k = 1$ leads to a known efficient method. For $k = 1$, the above characteristic function has the form

$$\exp(-K \cdot |\omega|).$$

By applying the inverse Fourier transform to this expression, we get the Cauchy distribution, with probability density

$$\rho_1(x) = \frac{1}{\pi \cdot K} \cdot \frac{1}{1 + \frac{x^2}{K^2}}.$$

Monte-Carlo methods based on the Cauchy distribution indeed lead to efficient estimation of first order uncertainty – e.g., bounds on mean; see, e.g., [6].

What about larger odd values k? Alas, for $k \geq 3$, we have a problem:

- when we apply the inverse Fourier transform to the characteristic function

$$\exp(-|K| \cdot |\omega|^k),$$

- the resulting function $\rho_1(\Delta x_1)$ takes negative values for some x, and thus, cannot serve as a usual probability density function; see, e.g., [12].

However:

- if negative probabilities are physically possible,
- then we can indeed use the same idea to speed up computation of Δ_k for odd values

$$k \geq 3.$$

If negative probabilities are physically possible, then we can speed up uncertainty propagation—namely, computation of Δ_k. If negative probabilities are indeed physically possible, then we can use the following algorithm to speed up the computation of Δ_k.

Let us assume that we are able to simulate a "random" variable η whose (sometimes negative) probability density function $\rho_1(x)$ is the inverse Fourier transform of the function

$$\chi_1(\omega) = \exp(-|\omega|^k).$$

We will use the corresponding "random" number generator for each variable x_i and for each iteration $\ell = 1, 2, \ldots, L$. The corresponding value will be denoted by $\eta_i^{(\ell)}$.

The value $\eta_i^{(\ell)}$ will corresponds to the value of the k-th cumulant equal to 1. To simulate a random variable corresponding to parameter Δ_{ik}, we use

$$(\Delta_{ik})^{1/k} \cdot \eta_i^{(\ell)}.$$

Thus, we arrive at the following algorithm:

Several times $\ell = 1, 2, \ldots, L$, we:

- simulate the values $\Delta x_i^{(\ell)}$ as

$$(\Delta_{ik})^{1/k} \cdot \eta_i^{(\ell)},$$

  and
- estimate

$$\Delta y^{(\ell)} = \widetilde{y} - f(\widetilde{x}_1 - \Delta x_1^{(\ell)}, \ldots, \widetilde{x}_n - \Delta x_n^{(\ell)}).$$

One can show that in this case, the resulting distribution for $\Delta y^{(\ell)}$ has the same distribution as η multiplied by the k-th root of the desired value

$$\Delta_k = \sum_{i=1}^{n} (c_i)^k \cdot \Delta_{ik}.$$

Thus, by computing the corresponding characteristic of the sample

$$\Delta y^{(1)}, \ldots, \Delta y^{(L)},$$

we can find the desired bound Δ_k on the k-th order cumulant.

So, we can indeed use fast Monte-Carlo methods to estimate both values $\widetilde{\kappa}_k$ and Δ_k – and thus, to speed up uncertainty propagation.

Acknowledgements This work was supported in part:
- by the National Science Foundation grants
  - HRD-0734825 and HRD-1242122 (Cyber-ShARE Center of Excellence) and
  - DUE-0926721, and
- by the award "UTEP and Prudential Actuarial Science Academy and Pipeline Initiative" from Prudential Foundation.

References

1. Aczél, J.: Lectures on Functional Equations and Their Applications. Dover, New York (2006)
2. Dirac, P.A.M.: The physical interpretation of quantum mechanics. Proc. Royal Soc. A: Math. Phys. Eng. Sci. **180**(980), 1–39 (1942)
3. Feynman, R.P.: Negative probability. In: Peat, F.D., Hiley, B. (eds.) Quantum Implications: Essays in Honour of David Bohm, pp. 235–248. Routledge & Kegan Paul Ltd., Abingdon-on-Thames, UK (1987)
4. Feynman, R., Leighton, R., Sands, M.: The Feynman Lectures on Physics. Addison Wesley, Boston, Massachusetts (2005)
5. Fishburn, P.C.: Utility Theory for Decision Making. Wiley, New York (1969)
6. Kreinovich, V., Ferson, S.: A new Cauchy-based black-box technique for uncertainty in risk analysis. Reliab. Eng. Syst. Saf. **85**(1–3), 267–279 (2004)
7. Luce, R.D., Raiffa, R.: Games and Decisions: Introduction and Critical Survey. Dover, New York (1989)
8. Nguyen, H.T., Kosheleva, O., Kreinovich, V.: Decision making beyond Arrow's 'impossibility theorem', with the analysis of effects of collusion and mutual attraction. Int. J. Intell. Syst. **24**(1), 27–47 (2009)
9. Nielsen, M.A., Chuang, I.L.: Quantum Computation and Quantum Information. Cambridge University Press, Cambridge, U.K. (2000)
10. Rabinovich, S.G.: Measurement Errors and Uncertainty. Theory and Practice. Springer Verlag, Berlin (2005)
11. Raiffa, H.: Decision Analysis. Addison-Wesley, Reading, Massachusetts (1970)
12. Samorodnitsky, G., Taqqu, M.S.: Stable Non-Gaussian Random Proceses Stochastic Models with Infinite Variance. Chapman & Hall, New York (1994)
13. Sheskin, D.J.: Handbook of Parametric and Nonparametric Statistical Procedures. Chapman and Hall/CRC, Boca Raton, Florida (2011)

New Method of Obtaining the Kochen-Specker Theorem

Koji Nagata, Tadao Nakamura and Ahmed Farouk

Abstract We derive a new type of no-hidden-variable theorem based on the assumptions proposed by Kochen and Specker. We consider N spin-1/2 systems. The hidden results of measurement are either $+1$ or -1 (in $\hbar/2$ unit). We derive some proposition concerning a quantum expected value under an assumption about the existence of the Bloch sphere in N spin-1/2 systems. However, the hidden variables theory violates the proposition with a magnitude that grows exponentially with the number of particles. Therefore, we have to give up either the existence of the Bloch sphere or the hidden variables theory. Also we discuss a two-dimensional no-hidden-variables theorem of the KS type. Especially, we systematically describe our assertion based on more mathematical analysis using raw data in a thoughtful experiment.

Keywords 03.65.Ud (Quantum non locality) · 03.65.Ta (Quantum measurement theory) · 03.65.Ca (Formalism)

1 Introduction

Quantum mechanics (cf. [1–6]) gives accurate and at times remarkably accurate numerical predictions. Much experimental data has fit to the quantum predictions for long time.

Kochen and Specker present the no-hidden-variables theorem (the KS theorem) [7]. The KS theorem says the non-existence of a real-valued function which is multiplicative and linear on commuting operators. The proof of the KS theorem relies on

K. Nagata (✉)
Department of Physics, Korea Advanced Institute of Science and Technology,
Daejeon 34141, Korea
e-mail: ko_mi_na@yahoo.co.jp

T. Nakamura
Department of Information and Computer Science, Keio University,
3-14-1 Hiyoshi, Kohoku-ku, Yokohama 223-8522, Japan

A. Farouk
Computer Sciences Department, Faculty of Computers and Information,
Mansoura University, Mansoura, Egypt

A.E. Hassanien et al. (eds.), *Quantum Computing: An Environment for Intelligent Large Scale Real Application*, Studies in Big Data 33,
https://doi.org/10.1007/978-3-319-63639-9_12

intricate geometric argument. Greenberger, Horne, and Zeilinger discover [8, 9] the so-called GHZ theorem for four-partite GHZ state. And, the KS theorem becomes very simple form (see also Refs. [10–14]).

It is begun to research the validity of the KS theorem by using inequalities (see Refs. [15–18]). To find such inequalities to test the validity of the KS theorem is particularly useful for experimental investigation [19]. One of the authors derives an inequality [18] as tests for the validity of the KS theorem. The quantum predictions violate the inequality when the system is in an uncorrelated state. An uncorrelated state is defined in Ref. [20]. The quantum predictions by n-partite uncorrelated state violate the inequality by an amount that grows exponentially with n.

Recently, Leggett-type non-local variables theory [21] is experimentally investigated [22–24]. The experiments report that quantum mechanics does not accept Leggett-type non-local variables interpretation. However there are debates for the conclusions of the experiments. See Refs. [25–27].

As for the applications of quantum mechanics, implementation of a quantum algorithm to solve Deutsch's problem [28–30] on a nuclear magnetic resonance quantum computer is reported firstly [31]. Implementation of the Deutsch-Jozsa algorithm for an ion-trap quantum computer is also reported [32]. There are several attempts to use single-photon two-qubit states for quantum computing. Oliveira *et al.* implement Deutsch's algorithm with polarization and transverse spatial modes of the electromagnetic field as qubits [33]. Single-photon Bell states are prepared and measured [34]. Also the decoherence-free implementation of Deutsch's algorithm is reported by using such a single-photon and by using two logical qubits [35]. More recently, a one-way based experimental implementation of Deutsch's algorithm is reported [36]. In 1993, the Bernstein-Vazirani algorithm was reported [37]. It can be considered as an extended Deutsch-Jozsa algorithm. In 1994, Simon's algorithm was reported [38]. Implementation of a quantum algorithm to solve the Bernstein-Vazirani parity problem without entanglement on an ensemble quantum computer is reported [39]. Fiber-optics implementation of the Deutsch-Jozsa and Bernstein-Vazirani quantum algorithms with three qubits is discussed [40]. A quantum algorithm for approximating the influences of Boolean functions and its applications is recently reported [41].

On the other hand, the double-slit experiment is an illustration of wave-particle duality. In it, a beam of particles (such as photons) travels through a barrier with two slits removed. If one puts a detector screen on the other side, the pattern of detected particles shows interference fringes characteristic of waves; however, the detector screen responds to particles. The system exhibits the behaviour of both waves (interference patterns) and particles (dots on the screen).

If we modify this experiment so that one slit is closed, no interference pattern is observed. Thus, the state of both slits affects the final results. We can also arrange to have a minimally invasive detector at one of the slits to detect which slit the particle went through. When we do that, the interference pattern disappears [42]. An analysis of a two-atom double-slit experiment based on environment-induced measurements is reported [43].

We try to implement the double-slit experiment. There is a detector just after each slit. Thus interference figure does not appear, and we do not consider such a pattern. The possible values of the result of measurements are ± 1 (in $\hbar/2$ unit). If a particle passes one side slit, then the value of the result of measurement is $+1$. If a particle passes through another slit, then the value of the result of measurement is -1. This is an easy detector model for a Pauli observable.

In this paper, we derive a new type of no-hidden-variables theorem based on the assumptions proposed by Kochen and Specker. We consider N spin-1/2 systems. The hidden results of measurement are either $+1$ or -1 (in $\hbar/2$ unit). We derive some proposition concerning a quantum expected value under an assumption about the existence of the Bloch sphere in N spin-1/2 systems. However, the hidden variables theory violates the proposition with a magnitude that grows exponentially with the number of particles. Therefore, we have to give up either the existence of the Bloch sphere or the hidden variables theory. Also we discuss two-dimensional no-hidden-variables theorem of the KS type, by using the double-slit experiment. Especially, we systematically describe our assertion based on more mathematical analysis using raw data in a thoughtful experiment.

Throughout this paper, we confine ourselves to the two-level (e.g., electron spin, photon polarizations, and so on) and the discrete eigenvalue case.

2 Notations and Preparation to Get New Type of No-Hidden-Variables Theorem of the KS Type

We consider a two-dimensional space H. Let $\mathbf{N}$ denote a set of the numbers

$$\{1, 2, \ldots, +\infty\} \tag{2.1}$$

that contains the countably infinite. Let S be $\{\pm 1\}$. We assume that every result of measurements lies in S. We assume that every time t lies in $\mathbf{N}$. Let $\mathbf{N}_1$ denote a set of the numbers

$$\{1, 5, 9, \ldots, +\infty\} \tag{2.2}$$

that contains the countably infinite. Here we introduce $t_1 \in \mathbf{N}_1$. Let $\mathbf{N}_2$ denote a set of the numbers

$$\{2, 6, 10, \ldots, +\infty\} \tag{2.3}$$

that contains the countably infinite. Here we introduce $t_2 \in \mathbf{N}_2$. Let $\mathbf{N}_3$ denote a set of the numbers

$$\{3, 7, 11, \ldots, +\infty\} \tag{2.4}$$

that contains the countably infinite. Here we introduce $t_3 \in \mathbf{N}_3$. Let $\mathbf{N}_4$ denote a set of the numbers

$$\{4, 8, 12, \ldots, +\infty\} \tag{2.5}$$

that contains the countably infinite. Here we introduce $t_4 \in \mathbf{N}_4$. Let $\vec{\sigma}$ be

$$(\sigma_x, \sigma_y, \sigma_z), \tag{2.6}$$

the vector of Pauli operators. The measurements (observables) of $\vec{n} \cdot \vec{\sigma}$ are parameterized by a unit vector $\vec{n}$ (its direction along which the spin component is measured). Here, $\cdot$ is the scalar product in $\mathbf{R}^3$. One measures an observable $\vec{n} \cdot \vec{\sigma}$. We define a notation $\theta(t)$ which represents predetermined result of measurements at time t. We assume that measurement of an observable $\vec{n} \cdot \vec{\sigma}$ at time t for a physical system in a state ψ yields a value $\theta(\psi, \vec{n} \cdot \vec{\sigma}, t) \in S$.

We consider the following:

Assumption: M (*predetermined measurement outcome*),

$$\theta(\psi, \vec{n} \cdot \vec{\sigma}, t) \in S. \tag{2.7}$$

Assumption: E (*quantum expected value*),

$$\mathrm{Tr}[\psi \vec{n} \cdot \vec{\sigma}] = \lim_{m \to \infty} \frac{\sum_{t=1}^{m} \theta(\psi, \vec{n} \cdot \vec{\sigma}, t)}{m}. \tag{2.8}$$

Assumption: T
If

$$\mathrm{Tr}[\psi \vec{n} \cdot \vec{\sigma}] = \lim_{m \to \infty} \frac{\sum_{t=1}^{m} \theta(\psi, \vec{n} \cdot \vec{\sigma}, t)}{m}, \tag{2.9}$$

then

$$\mathrm{Tr}[\psi \vec{n} \cdot \vec{\sigma}] = \lim_{m_1 \to \infty} \frac{\sum_{t_1=1}^{m_1} \theta(\psi, \vec{n} \cdot \vec{\sigma}, t_1)}{m_1} = \lim_{m_2 \to \infty} \frac{\sum_{t_2=2}^{m_2} \theta(\psi, \vec{n} \cdot \vec{\sigma}, t_2)}{m_2} \tag{2.10}$$

and

$$\mathrm{Tr}[\psi \vec{n} \cdot \vec{\sigma}] = \lim_{m_3 \to \infty} \frac{\sum_{t_3=3}^{m_3} \theta(\psi, \vec{n} \cdot \vec{\sigma}, t_3)}{m_3} = \lim_{m_4 \to \infty} \frac{\sum_{t_4=4}^{m_4} \theta(\psi, \vec{n} \cdot \vec{\sigma}, t_4)}{m_4}. \tag{2.11}$$

3 New Type of No-Hidden-Variables Theorem of the KS Type

In this section, we give new type of no-hidden-variables theorem of the KS type.

3.1 The Existence of the Bloch Sphere

We assume a pure spin-1/2 state ψ lying in the x-y plane. Let $\vec{\sigma}$ be $(\sigma_x, \sigma_y, \sigma_z)$, the vector of Pauli operators. The measurements (observables) on a spin-1/2 state lying in the x-y plane of $\vec{n} \cdot \vec{\sigma}$ are parameterized by a unit vector $\vec{n}$ (its direction along which the spin component is measured). Here, $\cdot$ is the scalar product in $\mathbf{R}^3$.

We have a quantum expected value $E_{\rm QM}^k$, $k = 1, 2$ as

$$E_{\rm QM}^k \equiv {\rm Tr}[\psi \vec{n}_k \cdot \vec{\sigma}],\ k = 1, 2. \tag{3.1}$$

We have $\vec{x} \equiv \vec{x}^{(1)}$, $\vec{y} \equiv \vec{x}^{(2)}$, and $\vec{z} \equiv \vec{x}^{(3)}$. They are the Cartesian axes relative to which spherical angles are measured. We write two unit vectors in the plane defined by $\vec{x}^{(1)}$ and $\vec{x}^{(2)}$ in the following way:

$$\vec{n}_k = \cos\theta_k \vec{x}^{(1)} + \sin\theta_k \vec{x}^{(2)}. \tag{3.2}$$

Here, the angle θ_k takes only two values:

$$\theta_1 = 0,\ \theta_2 = \frac{\pi}{2}. \tag{3.3}$$

We derive a necessary condition for the quantum expected value for the system in a pure spin-1/2 state lying in the x-y plane given in (3.1). We derive the possible values of the scalar product

$$\sum_{k=1}^{2} \left(E_{\rm QM}^k \times E_{\rm QM}^k\right) \equiv \|E_{\rm QM}\|^2. \tag{3.4}$$

$E_{\rm QM}^k$ is the quantum expected value given in (3.1). We see that

$$\|E_{\rm QM}\|^2 = \langle\sigma_x\rangle^2 + \langle\sigma_y\rangle^2. \tag{3.5}$$

We use the decomposition (3.2). We introduce simplified notations as

$$T_i = {\rm Tr}[\psi \vec{x}^{(i)} \cdot \vec{\sigma}] \tag{3.6}$$

and

$$(c_k^1, c_k^2,) = (\cos\theta_k, \sin\theta_k). \tag{3.7}$$

Then, we have

$$\|E_{\mathrm{QM}}\|^2 = \sum_{k=1}^{2}\left(\sum_{i=1}^{2} T_i c_k^i\right)^2 = \sum_{i=1}^{2} T_i^2 \le 1, \tag{3.8}$$

where we use the orthogonality relation

$$\sum_{k=1}^{2} c_k^\alpha c_k^\beta = \delta_{\alpha,\beta}. \tag{3.9}$$

From a proposition of the quantum theory, the Bloch sphere with the value of

$$\sum_{i=1}^{2} T_i^2 \tag{3.10}$$

is bounded as

$$\sum_{i=1}^{2} T_i^2 \le 1. \tag{3.11}$$

The reason of the condition (3.8) is the Bloch sphere

$$\sum_{i=1}^{3} (\mathrm{Tr}[\psi \vec{x}^{(i)} \cdot \vec{\sigma}])^2 \le 1. \tag{3.12}$$

Thus we derive a proposition concerning a quantum expected value under an assumption of the existence of the Bloch sphere (in a spin-1/2 system). The proposition is

$$\|E_{\mathrm{QM}}\|^2 \le 1. \tag{3.13}$$

This inequality is saturated iff ψ is a pure state lying in the x-y plane. That is,

$$\sum_{i=1}^{2} (\mathrm{Tr}[\psi \vec{x}^{(i)} \cdot \vec{\sigma}])^2 = 1. \tag{3.14}$$

Hence, we derive the following proposition concerning the existence of the Bloch sphere when the system is in a pure state lying in the x-y plane

$$\|E_{\rm QM}\|_{\max}^2 = 1. \tag{3.15}$$

3.2 *The Existence of Hidden Measurement Outcome Which Is* ± 1

We assign the truth value "1" for Assumption M, Assumption E, and Assumption T. Let A_k be $\vec{n}_k \cdot \vec{\sigma}$. We assume four gedanken experiments in the same state ψ. The value of $\theta(\psi, A_1, t_1)$ is independent of $\theta(\psi, A_1, t_2)$. We note that the measurement time is different from each other. Here, we assume $t_1 \in \mathbf{N}_1$ and $t_2 \in \mathbf{N}_2$. The value of $\theta(\psi, A_2, t_3)$ is independent of $\theta(\psi, A_2, t_4)$. We note that the measurement time is different from each other. Here, we assume $t_3 \in \mathbf{N}_3$ and $t_4 \in \mathbf{N}_4$. The values of $\theta(\psi, A_1, t_1)$, $\theta(\psi, A_1, t_2)$, $\theta(\psi, A_2, t_3)$, and $\theta(\psi, A_2, t_4)$ are independent of each other. We note that the measurement time is different from each other. We assume that the number of each of quantum measurements is the countably infinite. We know that a sum of 'four' countably infinite is the countably infinite. We do not have to assign definite values to non-commuting observables in the same time.

From Assumption E and Assumption T, the quantum expected value in (3.1) ($k = 1$), which is the average of the results of measurements, is given by

$$E_{\rm QM}^1 = \lim_{m_1 \to \infty} \frac{\sum_{t_1=1}^{m_1} \theta(\psi, A_1, t_1)}{m_1}. \tag{3.16}$$

From Assumption M, the possible values of the measured result $\theta(\psi, A_1, t_1)$ are ± 1.

From Assumption T, the same quantum expected value is given by

$$E_{\rm QM}^1 = \lim_{m_2 \to \infty} \frac{\sum_{t_2=2}^{m_2} \theta(\psi, A_1, t_2)}{m_2}. \tag{3.17}$$

From Assumption M, the possible values of the measured result $\theta(\psi, A_1, t_2)$ are ± 1. From Assumption T, we see

$$\begin{aligned} &\|\{t_1 | t_1 \in \mathbf{N}_1 \wedge \theta(\psi, A_1, t_1) = 1\}\| = \|\{t_2 | t_2 \in \mathbf{N}_2 \wedge \theta(\psi, A_1, t_2) = 1\}\|, \\ &\|\{t_1 | t_1 \in \mathbf{N}_1 \wedge \theta(\psi, A_1, t_1) = -1\}\| = \|\{t_2 | t_2 \in \mathbf{N}_2 \wedge \theta(\psi, A_1, t_2) = -1\}\|. \end{aligned} \tag{3.18}$$

From Assumption E and Assumption T, the quantum expected value in (3.1) ($k = 2$), which is the average of the results of measurements, is given by

$$E_{\rm QM}^2 = \lim_{m_3 \to \infty} \frac{\sum_{t_3=3}^{m_3} \theta(\psi, A_2, t_3)}{m_3}. \tag{3.19}$$

From Assumption M, the possible values of the measured result $\theta(\psi, A_2, t_3)$ are ± 1.

From Assumption T, the same quantum expected value is given by

$$E_{\text{QM}}^2 = \lim_{m_4 \to \infty} \frac{\sum_{t_4=4}^{m_4} \theta(\psi, A_2, t_4)}{m_4}. \tag{3.20}$$

From Assumption M, the possible values of the measured result $\theta(\psi, A_2, t_4)$ are ± 1. From Assumption T, we see

$$\begin{aligned}
&\|\{t_3 | t_3 \in \mathbf{N}_3 \wedge \theta(\psi, A_2, t_3) = 1\}\| = \|\{t_4 | t_4 \in \mathbf{N}_4 \wedge \theta(\psi, A_2, t_4) = 1\}\|, \\
&\|\{t_3 | t_3 \in \mathbf{N}_3 \wedge \theta(\psi, A_2, t_3) = -1\}\| = \|\{t_4 | t_4 \in \mathbf{N}_4 \wedge \theta(\psi, A_2, t_4) = -1\}\|.
\end{aligned} \tag{3.21}$$

We derive a necessary condition for the two quantum expected values for the system in a pure spin-1/2 state lying in the x-y plane given in (3.16) and (3.19). We derive the possible values of the scalar product $\|E_{\text{QM}}\|^2$ of the two quantum expected values, E_{QM}^k given in (3.16) and (3.19).

We introduce an assumption that Sum rule and Product rule commute with each other [44]. We do not pursue the details of the assumption. To pursue the details is an interesting point. It is suitable to the next step of researches. We have

$$\begin{aligned}
&\|E_{\text{QM}}\|^2 \\
&= \left(\lim_{m_1 \to \infty} \frac{\sum_{t_1=1}^{m_1} \theta(\psi, A_1, t_1)}{m_1} \times \lim_{m_2 \to \infty} \frac{\sum_{t_2=2}^{m_2} \theta(\psi, A_1, t_2)}{m_2} \right) \\
&+ \left(\lim_{m_3 \to \infty} \frac{\sum_{t_3=3}^{m_3} \theta(\psi, A_2, t_3)}{m_3} \times \lim_{m_4 \to \infty} \frac{\sum_{t_4=4}^{m_4} \theta(\psi, A_2, t_4)}{m_4} \right) \\
&= \left(\lim_{m_1 \to \infty} \frac{\sum_{t_1=1}^{m_1}}{m_1} \cdot \lim_{m_2 \to \infty} \frac{\sum_{t_2=2}^{m_2}}{m_2} \theta(\psi, A_1, t_1) \theta(\psi, A_1, t_2) \right) \\
&+ \left(\lim_{m_3 \to \infty} \frac{\sum_{t_3=3}^{m_3}}{m_3} \cdot \lim_{m_4 \to \infty} \frac{\sum_{t_4=4}^{m_4}}{m_4} \theta(\psi, A_2, t_3) \theta(\psi, A_2, t_4) \right) \\
&\leq \left(\lim_{m_1 \to \infty} \frac{\sum_{t_1=1}^{m_1}}{m_1} \cdot \lim_{m_2 \to \infty} \frac{\sum_{t_2=2}^{m_2}}{m_2} |\theta(\psi, A_1, t_1) \theta(\psi, A_1, t_2)| \right) \\
&+ \left(\lim_{m_3 \to \infty} \frac{\sum_{t_3=3}^{m_3}}{m_3} \cdot \lim_{m_4 \to \infty} \frac{\sum_{t_4=4}^{m_4}}{m_4} |\theta(\psi, A_2, t_3) \theta(\psi, A_2, t_4)| \right) \\
&= \left(\lim_{m_1 \to \infty} \frac{\sum_{t_1=1}^{m_1}}{m_1} \cdot \lim_{m_2 \to \infty} \frac{\sum_{t_2=2}^{m_2}}{m_2} \right) + \left(\lim_{m_3 \to \infty} \frac{\sum_{t_3=3}^{m_3}}{m_3} \cdot \lim_{m_4 \to \infty} \frac{\sum_{t_4=4}^{m_4}}{m_4} \right) = 2.
\end{aligned} \tag{3.22}$$

From Assumption M, we have

$$|\theta(\psi, A_1, t_1)\theta(\psi, A_1, t_2)| = +1, \ |\theta(\psi, A_2, t_3)\theta(\psi, A_2, t_4)| = +1. \tag{3.23}$$

The above inequality (3.22) is saturated when

$$\theta(\psi, A_1, t_1)\theta(\psi, A_1, t_2) = 1, \ \theta(\psi, A_2, t_3)\theta(\psi, A_2, t_4) = 1. \tag{3.24}$$

This implies

$$\theta(\psi, A_1, t_1) = \theta(\psi, A_1, t_2), \ \theta(\psi, A_2, t_3) = \theta(\psi, A_2, t_4). \tag{3.25}$$

The above condition (3.25) can be possible since, as we have said,

$$\begin{aligned} &\|\{t_1|t_1 \in \mathbf{N}_1 \wedge \theta(\psi, A_1, t_1) = 1\}\| = \|\{t_2|t_2 \in \mathbf{N}_2 \wedge \theta(\psi, A_1, t_2) = 1\}\|, \\ &\|\{t_1|t_1 \in \mathbf{N}_1 \wedge \theta(\psi, A_1, t_1) = -1\}\| = \|\{t_2|t_2 \in \mathbf{N}_2 \wedge \theta(\psi, A_1, t_2) = -1\}\|, \end{aligned} \tag{3.26}$$

and

$$\begin{aligned} &\|\{t_3|t_3 \in \mathbf{N}_3 \wedge \theta(\psi, A_2, t_3) = 1\}\| = \|\{t_4|t_4 \in \mathbf{N}_4 \wedge \theta(\psi, A_2, t_4) = 1\}\|, \\ &\|\{t_3|t_3 \in \mathbf{N}_3 \wedge \theta(\psi, A_2, t_3) = -1\}\| = \|\{t_4|t_4 \in \mathbf{N}_4 \wedge \theta(\psi, A_2, t_4) = -1\}\|. \end{aligned} \tag{3.27}$$

Thus we derive a proposition concerning the two quantum expected values under an assumption that we assign the truth value "1" for Assumption M, Assumption E, and Assumption T (in a spin-1/2 system). The proposition is $\|E_{\rm QM}\|^2 \le 2$. This inequality can be saturated. Hence we derive the following proposition concerning Assumption M, Assumption E, and Assumption T:

$$\|E_{\rm QM}\|^2_{\rm max} = 2. \tag{3.28}$$

3.3 Contradiction

We cannot assign the truth value "1" for two propositions (3.15) (concerning the existence of the Bloch sphere) and (3.28) (concerning Assumption M, Assumption E, and Assumption T), simultaneously, when the system is in a pure state lying in the x-y plane. Therefore, we are in the KS contradiction. We do not assign the truth value "1" for five assumptions

1. Assumption M
2. Assumption E
3. Assumption T
4. The existence of the Bloch sphere
5. Sum rule and Product rule commute with each other, simultaneously.

4 High Dimensional No-Hidden-Variables Theorem of the KS Type

In this section, we derive a proposition concerning a quantum expected value under an assumption of the existence of the Bloch sphere in N spin-1/2 systems ($1 \leq N < +\infty$). This assumption intuitively depictures our physical world. However, the hidden variables theory (the result of measurements is ± 1) violates the proposition with a magnitude that grows exponentially with the number of particles. We have to give up either the existence of the Bloch sphere or the hidden variables theory. Therefore, the hidden variables theory cannot depicture our physical world with a violation factor that grows exponentially with the number of particles.

4.1 *The Existence of the Bloch Sphere*

Assume that we have a set of N spins $\frac{1}{2}$. Each of them is a spin-$1/2$ pure state lying in the x-y plane. Let us assume that one source of N uncorrelated spin-carrying particles emits them in a state, which can be described as a multi spin-1/2 pure uncorrelated state. Let us parameterize the settings of the jth observer with a unit vector $\vec{n}_j$ (its direction along which the spin component is measured) with $j = 1, \ldots, N$. One can introduce the 'hidden variables' correlation function, which is the average of the product of the hidden results of measurement

$$E_{\mathrm{HV}}(\vec{n}_1, \vec{n}_2, \ldots, \vec{n}_N) = \langle r(\vec{n}_1, \vec{n}_2, \ldots, \vec{n}_N) \rangle_{\mathrm{avg}}, \tag{4.1}$$

where r is hidden result. We assume the value of r is ± 1 (in $(\hbar/2)^N$ unit), which is obtained if the measurement directions are set at $\vec{n}_1, \vec{n}_2, \ldots, \vec{n}_N$.

Also one can introduce a quantum correlation function with the system in such a pure uncorrelated state

$$E_{\mathrm{QM}}(\vec{n}_1, \vec{n}_2, \ldots, \vec{n}_N) = \mathrm{tr}[\rho \vec{n}_1 \cdot \vec{\sigma} \otimes \vec{n}_2 \cdot \vec{\sigma} \otimes \cdots \otimes \vec{n}_N \cdot \vec{\sigma}] \tag{4.2}$$

where $\otimes$ denotes the tensor product, $\cdot$ the scalar product in $\mathbf{R}^2$, $\vec{\sigma} = (\sigma_x, \sigma_y)$ is a vector of two Pauli operators, and ρ is the pure uncorrelated state,

$$\rho = \rho_1 \otimes \rho_2 \otimes \cdots \otimes \rho_N \tag{4.3}$$

with $\rho_j = |\Psi_j\rangle\langle\Psi_j|$ and $|\Psi_j\rangle$ is a spin-$1/2$ pure state lying in the x-y plane.

One can write the observable (unit) vector $\vec{n}_j$ in a plane coordinate system as follows:

$$\vec{n}_j(\theta_j^{k_j}) = \cos\theta_j^{k_j} \vec{x}_j^{(1)} + \sin\theta_j^{k_j} \vec{x}_j^{(2)}, \tag{4.4}$$

where $\vec{x}_j^{(1)} = \vec{x}$ and $\vec{x}_j^{(2)} = \vec{y}$ are the Cartesian axes. Here, the angle $\theta_j^{k_j}$ takes two values (two-setting model):

$$\theta_j^1 = 0,\ \theta_j^2 = \frac{\pi}{2}. \tag{4.5}$$

We derive a necessary condition to be satisfied by the quantum correlation function with the system in a pure uncorrelated state given in (4.2). In more detail, we derive the value of the product of the quantum correlation function, $E_{\rm QM}$ given in (4.2), i.e., $\|E_{\rm QM}\|^2$. We use the decomposition (4.4). We introduce simplified notations as

$$T_{i_1 i_2 \ldots i_N} = {\rm tr}[\rho \vec{x}_1^{(i_1)} \cdot \vec{\sigma} \otimes \vec{x}_2^{(i_2)} \cdot \vec{\sigma} \otimes \cdots \otimes \vec{x}_N^{(i_N)} \cdot \vec{\sigma}] \tag{4.6}$$

and

$$\vec{c}_j = (c_j^1, c_j^2) = (\cos\theta_j^{k_j}, \sin\theta_j^{k_j}). \tag{4.7}$$

Then, we have

$$\begin{aligned} &\|E_{\rm QM}\|^2 \\ &= \sum_{k_1=1}^{2} \cdots \sum_{k_N=1}^{2} \left(\sum_{i_1,\ldots,i_N=1}^{2} T_{i_1\ldots i_N} c_1^{i_1} \cdots c_N^{i_N} \right)^2 \\ &= \sum_{i_1,\ldots,i_N=1}^{2} T_{i_1\ldots i_N}^2 \leq 1, \end{aligned} \tag{4.8}$$

where we use the orthogonality relation $\sum_{k_j=1}^{2} c_j^{\alpha} c_j^{\beta} = \delta_{\alpha,\beta}$. The value of $\sum_{i_1,\ldots,i_N=1}^{2}$ $T_{i_1\ldots i_N}^2$ is bounded as $\sum_{i_1,\ldots,i_N=1}^{2} T_{i_1\ldots i_N}^2 \leq 1$. We have

$$\prod_{j=1}^{N} \sum_{i_j=1}^{2} ({\rm tr}[\rho_j \vec{x}_j^{(i_j)} \cdot \vec{\sigma}])^2 \leq 1. \tag{4.9}$$

From the convex argument, all quantum separable states must satisfy the inequality (4.8). Therefore, it is a separability inequality. It is important that the separability inequality (4.8) is saturated iff ρ is a multi spin-1/2 pure uncorrelated state such that, for every j, $|\Psi_j\rangle$ is a spin-$1/2$ pure state lying in the x-y plane. The reason of the inequality (4.8) is due to the following quantum inequality

$$\sum_{i_j=1}^{2} ({\rm tr}[\rho_j \vec{x}_j^{(i_j)} \cdot \vec{\sigma}])^2 \leq 1. \tag{4.10}$$

The inequality (4.10) is saturated iff $\rho_j = |\Psi_j\rangle\langle\Psi_j|$ and $|\Psi_j\rangle$ is a spin-$1/2$ pure state lying in the x-y plane. The inequality (4.8) is saturated iff the inequality (4.10) is saturated for every j. Thus we have the maximal possible value of the scalar product as a quantum proposition concerning the existence of the Bloch sphere

$$\|E_{\rm QM}\|_{\max}^2 = 1 \tag{4.11}$$

when the system is in such a multi spin-1/2 pure uncorrelated state.

4.2 The Hidden Variables Theory

On the other hand, a correlation function satisfies the hidden variables theory if it can be written as

$$E_{\rm HV}(\vec{n}_1, \vec{n}_2, \ldots, \vec{n}_N) = \lim_{m\to\infty} \frac{\sum_{l=1}^{m} r(\vec{n}_1, \vec{n}_2, \ldots, \vec{n}_N, l)}{m} \tag{4.12}$$

where l denotes a label and r is the result of measurement of the dichotomic observables parameterized by the directions of $\vec{n}_1, \vec{n}_2, \ldots, \vec{n}_N$.

Assume the quantum correlation function with the system in a pure uncorrelated state given in (4.2) admits the hidden variables theory. One has the following proposition concerning the hidden variables theory

$$E_{\rm QM}(\vec{n}_1, \vec{n}_2, \ldots, \vec{n}_N) = \lim_{m\to\infty} \frac{\sum_{l=1}^{m} r(\vec{n}_1, \vec{n}_2, \ldots, \vec{n}_N, l)}{m}. \tag{4.13}$$

In what follows, we show that we cannot assign the truth value "1" for the proposition (4.13) concerning the hidden variables theory.

Assume the proposition (4.13) is true. By changing the label l into l', we have the same quantum expected value as follows

$$E_{\rm QM}(\vec{n}_1, \vec{n}_2, \ldots, \vec{n}_N) = \lim_{m\to\infty} \frac{\sum_{l'=1}^{m} r(\vec{n}_1, \vec{n}_2, \ldots, \vec{n}_N, l')}{m}. \tag{4.14}$$

An important note here is that the value of the right-hand-side of (4.13) is equal to the value of the right-hand-side of (4.14) because we only change the label.

We abbreviate $r(\vec{n}_1, \vec{n}_2, \ldots, \vec{n}_N, l)$ to $r(l)$ and $r(\vec{n}_1, \vec{n}_2, \ldots, \vec{n}_N, l')$ to $r(l')$.

We introduce an assumption that Sum rule and Product rule commute with each other [44]. We have

$$
\begin{aligned}
&\|E_{\mathrm{QM}}\|^2 \\
&= \sum_{k_1=1}^{2} \cdots \sum_{k_N=1}^{2} \left(\lim_{m\to\infty} \frac{\sum_{l=1}^{m} r(l)}{m} \times \lim_{m\to\infty} \frac{\sum_{l'=1}^{m} r(l')}{m} \right) \\
&= \sum_{k_1=1}^{2} \cdots \sum_{k_N=1}^{2} \left(\lim_{m\to\infty} \frac{\sum_{l=1}^{m}}{m} \cdot \lim_{m\to\infty} \frac{\sum_{l'=1}^{m}}{m} r(l)r(l') \right) \\
&\le \sum_{k_1=1}^{2} \cdots \sum_{k_N=1}^{2} \left(\lim_{m\to\infty} \frac{\sum_{l=1}^{m}}{m} \cdot \lim_{m\to\infty} \frac{\sum_{l'=1}^{m}}{m} |r(l)r(l')| \right) \\
&= \sum_{k_1=1}^{2} \cdots \sum_{k_N=1}^{2} \left(\lim_{m\to\infty} \frac{\sum_{l=1}^{m}}{m} \cdot \lim_{m\to\infty} \frac{\sum_{l'=1}^{m}}{m} \right) = 2^N. \qquad (4.15)
\end{aligned}
$$

We use the following fact

$$|r(\vec{n}_1, \vec{n}_2, \ldots, \vec{n}_N, l) r(\vec{n}_1, \vec{n}_2, \ldots, \vec{n}_N, l')| = +1. \qquad (4.16)$$

The inequality (4.15) is saturated since we have

$$
\begin{aligned}
&\|\{l | r(\vec{n}_1, \vec{n}_2, \ldots, \vec{n}_N, l) = 1 \wedge l \in \mathbf{N}\}\| \\
&= \|\{l' | r(\vec{n}_1, \vec{n}_2, \ldots, \vec{n}_N, l') = 1 \wedge l' \in \mathbf{N}\}\|, \\
&\|\{l | r(\vec{n}_1, \vec{n}_2, \ldots, \vec{n}_N, l) = -1 \wedge l \in \mathbf{N}\}\| \\
&= \|\{l' | r(\vec{n}_1, \vec{n}_2, \ldots, \vec{n}_N, l') = -1 \wedge l' \in \mathbf{N}\}\|. \qquad (4.17)
\end{aligned}
$$

Hence one has the following proposition concerning the hidden variables theory.

$$\|E_{\mathrm{QM}}\|^2_{\max} = 2^N. \qquad (4.18)$$

4.3 Contradiction

Clearly, we cannot assign the truth value "1" for two propositions (4.11) (concerning the existence of the Bloch sphere) and (4.18) (concerning the hidden variables theory), simultaneously, when the system is in a multiparticle pure uncorrelated state. Of course, each of them is a spin-1/2 pure state lying in the x-y plane. Therefore, we are in the KS contradiction when the system is in such a multiparticle pure uncorrelated state. Thus, we cannot accept the validity of the proposition (4.13) (concerning the hidden variables theory) if we assign the truth value "1" for the proposition (4.11) (concerning the existence of the Bloch sphere). In other words, the hidden variables theory does not reveal our physical world.

5 Two-Dimensional No-Hidden-Variables Theorem of the KS Type

In this section, we consider the relation between the double-slit experiment and the hidden variables theory. We try to implement the double-slit experiment. There is a detector just after each slit. Thus interference figure does not appear, and we do not consider such a pattern. The possible values of the result of measurements are ± 1 (in $\hbar/2$ unit). If a particle passes one side slit, then the value of the result of measurement is $+1$. If a particle passes through another slit, then the value of the result of measurement is -1.

5.1 A Wave Function Analysis

Let (σ_z, σ_x) be Pauli vector. We assume that a source of spin-carrying particles emits them in a state $|\psi\rangle$, which can be described as an eigenvector of a Pauli observable σ_z. We consider a quantum expected value $\langle\sigma_x\rangle$ as

$$\langle\sigma_x\rangle = \langle\psi|\sigma_x|\psi\rangle = 0. \tag{5.1}$$

The above quantum expected value is zero if we consider only a wave function analysis.

We derive a necessary condition for the quantum expected value for the system in the pure spin-1/2 state $|\psi\rangle$ given in (5.1). We derive the possible value of the product $\langle\sigma_x\rangle \times \langle\sigma_x\rangle = \langle\sigma_x\rangle^2$. $\langle\sigma_x\rangle$ is the quantum expected value given in (5.1). We derive the following proposition

$$\langle\sigma_x\rangle^2 = 0. \tag{5.2}$$

Hence we have

$$\langle\sigma_x\rangle^2 \leq 0. \tag{5.3}$$

Thus,

$$(\langle\sigma_x\rangle^2)_{\max} = 0. \tag{5.4}$$

5.2 The Hidden Variables Theory

On the other hand, a mean value E admits the hidden variables theory if it can be written as

$$E = \frac{\sum_{l=1}^{m} r_l(\sigma_x)}{m} \tag{5.5}$$

where l denotes a label and r is the result of measurement of the Pauli observable σ_x. We assume the value of r is ± 1 (in $\hbar/2$ unit).

Assume the quantum mean value with the system in an eigenvector ($|\psi\rangle$) of the Pauli observable σ_z given in (5.1) admits the hidden variables theory. One has the following proposition concerning the hidden variables theory

$$\langle \sigma_x \rangle(m) = \frac{\sum_{l=1}^{m} r_l(\sigma_x)}{m}. \tag{5.6}$$

We can assume as follows by Strong Law of Large Numbers [45],

$$\langle \sigma_x \rangle(+\infty) = \langle \sigma_x \rangle = \langle \psi | \sigma_x | \psi \rangle. \tag{5.7}$$

In what follows, we show that we cannot assign the truth value "1" for the proposition (5.6) concerning the hidden variables theory.

Assume the proposition (5.6) is true. By changing the label l into l', we have the same quantum mean value as follows

$$\langle \sigma_x \rangle(m) = \frac{\sum_{l'=1}^{m} r_{l'}(\sigma_x)}{m}. \tag{5.8}$$

An important note here is that the value of the right-hand-side of (5.6) is equal to the value of the right-hand-side of (5.8) because we only change the label.

We introduce an assumption that Sum rule and Product rule commute with each other [44]. We have

$$\begin{aligned} &\langle \sigma_x \rangle(m) \times \langle \sigma_x \rangle(m) \\ &= \frac{\sum_{l=1}^{m} r_l(\sigma_x)}{m} \times \frac{\sum_{l'=1}^{m} r_{l'}(\sigma_x)}{m} \\ &= \frac{\sum_{l=1}^{m}}{m} \cdot \frac{\sum_{l'=1}^{m}}{m} r_l(\sigma_x) r_{l'}(\sigma_x) \\ &\leq \frac{\sum_{l=1}^{m}}{m} \cdot \frac{\sum_{l'=1}^{m}}{m} |r_l(\sigma_x) r_{l'}(\sigma_x)| \\ &= \frac{\sum_{l=1}^{m}}{m} \cdot \frac{\sum_{l'=1}^{m}}{m} = 1. \end{aligned} \tag{5.9}$$

We use the following fact

$$|r_l(\sigma_x) r_{l'}(\sigma_x)| = 1. \tag{5.10}$$

The inequality (5.9) is saturated since we have

$$\|\{l|r_l(\sigma_x) = 1 \wedge l \in \mathbf{N}\}\| = \|\{l'|r_{l'}(\sigma_x) = 1 \wedge l' \in \mathbf{N}\}\|,$$
$$\|\{l|r_l(\sigma_x) = -1 \wedge l \in \mathbf{N}\}\| = \|\{l'|r_{l'}(\sigma_x) = -1 \wedge l' \in \mathbf{N}\}\|. \tag{5.11}$$

Thus we derive a proposition concerning the quantum mean value under an assumption that the hidden variables theory is true (in a spin-1/2 system), that is

$$(\langle\sigma_x\rangle(m) \times \langle\sigma_x\rangle(m))_{\max} = 1. \tag{5.12}$$

From Strong Law of Large Numbers, we have

$$(\langle\sigma_x\rangle \times \langle\sigma_x\rangle)_{\max} = 1. \tag{5.13}$$

Hence we derive the following proposition concerning the hidden variables theory

$$(\langle\sigma_x\rangle^2)_{\max} = 1. \tag{5.14}$$

5.3 *Contradiction*

We do not assign the truth value “1” for two propositions (5.4) (concerning a wave function analysis) and (5.14) (concerning the hidden variables theory), simultaneously. We are in the KS contradiction.

We cannot accept the validity of the proposition (5.6) (concerning the hidden variables theory) if we assign the truth value “1” for the proposition (5.4) (concerning a wave function analysis). In other words, the hidden variables theory does not meet the detector model for the spin observable σ_x.

6 Conclusions

In conclusion, we have derived a new type of no-hidden-variable theorem based on the assumptions proposed by Kochen and Specker. We have considered N spin-1/2 systems. The hidden results of measurement have been either $+1$ or -1 (in $\hbar/2$ unit). We have derived some proposition concerning a quantum expected value under an assumption about the existence of the Bloch sphere in N spin-1/2 systems. However, the hidden variables theory has violated the proposition with a magnitude that grows exponentially with the number of particles. Therefore, we have to have given up either the existence of the Bloch sphere or the hidden variables theory. Also we have discussed two-dimensional no-hidden-variables theorem of the KS type. Especially, we have systematically described our assertion based on more mathematical analysis using raw data in a thoughtful experiment.

Acknowledgements We would like to thank Professor Niizeki and Dr. Ren for valuable comments.

References

1. Sakurai, J.J.: Modern Quantum Mechanics. Revised ed. Addison-Wesley Publishing Company (1995)
2. Peres, A.: Quantum Theory: Concepts and Methods. Kluwer Academic, Dordrecht, The Netherlands (1993)
3. Redhead, M.: Incompleteness, Nonlocality, and Realism. 2nd ed. Clarendon Press, Oxford (1989)
4. Nielsen, M.A., Chuang, I.L.: Quantum Computation and Quantum Information. Cambridge University Press (2000)
5. von Neumann, J.: Mathematical Foundations of Quantum Mechanics. Princeton University Press, Princeton, New Jersey (1955)
6. Feynman, R.P., Leighton, R.B., Sands, M.: Lectures on Physics, Volume III, Quantum mechanics. Addison-Wesley Publishing Company (1965)
7. Kochen, S., Specker, E.P.: J. Math. Mech. **17**, 59 (1967)
8. Greenberger, D.M., Horne, M.A., Zeilinger, A.: In: Kafatos, M. (ed.) Bell's Theorem, Quantum Theory and Conceptions of the Universe, pp. 69–72. Kluwer Academic, Dordrecht, The Netherlands (1989)
9. Greenberger, D.M., Horne, M.A., Shimony, A., Zeilinger, A.: Am. J. Phys. **58**, 1131 (1990)
10. Pagonis, C., Redhead, M.L.G., Clifton, R.K.: Phys. Lett. A **155**, 441 (1991)
11. Mermin, N.D.: Phys. Today **43**(6), 9 (1990)
12. Mermin, N.D.: Am. J. Phys. **58**, 731 (1990)
13. Peres, A.: Phys. Lett. A **151**, 107 (1990)
14. Mermin, N.D.: Phys. Rev. Lett. **65**, 3373 (1990)
15. Simon, C., Brukner, Č., Zeilinger, A.: Phys. Rev. Lett. **86**, 4427 (2001)
16. Larsson, J.Å.: Europhys. Lett. **58**, 799 (2002)
17. Cabello, A.: Phys. Rev. A **65**, 052101 (2002)
18. Nagata, K.: J. Math. Phys. **46**, 102101 (2005)
19. Huang, Y.-F., Li, C.-F., Zhang, Y.-S., Pan, J.-W., Guo, G.-C.: Phys. Rev. Lett. **90**, 250401 (2003)
20. Werner, R.F.: Phys. Rev. A **40**, 4277 (1989)
21. Leggett, A.J.: Found. Phys. **33**, 1469 (2003)
22. Gröblacher, S., Paterek, T., Kaltenbaek, R., Brukner, Č., Żukowski, M., Aspelmeyer, M., Zeilinger, A.: Nature (London) **446**, 871 (2007)
23. Paterek, T., Fedrizzi, A., Gröblacher, S., Jennewein, T., Żukowski, M., Aspelmeyer, M., Zeilinger, A.: Phys. Rev. Lett. **99**, 210406 (2007)
24. Branciard, C., Ling, A., Gisin, N., Kurtsiefer, C., Lamas-Linares, A., Scarani, V.: Phys. Rev. Lett. **99**, 210407 (2007)
25. Suarez, A.: Found. Phys. **38**, 583 (2008)
26. Żukowski, M.: Found. Phys. **38**, 1070 (2008)
27. Suarez, A.: Found. Phys. **39**, 156 (2009)
28. Deutsch, D.: Proc. Roy. Soc. London Ser. A **400**, 97 (1985)
29. Deutsch, D., Jozsa, R.: Proc. Roy. Soc. London Ser. A **439**, 553 (1992)
30. Cleve, R., Ekert, A., Macchiavello, C., Mosca, M.: Proc. Roy. Soc. London Ser. A **454**, 339 (1998)
31. Jones, J.A., Mosca, M.: J. Chem. Phys. **109**, 1648 (1998)
32. Gulde, S., Riebe, M., Lancaster, G.P.T., Becher, C., Eschner, J., Häffner, H., Schmidt-Kaler, F., Chuang, I.L., Blatt, R.: Nature (London) **421**, 48 (2003)
33. de Oliveira, A.N., Walborn, S.P., Monken, C.H.: J. Opt. B: Quantum Semiclass. Opt. **7**, 288–292 (2005)

34. Kim, Y.-H.: Phys. Rev. A **67**, 040301(R) (2003)
35. Mohseni, M., Lundeen, J.S., Resch, K.J., Steinberg, A.M.: Phys. Rev. Lett. **91**, 187903 (2003)
36. Tame, M.S., Prevedel, R., Paternostro, M., Böhi, P., Kim, M.S., Zeilinger, A.: Phys. Rev. Lett. **98**, 140501 (2007)
37. Bernstein, E., Vazirani, U.: Proceedings of the Twenty-Fifth Annual ACM Symposium on Theory of Computing (STOC '93), pp. 11–20 (1993). doi:10.1145/167088.167097; SIAM J. Comput. 26–5, pp. 1411–1473 (1997)
38. Simon, D.R.: Foundations of Computer Science. In: Proceedings of the 35th Annual Symposium on: 116-123, retrieved 2011-06-06 (1994)
39. Du, J., Shi, M., Zhou, X., Fan, Y., Ye, B.J., Han, R., Wu, J.: Phys. Rev. A **64**, 042306 (2001)
40. Brainis, E., Lamoureux, L.-P., Cerf, N.J., Emplit, P., Haelterman, M., Massar, S.: Phys. Rev. Lett. **90**, 157902 (2003)
41. Li, H., Yang, L.: Quantum Inf. Process. **14**, 1787 (2015)
42. De Broglie-Bohm theory—Wikipedia, the free encyclopedia
43. Schon, C., Beige, A.: Phys. Rev. A **64**, 023806 (2001)
44. Nagata, K., Nakamura, T.: Physics Journal **1**(3), 183 (2015)
45. In probability theory, the law of large numbers is a theorem that describes the result of performing the same experiment a large number of times. According to the law, the average of the results obtained from a large number of trials should be close to the expected value, and will tend to become closer as more trials are performed. The strong law of large numbers states that the sample average converges almost surely to the expected value

Proposal for a Quantum-Based Memory for Storing Classical Information and the Connection Between Molecular Dynamics Simulations and the Landauer's Principle

Josep Batle, Mohamed Elhoseny and Ahmed Farouk

Abstract The development of high-capacity memory devices plays an increasingly important role in modern society. High capacities in information storage constitutes a key resource for dealing with the everyday generation of information, as well as for handling the so called Big Data generated in different scientific and technological scenarios. By combining precision metrology and quantum devices such as quantum dots and quantum wires, we propose a quantum memory whose capacity depends on the particular architecture chosen, namely, linear or planar. We show that the geometric disposition of minimal quantum cells or chips is critical in having similar or dramatically outperformed information capacities as compared to current devices. This information is stored in the form of classical bits, though. Realization of such a quantum memory may solve a two-fold problem at the same time: unprecedented higher information capacity with undefined longevity.

We shall obtain as well, by rigorously applying the definition of the exponentiation of a Hermitian matrix, the set of Hamiltonians whose evolution corresponds to the set of universal gates.

Also, Landauer's principle is a fundamental link between thermodynamics and information theory, which implies that the erasure of information comes at an energetic price, either in classical or quantum computation. In the present contribution we analyze to what extend the usual molecular dynamics (MD) simulation formalism can handle the Landauer's bound $k_B T \ln 2$ in the simplest case of one particle treated classically. The erasure of one bit of information is performed by adiabatically varying the shape of a bistable potential in a full cycle. We will highlight the inadequacy

J. Batle (✉)
Departament de Física, Universitat de Les Illes Balears, Balearic Islands, 07122 Palma de Mallorca, Spain
e-mail: jbv276@uib.es

M. Elhoseny
Faculty of Computers and Information, Mansoura University, Mansoura, Egypt
e-mail: mohamed_elhoseny@mans.edu.eg

A. Farouk
Scientific Research Group in Egypt (SRGE), Cairo, Egypt
e-mail: dr.ahmedfarouk85@yahoo.com
URL: http://www.egyptscience.net

A.E. Hassanien et al. (eds.), *Quantum Computing: An Environment for Intelligent Large Scale Real Application*, Studies in Big Data 33,
https://doi.org/10.1007/978-3-319-63639-9_13

of either the microcanonical or canonical ensemble treatments currently employed in MD simulations and propose potential solutions.

1 Introduction

Ever since the introduction of writing, several supports have been used in order to store information (wood, rock, paper, etc.). These physical supports have not prevented the corruption of the information contained due to erosion or other physical/chemical processes acting on the organic materials employed, although some writings have remained intelligible for centuries and even millennia. On the other hand, phonograph records such as The Golden Record carried in the Voyager missions [1]–which do not dramatically differ from previous writing methods–constitute examples of extremely well-designed systems were longevity is expected to be of the order of several million years [1]. All these previous examples constitute analog systems of storing information, differing only in their longevity scale.

In recent years, it has become the subject of universal concern the study of the storage of information in a way that it can endure being uncorrupted. Modern systems of magnetic storage of digital information have a finite lifespan, and devices commonly used to store information such as CDs or DVDs cannot offer unlimited endurance. Due to this fact, it is not nonsensical to assert the legacy of Mankind is at a stake if present generations are not able to transmit the current knowledge to the future ones.

In point of fact, the scenario of modern times is two-fold problematic: (i) on the one hand, the everyday Big Data generation of information is enormous, which needs to be stored and therefore requires a physical support, whereas on the other hand (ii) it has to possess a reasonable longevity. These two concepts, big capacity and longevity, will be required in the near future to prevent an informational collapse.

Along these lines, and avoiding the constraints of current state-of-the-art classical information storage, a large number of proposals have been developed for how one can construct quantum memories based on minimal physical systems such as atomic ensembles [2]. A large class of quantum memory approaches based on atomic ensembles uses classical laser fields to control the memory process in such a way that the incoming field is mapped into the ground-state coherence of the atoms (see for instance [3–9]). However, the need for quantum coherence requires these quantum systems to have very specific conditions such as low temperatures and a high control on the environment. This extreme control is nowadays impossible to reach on demand, for otherwise quantum memories would be already a reality.

Here we shall pursue a different approach. We shall try to find a solution to the storage of large amounts of information and, at the same time, make the supporting system to be robust against the action of the environment. As we shall see, the abstract concept of bits of information will remain classical, but the underlying physical system will be fully quantum. In this way, we believe that our proposal can constitute a mid-term solution to the problem, employing the current technology available.

However, we are positive about the fact that the final solution will be purely quantum in the future.

Ever since Deutsch [24] introduced a universal gate for three qubits, universal gates for qubits have been extensively studied [25–28]. Widely speaking, a set of universal quantum gates is any set of gates to which any operation can be reduced. In other words, any other unitary operation can be expressed as a finite approximate sequence of gates.

The actual realization of a universal quantum computer must fulfill several requirements, such as the storage of quantum information into two-level systems or *qubits*, readout, etc. DiVincenzo [29] actually put together a set of rules that generalized the task of the physical realization of a quantum computer.

Here we shall be concerned with those Hamiltonians that lead to the set of universal gates introduced by Barenco [30]. We shall provide an approach to quantum gates, vital in quantum computation, from the operational perspective. That is, we introduce a way to characterize them that involves the usual Pauli operators, which can easily be implemented say, for instance, in quantum optics.

The fundamental bound on the minimum thermodynamic energy cost of information processing –quantum computing being one part of it, has been a topic of active research [31–38, 43, 44]. According to Landauer's principle [34], an average of at least $k_{\mathrm{B}}T\ln 2$ of work is required to delete one bit of information from a physical system. A direct consequence of this logically irreversible transformation is that the entropy of the environment increases by a finite amount.

Recent technological developments have enabled the direct measurement of such small amounts of work for small non-equilibrium thermodynamic systems [47]. In particular, Bérut et al. [48] recently showed experimentally the Landauer's principle linking information and thermodynamics.

The fact that erasure of information is a process which costs free energy has interesting echoes in quantum information theory as well. To be more precise, if one is able to efficiently erase information, which is tantamount as to saturate Landauer's bound $k_BT\ln 2$, then one can provide a physical interpretation [49, 50] of the so called Holevo bound [51], which is related to the information capacity in quantum channels.

In the quantum realm, it was pointed out [52], however, that erasure of a single unknown state may be thought of as swapping it with some standard state and then dumping it into the environment. This kind of deletion led to the conclusion that there is no quantum eraser that can delete one unknown state against a copy in either a reversible or an irreversible manner. This fact would imply that a failure of Landauer's erasure principle deep in the quantum domain would therefore have far-reaching consequences. Fortunately, this conundrum was resolved in [53], and hence establishing the validity of the erasure principle in classical and quantum physics for arbitrary reservoir interaction strengths.

A topic which is also addressed here is that MD. MD is a technique for computer simulation of complex systems, modeled at the atomic level. The equations of motion are solved numerically to follow the time evolution of the system, allow-ing the derivation of kinetic and thermodynamic properties computationally. MD emerged as one of the first simulation methods from the pioneering applications to the dynamics of liquids. Early on, Alder and Wainwright [54] applied it to study the hard sphere fluid. Afterwards, many authors analyzed Lennard-Jones fluids using this method [55, 60]. For references on basic MD implementation see [61].

Macroscopic quantities are computed using time averages. This is supported by the ergodic hypothesis, which states that the time average equals the ensemble average. In the microcanonical ensemble, there is no defined temperature, as opposite to the canonical ensemble. This implies that the former is not embedded in a thermal bath, whereas the latter is, giving rise to the Gibbs factor.

The purpose of the present work is also to elucidate to what extend one can use the Landauer's principle in an scenario where no heat bath is contemplated, that is, in physical systems where the temperature T has no thermodynamical meaning. We shall also discuss the erasure of one bit of information when MD are approached from the canonical ensemble view, that is, for a fixed temperature T.

The concept of quantum memory with classical bits

A quantum memory has to necessarily make use of quantum bits or *qubits* encoded in the state of a quantum system. In the Bloch representation, they constitute a linear superposition of states $|0\rangle$ and $|1\rangle$ in the form $\sin\theta|0\rangle + \cos\theta e^{i\phi}|1\rangle$. Classical information only uses the poles $\{|0\rangle, |1\rangle\}$ to perform abstract operations. A memory being quantum implies that its constituents follow the tenets of quantum mechanics. Therefore we shall treat the evolution of them according to some unitary transformation.

In our case, the quantum memory will consist of an string of minimal memory chips. The different spatial distribution of the array of memory chips will define a particular architecture. Our proposal will contain a quantum chip as the one depicted in Fig. 1. There we can see that one positive charge is stored in a quantum dot, which in turn is connected through a quantum wire to another quantum dot where a negative charge is confined (present techniques may soon allow quantum dots to operate at room temperature). If the diameter of a wire is sufficiently small, the confined charged particle inside (another electron) will experience quantum confinement in the transverse direction. As a result, their transverse energy will be quantized into a series of discrete values. But these energies are not relevant here, because we shall focus only in the longitudinal direction. The confined electron is thus under the action of the Coulomb electrostatic interaction between the positive charge in one end, and the negative one at the other end. The effective potential of this chip will be that of the 1D Hydrogen atom plus a repulsion. That is,

$$V_c(x) = \begin{cases} \infty & \text{if } x \leq 0 \\ -\frac{1}{x} + \frac{1}{l_0 - x} & \text{if } x > 0 \end{cases},$$

where l_0 is the length between endpoints. The corresponding eigenfunctions after solving the corresponding Schrödinger equation are similar to the analytic ones obtained from the 1D Hydrogen atom, although the eigenenergies differ considerably

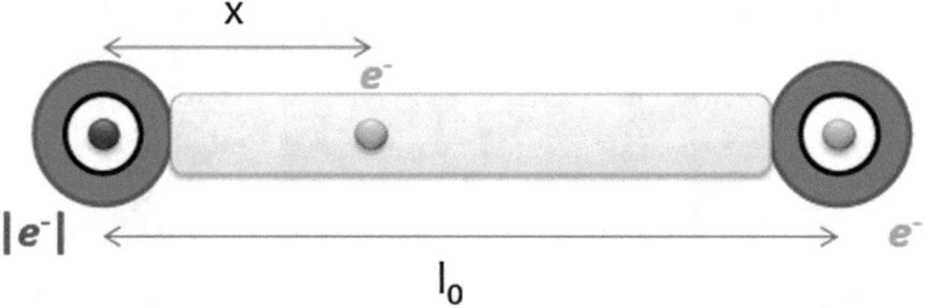

Fig. 1 **Structure of the minimal quantum memory chip**. Two quantum dots placed at the endpoints of a nanowire of lenght l_0, each one containing a proton and an electron, respectively, confine an electron inside the quantum wire. Although wavefunctions have similar shape as compared to the analytic 1D Hydrogen atom, eigenenergies differ absolutely (in the quantum chip, they are monotonically increasing with the quantum number N). See text for details

(in the 1D-Hydrogen atom, E_n are proportional to $1/n^2$, whereas in the present case, they go as $An^2 - B$, where A and B are positive constants).

The Schrödinger equation describing the electron confined between a proton and another electron in the 1D potential $V(x) = \frac{-1}{x} + \frac{1}{l_0 - x}$ is described (in atomic units) by

$$-\frac{1}{2}\nabla^2\psi(x) + V(x)\psi(x) = E_N\psi(x). \tag{1}$$

The corresponding solved first three states $\psi_N(x)$ are depicted in Fig. 2 for $l_0 = 10$ a.u. They are almost identical to the 1D Hydrogen atom [10], except for the fact that go to zero sharply at $x = l_0$. Interest in this particular potential is not merely mathematical. The reason for the great deal of interest in discussing the 1D hydrogen atom potential is partly due to its wide applications to different physical topics involving excitons in high-temperature superconductivity [11], semiconductors [13, 25], polymers [26] or 1D electron gas at the helium surface [27] and the so-called Wigner crystal [17, 28].

In our case, the action of the background has to be fully considered considered if one has to be consistent with all electrostatic interactions taking place. The original potential is $V(x) = \frac{-1}{x} + \frac{1}{l_0 - x} = (2x - l_0)\frac{1}{(x)(l_0 - x)}$, which is clearly asymmetric function around $x = l_0/2$. If take into account the action of all the rest of charges, we then have a potential of the type

$$\begin{aligned} V_T(x) &= (2x - l_0) \sum_{i=0}^{\infty} \frac{(-1)^i}{(x + l_0 i)(l_0 - x + l_0 i)} \\ &\equiv \Phi(-1, 1, x) - \Phi(-1, 1, l_0 - x), \end{aligned} \tag{2}$$

where $\Phi(z, s, a)$ is the Hurwitz Lerch transcendent, defined by $\sum_{k=0}^{\infty} \frac{z^k}{(a+k)^s}$ [18]. This total potential converges and the asymmetry around $x = 1_0/2$ is clearly observed there. This total confining potential resembles that of an infinite square well: it can retain an infinite number of bound states and the tails at the endpoints are responsible

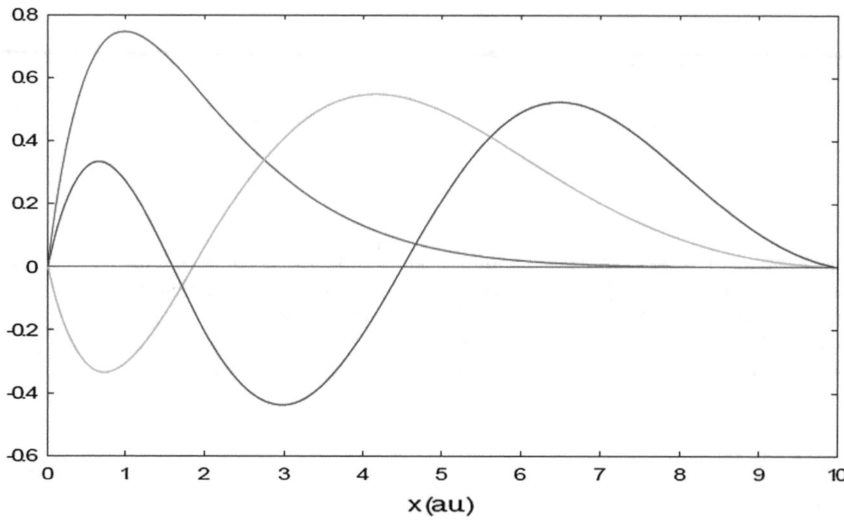

Fig. 2 First three eigenfunctions for the confining potential of the chip with no background. Even the action of linear or quadratic backgrounds do not change much the corresponding. That is why expectation values such as the position of the confined electron are not substantially altered

for the eigenfunctions to be null there. As a matter of fact, the spectrum of energies looks like a quadratic function with the quantum number N. The first four eigenenergies obtained without the background are $\{-0.3813, 0.0911, 0.3875, 0.7528\}$, whereas with the all iterations we obtain $\{-0.3358, 0.1014, 0.3857, 0.7489\}$.

In the planar case, we obtain the same $x = 1_0/2$-asymmetry. In this case, the outcome of the final potential is not analytic, but it converges. When computing the corresponding energies solving the Schrödinger equation (1), neither the energies nor the average positions change much. Converge for the total potential is reached already for plaquettes of 100 l_0 side length.

Here lies our proposal: all the interactions present in the system will be of electrostatic nature. In this way, we are avoiding having to maintain the system at low temperatures, or having to isolate it by complex means from the environment. Now that we have our minimal quantum system settled, we have to define the abstract notion of classical bits 0 and 1. Since we are assuming all our interactions to be of electrostatic nature, *the presence or absence of the confined electron will be our 1 and 0, respectively*. This is done in such a way that the readout of information will be solely based in determining the presence of the confined electron. Let us summarize the steps that have to be followed in our proposed quantum memory (we assume that each individual chip is independent of the next one):

- We load all the electrons of the chips in the memory. This injection could be performed by the emission of the single electrons from a condensed phase of metallic or semiconducting materials that are subjected to high electric field into the vacuum [19]. Noteworthy, emission occurs by electrons tunneling through a surface potential barrier that requires no supply of energy to electrons (making the process temperature independent). Sharp tips of a few hundred nanometres in radii have been prepared by electrochemical etching of fine tungsten wires of high purity (99.9%) [20].

- Although writing is a process that is conceived to be repeated many times, in the present approach it is only done once, for the information has to be stored and read only. Depending on the technology employed, confined electrons may be either drained using standard semiconductor-related techniques or promoted to the continuum by shining a laser on them. In the latter case, they have to be collected from an electrode plate placed above the whole system.
- Readout, as stressed before, must rely on the detection of single electrons confined inside each quantum wire. One way is to use an electrode tip positioning with an accuracy in the nanometer range, so that it can detect the presence of a single electron. Needless to say, low drift, low backlash, and insusceptibility to vibrations of the micro-probing setup are indispensable for a successful use of this method. On the contrary, one can use other types of technologies available for detecting a single charge. One such candidate could be a quantum point contact (QPC) placed at the end of a nanometer tip. A QPC is a narrow constriction between two wide electrically conducting regions, of a width comparable to the electronic wavelength. Since the conductance through the contact strongly depends on the size of the constriction, any potential fluctuation (for instance, created by other electrons) in the vicinity will influence the current through the QPC, which makes it good candidate for a readout device.

In view of the previous steps for storing and reading information, it is plain that what we mean by "quantum memory with classical bits" is nothing but the usage of small quantum devices, and regard the'presence' or'absence' of an elementary charge during readout as classical '1' or'0'. This is not a fully quantum treatment, although we do exploit the low size of them to increase the total physical memory.

One- and two-dimensional architectures. Memory capacity

The disposition in space of each quantum chip will define a precise architecture for the quantum memory. The fact of aligning all chips in one dimension, in a zig-zag shape or in two dimensions has a considerable implication in the amount of information being stored. In point of fact, the existence of the Coulombian long-range interaction affects the effective potential of the trapped electron. This fact, in turn, becomes relevant for the eigenenergies and eigenfunctions of the confined electron (as explained in the Supplementary Information, the effect on the previous ones is not significant). However, we have to consider that our approach of growing the quantum memory could not be considered if all charges had the same sign. In these case, it would not be difficult to prove that the electronic background would make a huge contribution to every quantum chip, rendering the whole system explosively unstable.

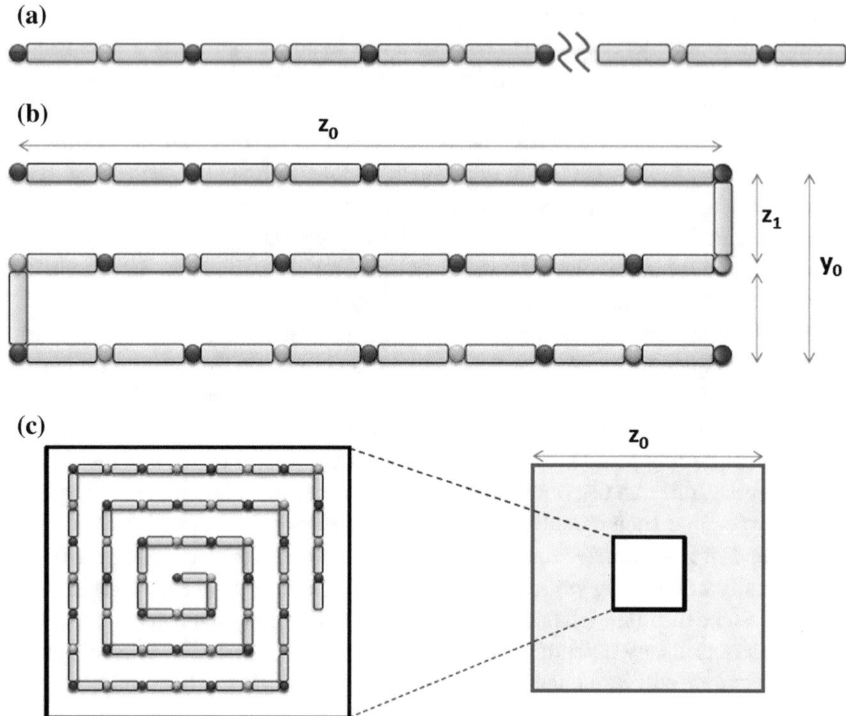

Fig. 3 **Different architectures of the quantum memory**. **a** Rod structure of the quantum chips. Due to the one-directional design, it is the simplest possible functional design. **b** Zig-zag disposition of the linear design in **a**. Longitudinal length has to be much greater than the transverse one. **c** Fully two-dimensional quantum memory. Readout in this square spiral has to have maximum precision and stability

In Fig. 3 we can appreciate the different possible geometries for the proposed quantum memory. In all cases, the distance of quantum wires is l_0. Let us now discuss the memory capacity for every architecture. In the one-directional design, every chip can be either occupied or empty, which is represented by one bit of information. In other words, for a total length of the rod z_0, we have $z_0 = n\, l_0$, where n is the number of stored bits. Thus, the total information in the length z_0 is given by n bits.

In the case of a zig-zig configuration, we define the lateral size $z_1 = \nu l_0$, in addition to $z_0 = nl_0$. Besides, we want the total lateral size y_0 to be a fraction of z_0, so that the main contribution is longitudinal. Thus, $y_0 = \frac{z_0}{f}$. By adding all lengths, we finally obtain a total length

$$L = \left(\frac{n^2}{\nu f} + (\frac{1}{f} + 1)n\right) l_0 \equiv S_n\, l_0, \tag{3}$$

where S_n is the number of bits stored in this configuration. Although the leading terms seems to be quadratic with n, it is in fact proportional to n. In practice, in order to make the system look like almost one-dimensional, we shall take f between 4 and 10.

Now, the case where the system grows starting from a square spiral is a bit more involved. Like in the previous configuration, the micro-electrode tip needs more precise movements. This fact may slow-down readout. However, this planar configurations will provide a better storage capacity. Every turn of the spiral is increased by 8 units. Staring from the center, we get the following series 8+16+24+..., which is an arithmetic series for a general term of the kind $a_n = 8 + 8(n-1) = 8n$. After n turns, we have a total length of

$$L = 4n(n+1)\, l_0 \approx 4n^2\, l_0 = S_n\, l_0. \tag{4}$$

Regarding z_0 as the side of a the square where the spiral is embedded, we thus have $z_0 = 2n\, l_0$. Thus, the information capacity in terms of the size of a typical square would be given by z_0^2/l_0^2. Among the three cases considered, it is apparent that the last one is capable of storing much more information by simply folding the line of quantum chips forming a spiral. As in usual CDs and DVDs, readout would be done outwards. Of course, this problem can be avoided if writing is done in rows, in which case the micro-electrode tip just has to read longitudinally.

Here we shall provide a comparison of information capacities among the three cases. For the sake of simplicity, we shall consider the linear and the planar structures. For such a purpose, we shall define the length l_0 to be 10 atomic units, that is to say, $5.292\, 10^{-10}$ m. If we assume a reasonable size $z_0 = 0.1$ meters, then we have $n \approx 2\, 10^8$ bits, which is $1/1000$ times the capacity of a typical one-sided Blu-ray disc. Changing l_0 and z_0 according to present state-of-the-art technology, capacity definitely can be definitely increased at will. The planar structure of the square spiral contains a capacity that goes as $(z_0/l_0)^2$, that is, the square of the linear one. For the same $\{l_0, z_0\}$, we thus have $n \approx (2\, 10^8)^2 = 4\, 10^{16}$ bits, which corresponds to 200000 times the capacity of a one-sided Blu-ray disc.
Depending of the purpose of the quantum memory, that is, storing large amounts of data like a library, one can easily compare this capacity with that of the Library of Congress, which has more than 24 petabits of digital collections [21]. Both capacities are almost equal. In other words, a square z_0 of 10 cm containing spiral chips of length $l_0 = 10$ a.u. can store as much information as the whole Library of the Congress of the United States. Of course, limitations in size due to present technology can change considerably the eventual size, but the ultimate outcome of storing large amounts of information could in principle be achieved.
The intermediate solution provided by the zig-zag structure could provide an intermediate solution, easy to handle and with a reasonable capacity.

Stability analysis

Suppose that we want to operate at room temperature. The thermal energy induced in the confined electron is of the order of a thousand Hartrees, which from Fig. 4 we can appreciate that it corresponds to several hundreds units in the principal number N. In turn, the average position $\langle x \rangle$ for those energies is around $l_0/2$. It is remarkable to compare this result with the one obtained with the infinite square well, where the average position is always $l_0/2$, regardless of the state of the particle. Realization of the full-scale network of the quantum memory described here will require a number of technological advances in both metrology and experimental maintenance of quantum properties at room temperature. Promising new experimental graphene quantum dots endowed with addition energies as large as 1.6 eV (room temperature) fabricated by the controlled rupture of a graphene sheet have been obtained recently [22]. Atomic and molecular junctions are an emerging class materials that exploit quantum confinement effects to obtain an enhanced figure of merit. The case of QPC is still a bit far of being operational at room temperature, although some advances have been achieved along these lines [23].

The quantum memory, regardless of the corresponding structure, also has to prevent the action of the environment from the classical point of view. We do not have to worry about quantum coherence as is the fully quantum memory. However, the most challenging issue here will be that of preventing the leakage of charges. An additional ultra-thing metallic layer should cover all quantum chips and additional devices once the information has been stored.

2 Class of Universal Gates

Consider any two qubit gate **A** whose action is given in the computation basis $\{|00\rangle, |01\rangle, |10\rangle, |11\rangle\}$ by the unitary matrix

$$A(\phi, \alpha, \theta) = \begin{pmatrix} 1\,0 & 0 & 0 \\ 0\,1 & 0 & 0 \\ 0\,0 & e^{i\alpha}\cos\theta & -ie^{i(\alpha-\phi)}\sin\theta \\ 0\,0 & -ie^{i(\alpha+\phi)}\sin\theta & e^{i\alpha}\cos\theta \end{pmatrix}, \qquad (5)$$

where ϕ, α and θ are fixed irrational multiples of π and of each other. Barenco [30] showed that any such gate is universal.

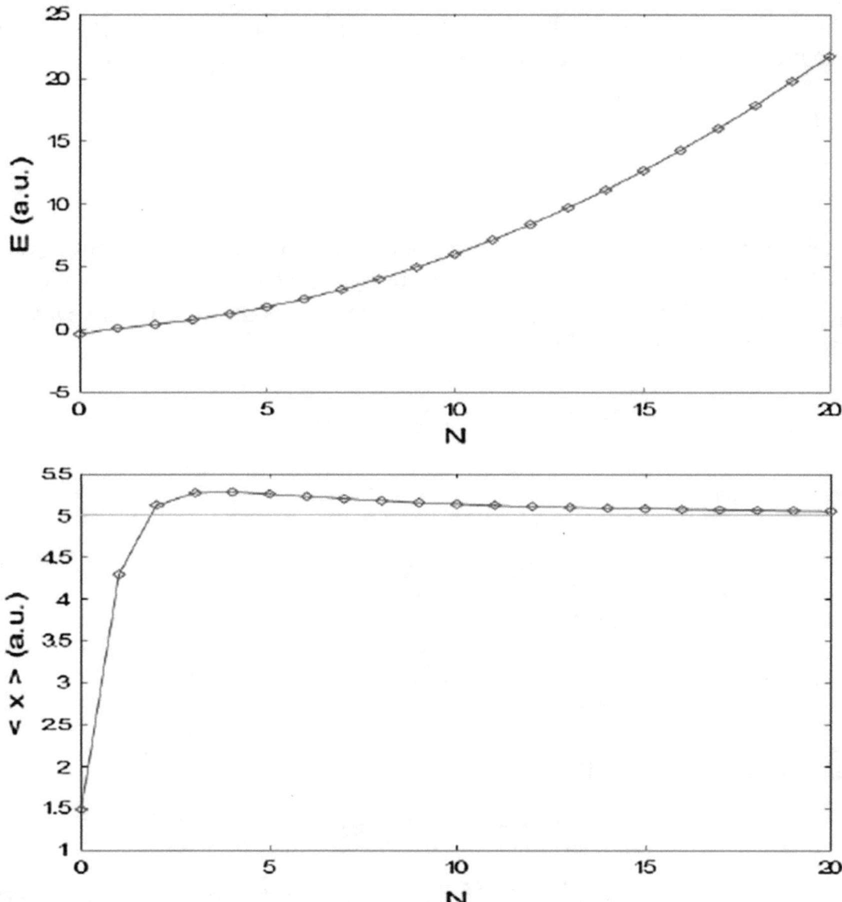

Fig. 4 Energy and average position of the confined electron versus the quantum main quantum number N for $l_0 = 10$ **a.u**. Even though the background contributes to confining the particle even more (see Supplementary Information for details), energies and mean positions do not change significantly. **a** Spectrum of the electron inside the quantum chip. Energies are given in Hartrees. The functional dependency is almost quadratic. **b** Computed mean positions of confined electrons versus N. We can appreciate that it quickly reaches the value $l_0/2$

3 Class of Universal Hamiltonians

After some algebra, we obtain that any universal computational quantum gate $A(\phi, \alpha, \theta)$ (5) is written in terms of spin-$\frac{1}{2}$ operators as

$$\frac{1}{2}\left((I_{2\times2} + \sigma_z) \otimes I_{2\times2} + (I_{2\times2} - \sigma_z) \otimes e^{i[\alpha - \theta \mathbf{n}(\phi)\cdot\sigma]} \right), \tag{6}$$

where $\mathbf{n}(\phi) = (\cos\phi, \sin\phi, 0)$ and $\sigma = (\sigma_x, \sigma_y, \sigma_z)$. The rotation nature of (5) is only mentioned in the original work [30]. Here we show it explicitly.

Now let us apply the following:

$$A = e^{iH} \longrightarrow H = -i \log A. \tag{7}$$

Notice that we have to take the logarithm of a matrix, and logarithm of complex numbers are multivalued. Either employing the direct diagonalization $\log A = \sum_i \log \lambda_i |\lambda_i\rangle\langle\lambda_i|$, or using straightforward computation of the expansion in the Mercator series for a matrix X from (6)

$$\log(I + X) = X - \frac{X^2}{2} + \frac{X^3}{3} - \frac{X^4}{4} + \cdots \tag{8}$$

we obtain the result

$$H = h(\phi, \pi n, -\pi n') + h(\phi, \alpha, \theta), \tag{9}$$

n and n' being integers, with

$$h(\phi, \alpha, \theta) = \frac{I_{2\times2} - \sigma_z}{2} \otimes [\alpha I_{2\times2} - \theta \mathbf{n}(\phi) \cdot \sigma]. \tag{10}$$

Expanding the terms in (10), we obtain

$$\begin{aligned} &\frac{\alpha}{2} I_{4\times4} - \frac{\theta\cos\phi}{2} I_{2\times2} \otimes \sigma_x - \frac{\theta\sin\phi}{2} I_{2\times2} \otimes \sigma_y - \\ &\frac{\alpha}{2}\sigma_z \otimes I_{2\times2} + \frac{\theta\cos\phi}{2}\sigma_z \otimes \sigma_x + \frac{\theta\sin\phi}{2}\sigma_z \otimes \sigma_y. \end{aligned} \tag{11}$$

The final expression contains the basic ingredients for implementing Hamiltonians leading to universal quantum gates. Notice that the effective Hilbert space involve is a subspace of the computational basis for two qubits, namely, $\{|10\rangle, |11\rangle\}$

4 A Classical Model for Erasing One Bit of Information

One bit of information stored in the position of a particle in a bistable potential is erased by modifying its curvature and hence the positions of the extremal points, first lowering the initial position of the right minimum, which lifts the central barrier. The ensuing tilting force causes the particle to move to the right. Once the particle is confined there, the process is reversed until a whole cycle in time τ is completed. The cycle is schematically depicted in Fig. 5. The memory (position of the particle either

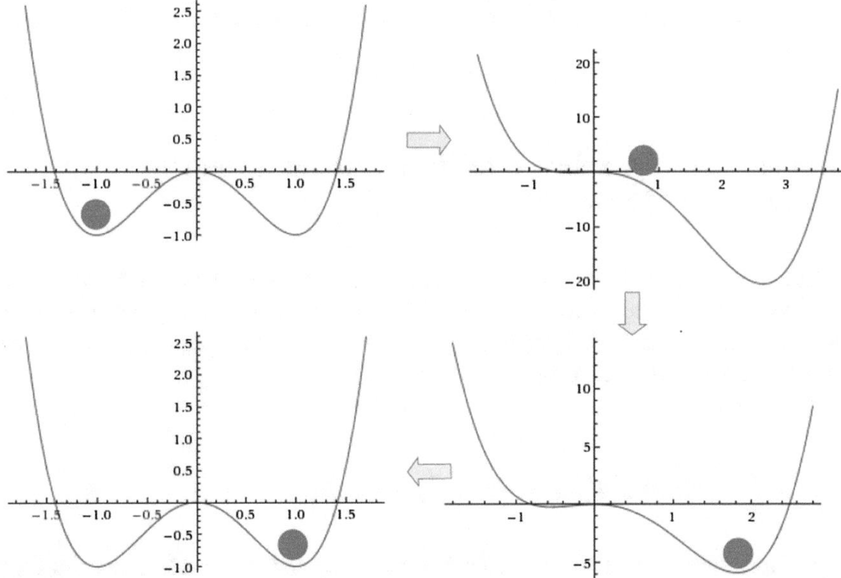

Fig. 5 (Color online) Evolution of the confining potential $V(x, t)$ (12). The process ends after the time cycle τ. In this fashion, the position of the particle is changed from left to right, which encodes one bit of information. See text for details

on the left or the right) is then said to be erased when its state is reset irrespective of its initial state.

The potential we are going to employ is given by

$$V(x, t) = x^4 - c(t)x^3 - 2x^2. \tag{12}$$

The asymmetry parameter $c(t)$ is chosen so that the evolution is adiabatic, which implies that $c(t)$ grows slowly enough as compared to a typical time scale, and that it is symmetric with respect to $\tau/2$, which is tantamount as saying that the potential returns to its original shape at time $t = \tau$. In our case, we use a Gaussian curve centered at $\tau/2$.

The closest quantum counterpart of the previous situation was considered by Zurek [62] when extending Szilard's model of a particle in a box to that on and infinite well. However, there was no evolution of the kind depicted in Fig. 5. Instead, a partition was placed at some point between the walls and the thermodynamics of a single particle were carried out in the usual canonical ensemble, obviously implying a fixed temperature T. Notice here that no temperature is defined.

In point of fact, a quantum analogue (with no temperature involved) of the evolution depicted in Fig. 5 could be conceived by raising and lowering left and right wells inside a bigger infinite one. It can be shown that the overall process only induces a global phase on the final position eigenstate of the particle with respect to the initial one.

5 Molecular Dynamics, Statistical Ensembles and Landauer's Principle

Thermodynamic quantities would be stochastic variables at the microscopic level if a temperature was fixed, because thermal fluctuations cannot be neglected. Even in the case where there is no temperature defined, a mechanical equivalent of power released (heat) can be conceived.

5.1 *Microcanonical Ensemble Approach*

The temperature T in MD simulations is directly related to the total kinetic energy by the well-known equipartition formula, assigning an average kinetic energy $k_BT/2$ per degree of freedom $K = \frac{3}{2}Nk_BT$. Notice that if we consider a one dimensional problem, the factor is $\frac{1}{2}$ instead of the usual $\frac{3}{2}$ in three dimensions. In MD therefore the temperature is *estimated* directly from the average total kinetic energy K.

Hence, we can define an *instantaneous* temperature $T(t)$ from the relation $\frac{1}{2}m\dot{x}^2 = \frac{1}{2}k_BT(t)$, that is, $k_BT(t) = m\dot{x}^2$. The heat released along the trajectory $x(t)$ is given by the expression

$$Q = -\int_0^{\tau} dt\, m\,\dot{x}(t)\,\frac{\partial V(x,t)}{\partial x}, \tag{13}$$

where τ is the total time of the cycle in Fig. 5. Averaging over many trajectories will return the mean dissipated heat $\langle Q \rangle$, which fulfills the entropic inequality $\langle Q \rangle \geq -T\Delta S = Tk_B \ln 2 = m\dot{x}^2 \ln 2 = \langle Q \rangle_{\text{Landauer}}$. The critical figure of merit will be then be the ratio $\langle Q \rangle / \langle Q \rangle_{\text{Landauer}}$.

5.2 *Canonical Ensemble Approach*

As we have presented it so far, MD is performed in the microcanonical ensemble. We saw that it is possible to estimate the temperature using thermodynamic averages. It is often desirable, however, to specify the temperature a priori and perform a simulation in the canonical ensemble.

The aim of MD here is (i) to preserve the correct thermodynamics, i.e., the correct microstate distribution in the canonical ensemble, as well as (ii) a realistic dynamics of the equations of motion. One popular method is given by the Nosé-Hoover thermostat [63]. The time-evolution of a particle's position and momentum in one dimension is governed by the following set of equations:

$$\begin{aligned}\frac{dx(t)}{dt} &= v(t)\\ \frac{dv(t)}{dt} &= -\frac{\partial V(x,t)}{\partial x} - \zeta(t)\,v(t)\\ \frac{d\zeta(t)}{dt} &= \left(mv(t)^2 - k_B T\right)/Q_{\text{mass}}.\end{aligned} \tag{14}$$

Here ζ acts as a friction coefficient, and Q_{mass} is the so-called heat bath mass. As we shall see, it constitutes an important parameter to be chosen. Notice that when ζ is constant, one instantaneously recovers the microcanonical case. Given a fixed temperature T, the Landauer analysis is exactly the same as in the previous case.

6 Results

In the same way that ergodicity of dynamics is a pre-requisite for obtaining statistical properties of a system simulated via MD, it should also be the case for the Landauer's principle. The former has been shown to fail [64] in the widely used Nosé-Hoover thermostat in MD, generally believed to impart the canonical distribution on the thermostatted physical system. Therefore if a canonical ensemble-based MD simulation of a physical system failed within the context of the most intimate link between information erasure and thermodynamics, then we would be led to conclude that no MD approach in the canonical ensemble has yet been able to reproduce *exactly* a physical system in a thermal bath. We shall see now that both the microcanonical and the canonical approach to MD violates Landauer's principle, although due to different reasons.

From now on we shall use atomic units, with $k_B = m = 1$. Let us denote rand_u() a random number uniformly distributed between 0 an 1. Different trajectories are defined by different initial conditions. Therefore, at time $t = 0$, we shall generate M initial positions and velocities $\{x_0^k, v_0^k\}$ $k \in [1, M]$ in such a way that, for small departures (Δ_x, Δ_v) of the left equilibrium position and velocity $(-1, 0)$ in (12), we shall have $x_0^k = (-1 - \Delta_x) + 2\Delta_x \text{rand\_u}()$ and $v_0^k = (0 - \Delta_v) + 2\Delta_v \text{rand\_u}()$. The concomitant average of a quantity A will simply read as $\langle A \rangle = \frac{1}{M}\sum_i A_i$.

Let us have then a particle that undergoes M repetitions of the cycle τ in Fig. 5, all of them being different trajectories. We depict in Fig. 6 a typical evolution of both the position and velocity of the particle, the position of which has been moved from left to right adiabatically. The equation of motion of a particle under the action of potential (12) is solved numerically using a fourth order Runge-Kutta algorithm. The position erasure corresponding to one bit of information is better shown on the average position and velocity. However, the small oscillations along the evolution

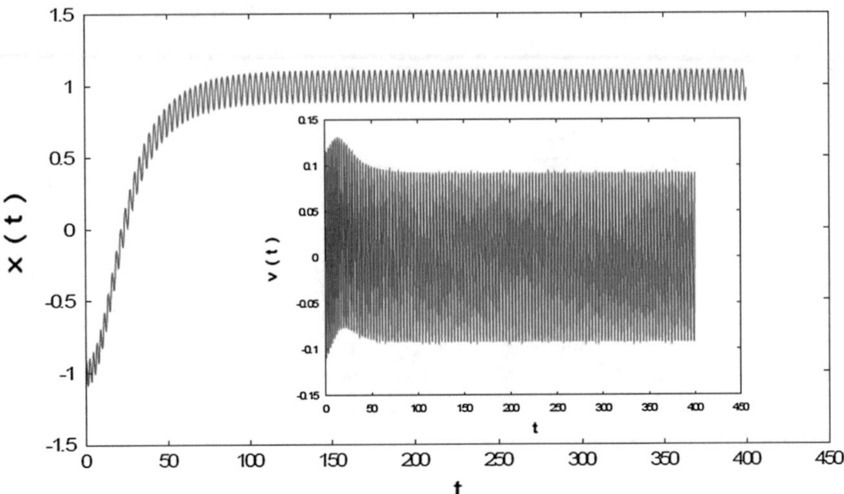

Fig. 6 (Color online) Evolution of the position and velocity for a particle confined to the potential $V(x, t)$ (12) after a complete cycle τ. See text for details

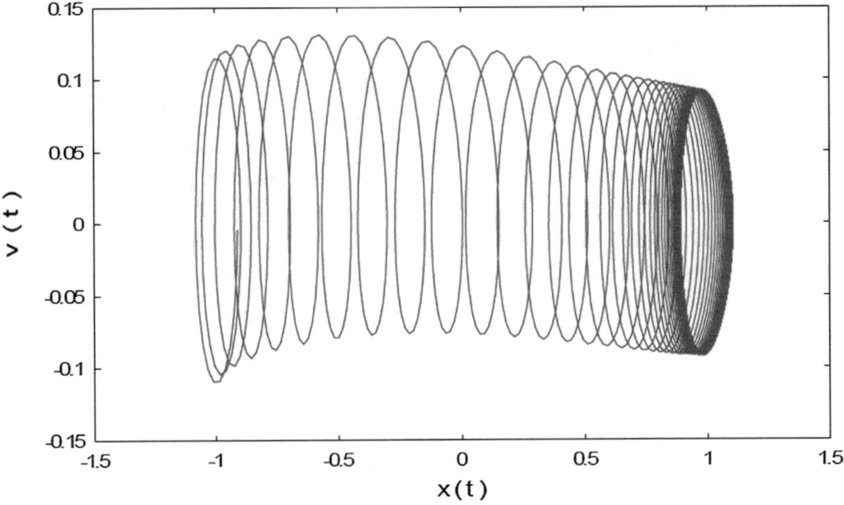

Fig. 7 (Color online) Phase portrait of the evolution depicted in Fig. 6. See text for details

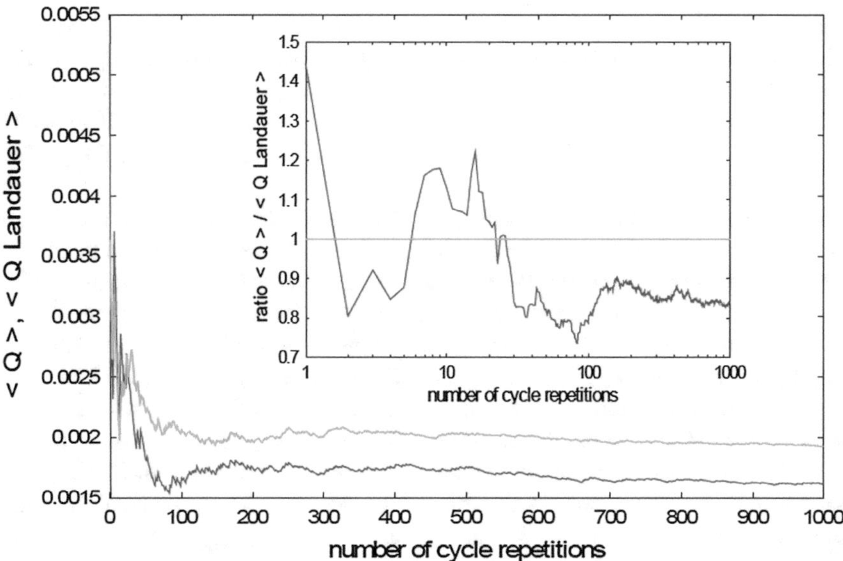

Fig. 8 (Color online) Evolution of the mean values for the heat released by the particle (lower red curve) and the corresponding Landauer's lower bound (upper green curve). The ratio between the two depicted in the inset clearly shows that there is a violation of Landauer's principle. See text for details

do not really depart considerably. Figure 7 shows the phase portrait of the particle's evolution.

However, these results do not imply by themselves any judgement on the validity of any thermodynamic quantity computation. To this extend, MD is providing the dynamics of the system, even if it is as simple as that of a single particle. The critical results have to compare the mean heat released by the particle along the different trajectories, and the heat involved in erasing one bit of information. The ensuing results after calculations as the number of repetitions M approaches 1000 show in Fig. 8 that there is indeed a violation of the Landauer's principle.

The previous results are based entirely on numerical calculations. As long as the numerics are carried out in utmost detail, as we have done here, these results are correct. Admittedly, the evolution for a particle inside the particular double-well (12), the non-analytic nature of the concomitant equations of motion, the form of the computed quantities such as (13) and the average over many initial condition situations may induce to think that some kind of mistake can be made in several possible forms.

To further elucidate the situation, we shall consider an almost analytic example of a particle in a box of width L. Let us suppose that we initially have a particle on the left of a potential where a barrier of width $\delta \ll L$ has been inserted right in the middle. In our description it does not bare much significance whether we remove the barrier adiabatically or not. In any case, we lower the middle barrier adiabatically while the right hand side of the potential is tilted downwards. This last step induces a constant positive force F that drags the particle to the right. After some time (which is irrelevant), the middle barrier of width δ is raised again and $F \to 0$. The overall process has removed one bit of information encoded in the position of the particle from *"left"* to *"right"*.

In the following we shall focus in the part where the particle remains to the right of the middle-wall (all the other steps in the previous bit erasure involve a small, but non-zero total Q (13)). The particle bounces back and forth against the right wall of the box. Since F is constant, the dependence of $\dot{x}(t)$ is linear with t, and changes from its maximum value $v_{\max} > 0$ to $-v_{\max}$ almost instantaneously during the collision with the wall. Hence, $\dot{x}(t)$ will have a periodic shape (a sawtooth wave). Therefore, Q (13) will be proportional to the time-average of $\dot{x}(t)$ over the cycle, which shall be zero. Thus, in this situation, the ensuing averaged $\langle Q \rangle$ will be quite close to zero.

As far as the Landauer's bound is concerned, the instantaneous value for $\dot{x}(t)^2$ will be non-zero (its time-average will certainly be positive). Then, we shall have that the inequality $\langle Q \rangle \geq -T\Delta S = Tk_B \ln 2 = m\dot{x}^2 \ln 2 = \langle Q \rangle_{\text{Landauer}}$ will be flagrantly violated most of the times, and only respected when $\dot{x}(t)$ is actually zero, which can happen for a very specific subset of initial conditions (definite times for those initial conditions, to be more precise). Summing up, we have found a simple situation where the erasure of one bit can violate the bound imposed by the Landauer's principle.

In the light of these results, we have seen that the violation of the Landauer's principle is possible in the approach given by the microcanonical ensemble. We shall discuss later on the possible implications and the compatibility of this approach in the framework of MD simulations. We cannot conclude, however, that this violation is universal for any means of erasing the information encoded in the position of a single particle. By the same token, we cannot rule out the existence of similar procedures such that an eventual overall erasure process respects the Landauer's bound.

Finding the conditions for which a system can or cannot respect the bound imposed by the Landauer's principle is not an easy task, and, in a way, it goes beyond the scope of the present work. The message behind the previous results for the microcanonical ensemble scenario in MD is that there is no inherent reason why MD should respect (or not) the Landauer's principle.

Let us now consider a canonical ensemble approach to MD simulations. The set of equations that one has to solve is now the one corresponding to (14). In this new setting, and as opposed to the microcanonical case, we have to deal with the new set of parameters $\{T, Q_{\text{mass}}\}$ and a new variable, the friction $\zeta(t)$. Q_{mass} is important in determining the rate of exchange of the system energy with an imaginary heat bath. Since it is a constant, we shall use different values in order to illustrate the corresponding results. In any case, it is expected to play a paramount role in our discussion. This new parameter makes a huge difference with respect to the previous results obtained in the microcanonical ensemble, for it provides with a real adjustable

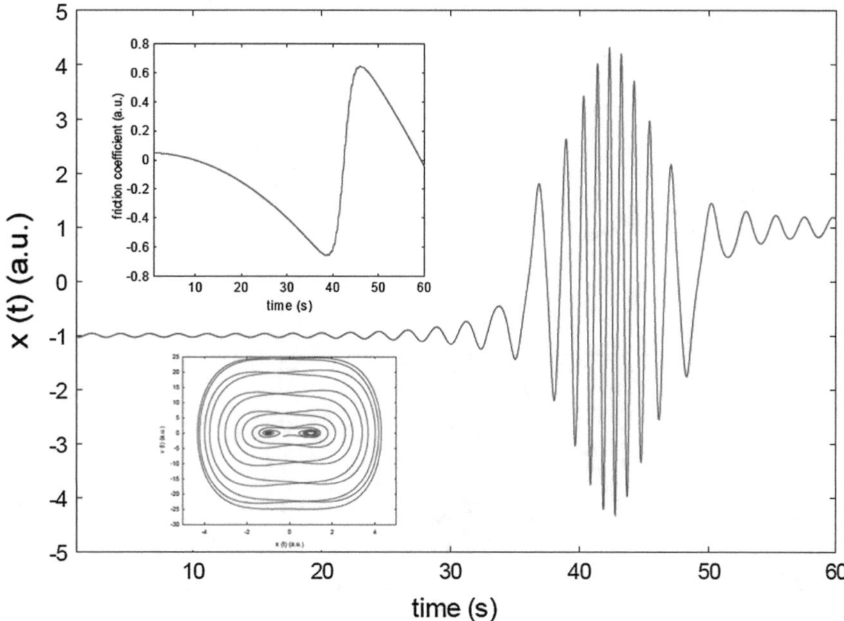

Fig. 9 (Color online) Time evolution of the position of the particle in the canonical ensemble approach, with $Q_{\text{mass}} = 10^3$. The concomitant phase portrait is shown in the lower inset. The upper inset shows the time evolution of the friction coefficient $\zeta(t)$. Notice that it abruptly changes in the "turbulent" times until the particle is accommodated on the right hand side of the double well. See text for details

parameter that will definitely play a role in the description of the Landauer's principle within the framework of MD.

We shall consider a fixed value for the temperature $T = 10K$ in all the forthcoming calculations. In Fig. 9 we depict the results corresponding to $Q_{\text{mass}} = 10^3$. Although the evolution is governed by a different set of equations, the particle eventually reaches an equilibrium position on the right side of the potential (12). The position of the particle evolves around the initial values at first and then some very fast changes occur. This fact is also reflected in the phase portrait shown in the lower inset of Fig. 9.

Increasing Q_{mass} up to 10^5 has definitely a huge impact in the evolution of the particle. As seen from Fig. 10, friction coefficient shrinks—although keeping a similar evolution—and the transition from one equilibrium position to the other is much more abrupt. As a consequence, the corresponding phase diagram has a richer structure as compared to the one in Fig. 9.

The difference between these evolutions will entail that the heats exchanged, once averaged over many trajectories, will be rather different. Indeed, the rapid changes in the position imply greater values of the velocity, which is certainly relevant in the set differential equations (14). However, since this set of equations is coupled and

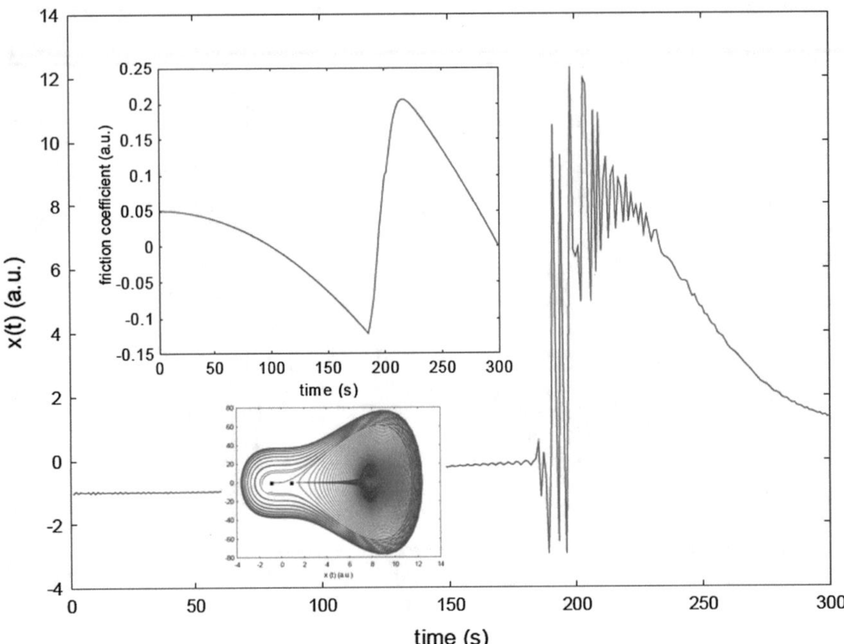

Fig. 10 (Color online) Same as in Fig. 9, but with $Q_{\text{mass}} = 10^5$. Increasing Q_{mass} has an immediate effect in reducing the absolute value of $\zeta(t)$ and in making the transition to the right position longer, as well as more fluctuating in time. As a consequence, the phase portrait presents a richer structure (equilibrium positions are shown by the two dots in the lower inset). See text for details

non-linear, it is not so obvious from them to draw conclusions. Therefore, we have computed in Fig. 11 the same as in Fig. 8 but only for the ratio $\langle Q\rangle/\langle Q\rangle_{\text{Landauer}}$ for different values of Q_{mass}. The corresponding results show that there is a threshold beyond which there is always a violation of Landauer's principle. Therefore, the framework of the celebrated Nosé-Hoover thermostat in MD seems to not account for the intimate connection between information erasure and thermodynamics. In other words, when a finite system is at equilibrium with a heat bath, the equilibrium temperature is dictated by the heat bath, and not by the intrinsic thermostatistics of the finite system. And if it is not sufficiently large, it may be necessary for the finite system to change its thermostatistics to be in equilibrium again.

It is indeed the fact that one can change at will the size of the heat reservoir that has an immediate impact of the heat released by the particle via the set of equations (14). In point of fact, the last equation in (14), that is, $\frac{d\zeta(t)}{dt} = \left(mv(t)^2 - k_B T\right)/Q_{\text{mass}}$ is designed such that the friction disappears when one reaches "equilibrium" $mv(t)^2 = k_B T$ given by the equipartition theorem. However, one can artificially reach the same goal by changing the finite heat reservoir. Remarkably enough, one recovers the occurrence of the violation of the Landauer's principle in the microcanonical case for Q_{mass} sufficiently large.

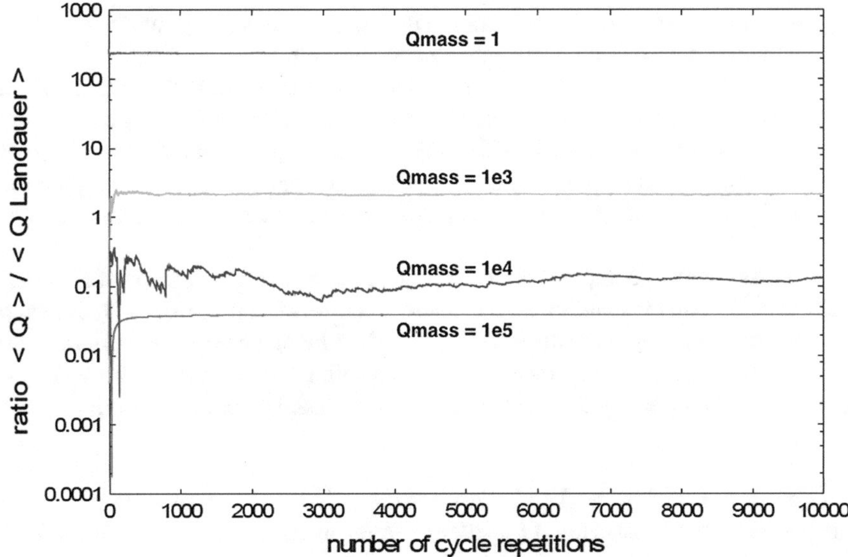

Fig. 11 (Color online) Calculation of the ratio of the mean values of $\langle Q \rangle$ and $\langle Q \rangle_{\text{Landauer}}$ (recall that the Landauer's bound is equal to $k_B T \ln 2$). Choosing different couplings to an environment may induce a violation of the Landauer's principle. See text for details

Therefore, we are able to introduce an *ad hoc* **criterion** for fulfilling the Landauer's bound and at the same time simulate a system of particles in order to find correct thermodynamic averages: simulate the system of N particles and find the critical value of Q^*_{mass} for which $\langle Q \rangle = -T\Delta S = NTk_B \ln 2$. Once obtained, only MD simulations will be accepted provided $Q_{\text{mass}} < Q^*_{\text{mass}}$.

7 Discussion

One of the advantages of the proposed quantum memory involves its ability to maintain information over time (longevity). Unlike the current magnetic memories, where the individual bits can disappear due to the action of several external effects, our proposal can be made to be robust against these effects for its only interaction is of electrostatic nature.

The writing of information is done by promoting confined electrons in the quantum wires to the continuum after the whole system has been charge-injected, whereas readout of information is performed by detecting the presence of single charge carriers.

The enhanced capacity achieved outperforms any classical device. We acknowledge the fact that we propose quantum devices to stored classical bits. This is the main advantage. Ee believe that our proposal constitutes a half-way solution to storing big quantities of information until the drawbacks related to quantum coherence are completely solved in the fully quantum counterpart.

If realized, such a quantum network of chips constituting a quantum memory can have important technological and social consequences. Besides creating a new solid proposal for storing information with undetermined longevity, it solves the issue of keeping the concomitant information safe and, at the same time, solves the situation creating by handling the Big Data generated at present times by society.

As far as quantum computation is concerned, we described what type of Hamiltonians lead directly to quantum gates considered universal and proved so by Barenco [30]. We have employed the direct definition of e^{iH} being unitary if H is Hermitian. This result may constitute a practical way of unveiling the structure of those Hamiltonians capable of entangling and evolving to universal gates in the form of Pauli matrices.

Landauer's principle can be checked –never derived– from examples employing one particle either classically *à la Szilard's* or quantum mechanically *à la Zurek's*. So treating one particle only should not have thermodynamic implications. In any case, either considering one or many particles, Landauer's bound cannot be derived from thermodynamics alone. It is the connection with information theory that gives that bound the status of *principle*. This fact might have implications in designing thermostats. MD simulations using a rigorous approach to canonical ensembles, such as the widely used Nosé-Hoover thermostat, but at the same time treating the ensuing dynamics in a proper way (energy preserved, total momentum, etc.), seem to provide a reasonable framework for chemical physics. We do believe that. The main issue here is that if these kind of thermostats could sustain the preservation of a fundament energy bound related to entropy and information.

In the case where no temperature is defined strictly speaking, where there is an instantaneous temperature, it may seem that the approach we have carried out is bound to fail. However, there is nothing against that: we have a system where a time-dependent force drives a particle from one side of a bistable potential to the other. This implies the erasing of one bit of information. Employing the tools provided by the bound $k_B T(t) \ln 2$ may constitute a mistake, but only conceptually. In addition, one could argue that the"heat" dissipated by the particle during a given trajectory $x(t)$ has nowhere to be released, since there is no (finite or infinite) heat reservoir. This should explain why Landauer's principle fails in that context. Also, there are other issues at play, like the following ones:

- Considering more particles is closer to reality. Landauer's principle might hold for some other system where more than one single bit of information is deleted
- Instantaneous temperatures $T(t)$ constitute a mere guide in usual MD simulations, where N particles interact. Averages with time are able to obtain macroscopic quantities via the ergodic theorem. Therefore the actual use of $T(t)$ bears no thermodynamic relevance

- There is always an environment, therefore Landauer's lower bound cannot apply in the microcanonical ensemble

In the more realistic case where a fixed temperature is given, adding the constraint of fulfilling the Landauer's principle may constitute an impossible task. In any case, and in the light of our results, one has to admit that the action of an interaction of the system with a virtual environment via the constant parameter Q_{mass} is rather *the* critical issue here.

In view of the problem of connecting a heat reservoir—virtual or not—to the system might be overcome by considering finite instances and/or broader approaches, which may imply for instance replacing the usual Gibbs factor by another one related to non-extensive statistical mechanics [65]. Since the goal in MD simulations is to reproduce certain thermodynamic quantities, a good opportunity to explore the implications of our study would be to concoct the extension of the present formalism to that of Tsallis' formalism.

Some criticism to the present study could be made on the basis of the very definition of the temperature T of the system, which is basically what is done in MD simulations. The temperature set in the system does not appear to follow the definition $T = \frac{dS}{dQ}$, where S stands for entropy. Hence, one might think that therefore it is incorrect to state either that Landauer's principle were violated in the microcanonical ensemble, or that the microcanonical ensemble model were incorrect. Conclusion: what can only be stated as such is that the temperature cannot be trivially treated to be equal to the kinetic energy of a single particle? This criticism would be certainly against the very nature of MD simulations, not the present study. What we present here are the sheer consequences of applying MD simulations, with an artificial definition of the temperature T that has, as an immediate consequence, the violation of the Landauer's principle, which ultimately addresses the Second law of thermodynamics.

Summing up, we have tried to shed new light on the implications of MD simulations regarding a profound information-theoretical bound relating minimal amounts of heat transfer. One question remains, though: could we expect that a quantum approach to the present work had no flaws? The quantum information treatment of other systems usually operates at zero temperature, and entropy differences can usually be calculated. There also exists an environment, of course, but the system is coupled to it by means that may or may not have a classical counterpart. In other words, we can have an environment at zero temperature, as opposed to a classical treatment. If one wants to take into account the thermodynamics of a quantum system for a given temperature T, then the environment has to be modeled accordingly. This implies an evolution of the density matrix describing the entire system, including the environment. However, resorting to the quantum case may not be a solution. In the classical regime, experiments have been devised that accurately confirm many of the modern relations involving heat and work statistics in stochastic thermodynamics. However, in quantum systems, measuring such quantities shall induce new questions according to general principles of quantum mechanics. This situation is currently the subject of intense reserach [66].

To wether or not the Landauer's principle is violated in the *simulation* of a specific quantum system + environment remains an open problem. In any case, we have provided in the present study that MD simulation are not designed to follow the tenets of the Landauer's principle, hence we can provide situations were the physical system can be tailored to respect or violate this fundamental relation between thermodynamics and information theory.

Acknowledgements J. Batle acknowledges fruitful discussions with J. Rosselló, Maria del Mar Batle and Regina Batle.

References

1. Voyager NASA mission web site: http://voyager.jpl.nasa.gov/
2. Hammerer, K., Sorensen, A.S., Polzik, E.S.: 2010 Quantum interface between light and atomic ensembles. Rev. Mod. Phys. **82**, 1041–93 (2010)
3. Choi, K.S., Deng, H., Laurat, J., Kimble, H.J.: Mapping photonic entanglement into and out of a quantum memory. Nature **452**, 67–71 (2008)
4. Zhao, B., Chen, Y-A, Bao, X-H, Strassel, T., Chuu, C-S, Jin, X-M, Schmiedmayer, J., Yuan, Z-S, Chen, S. & Pan, J.W. A millisecond quantum memory for scalable quantum networks. Nature Phys. **5**, 95–9 (2008)
5. Reim, K.F., Nunn, J., Lorenz, V.O., Sussman, B.J., Lee, K.C., Langford, N.K., Jaksch, D., Walmsley, I.A.: Towards high-speed optical quantum memories. Nature Photon. **4**, 218–21 (2010)
6. Schnorrberger, U., Thompson, J.D., Trotzky, S., Pugatch, R., Davidson, N., Kuhr, S., Bloch, I.: Electromagnetically induced transparency and light storage in an atomic Mott insulator. Phys. Rev. Lett. **103**, 033003 (2009)
7. Liu, C., Dutton, Z., Behroozi, C. H. & Hau, L. V. Observation of coherent optical information storage in an atomic medium using halted light pulses. *Nature* **409**, 490–3 (2001)
8. Julsgaard, B., Sherson, J., Ignacio Cirac, J., Fiurasek, J., Polzik, E.S.: Experimental demonstration of quantum memory for light. Nature **432**, 482–6 (2004)
9. Eisaman, M.D., André, A., Massou, F., Fleischhauer, M., Zibrov, A.S., Lukin, M.D.: Electromagnetically induced transparency with tunable single-photon pulses. Nature **438**, 837–41 (2005)
10. Ran, Y., Xue, L., Hu, S., Su, R.-K.: On the Coulomb-type potential of the one-dimensional Schrödinger equation. J. Phys. A: Math. Gen. **33**, 9265–9272 (2000)
11. Ginzburg, V.L.: Once again about high-temperature superconductivity. Contemp. Phys. **33**, 15 (1992)
12. Brown, J.W., Spector, H.N.: Exciton binding energy in a quantum-well wire. Phys. Rev. B **35**, 3009 (1987)
13. Reyes, J.A., del Castillo-Mussot, M.: Wannier-Mott exciton formed by electron and hole separated in parallel quantum wires. Phys. Rev. B **57**, 1690 (1998)
14. Heeger, A.J., Kivelson, S., Schrieffer, J.R., Su, W.P.: Solitons in conducting polymers. Rev. Mod. Phys. **60**, 731 (1988)
15. Abe, S., Su, W.P.: Excitons and Charge Transfer States in One-Dimensional Semiconductors. Mol. Cryst. Liq. Cryst. **194**, 357–362 (1991)
16. Wigner, E.P.: Effects of the electron interaction on the energy levels of electrons in metals. Trans. Faraday Soc. **34**, 678 (1938)
17. Carr Jr., W.J.: Energy, specific heat, and magnetic properties of the low-density electron gas. Phys. Rev. **122**, 1437 (1961)
18. Andrews, G.E., Askey, R., Roy, R.: Special Functions. Cambridge University Press (1999)

19. Fowler, R.H., Nordheim, L.W.: Electron emission in intense electric fields. Proc. Roy. Soc. (London) A **119**, 173–181 (1928)
20. Melmed, A.J.: The art of science and other aspects of making sharps tips. J. Vac. Sci. Technol. B **9**, 601–608 (1991)
21. Library of Congress web site: http://www.loc.gov/
22. Barreiro, A., van der Zant, Herre S.J., Vandersypen, L.M.K.: Quantum Dots at Room Temperature carved out from Few-Layer Graphene. *Nano Lett.* **12** 6096 (2012)
23. Tsutsui, M., Morikawa, T., Arima, A., Taniguchi, M.: Thermoelectricity in atom-sized junctions at room temperatures. Sci Rep. **3**, 3326 (2013)
24. Deutsch, D.: Proc. Royal Soc. London A **425**, 73 (1989)
25. Barenco, A., Bennett, C.H., Cleve, R., DiVincenzo, D.P., Margolus, N., Shor, P., Sleator, T., Smolin, J.A., Weinfurter, H.: Phys. Rev. A **52**, 3457 (1995)
26. Deutsch, D., Barenco, A., Ekert, A.: Proc. Royal Soc. London **449**, 669 (1995)
27. DiVincenzo, D.P.: Phys, Rev. A **51**, 1015 (1995)
28. Lloyd, S.: Phys. Rev. Lett. **75**, 346 (1995)
29. DiVincenzo, D.P.: Fortschr. Phys. **48**, 771 (2000)
30. Barenco, A.: Proc. R. Soc. Lond. A **449**, 679 (1995)
31. Leff, H.S., Rex, A.F. (eds.): Maxwell's demon 2: Entropy, Classical and Quantum Information, Computing. Princeton University Press, New Jersey (2003)
32. Szilard, L.: Z. Phys. **53**, 840 (1929)
33. Brillouin, L.: J. Appl. Phys. **22**, 334 (1951)
34. Landauer, R.: IBM J. Res. Dev. **5**(183) (1961); Landauer, R.: Nature **335**, 779 (1988); Landauer, R.: Science **272**, 1914 (1996)
35. Bennett, C.H.: Int. J. Theor. Phys. **21**, 905 (1982)
36. Piechocinska, B.: Phys. Rev. A **61**, 062314 (2000)
37. Barkeshli, M.M. (2005). arXiv:cond-mat/0504323
38. Maroney, O.J.E.: Phys. Rev. E **79**, 031105 (2009)
39. Metawa, N., Elhoseny, M., Kabir Hassan, M., Hassanien, A.: Loan portfolio optimization using genetic algorithm: a case of credit constraints. In: 12th International Computer Engineering Conference (ICENCO), IEEE, 59–64 (2016). doi:10.1109/ICENCO.2016.7856446
40. Metawa, N., Hassan, M.K., Elhoseny, M.: Genetic algorithm based model for optimizing bank lending decisions, Expert Systems with Applications, vol. 80, 1 September 2017, pp. 75–82, ISSN 0957-4174. doi:10.1016/j.eswa.2017.03.021
41. Elhoseny, M., Elminir, H., Riad, A., Yuan, X.: Recent advances of secure clustering protocols in wireless sensor networks. Int. J. Comput. Netw. Commun. Secur. **2**(11), 400–413 (2014)
42. Elhoseny, M., Yuan, X., El-Minir, H.K., Riad, A.M.: Riad, an energy efficient encryption method for secure dynamic WSN. Secur. Commun. Netw. **9**, 2024–2031 (2016)
43. Sagawa, T., Ueda, M.: Phys. Rev. Lett. **100**, 80403 (2008)
44. Bremermann, B.: Int. J. Theor. Phys. **21**(203) (1982); Lloyd, S., Zurek, W.H.: J. Stat. Phys. **62**(819) (1991); Caves, C.M., Drummond, P.M.: Rev. Mod. Phys. **66**(481) (1994); Magnasco, M.O.: Europhys. Lett. **33**(583) (1996); Zurek, W.H.: arXiv:quant-ph/0301076 (2003); Scully, M.O. et al. Science **299**(862) (2003); Kieu, T.D.: Phys. Rev. Lett. **93**, 140–403 (2004); Allahverdyan, A.E., et al.: J. Mod. Optics **51**(2703) (2004); Maruyama, K., et al.: J. Phys. A **38**(7175) (2005); Quan, H.T. et al.: Phys. Rev. Lett. **97**(180402) (2006); Maruyama, K. et al.: Rev. Mod. Phys. **81**(1) (2009)
45. Elhoseny, M., Yuan, X., Yu, Z., Mao, C., El-Minir, H.K., Riad, A.M.: Balancing energy consumption in heterogeneous wireless sensor networks using genetic algorithm. IEEE Commun. Lett. (99), 1–4 (2014)
46. Riad, A.M., El-minir, H.K., Elhoseny, M.: Secure routing in wireless sensor network: a state of the art. Int. J. Comput. Appl. **67**(7) (2013)
47. Jarzynski, C.: Phys. Rev. Lett. **78**(2690) (1997); Crooks, G.E.: Phys. Rev. E **60**(2721) (1999); Mukamel, S.: Phys. Rev. Lett. **90**(170604) (2003); Kawai, R.: et al. Phys. Rev. Lett. **98**(80602) (2007); J. Liphardt et al. Science **296**(1832) (2002); Collin, M. et al.: Nature **437**(231) (2005)
48. Bérut, A., et al.: Nature **483**, 187 (2012)

49. Vedral, V.: Proc. Roy. Soc. Lond. **456**, 969 (1996)
50. Plenio, M.B.: Phys. Lett. A **263**, 281 (1999)
51. Holevo, A.S.: Probl. Inf. Transm. **9**, 3 (1973)
52. Pati, A.K., Braunstein, S.L.: Nature **404**, 164 (2000)
53. Hilt, S., Shabbir, S., Anders, J., Lutz, E.: Phys. Rev. E **83**, 030102 (2011)
54. Alder, B.J., Wainwright, T.E.: J. Chem. Phys. **27**(1208) (1957); Alder, B.J., Wainwright, T.E.: J. Chem. Phys. **31**(459) (1959)
55. Rahman, A.: Phys. Rev. **136**, 405 (1964)
56. Elhoseny, M., Farouk, A., Zhou, N., Wang, M.-M., Abdalla, S., Batle, J.: Dynamic multi-hop clustering in a wireless sensor network: Performance improvement. Wirel. Pers. Commun., 121 (2017)
57. Elhoseny, M., Yuan, X., Yu, Z., Mao, C., El-Minir, H.K., Riad, A.M.: Balancing energy consumption in heterogeneous wireless sensor networks using genetic algorithm. IEEE Commun. Lett. (99), 1–4 (2014)
58. Yuan, X., Elhoseny, M., Minir, H., Riad, A.: A genetic algorithm-based, dynamic clustering method towards improved WSN longevity. J. Netw. Syst. Manage., 1–26, Springer US (2016). doi:10.1007/s10922-016-9379-7
59. Elhoseny, M., Yuan, X., El-Minir, H.K., Riad, A.M.: An energy efficient encryption method for secure dynamic WSN. Secur. Commun. Netw. (9), 2024–2031 (2016)
60. Verlet, L.: Phys. Rev. **159**(98) (1964); Verlet, L.: Phys. Rev. **165**(201) (1968)
61. Frenkel, D., Smit, B.: Understanding Molecular Simulation (Academic Press, San Diego, 1996). Computer Simulation of Liquids (Claredon Press, Oxford, M. P. Allen and D. J. Tildesley (1986)
62. Zurek, W.H.: Maxwell's Demon, Szilard's Engine and Quantum Measurements, Frontiers of Nonequilibrium Statistical Physics 135, 151. Plenum Press, New York (1986)
63. Nosé, S.: J. Chem. Phys. **81**(511) (1984); Hoover, W.G.: Phys. Rev. A **31**(1695) (1985)
64. Kumar Patra, P., Bhattacharya, B.: Phys. Rev. E **90**, 43304 (2014)
65. Tsallis, C.: J. Stat. Phys. **52**(479) (1988); Gell-Mann, M.: C. Tsallis (Eds.), Nonextensive Entropy: Interdisciplinary Applications. Oxford University Press, New York (2004); C. Tsallis, Introduction to Nonextensive Statistical Mechanics: Approaching a Complex World. Springer, New York (2009)
66. Pekola, J.P., Suomela, S., Galperin, Y.M.: J. Low Temp. Phys. **184**, 1015 (2016)

Morphogenetic Sources in Quantum, Neural and Wave Fields: Part 1

G. Resconi, K. Nagata, O. Tarawneh and Ahmed Farouk

Abstract Neural network and quantum computer have the same conceptual structure similar to Huygens sources in the wave field generation. Any point of the space is a source with different intensity of waves that transport information in all the space where are superposed in a complex way to generate the wave field. In wave theory this sources are denoted Huygens sources. The morphogenetic field is the wave field generate by computed sources that are designed in a way to transform the original field in a wanted field to satisfy wanted property. The morphogenetic computation is this type of global computation by sources like Huygens sources that in parallel and synchronic way give us the designed field. So the intensity of the sources must be computed a priory before the morphogenetic effective computation in a way to have an entanglement of the sources that in the same time compute the field. If we cannot design the sources a priory and we want generate the field by a recursion process we enter easily in a deadlock state for which one source generate local wanted field that destroy the generation of another local field. So we have a contradiction between the action of different non entangled sources that cannot generate all the wanted field. In neural network we have the superposition of the input vectors in quantum mechanics we have the superposition of the states. In the neural network the intensity of the sources are the neural weights and the threshold. In quantum mechanics the intensity of the sources are the coefficients of the quantum states superposition. To design neural sources intensity

G. Resconi
Department of Mathematics and Physics, Catholic University, I-25121 Brescia, Italy

K. Nagata (✉)
Department of Physics, Korea Advanced Institute of Science and Technology,
Daejeon 34141, Korea
e-mail: ko_mi_na@yahoo.co.jp

O. Tarawneh
Information Technology Department, Al-Zahra College for Women,
P.O.Box 3365, Muscat, Oman

A. Farouk
Faculty of Computers and Information, Computer Sciences Department,
Mansoura University, Mansoura, Egypt

A.E. Hassanien et al. (eds.), *Quantum Computing: An Environment for Intelligent Large Scale Real Application*, Studies in Big Data 33,
https://doi.org/10.1007/978-3-319-63639-9_14

(weights) we use the matrix of all possible inputs by which we can define all possible outputs. In the design neural network we cannot use the simple theory of input output but all the past or future input output are used. Space and time is not important in the design the network more important is to use the space of all possible input and output. The same in the quantum computer where we must design the unitary transformation for which only one wanted state coefficient is different from zero all the other coefficients are put to zero. In this way we can select among a huge possible states any one wanted state solution of our problem. In this scheme we include Deutsch problems, Berstein Varizani theorem and Nagata parallel function computing. The difference between quantum computer and neural network is that in quantum computer the basis is the oracle square matrix without any threshold and contradiction. In neural network the basis is a rectangular matrix of possible input with possible contradiction and threshold. So in neural network is necessary first to enlarge the basis in a way to solve with the minimum enlargement the contradiction and after use the threshold to reduce the complexity of the input basis. In the one step neural method we compute the parameters in one step as in quantum computer we use one query to generate the wanted result by a unitary matrix. To select wanted result in quantum computer and to obtain the wanted function in neural network, we use the projection operator method for non orthogonal states as oracle and inputs in quantum computer and neural network. Coker Specher theorem is revised in the light of the projection operator. In fact projection operator can select in a superposition one and only one element. Now when we have many basis with elements in common the local projection can enter in conflict with other connected basis projections. This put up in evidence that quantum computer and neural computer include contradiction or conflict. So before any computation we must solve the contradiction itself by the entangled projection method.

1 Introduction

Quantum computer has no contradiction because any time transformation (unitary transformation) of the Hilbert space the input and the output can be reversed. In quantum computer we use the projection operator into a space that include the wave function that for Coker Specher theorems quantum computer can have conflicts only for a set of connected set of basis. Neural computer has contradiction because we cannot reverse the input output transformation (projection operator). So the quantum computer and the neural computation are complementary one with the other. When we try to reduce the quantum computer to an ordinary computer or a neural computer we must introduce hidden variables that create contradictions shown in KS and Bell paradoxes. So quantum computer has no locality (entanglement) no direction in a particular space (in fact is all the universe that change in time but not an individual particle). All cells or states in the quantum computer are computed as one entity in a superposition state (coherent cells state as in laser). Probability to found the same cell is in any place different from zero (non locality). In the superposition state that cells (states) can interfere in a negative way to eliminate any probability to see the cell itself. In classical computer the superposition and interference are an intrinsic contradictory condition because how is the meaning to have the same particle or cell in two different positions or how is the meaning of two memory cells to have. By way of 2017, the improvement and growth of a real quantum computer is still in early stages but many poetical and theoretical experimentations were implemented by many research groups [1–23]. In conclusion the identity of one cell is not sure in the quantum computer.

2 Generation of Morphogenetic Field by Computation of the Sources and Superposition of Fields

Given the general function of one element of the basin in one point

$$F(x, y) = S\,[e^{-h((x-x_0)^2+(y-y_0)^2)}] \tag{1}$$

where S and h are parameters and x_0, y_0 is the position of the maximum value of the fields in the space x, y. The example of two basis fields is the parameters of the fields in Fig. 1 are for F_1 S = 1 h = 2 and $x_0 = -0.5$ and $y_0 = -0.5$, the parameters of the field F_2 are S = 1 h = 2 and $x_0 = 0.5$ and $y_0 = 0.5$.

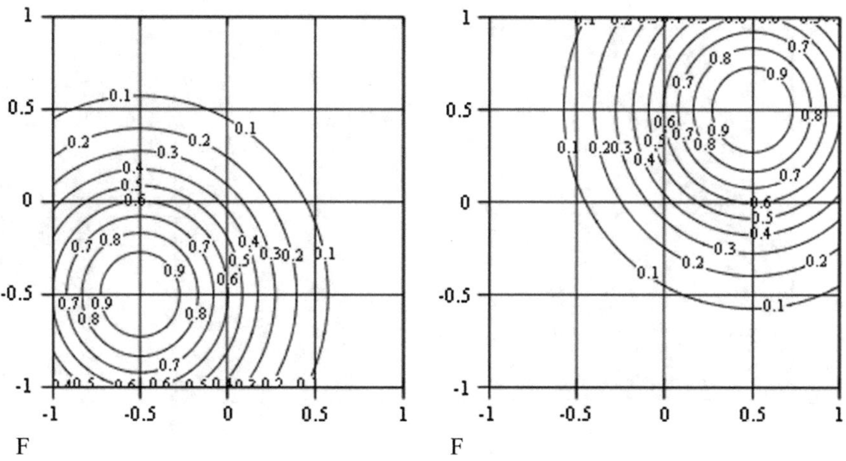

Fig. 1 Two basis fields or states

The matrix representation of the basis vectors is

$$A = \begin{bmatrix} F_1(p_1) & F_2(p_1) & \dots & F_m(p_1) \\ F_1(p_2) & F_2(p_2) & \dots & F_m(p_2) \\ \dots & \dots & \dots & \dots \\ F_1(p_n) & F_2(p_n) & \dots & F_m(p_n) \end{bmatrix} \quad (2)$$

Where any colon j are the values of the intensity of the field F_j *in the positions* $p_1, p_2, \ldots, p_n$ *in the reference space (x, y). We show an example in Fig. 1 of two different basic fields in a two dimensional reference space (x, y). The two fields are represented by two colons in the matrix A.*

The superposition field F that is shown in Fig. 2 is $F = F_1 + F_2$. For the sources $S_1 = 1$ and $S_2 = 2$, the superposition field F that is shown again in Fig. 2 is $F = F_1 + 2\,F_2$.

Given the field F in the two dimension x, y with one colon vector

$$F = \begin{bmatrix} F(p_1) \\ F(p_2) \\ \dots \\ F(p_n) \end{bmatrix} \quad (3)$$

The sources are the coefficients of the basis fields to rebuilt the field F by the basis fields A. So we have

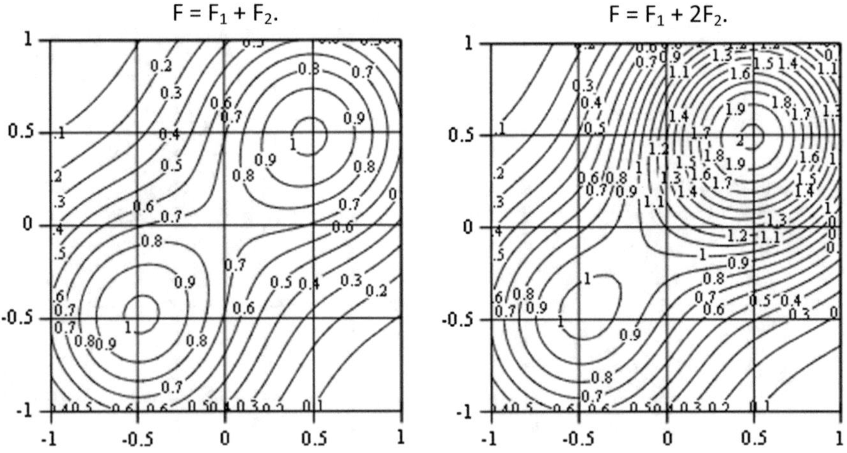

Fig. 2 Two different basis superposition

$$S = \begin{bmatrix} s_1 \\ s_2 \\ \cdots \\ s_m \end{bmatrix} = (A^T A)^{-1} A^T F = OPF \tag{4}$$

And the rebuilt field by the sources is

$$F' = s_1 F_1 + \cdots + s_m F_m \tag{5}$$

or

$$F' = QF = AS = A(A^T A)^{-1} A^T F \tag{6}$$

The operator Q is a projection operator of F into the space of the basis fields $F_1, \ldots, F_m$. Now when we change the original field by the operator U we have

$$S_U = (A^T A)^{-1} A^T (UF) = OPUF \tag{7}$$

where

$$OP = (A^T A)^{-1} A^T \tag{8}$$

A. Compensation and contradiction in superposition of basis fields

Given

$$OPUF = UOPF + (OPU - UOP)F = US + [OP, U]F \tag{9}$$

In the graph way we have (Fig. 3)

We remark that when the sources S change in the same way as the field (homotopy) the system is consistent but when the sources change in a different way as the field we have an inconsistent or contradictory source field system. To solve the contradiction we adjoin new sources that generate compensatory fields *that eliminate the contradictory or inconsistent state*. The compensatory sources are

$$S(F) = [OP, U]F \tag{10}$$

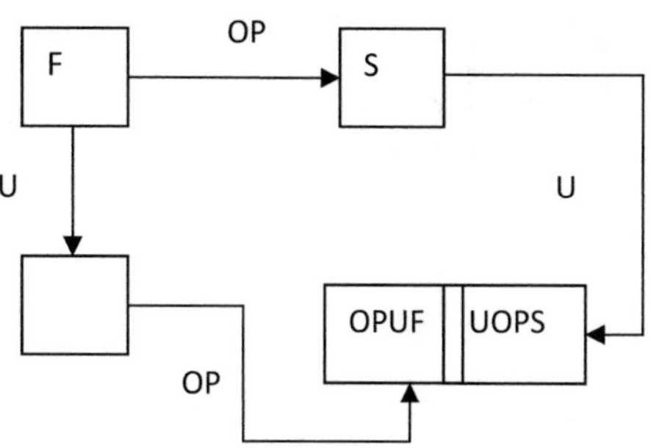

Fig. 3 Non commutative diagram for the U transformation

We remark that S(F) are sources that are function of the field so are not constant values but must be change in agreement with the local field for any point. This create the inconsistent problems- in fact this inconsistent problem can be solved only by a synchronic change of sources in any point of the reference point. Given a point with the previous operation we can create locally the wanted field by the knowledge of the previous field. Now with the compensatory field we change the local but also the global field (non locality). Now we can try to use another point to obtain local wanted field but in this way we destroy the wanted field in the first point. So we can move between one point to another without to obtain the wanted field in the two points this is the *inconsistent time problem* that we can solve only by a total synchronic activation of the compensatory sources. The field F can be represented by the function $\psi(x)$ where $x = \{x_1, x_2, \ldots, x_n\}$ is the n dimensional space. In this case we have again the same diagram (Fig. 4)

Where

$$\begin{aligned} OPU\psi &= UOP\psi + (OPU - UOP)\psi \\ &= US + [OP, U]\psi = US + S(\psi) \end{aligned} \tag{11}$$

The contradiction is connected with the commutative difference

$$L\psi = (OPU - UOP)\psi = [OP, U]\psi \tag{12}$$

When L = 0 the diagram is a commutative diagram for which the transformation U is the same before and after the application of the operator OP. This is consistent condition. When L is different from zero we are in the inconsistent condition because after the application of OP the same transformation U can have different values so if one is true the other is false and true and false are associate to the same transformation and true and false can be true but this is the contradiction that is impossible in the classical logic. We remark that the contradiction is compensate by a new type of sources or $S(\psi)$ (secondary sources) that generate a compensation (expansion) that solve the contradiction. The source $S(\psi)$ is locate in all the spaces and is dependent on the initial field ψ that synchronic way ACTIVATE the secondary sources that change the original field into a new wanted field $U\psi$. Example of the secondary sources or compensatory sources

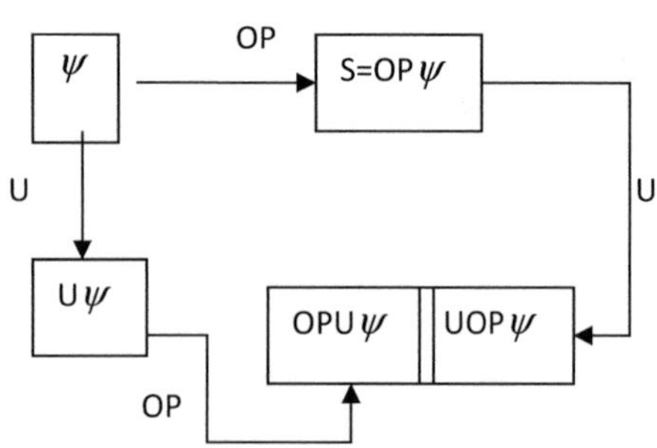

Fig. 4 Non commutative diagram for the U transformation where the field F is represented by the function $\psi(x)$ where $x = \{x_1, x_2, \ldots, x_n\}$

$$S(\psi) = [OP, U]\psi \tag{13}$$

Where

$$[\frac{\partial^2}{\partial t^2} - c^2(\frac{\partial^2}{\partial x^2} + \frac{\partial^2}{\partial y^2})]\psi = OP\psi = S \tag{14}$$

Is the wave function and S is the source of the wave field $\psi(x, y, t)$. Given the diagram

The diagram in Fig. 5 is a diagram where U is invariant (symmetry or coherence). In fact we have

$$\begin{aligned} OPU\psi &= UOP\psi + (OPU - UOP)\psi \\ &= US + [OP, U]\psi = US + S(\psi) = US \end{aligned} \tag{15}$$

The transformation U change in the same way the sources and the field. So when we want to change the field in this way $\psi \to U\psi$ we change the source in the same way $S \to US$. Now we want to found the transformation U for which $[OP, U]\psi = 0$ and OP is

$$\frac{\partial^2}{\partial t^2} - c^2(\frac{\partial^2}{\partial x^2} + \frac{\partial^2}{\partial y^2}) = OP \tag{16}$$

and

$$\begin{aligned} (OPU - UOP)\psi &= (\frac{\partial^2}{\partial t^2} - c^2(\frac{\partial^2}{\partial x^2} + \frac{\partial^2}{\partial y^2}))U\psi - U(\frac{\partial^2}{\partial t^2} - c^2(\frac{\partial^2}{\partial x^2} + \frac{\partial^2}{\partial y^2}))\psi \\ &= (\frac{\partial^2 U\psi}{\partial t^2} - c^2(\frac{\partial^2 U\psi}{\partial x^2} + \frac{\partial^2 U\psi}{\partial y^2})) - U(\frac{\partial^2}{\partial t^2} - c^2(\frac{\partial^2}{\partial x^2} + \frac{\partial^2}{\partial y^2}))\psi \\ &= \psi(\frac{\partial^2 U}{\partial x^2} - c^2(\frac{\partial^2 U}{\partial x^2} + \frac{\partial^2 U}{\partial y^2})) + 2(\frac{\partial U}{\partial t}\frac{\partial \psi}{\partial t} - c^2(\frac{\partial U}{\partial x}\frac{\partial \psi}{\partial x} + \frac{\partial U}{\partial y}\frac{\partial \psi}{\partial y}) \\ &\quad + U(\frac{\partial^2}{\partial t^2} - c^2(\frac{\partial^2}{\partial x^2} + \frac{\partial^2}{\partial y^2}))\psi - U(\frac{\partial^2}{\partial t^2} - c^2(\frac{\partial^2}{\partial x^2} + \frac{\partial^2}{\partial y^2}))\psi \\ &= \psi(\frac{\partial^2 U}{\partial x^2} - c^2(\frac{\partial^2 U}{\partial x^2} + \frac{\partial^2 U}{\partial y^2})) + 2(\frac{\partial U}{\partial t}\frac{\partial \psi}{\partial t} - c^2(\frac{\partial U}{\partial x}\frac{\partial \psi}{\partial x} + \frac{\partial U}{\partial y}\frac{\partial \psi}{\partial y}) = 0 \end{aligned} \tag{17}$$

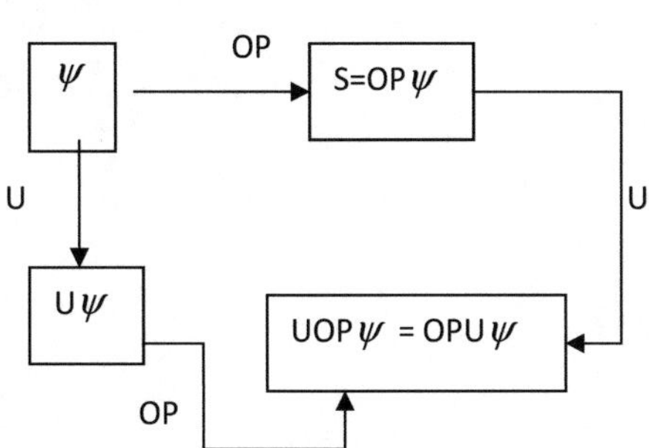

Fig. 5 Non commutative diagram for the U and constraint UOP ψ = OPU ψ

In a tensor form and in the space time reference space we have

$$
\begin{aligned}
&(OPU - UOP)\psi = \psi\partial^{\mu}\partial_{\mu}U + 2\partial^{\mu}\psi\partial_{\mu}U = 0 \\
&\partial_{\mu} = (\frac{\partial}{\partial ct}, \frac{\partial}{\partial x}, \frac{\partial}{\partial y}, \frac{\partial}{\partial z})
\end{aligned} \tag{18}
$$

In this condition the system is closed and coherent, No contradiction exist and any transformation U cannot change the properties or invariant of the system. In fact the invariant property or constrain can be seen in this computation

$$
\begin{aligned}
&\psi^2\partial^{\mu}\partial_{\mu}U + 2\psi\partial^{\mu}\psi\partial_{\mu}U \\
&= \partial^{\mu}(\psi^2\partial_{\mu}U) = \partial^{\mu}J_{\mu} = 0
\end{aligned} \tag{19}
$$

That is the continuity equation that now assume a new meaning as the condition to avoid contradiction and inconsistency. We remark that.

$$
\begin{aligned}
J_{\mu} &= \psi^2\partial_{\mu}U \\
&= \psi^2(\frac{\partial U}{\partial ct}, \frac{\partial U}{\partial x}, \frac{\partial U}{\partial y}, \frac{\partial U}{\partial z}) = (c\rho, j_x, j_y, j_z)
\end{aligned} \tag{20}
$$

The divergence of the four vector of the current is equal to zero. So when we have zero value of commutator or of Huygens sources S (active sources) we have the conservation of the charge

$$
\partial^{\mu}J_{\mu} = \frac{\partial\rho}{\partial t} + \nabla j = 0 \tag{21}
$$

When the probability density $\rho = R^2$ we have

$$
\begin{aligned}
&\frac{\partial R^2}{\partial t} + \nabla\cdot(R^2 v) = 2R\frac{\partial R}{\partial t} + \nabla\cdot(R^2 v) \\
&= 2R\frac{\partial R}{\partial t} + 2R\nabla Rv + R^2\nabla v = 0
\end{aligned} \tag{22}
$$

For $R \neq 0$ we have

$$
\begin{aligned}
&\frac{\partial R}{\partial t} + \nabla Rv + \frac{1}{2}R\nabla v \\
&= \frac{\partial R}{\partial t} + \frac{1}{2m}(2\nabla Rmv + R\nabla mv)
\end{aligned} \tag{23}
$$

Now in classical mechanics we have for the action S the relation

$$\nabla S = mv = p \tag{24}$$

So

$$\frac{\partial R}{\partial t} + \frac{1}{2m}(R\nabla^2 S + 2\nabla R \nabla S) = 0 \tag{25}$$

where S is the action and for the wave the phase of the wave. Given the system of the conditional probabilities.

when R is independent on time we have

$$\begin{aligned} &R\nabla^2 S + 2\nabla R\nabla S = 0 \\ &R^2\nabla^2 S + 2R\nabla R\nabla S = 0 \\ &\partial_\mu(R^2\nabla S) = \partial_\mu(P\nabla S) = \partial_\mu J^\mu = 0 \end{aligned} \tag{26}$$

where P is the probability. Now we show the connection between the action S and the refraction index and wave movement

$$\begin{aligned} \nabla S \cdot \nabla S &= (\frac{\partial S}{\partial x_1})^2 + (\frac{\partial S}{\partial x_2})^2 + (\frac{\partial S}{\partial x_3})^2 \\ &= \frac{1}{v(x)^2} = n^2 \end{aligned} \tag{27}$$

So the travel time T is equivalent to the electrostatic potential V and the vector of slowness is equivalent to the electrical field. For the phase $S = cT$ we have the image (Fig. 6)

In a more explicit way we have (Fig. 7)

In electromagnetic field we have the eikonale

$$J = \rho E = -\rho\nabla V \approx \psi^2 \nabla U \tag{28}$$

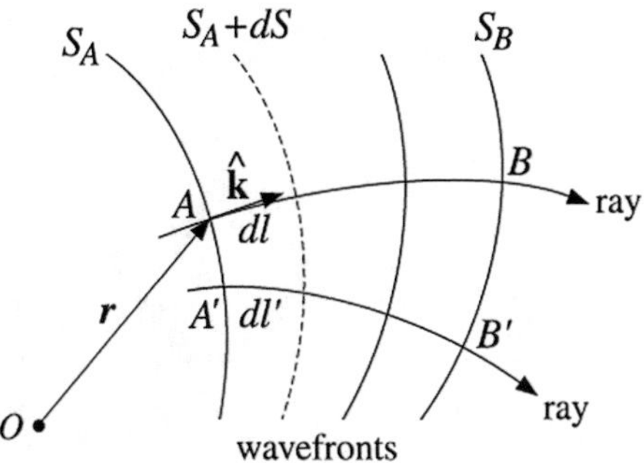

Fig. 6 Optical trajectory phase wave S and Eikonal ray

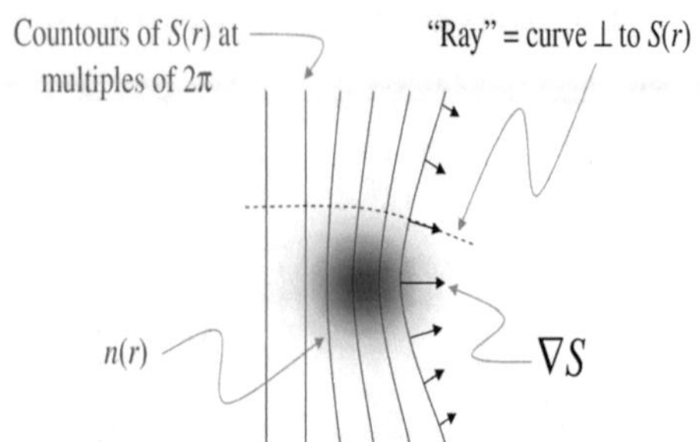

Fig. 7 Deformation of the ray or Eikonal for the refraction index that in this way control the optical ray and possible images (Eikonal)

The zero divergence property is the symmetric (non contradictory system) property for which the transformation U does not change when we move from the field to the sources so the commutator is equal to zero. The system is isolate and conservative (The source change in the same way as the field). The current is orthogonal to the equipotential surface. The intensity of the current is the density of conductance ρ.

Examples of potential field and currents (Figs. 8, 9)

And

B. Morphogenetic Sources by contradiction and inconsistency

In contradictory and inconsistent system we break the symmetry and new sources appear that we denote Morphogenetic Sources $[OP, U]\psi = S(\psi)$ that can compute by the expression

$$\begin{aligned} OPU\psi &= UOP\psi + (OPU - UOP)\psi \\ &= US + [OP, U]\psi = US + S(\psi) \end{aligned} \tag{29}$$

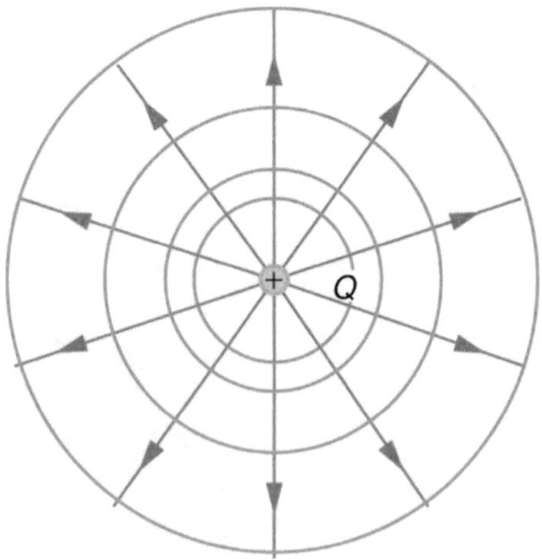

Fig. 8 Central potential field or monopole and currents that are orthogonal to equipotential (Eikonal)

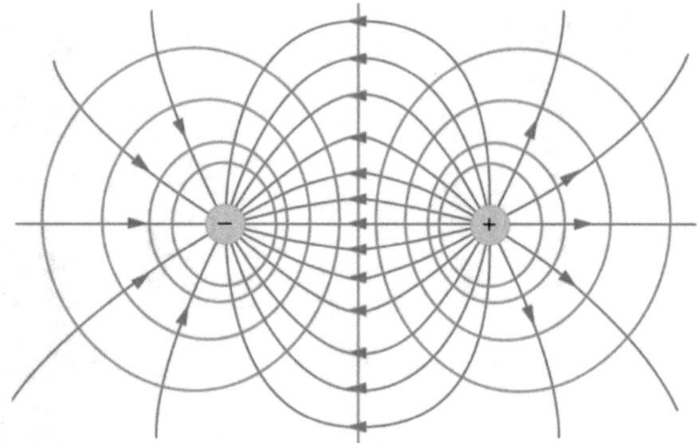

Fig. 9 Bipole potemtial field and orthogonal currents

where U is the goal and morphogenetic sources compensate the conservative sources *US* to generate any type of field different from conservative field or Eikonale.

Examples of Morphogenetic sources to generate designed fields or create Fields for special purpose (Figs. 10, 11, 12, 13).

Morphogenetic sources controlled by the forced function $f(x)$, and transformation $U(x)$ to compute the Morphogenetic field $u(x)$ and its transformation $U(x)u(x)$ for fixed basis functions $\varphi_n(x)$

Given the differential operator L and the equation

$$OPu(x) = f(x) \tag{30}$$

where $f(x)$ is the forced function and for example can be the wave differential operator

$$OP = \frac{\partial^2}{\partial T^2} - c^2\left(\frac{\partial^2}{\partial X^2} + \frac{\partial^2}{\partial Y^2}\right), x = (X, Y, T) \tag{31}$$

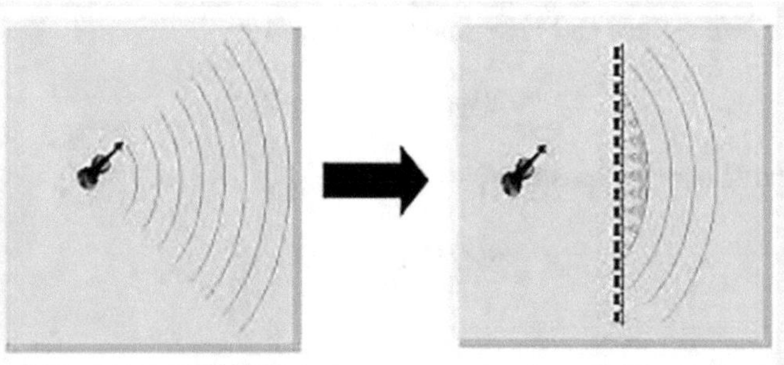

Fig. 10 Sound reproduction at the right open system of the left sound by a set of morphogenetic sources computed and reproduce by loudspeakers

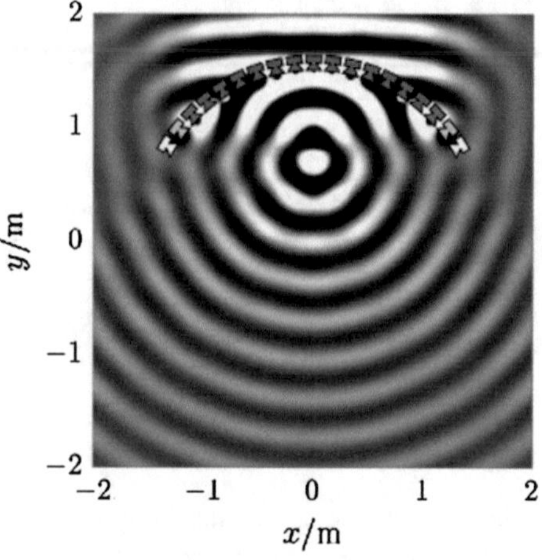

Fig. 11 Morphognetic sources to deform and control wave functions

Fig. 12 Synthesis of new type of wave by morphognetic sources that expand the space of the possible waves and select (projection) one particular case of the new possible wave

With the eigenvalues and eigenfunctions equation

$$OP\varphi(x) + \lambda\varphi(x) = 0 \tag{32}$$

Given the linear superposition

$$u(x) = \sum_n a_n\varphi_n(x) \tag{33}$$

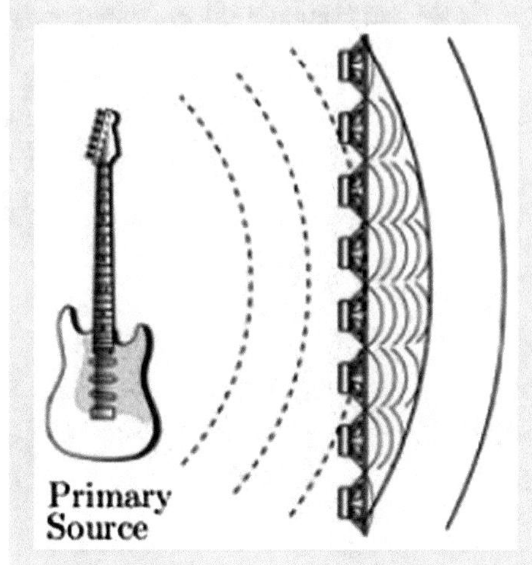

Fig. 13 More clear representation of the control by loudspeakers of the sound. We open the system by morphogenetic sources to create wanted behavior of the sound in the closed system

We have

$$OPu(x) = \sum_n a_n OP\varphi_n(x) = \sum_n a_n OP\varphi_n(x) = -\sum_n a_n \lambda_n \varphi_n(x) = f(x) \tag{34}$$

Now we integrate the two and we multiply with $\varphi_m(x)$

$$\int_a^b \varphi_m(x) f(z) dz = -\int_a^b \sum_n a_n \lambda_n \varphi_n(z) \varphi_m(z) dz \tag{35}$$

For the orthonormality of the eigenfunction we have

$$\int_a^b \varphi_n(x) f(x) dx = -\int_a^b \sum_n a_n \lambda_n \varphi_n(x) \varphi_n(x) dx = -a_n \lambda_n \int_a^b \sum_n \varphi_n(x)^2 dx \tag{36}$$

So we have

$$a_n \lambda_n = -\frac{\int_a^b \varphi_n(z) f(z) dz}{\int_a^b \sum_n \varphi_n(z)^2 dz} \tag{37}$$

So we can compute the coefficients a_n of (32)

$$a_n = -\frac{1}{\lambda_n}\frac{\int_a^b \varphi_n(z)f(z)dz}{\int_a^b \sum_n \varphi_n(z)^2 dz} \tag{38}$$

So we have

$$u(x) = \sum_n \left(-\frac{1}{\lambda_n}\frac{\int_a^b \varphi_n(z)f(z)dz}{\int_a^b \sum_n \varphi_n(z)^2 dz}\right)\varphi_n(x) = \int_a^b \sum_n \left(-\frac{1}{\lambda_n}\frac{\int_a^b \varphi_n(z)\varphi_n(x)dz}{\int_a^b \sum_n \varphi_n(z)^2 dz}\right)f(z) = \int_a^b G(x,z)f(z)dz \tag{39}$$

where $G(x, z)$ is the Green function and

$$S_n = \frac{-1}{\lambda_n}\frac{\int_a^b \varphi_n(z)f(z)dz}{\int_a^b \sum_n \varphi_n(z)^2 dz} \tag{40}$$

Are the Morphogenetic sources controlled by the forced function $f(x)$. Because we have

$$\begin{aligned} OPU(x)u(x) &= U(x)OPu(x) - U(x)OPu(x) + OPU(x)u(x) \\ &= U(x)f(x) + (OPU(x) - U(x)OP)u(x) = U(x)f(x) + [OP, U(x)]u(x) \end{aligned} \tag{41}$$

We obtain

$$\begin{aligned} U(x)u(x) &= \sum_n \left(-\frac{1}{\lambda_n}\frac{\int_a^b \varphi_n(z)(U(z)f(z) + [OP, U(z)]u(z))dz}{\int_a^b \sum_n \varphi_n(z)^2 dz}\right)\varphi_n(x) \\ &= -\sum_n \frac{1}{\lambda_n}\left(\frac{\int_a^b \varphi_n(z)U(z)f(z)dz}{\int_a^b \sum_n \varphi_n(z)^2 dz}\right) + \frac{\int_a^b [OP, U(z)]u(z))dz}{\int_a^b \sum_n \varphi_n(z)^2 dz}\Bigg)\varphi_n(x) \end{aligned} \tag{42}$$

Fourier transformation as unitary transformation to compute Morphogenetic sources and in quantum computer to generate designed field.

Given the equation

$$\begin{aligned} x^2 &= 1 \\ x &= 1, x = -1 \end{aligned} \tag{43}$$

For the generator $w = -1$ we have the cycle group composition low

$$\begin{bmatrix} & w & w^2 \\ w & w^2 & w \\ w^2 & w & w^2 \end{bmatrix} \tag{44}$$

Given the colon space (one qubit)

$$\begin{bmatrix} 0 \\ 1 \end{bmatrix} \tag{45}$$

We can compute the transformation matrix

$$\begin{bmatrix} 0 \\ 1 \end{bmatrix} \begin{bmatrix} 0 \\ 1 \end{bmatrix}^T = \begin{bmatrix} 0 & 0 \\ 0 & 1 \end{bmatrix} \tag{46}$$

Now we know that the eigenvalue of the unitary matrix are roots of the unity. In our case we have

$x^2 = 1$ which roots are 1 and -1. The generator of the cycle group of the roots is $\omega = -1$. So we have the discrete Fourier matrix

$$F = \begin{bmatrix} (\omega^0)^0 & (\omega^0)^1 \\ (\omega^1)^0 & (\omega^1)^1 \end{bmatrix} = \begin{bmatrix} 1 & 1 \\ 1 & -1 \end{bmatrix} \tag{47}$$

And the discrete Fourier transform of the vector p_k into a vector of Morphogenetic sources P^i

$$P^i = \sum_{k=0}^{n-1} p_k \omega^{ik} \tag{48}$$

With the inverse discrete Fourier transform where we use the vector of Morphogenetic sources P^i as coefficients to rebuilt the original field given by the vector p_k.

$$p_k = \sum_{i=0}^{n-1} P^i \omega_{ik}, \omega_{ik} = \omega^{-ik} \tag{49}$$

Where $\omega = e^{i\frac{2\pi}{2}} = -1$. So

$$F = \begin{bmatrix} 1 & 1 \\ 1 & -1 \end{bmatrix} \quad (50)$$

With the same generator the Fourier matrix can be obtained by the colon space in this way.

$$(-1)^{\begin{bmatrix} 0 & 0 \\ 0 & 1 \end{bmatrix}} = \begin{bmatrix} 1 & 1 \\ 1 & -1 \end{bmatrix} = F \quad (51)$$

Remark We remark that −1 is a generator of a cycle group for which we have the product table

$$\begin{bmatrix} & 1 & -1 \\ 1 & 1 & -1 \\ -1 & -1 & 1 \end{bmatrix} \quad (52)$$

The matrix F is generate by exponential method by the generator −1. So F is a generator of a group with the same product table of −1. The group generate by F is a new representation of the group with generator −1. So F is an unitary operator that give us a new representation of the group with −1 as generator.

For the impulse function $\begin{bmatrix} 1 \\ 0 \end{bmatrix}$ we have the discrete Fourier transformation

$$\begin{bmatrix} 1 & 1 \\ 1 & -1 \end{bmatrix} \begin{bmatrix} 1 \\ 0 \end{bmatrix} = \begin{bmatrix} 1 \\ 1 \end{bmatrix} \quad (53)$$

In the impulse function are present all the two frequencies with the same intensity. The values $\begin{bmatrix} 1 \\ 1 \end{bmatrix}$ are the coefficients of the discrete Fourier series that we denote as Morphogenetic Sources.

For

$$\begin{bmatrix} 0 & 0 \\ 1 & 0 \\ 0 & 1 \\ 1 & 1 \end{bmatrix} \begin{bmatrix} 0 & 0 \\ 1 & 0 \\ 0 & 1 \\ 1 & 1 \end{bmatrix}^T = \begin{bmatrix} 0 & 0 & 0 & 0 \\ 0 & 1 & 0 & 1 \\ 0 & 0 & 1 & 1 \\ 0 & 1 & 1 & 2 \end{bmatrix} \quad (54)$$

We have

$$F \otimes F = (-1)^{\begin{bmatrix} 0 & 0 & 0 & 0 \\ 0 & 1 & 0 & 1 \\ 0 & 0 & 1 & 1 \\ 0 & 1 & 1 & 2 \end{bmatrix}} = \begin{bmatrix} 1 & 1 & 1 & 1 \\ 1 & 1 & -1 & -1 \\ 1 & -1 & 1 & -1 \\ 1 & -1 & -1 & 1 \end{bmatrix} \tag{55}$$

Given the Boolean function AND that can write by the Boolean vector

$$xANDy = \begin{bmatrix} 0 & 0 \\ 1 & 0 \\ 0 & 1 \\ 1 & 1 \end{bmatrix} \rightarrow \begin{bmatrix} 0 \\ 0 \\ 0 \\ 1 \end{bmatrix} \tag{56}$$

The Fourier discrete transformation is

$$\begin{bmatrix} 1 & 1 & 1 & 1 \\ 1 & 1 & -1 & -1 \\ 1 & -1 & 1 & -1 \\ 1 & -1 & -1 & 1 \end{bmatrix} \begin{bmatrix} 0 \\ 0 \\ 0 \\ 1 \end{bmatrix} = \begin{bmatrix} 1 \\ -1 \\ -1 \\ 1 \end{bmatrix} \tag{57}$$

For XOR we have

$$\begin{bmatrix} 1 & 1 & 1 & 1 \\ 1 & 1 & -1 & -1 \\ 1 & -1 & 1 & -1 \\ 1 & -1 & -1 & 1 \end{bmatrix} \begin{bmatrix} 0 \\ 1 \\ 1 \\ 0 \end{bmatrix} = \begin{bmatrix} 2 \\ 0 \\ 0 \\ -2 \end{bmatrix} \tag{58}$$

Remark At any Boolean function we can give a Fourier transformation. For example the Boolean function NOT have this Fourier transformation

$$\begin{bmatrix} 1 & 1 \\ 1 & -1 \end{bmatrix} \begin{bmatrix} 0 \\ 1 \end{bmatrix} = \begin{bmatrix} 1 \\ -1 \end{bmatrix} \tag{59}$$

For more complex Fourier transformation we have

$$F = \begin{bmatrix} (\omega^0)^0 & (\omega^0)^1 & (\omega^0)^2 & (\omega^0)^3 \\ (\omega^1)^0 & (\omega^1)^1 & (\omega^1)^2 & (\omega^1)^3 \\ (\omega^2)^0 & (\omega^2)^1 & (\omega^2)^2 & (\omega^2)^3 \\ (\omega^3)^0 & (\omega^3)^1 & (\omega^3)^2 & (\omega^3)^3 \end{bmatrix} \tag{60}$$

For

Where $\omega = e^{i\frac{2\pi}{4}} = i$ we have

$$F = \begin{bmatrix} 1 & 1 & 1 & 1 \\ 1 & i & -1 & -i \\ 1 & -1 & 1 & -1 \\ 1 & -i & -1 & i \end{bmatrix} \tag{61}$$

That can be obtained also in this way

$$A = \begin{bmatrix} 0 \\ 1 \\ 2 \\ 3 \end{bmatrix} \begin{bmatrix} 0 \\ 1 \\ 2 \\ 3 \end{bmatrix}^T = \begin{bmatrix} 0 & 0 & 0 & 0 \\ 0 & 1 & 2 & 3 \\ 0 & 2 & 4 & 6 \\ 0 & 3 & 6 & 9 \end{bmatrix} \tag{62}$$

So

$$F = (i)^{\begin{bmatrix} 0 & 0 & 0 & 0 \\ 0 & 1 & 2 & 3 \\ 0 & 2 & 4 & 6 \\ 0 & 3 & 6 & 9 \end{bmatrix}} = \begin{bmatrix} 1 & 1 & 1 & 1 \\ 1 & i & -1 & -i \\ 1 & -1 & 1 & -1 \\ 1 & -i & -1 & i \end{bmatrix} \tag{63}$$

Given the Boolean function AND that can write by the Boolean vector

$$AND \begin{bmatrix} 0 & 0 \\ 1 & 0 \\ 0 & 1 \\ 1 & 1 \end{bmatrix} = \begin{bmatrix} 0 \\ 0 \\ 0 \\ 1 \end{bmatrix} \tag{64}$$

The Fourier discrete transformation is

$$\begin{bmatrix} 1 & 1 & 1 & 1 \\ 1 & i & -1 & -i \\ 1 & -1 & 1 & -1 \\ 1 & -i & -1 & i \end{bmatrix} \begin{bmatrix} 0 \\ 0 \\ 0 \\ 1 \end{bmatrix} = \begin{bmatrix} 1 \\ -i \\ -1 \\ i \end{bmatrix} \tag{65}$$

For XOR we have

$$\begin{bmatrix} 1 & 1 & 1 & 1 \\ 1 & i & -1 & -i \\ 1 & -1 & 1 & -1 \\ 1 & -i & -1 & i \end{bmatrix} \begin{bmatrix} 0 \\ 1 \\ 1 \\ 0 \end{bmatrix} = \begin{bmatrix} 2 \\ i-1 \\ 0 \\ -(i+1) \end{bmatrix} \tag{66}$$

For two qubits we have the colon space representation of the entanglement (rows) and superposition (colons)

$$\begin{bmatrix} 0 & 0 \\ 1 & 0 \\ 0 & 1 \\ 1 & 1 \end{bmatrix} \equiv \alpha_0|00\rangle + \alpha_1|10\rangle + \alpha_2|01\rangle + \alpha_3|11\rangle \tag{67}$$

And

$$\begin{bmatrix} 0 & 0 \\ 1 & 0 \\ 0 & 1 \\ 1 & 1 \end{bmatrix} \begin{bmatrix} 0 & 0 \\ 1 & 0 \\ 0 & 1 \\ 1 & 1 \end{bmatrix}^T = \begin{bmatrix} 0 & 0 & 0 & 0 \\ 0 & 1 & 0 & 1 \\ 0 & 0 & 1 & 1 \\ 0 & 1 & 1 & 2 \end{bmatrix} \tag{68}$$

For $\omega = -1$ we have the expression

$$(-1)^{rs} = (-1)^{\begin{bmatrix} 0 & 0 & 0 & 0 \\ 0 & 1 & 0 & 1 \\ 0 & 0 & 1 & 1 \\ 0 & 1 & 1 & 2 \end{bmatrix}} = \begin{bmatrix} 1 & 1 & 1 & 1 \\ 1 & -1 & 1 & -1 \\ 1 & 1 & -1 & -1 \\ 1 & -1 & -1 & 1 \end{bmatrix} = \begin{bmatrix} 1 & 1 \\ 1 & -1 \end{bmatrix} \otimes \begin{bmatrix} 1 & 1 \\ 1 & -1 \end{bmatrix} = F \otimes F \tag{69}$$

3 Quantum Computer Theorems and Connection with the Projection Operator

Given the input output structure (Fig. 14)

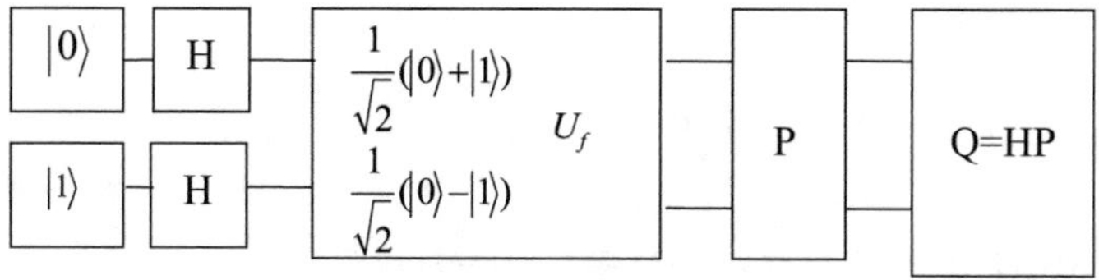

Fig. 14 Quantum computer unitary transformation with one qubit $|0\rangle$

$$\begin{aligned}
&H|0\rangle = \tfrac{1}{\sqrt{2}}(|0\rangle + |1\rangle)\\
&H|1\rangle = \tfrac{1}{\sqrt{2}}(|0\rangle - |1\rangle)\\
&\textit{for}\\
&U_f|x\rangle|y\rangle = |x\rangle|y \oplus f(x)\rangle\\
&P = U_f(\tfrac{1}{\sqrt{2}}(|0\rangle + |1\rangle)\tfrac{1}{\sqrt{2}}(|0\rangle - |1\rangle))\\
&= \tfrac{1}{2}U_f((|0\rangle|0\rangle - |0\rangle|1\rangle + |1\rangle|0\rangle - |1\rangle|1\rangle) = \tfrac{1}{2}|0\rangle|0 \oplus f(0)\rangle - |0\rangle|0 \oplus f(1)\rangle + |1\rangle|1 \oplus f(0)\rangle - |1\rangle|1 \oplus f(1)\rangle
\end{aligned} \tag{70}$$

We remark that

$$D = |0 \oplus f(1)\rangle - |1 \oplus f(0)\rangle = f(1) - (1 - f(0)) = f(1) + f(0) - 1 \tag{71}$$

Now we have that

$$\begin{aligned}
&if(0) \neq f(1) \ \textit{then} \ \mathrm{D} \ = f(1) + f(0) - 1 = 0\\
&f(0) = f(1) \ \textit{then} \ \mathrm{D} \ = f(1) + f(0) - 1 = 1
\end{aligned} \tag{72}$$

When we commute x with y the difference commute when $if(0) \neq f(1)$ and does not commute when $if(0) = f(1)$. Now with another H transformation we have

$$\begin{aligned}
Q = HP &= \frac{1}{2}H|0\rangle|0 \oplus f(0)\rangle - H|0\rangle|0 \oplus f(1)\rangle + H|1\rangle|1 \oplus f(0)\rangle - H|1\rangle|1 \oplus f(1)\rangle\\
&= \frac{1}{2\sqrt{2}}[(|0\rangle + |1\rangle)|0 \oplus f(0)\rangle - (|0\rangle + |1\rangle)|0 \oplus f(1)\rangle + (|0\rangle - |1\rangle)|1 \oplus f(0)\rangle\\
&\quad - (|0\rangle - |1\rangle)|1 \oplus f(1)\rangle\\
&= \frac{1}{2\sqrt{2}}[|0\rangle|0 \oplus f(0)\rangle - |0\rangle|0 \oplus f(1)\rangle + |0\rangle|1 \oplus f(0)\rangle - |0\rangle|1 \oplus f(1)\rangle\\
&\quad + \frac{1}{2\sqrt{2}}[|1\rangle|0 \oplus f(0)\rangle - |1\rangle|0 \oplus f(1)\rangle - |1\rangle|1 \oplus f(0)\rangle + |1\rangle|1 \oplus f(1)\rangle\\
&= \frac{|0\rangle}{2\sqrt{2}}[|0 \oplus f(0)\rangle - |0 \oplus f(1)\rangle + |1 \oplus f(0)\rangle - |1 \oplus f(1)\rangle\\
&\quad + \frac{|1\rangle}{2\sqrt{2}}[|0 \oplus f(0)\rangle - |0 \oplus f(1)\rangle - |1 \oplus f(0)\rangle + |1 \oplus f(1)\rangle
\end{aligned} \tag{73}$$

Now as we know if $f(0) = f(1) = 0$ we have

$$\begin{aligned}
&\frac{|0\rangle}{2\sqrt{2}}[|0 \oplus f(0)\rangle - |0 \oplus f(1)\rangle + |1 \oplus f(0)\rangle - |1 \oplus f(1)\rangle\\
&\quad + \frac{|1\rangle}{2\sqrt{2}}[|0 \oplus f(0)\rangle - |0 \oplus f(1)\rangle - |1 \oplus f(0)\rangle + |1 \oplus f(1)\rangle = -\frac{1}{\sqrt{2}}|0\rangle
\end{aligned} \tag{74}$$

Now as we know if $f(0) = f(1) = 1$ we have

$$\begin{aligned}&\frac{|0\rangle}{2\sqrt{2}}[|0 \oplus f(0)\rangle - |0 \oplus f(1)\rangle + |1 \oplus f(0)\rangle - |1 \oplus f(1)\rangle \\ &\quad + \frac{|1\rangle}{2\sqrt{2}}[|0 \oplus f(0)\rangle - |0 \oplus f(1)\rangle - |1 \oplus f(0)\rangle + |1 \oplus f(1)\rangle = \frac{1}{\sqrt{2}}|0\rangle\end{aligned} \tag{75}$$

Now as we know if $f(0) = 0, f(1) = 1$ we have

$$\begin{aligned}&\frac{|0\rangle}{2\sqrt{2}}[|0 \oplus f(0)\rangle - |0 \oplus f(1)\rangle + |1 \oplus f(0)\rangle - |1 \oplus f(1)\rangle \\ &\quad + \frac{|1\rangle}{2\sqrt{2}}[|0 \oplus f(0)\rangle - |0 \oplus f(1)\rangle - |1 \oplus f(0)\rangle + |1 \oplus f(1)\rangle = -\frac{1}{\sqrt{2}}|1\rangle\end{aligned} \tag{76}$$

Now as we know if $f(0) = 1, f(1) = 0$ we have

$$\begin{aligned}&\frac{|0\rangle}{2\sqrt{2}}[|0 \oplus f(0)\rangle - |0 \oplus f(1)\rangle + |1 \oplus f(0)\rangle - |1 \oplus f(1)\rangle \\ &\quad + \frac{|1\rangle}{2\sqrt{2}}[|0 \oplus f(0)\rangle - |0 \oplus f(1)\rangle - |1 \oplus f(0)\rangle + |1 \oplus f(1)\rangle = \frac{1}{\sqrt{2}}|1\rangle\end{aligned} \tag{77}$$

Colon space image of the superposition and entangled in quantum Hilbert space
Given the colon space

$$\begin{aligned}S = \begin{bmatrix} a_{11} & a_{12} & \ldots & a_{1n} \\ a_{21} & a_{22} & \ldots & a_{2n} \\ \ldots & \ldots & \ldots & \ldots \\ a_{m-1,1} & a_{m-1,2} & \ldots & a_{m-1,n} \\ a_{m,1} & a_{m,2} & \ldots & a_{m,n} \end{bmatrix} &\equiv \alpha_1 |a_{11} a_{12}, \ldots, a_{1n}\rangle \\ &+ \alpha_2 |a_{21} a_{22}, \ldots, a_{2n}\rangle + \cdots + \alpha_m |a_{m1} a_{m2}, \ldots, a_{mn}\rangle\end{aligned} \tag{78}$$

Deutsch problem by colon space

(1) $$S = [0 \quad 1] \equiv |01\rangle = |0\rangle|1\rangle \tag{79}$$

(2) $$S = \begin{bmatrix} 0 & 0 \\ 0 & -1 \\ 1 & 0 \\ 1 & -1 \end{bmatrix} \equiv |00\rangle - |01\rangle + |10\rangle - |11\rangle \tag{80}$$

$$(3) \quad S = \begin{bmatrix} 0 & 0 \oplus f(0) \\ 0 & -(1 \oplus f(0)) \\ 1 & 0 \oplus f(1) \\ 1 & -(1 \oplus f(1)) \end{bmatrix} \equiv |0\rangle|0 \oplus f(0)\rangle - |0\rangle|1 \oplus f(0)\rangle + |1\rangle|0 \oplus f(1)\rangle - |1\rangle|1 \oplus f(1)\rangle \tag{81}$$

$$(4) \quad S = \begin{bmatrix} 0 & 0 \oplus f(0) \\ 1 & 0 \oplus f(0) \\ 0 & -(1 \oplus f(0)) \\ 1 & -(1 \oplus f(0)) \\ 0 & 0 \oplus f(1) \\ -1 & 0 \oplus f(1) \\ 0 & -(1 \oplus f(1)) \\ -1 & -(1 \oplus f(1)) \end{bmatrix} \equiv H|0\rangle|0 \oplus f(0)\rangle - H|0\rangle|1 \oplus f(0)\rangle + H|1\rangle|0 \oplus f(1)\rangle$$
$$-H|1\rangle|1 \oplus f(1)\rangle = = (|0\rangle + |1\rangle)|0 \oplus f(0)\rangle - (|0\rangle + |1\rangle)|1 \oplus f(0)\rangle$$
$$+ (|0\rangle - |1\rangle)|0 \oplus f(1)\rangle - (|0\rangle - |1\rangle)|1 \oplus f(1)\rangle = w_1|0\rangle + w_2|1\rangle \tag{82}$$

Where

$$\begin{aligned} w_1 &= |0 \oplus f(0)\rangle - |1 \oplus f(0)\rangle + |0 \oplus f(1)\rangle - |1 \oplus f(1)\rangle = f(0) - (1 - f(0)) + f(1) - (1 - f(1)) \\ w_2 &= |0 \oplus f(0)\rangle - |1 \oplus f(0)\rangle - |0 \oplus f(1)\rangle + |1 \oplus f(1)\rangle = f(0) - (1 - f(0)) - f(1) + (1 - f(1)) \end{aligned} \tag{83}$$

But

$$\begin{aligned} w_1 &= f(0) - (1 - f(0)) + f(1) - (1 - f(1)) = 2(f(0) + f(1) - 1) \\ w_2 &= f(0) - (1 - f(0)) - f(1) + (1 - f(1)) = 2(f(0) - f(1)) \end{aligned} \tag{84}$$

Reverse of the Deutsch problem
Given the wanted weights or Morphogenetic Sources

$$2\begin{bmatrix} 1 \\ 0 \\ (f(0) + f(1) - 1 \\ f(0) - f(1) \end{bmatrix} \tag{85}$$

For which we can separate the situation where $f(0) = f(1), from f(0) \neq f(1)$ in fact we have for $f(0) = f(1)$

$$2\begin{bmatrix} 1 \\ 0 \\ (f(0)+f(1)-1 \\ f(0)-f(1) \end{bmatrix} = 2\begin{bmatrix} 1 \\ 0 \\ 1 \\ 0 \end{bmatrix}, 2\begin{bmatrix} 1 \\ 0 \\ (f(0)+f(1)-1 \\ f(0)-f(1) \end{bmatrix} = 2\begin{bmatrix} 1 \\ 0 \\ -1 \\ 0 \end{bmatrix} \tag{86}$$

For $f(0) \neq f(1)$

$$2\begin{bmatrix} 1 \\ 0 \\ (f(0)+f(1)-1 \\ f(0)-f(1) \end{bmatrix} = 2\begin{bmatrix} 1 \\ 0 \\ 0 \\ 1 \end{bmatrix}, 2\begin{bmatrix} 1 \\ 0 \\ (f(0)+f(1)-1 \\ f(0)-f(1) \end{bmatrix} = 2\begin{bmatrix} 1 \\ 0 \\ 0 \\ -1 \end{bmatrix} \tag{87}$$

Now given the aim we want to come back to the vector g for which

$$\begin{bmatrix} 1 & 1 & 1 & 1 \\ 1 & 1 & -1 & -1 \\ 1 & -1 & 1 & -1 \\ 1 & -1 & -1 & 0 \end{bmatrix}\begin{bmatrix} g_1 \\ g_2 \\ g_3 \\ g_4 \end{bmatrix} = 2\begin{bmatrix} 1 \\ 0 \\ 1-(f(0)+f(1)) \\ f(0)-f(1) \end{bmatrix} \tag{88}$$

So we have

$$\begin{bmatrix} 1 & 1 & 1 & 1 \\ 1 & 1 & -1 & -1 \\ 1 & -1 & 1 & -1 \\ 1 & -1 & -1 & 0 \end{bmatrix}\begin{bmatrix} g_1 \\ g_2 \\ g_3 \\ g_4 \end{bmatrix} = 2\begin{bmatrix} 1 & 1 & 1 & 1 \\ 1 & 1 & -1 & -1 \\ 1 & -1 & 1 & -1 \\ 1 & -1 & -1 & 0 \end{bmatrix}^{-1}\begin{bmatrix} 1 \\ 0 \\ (f(0)+f(1)-1 \\ f(0)-f(1) \end{bmatrix} =$$

$$2\begin{bmatrix} 1 & 1 & 1 & 1 \\ 1 & 1 & -1 & -1 \\ 1 & -1 & 1 & -1 \\ 1 & -1 & -1 & 0 \end{bmatrix}^{T}\begin{bmatrix} 1 \\ 0 \\ (f(0)+f(1)-1 \\ f(0)-f(1) \end{bmatrix} = \begin{bmatrix} f(0) \\ 1-f(0) \\ f(1) \\ 1-f(1) \end{bmatrix} = \begin{bmatrix} 0\oplus f(0) \\ 1\oplus f(0) \\ 0\oplus f(1) \\ 1\oplus f(1) \end{bmatrix} \tag{89}$$

In the previous expression we use the well known expression $a \oplus b = a\bar{b} + \bar{a}b$. With the previous reverse proves we give a motivation for the Deutsch oracle that we use in quantum computer. For a more general case we have (Fig. 15)

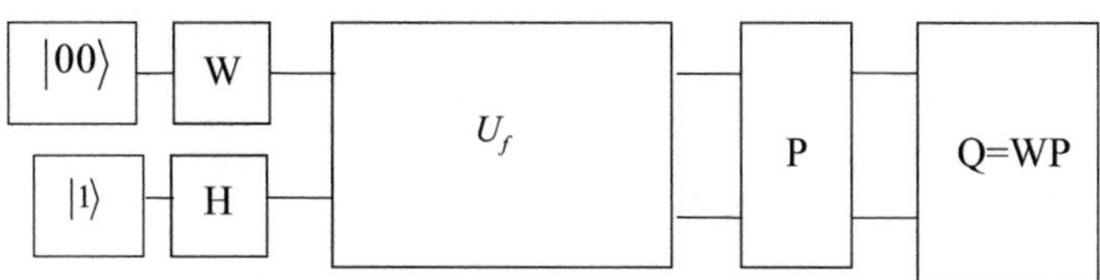

Fig. 15 Quantum computer unitary transformation with two qubits $|00\rangle$

$$W = H \otimes H = \begin{bmatrix} 1 & 1 & 1 & 1 \\ 1 & 1 & -1 & -1 \\ 1 & -1 & 1 & -1 \\ 1 & -1 & -1 & 0 \end{bmatrix} \begin{bmatrix} |00\rangle \\ |10\rangle \\ |01\rangle \\ |11\rangle \end{bmatrix} = \begin{bmatrix} |00\rangle + |10\rangle + |01\rangle + |11\rangle \\ |00\rangle + |10\rangle - |01\rangle - |11\rangle \\ |00\rangle - |10\rangle + |01\rangle - |11\rangle \\ |00\rangle - |10\rangle - |01\rangle + |11\rangle \end{bmatrix}$$

$$W|00\rangle = |00\rangle + |10\rangle + |01\rangle + |11\rangle$$
$$H|1\rangle = |0\rangle - |1\rangle$$
$$for$$
$$|x\rangle|y\rangle = |00\rangle|1\rangle$$
$$P = W|00\rangle H|1\rangle = (\frac{1}{2\sqrt{2}}|00\rangle + |10\rangle + |01\rangle + |11\rangle)(|0\rangle - |1\rangle))$$
$$\frac{1}{2\sqrt{2}} U_f(|00\rangle|0\rangle + |10\rangle|0\rangle + |01\rangle|0\rangle + |11\rangle|0\rangle) - (|00\rangle|1\rangle + |10\rangle|1\rangle + |01\rangle|1\rangle + |11\rangle|1\rangle)$$
$$= \frac{1}{2\sqrt{2}}|00\rangle|0 \oplus f(0,0)\rangle + |10\rangle|0 \oplus f(1,0)\rangle + |01\rangle|0 \oplus f(0,1)\rangle + |11\rangle|0 \oplus f(1,1)\rangle)$$
$$- (|00\rangle|1 \oplus f(0,0)\rangle +$$
$$+ |10\rangle|1 \oplus f(1,0)\rangle + |01\rangle|1 \oplus f(0,1)\rangle + |11\rangle|1 \oplus f(1,1)\rangle) \quad (90)$$

And

$$\frac{1}{2\sqrt{2}} W|00\rangle|0 \oplus f(0,0)\rangle + W|10\rangle|0 \oplus f(1,0)\rangle + W|01\rangle|0 \oplus f(0,1)\rangle$$
$$+ W|11\rangle|0 \oplus f(1,1)\rangle) - (W|00\rangle|1 \oplus f(0,0)\rangle + \quad (91)$$
$$+ W|10\rangle|1 \oplus f(1,0)\rangle + W|01\rangle|1 \oplus f(0,1)\rangle + W|11\rangle|1 \oplus f(1,1)\rangle)$$

Is

$$\frac{1}{2\sqrt{2}}(|00\rangle + |10\rangle + |01\rangle + |11\rangle)|0 \oplus f(0,0)\rangle + (|00\rangle + |10\rangle - |01\rangle - |11\rangle)|0 \oplus f(1,0)\rangle$$
$$+ (|00\rangle - |10\rangle + |01\rangle - |11\rangle)|0 \oplus f(0,1)\rangle + (|00\rangle - |10\rangle - |01\rangle + |11\rangle|0 \oplus f(1,1)\rangle)$$
$$- (|00\rangle + |10\rangle + |01\rangle + |11\rangle)|1 \oplus f(0,0)\rangle - (|00\rangle + |10\rangle - |01\rangle - |11\rangle)|1 \oplus f(1,0)\rangle +$$
$$- (|00\rangle - |10\rangle + |01\rangle - |11\rangle)|1 \oplus f(0,1)\rangle - (|00\rangle - |10\rangle - |01\rangle + |11\rangle)|1 \oplus f(1,1)\rangle) \quad (92)$$

So we have

$$
\begin{aligned}
S = \frac{1}{2\sqrt{2}} & |00\rangle[|0 \oplus f(0,0)\rangle + |0 \oplus f(1,0)\rangle + |0 \oplus f(0,1)\rangle + |0 \oplus f(1,1)\rangle) - |1 \oplus f(0,0)\rangle + \\
& - |1 \oplus f(1,0)\rangle - |1 \oplus f(0,1)\rangle - |1 \oplus f(1,1)\rangle] + \\
& |10\rangle[|0 \oplus f(0,0)\rangle + |0 \oplus f(1,0)\rangle - |0 \oplus f(0,1)\rangle - |0 \oplus f(1,1)\rangle) \\
& - |1 \oplus f(0,0)\rangle - |1 \oplus f(1,0)\rangle + |1 \oplus f(0,1)\rangle + |1 \oplus f(1,1)\rangle] + \\
& |01\rangle[|0 \oplus f(0,0)\rangle - |0 \oplus f(1,0)\rangle + |0 \oplus f(0,1)\rangle - |0 \oplus f(1,1)\rangle) \\
& - |1 \oplus f(0,0)\rangle |1 \oplus f(1,0)\rangle + \\
& - |1 \oplus f(0,1)\rangle + |1 \oplus f(1,1)\rangle) + \\
& |11\rangle[|0 \oplus f(0,0)\rangle - |0 \oplus f(1,0)\rangle - |0 \oplus f(0,1)\rangle + |0 \oplus f(1,1)\rangle) \\
& - |1 \oplus f(0,0)\rangle + |1 \oplus f(1,0)\rangle + |1 \oplus f(0,1)\rangle - |1 \oplus f(1,1)\rangle)
\end{aligned}
\tag{93}
$$

For

$$
\begin{bmatrix}
0 \oplus f(0,0) \\
1 \oplus f(0,0) \\
0 \oplus f(1,0) \\
1 \oplus f(1,0) \\
0 \oplus f(0,1) \\
1 \oplus f(0,1) \\
0 \oplus f(1,1) \\
1 \oplus f(1,1)
\end{bmatrix}
=
\begin{bmatrix}
f(0,0) \\
1 - f(0,0) \\
f(1,0) \\
1 - f(1,0) \\
f(0,1) \\
1 - f(0,1) \\
f(1,1) \\
1 - f(1,1)
\end{bmatrix}
\tag{94}
$$

We have

$$
\begin{aligned}
S = \frac{1}{2\sqrt{2}} & [(-4 + 2(f(0,0) + f(1,0) + f(0,1) + f(1,1)|00\rangle + 2(f(0,0) \\
& - f(1,0) - f(0,1) + f(1,1)|10\rangle \\
& 2(-f(0,0) + f(1,0) + f(0,1) - f(1,1)|01\rangle + 2(-f(0,0) - f(1,0) + f(0,1) + f(1,1)|11\rangle]
\end{aligned}
\tag{95}
$$

That can write in this direct way

$$W = (H \otimes H \otimes H)|y \oplus f(x_1.x_2)\rangle = \begin{bmatrix} 1 & 1 & 1 & 1 & 1 & 1 & 1 & 1 \\ 1 & -1 & 1 & -1 & 1 & -1 & 1 & -1 \\ 1 & 1 & -1 & -1 & 1 & 1 & -1 & -1 \\ 1 & -1 & -1 & 1 & 1 & -1 & -1 & 1 \\ 1 & 1 & 1 & 1 & -1 & -1 & -1 & -1 \\ 1 & -1 & 1 & -1 & -1 & 1 & -1 & 1 \\ 1 & 1 & -1 & -1 & -1 & -1 & 1 & 1 \\ 1 & -1 & -1 & 1 & -1 & 1 & 1 & -1 \end{bmatrix} \begin{bmatrix} 0 \oplus f(0,0) \\ 1 \oplus f(0,0) \\ 0 \oplus f(1,0) \\ 1 \oplus f(1,0) \\ 0 \oplus f(0,1) \\ 1 \oplus f(0,1) \\ 0 \oplus f(1,1) \\ 1 \oplus f(1,1) \end{bmatrix}$$

$$= \begin{bmatrix} f(0,0) \\ 1 - f(0,0) \\ f(1,0) \\ 1 - f(1,0) \\ f(0,1) \\ 1 - f(0,1) \\ f(1,1) \\ 1 - f(1,1) \end{bmatrix}$$

$$= \begin{bmatrix} 4 \\ 2f(0,0) + 2f(1,0) + 2f(0,1) + 2f(1,1) - 4 \\ 0 \\ 2f(0,0) - 2f(1,0) + 2f(0,1) - 2f(1,1) \\ 0 \\ 2f(0,0) + 2f(1,0) - 2f(0,1) - 2f(1,1) \\ 0 \\ 2f(0,0) - 2f(1,0) - 2f(0,1) + 2f(1,1) \end{bmatrix} \tag{96}$$

where we can see the transformation and functions. Boolean functions in quantum computer. Boolean functions and Deutsch problem

$$For \begin{bmatrix} f(0,0) \\ f(1,0) \\ f(0,1) \\ f(1,1) \end{bmatrix} = \begin{bmatrix} 1 \\ 1 \\ 1 \\ 1 \end{bmatrix} we \text{ have } \begin{bmatrix} 2f(0,0)+2f(1,0)+2f(0,1)+2f(1,1)-4 \\ 2f(0,0)-2f(1,0)+2f(0,1)-2f(1,1) \\ 2f(0,0)+2f(1,0)-2f(0,1)-2f(1,1) \\ 2f(0,0)-2f(1,0)-2f(0,1)+2f(1,1) \end{bmatrix} = \begin{bmatrix} 4 \\ 0 \\ 0 \\ 0 \end{bmatrix}$$

$$For \begin{bmatrix} f(0,0) \\ f(1,0) \\ f(0,1) \\ f(1,1) \end{bmatrix} = \begin{bmatrix} 0 \\ 0 \\ 0 \\ 0 \end{bmatrix} we \text{ have } \begin{bmatrix} 2f(0,0)+2f(1,0)+2f(0,1)+2f(1,1)-4 \\ 2f(0,0)-2f(1,0)+2f(0,1)-2f(1,1) \\ 2f(0,0)+2f(1,0)-2f(0,1)-2f(1,1) \\ 2f(0,0)-2f(1,0)-2f(0,1)+2f(1,1) \end{bmatrix} = \begin{bmatrix} -4 \\ 0 \\ 0 \\ 0 \end{bmatrix}$$

$$For \begin{bmatrix} f(0,0) \\ f(1,0) \\ f(0,1) \\ f(1,1) \end{bmatrix} = \begin{bmatrix} 0 \\ 1 \\ 0 \\ 1 \end{bmatrix} we \text{ have } \begin{bmatrix} 2f(0,0)+2f(1,0)+2f(0,1)+2f(1,1)-4 \\ 2f(0,0)-2f(1,0)+2f(0,1)-2f(1,1) \\ 2f(0,0)+2f(1,0)-2f(0,1)-2f(1,1) \\ 2f(0,0)-2f(1,0)-2f(0,1)+2f(1,1) \end{bmatrix} = \begin{bmatrix} 0 \\ -4 \\ 0 \\ 0 \end{bmatrix}$$

$$For \begin{bmatrix} f(0,0) \\ f(1,0) \\ f(0,1) \\ f(1,1) \end{bmatrix} = \begin{bmatrix} 0 \\ 0 \\ 1 \\ 1 \end{bmatrix} we \text{ have } \begin{bmatrix} 2f(0,0)+2f(1,0)+2f(0,1)+2f(1,1)-4 \\ 2f(0,0)-2f(1,0)+2f(0,1)-2f(1,1) \\ 2f(0,0)+2f(1,0)-2f(0,1)-2f(1,1) \\ 2f(0,0)-2f(1,0)-2f(0,1)+2f(1,1) \end{bmatrix} = \begin{bmatrix} 0 \\ 0 \\ -4 \\ 0 \end{bmatrix}$$

$$For \begin{bmatrix} f(0,0) \\ f(1,0) \\ f(0,1) \\ f(1,1) \end{bmatrix} = \begin{bmatrix} 1 & 0 & 0 & 0 \\ 0 & 1 & 0 & 0 \\ 0 & 0 & 0 & 1 \\ 0 & 0 & 1 & 0 \end{bmatrix} \begin{bmatrix} 0 \\ 1 \\ 0 \\ 1 \end{bmatrix} = \begin{bmatrix} 0 \\ 1 \\ 1 \\ 0 \end{bmatrix} = x \oplus y \, we \text{ have}$$

$$\begin{bmatrix} 2f(0,0)+2f(1,0)+2f(0,1)+2f(1,1)-4 \\ 2f(0,0)-2f(1,0)+2f(0,1)-2f(1,1) \\ 2f(0,0)+2f(1,0)-2f(0,1)-2f(1,1) \\ 2f(0,0)-2f(1,0)-2f(0,1)+2f(1,1) \end{bmatrix} = \begin{bmatrix} 0 \\ 0 \\ 0 \\ -4 \end{bmatrix} \tag{97}$$

$$\text{Given the strings } u = \begin{bmatrix} 0 \\ 0 \end{bmatrix}, \begin{bmatrix} 1 \\ 0 \end{bmatrix}, \begin{bmatrix} 0 \\ 1 \end{bmatrix}, \begin{bmatrix} 1 \\ 1 \end{bmatrix} \tag{98}$$

And the functions

$f_u = xu$

So

$$f_0 = \begin{bmatrix} 0 & 0 \\ 1 & 0 \\ 0 & 1 \\ 1 & 1 \end{bmatrix} \begin{bmatrix} 0 \\ 0 \end{bmatrix} = \begin{bmatrix} 0 \\ 0 \\ 0 \\ 0 \end{bmatrix}, f_1 = \begin{bmatrix} 0 & 0 \\ 1 & 0 \\ 0 & 1 \\ 1 & 1 \end{bmatrix} \begin{bmatrix} 1 \\ 0 \end{bmatrix} = \begin{bmatrix} 0 \\ 1 \\ 0 \\ 1 \end{bmatrix}$$
$$f_2 = \begin{bmatrix} 0 & 0 \\ 1 & 0 \\ 0 & 1 \\ 1 & 1 \end{bmatrix} \begin{bmatrix} 0 \\ 1 \end{bmatrix} = \begin{bmatrix} 0 \\ 0 \\ 1 \\ 1 \end{bmatrix}, f_3 = \begin{bmatrix} 0 & 0 \\ 1 & 0 \\ 0 & 1 \\ 1 & 1 \end{bmatrix} \begin{bmatrix} 1 \\ 1 \end{bmatrix} = \text{mod}\,(2) \begin{bmatrix} 0 \\ 1 \\ 1 \\ 2 \end{bmatrix} \equiv \begin{bmatrix} 0 \\ 1 \\ 1 \\ 0 \end{bmatrix} \tag{99}$$

So given for example

$$u = \begin{bmatrix} 1 \\ 0 \end{bmatrix} \tag{100}$$

We have

$$For \begin{bmatrix} f(0,0) \\ f(1,0) \\ f(0,1) \\ f(1,1) \end{bmatrix} = \begin{bmatrix} 0 \\ 0 \\ 1 \\ 1 \end{bmatrix} \textit{we}\text{ have} \begin{bmatrix} 2f(0,0)+2f(1,0)+2f(0,1)+2f(1,1)-4 \\ 2f(0,0)-2f(1,0)+2f(0,1)-2f(1,1) \\ 2f(0,0)+2f(1,0)-2f(0,1)-2f(1,1) \\ 2f(0,0)-2f(1,0)-2f(0,1)+2f(1,1) \end{bmatrix} = \begin{bmatrix} 0 \\ 0 \\ -4 \\ 0 \end{bmatrix}$$

and

$$\psi_u = -4|01\rangle \tag{101}$$

So with one query we know the value of u. This is the explicit expression of the Bernstein Varizani Theorm.

Another extension can be the **Nagata multifunction parallel computation**. For

$$f(x,y) = xg(a) \tag{102}$$

Where

$$g: R \to \{0,1\} \tag{103}$$

Among the Boolean function we have this basis elements

$$|00\rangle = |g_1 g_1\rangle = \begin{bmatrix} 1 \\ 1 \\ 1 \\ 1 \end{bmatrix}, |10\rangle = |g_2 g_1\rangle \begin{bmatrix} 0 \\ 1 \\ 0 \\ 1 \end{bmatrix}, |01\rangle = |g_1 g_2\rangle = \begin{bmatrix} 0 \\ 0 \\ 1 \\ 1 \end{bmatrix}, |11\rangle = |g_2 g_2\rangle = \begin{bmatrix} 0 \\ 1 \\ 1 \\ 0 \end{bmatrix} \quad (104)$$

The superposition of the four states we have

$$\psi(x, y) = c_1|00\rangle + c_2|10\rangle + c_3|10\rangle + c_4|10\rangle \quad (105)$$

For the non balance Boolean function with odd number of 1 we cannot reduce the wave function to one state but we have a superposition of the states.

Projection operator

The first idea is to transform the n inputs in n vectors which have at the maximum the dimension 2^n and a output vector that give us all possible output in one vector. For example given two inputs with values one or zero, we have all possible inputs in this two vectors in $A = \begin{bmatrix} 0 & 0 \\ 1 & 0 \\ 0 & 1 \\ 1 & 1 \end{bmatrix}$, the matrix A is denoted as the colon space S spanned by two vectors. The colon space will be given by the superposition of the input vectors

$$S = w_1 \begin{bmatrix} 0 \\ 1 \\ 0 \\ 1 \end{bmatrix} + w_2 \begin{bmatrix} 0 \\ 0 \\ 1 \\ 1 \end{bmatrix} \quad (106)$$

where w_1, w_2 are the coordinates of one point inside the plane defined by the two vectors in A. We remark that because n = 2 the dimension of the input vectors is $2^2 = 4$. The output is again a vector in the same 4 dimensional space 4. The output is a point in the 4 dimensional space. For example with $Y = \begin{bmatrix} 0 \\ 0 \\ 0 \\ 1 \end{bmatrix}$, this vector in the digital computer is the AND that is true only for the input (1, 1) and zero for the other three inputs. So we have

$$\begin{bmatrix} 0 & 0 \\ 1 & 0 \\ 0 & 1 \\ 1 & 1 \end{bmatrix} \rightarrow \begin{bmatrix} 0 \\ 0 \\ 0 \\ 1 \end{bmatrix} = X \wedge Y \tag{107}$$

In the digital gate theory we define basic Boolean operators as AND, OR, and NOR or in more simple way we can define only one Boolean operator as NAND by which we can built all possible complex expressions for any Boolean function. Now given a Boolean function to found the Boolean expression by the basic element is a very difficult problem that in general is a non polynomial complexity or PN. Now we change the point of view and we establish an algebraic connection between the Inputs vectors A and the output Y in this way.

$$Aw = Y \tag{108}$$

where w are the parameters that from the input A generate the output Y. After we will show that w are the neural weights. The Eq. (107) give the desired vector Y as a linear combination of the inputs. For the AND logic operator we have

$$\begin{bmatrix} 0 & 0 \\ 1 & 0 \\ 0 & 1 \\ 1 & 1 \end{bmatrix} \begin{bmatrix} w_1 \\ w_2 \end{bmatrix} = \begin{bmatrix} 0 \\ 0 \\ 0 \\ 1 \end{bmatrix} \tag{109}$$

As we can see the (107) and (108) cannot be solved to obtain the wanted parameters w. So we must change the (107) in this way

$$\begin{aligned} A^T A w &= A^T Y, \\ w &= (A^T A)^{-1} A^T Y \end{aligned} \tag{110}$$

For the case (108) we have

$$w = (A^T A)^{-1} A^T Y = \begin{bmatrix} \frac{1}{3} \\ \frac{1}{3} \end{bmatrix} \tag{111}$$

With w we can compute the projection QY of Y into the input world A by (107).

$$A w = A(A^T A)^{-1} A^T Y = QY \tag{112}$$

The operator Q is the projection because we have

$$Q^2 Y = A(A^T A)^{-1} A^T A (A^T A)^{-1} A^T Y = A(A^T A)^{-1} A^T Y \tag{113}$$

The projection of the projection is again a projection. Geometric example

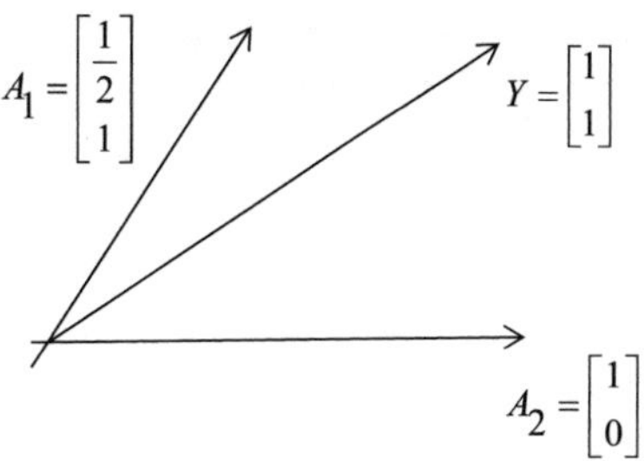

Fig. 16 Colon space of the two vectors A_1, A_2 and the Boolean function Y. Y is projected into the two dimensional space by the projection operator

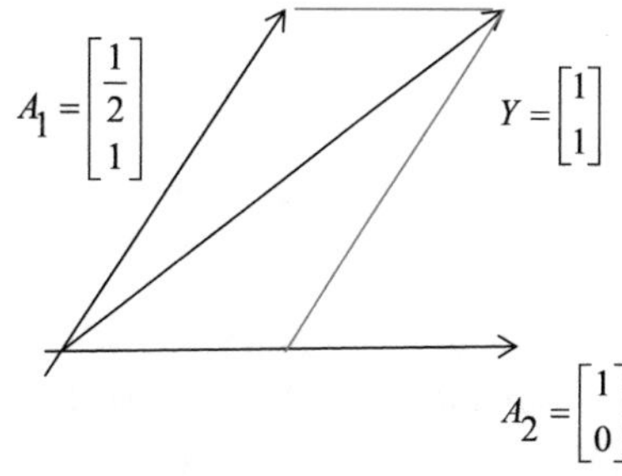

Fig. 17 Projection of Y into the colon space

For

$$A_1 = \begin{bmatrix} \frac{1}{2} \\ 1 \end{bmatrix}, A_2 = \begin{bmatrix} 1 \\ 0 \end{bmatrix}, A = \begin{bmatrix} \frac{1}{2} & 1 \\ 1 & 0 \end{bmatrix}, Y = \begin{bmatrix} 1 \\ 1 \end{bmatrix} \tag{114}$$

We have (Fig. 16)
For the projection operator we have

$$w = (A^T A)^{-1} A^T Y = \left(\begin{bmatrix} \frac{1}{2} & 1 \\ 1 & 0 \end{bmatrix}^T \begin{bmatrix} \frac{1}{2} & 1 \\ 1 & 0 \end{bmatrix} \right)^{-1} \begin{bmatrix} \frac{1}{2} & 1 \\ 1 & 0 \end{bmatrix}^T \begin{bmatrix} 1 \\ 1 \end{bmatrix} = \begin{bmatrix} 1 \\ \frac{1}{2} \end{bmatrix} \tag{115}$$

So we have

$$w = (A^T A)^{-1} A^T Y = QY = \begin{bmatrix} 1 \\ \frac{1}{2} \end{bmatrix} \tag{116}$$

and

$$Aw = A(A^T A)^{-1} A^T Y = QY = \begin{bmatrix} \frac{1}{2} \\ 1 \end{bmatrix} + \frac{1}{2} \begin{bmatrix} 1 \\ 0 \end{bmatrix} = \begin{bmatrix} 1 \\ 1 \end{bmatrix} \tag{117}$$

For the vector sum we have that (Fig. 17)

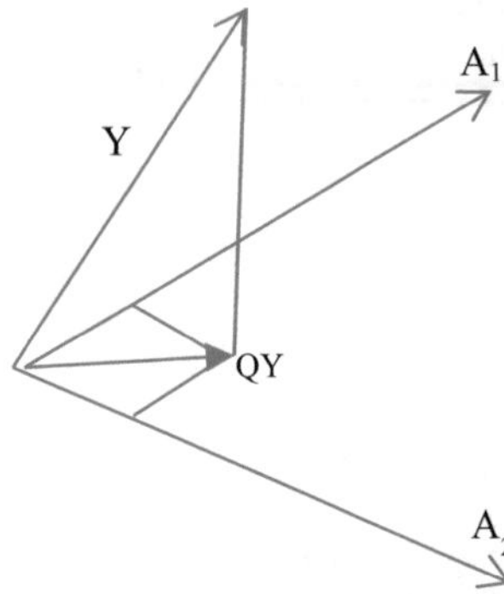

Fig. 18 Projection of Y into the two dimensional space

$$Y = w_1 \begin{bmatrix} \frac{1}{2} \\ 1 \end{bmatrix} + w_2 \begin{bmatrix} 1 \\ 0 \end{bmatrix} = 1 \begin{bmatrix} \frac{1}{2} \\ 1 \end{bmatrix} + \frac{1}{2} \begin{bmatrix} 1 \\ 0 \end{bmatrix} = \begin{bmatrix} 1 \\ 1 \end{bmatrix} \tag{118}$$

In three dimension we have (Fig. 18)

The inputs A define a subspace of the vectors space. Any Y in the vector space is projected into the space of the inputs. For the example 1 we have

$$QY = A\,w = \begin{bmatrix} 0 & 0 \\ 1 & 0 \\ 0 & 1 \\ 1 & 1 \end{bmatrix} \begin{bmatrix} \frac{1}{3} \\ \frac{1}{3} \end{bmatrix} = \frac{1}{3} \begin{bmatrix} 0 \\ 1 \\ 0 \\ 1 \end{bmatrix} + \frac{1}{3} \begin{bmatrix} 0 \\ 0 \\ 1 \\ 1 \end{bmatrix} = w_1 A_1 + w_2 A_2 = \begin{bmatrix} 0 \\ \frac{1}{3} \\ \frac{1}{3} \\ \frac{2}{3} \end{bmatrix}, \mathrm{Y} = \begin{bmatrix} 0 \\ 0 \\ 0 \\ 1 \end{bmatrix} \tag{119}$$

Where the vector space has four dimensions and the input space has two dimensions embedded in the four dimensional space. The two weights are the two component of the projection of Y into the two dimensional subspace of the inputs.

4 Conclusion

In Turing Machine input output and states form a system. When system is in one state the input is transformed in output. Local and sequential process is the ordinary computation in Turing machine and in the language for the digital computer. The software is built to connect all different sequential computations (statements) in a way that the final result will be the solution of our problem. The problem with digital computer is to build a code that will be useful for our purpose. Also if the digital computer can solve a lot of problems when we compare the efficiency of the digital computer with the efficiency of the brain we remark a huge difference in the two type of computation. Brain can solve problems that digital computer cannot. But how is the explanation of this very high brain efficiency? When we study the brain we can see that is compose by elements denoted neurons. Any neurons receive a lot of inputs, make weighted linear combination, with threshold give its reply. Any input can be one or zero n inputs have all possible value of the Boolean vector with dimension n. For two inputs we have (0,0), (1,0), (0,1), (1,1) possible inputs. So for first input has this possible inputs 0101, the second inputs has the possible inputs 0011. The first input has the morphogenetic field 0101, the second has the morphogenetic field 0011. The weights are the Morphogenetic sources or strength of the fields. Given the wanted field in a Boolean form, we project this field into the inputs fields and we compute the weights or morphogenetic sources. The superposition of the input fields generate the neuron field that with the threshold is can be the wanted field. If the wanted field is incompatible with the input field we have a method to enlarge the number of inputs in a way to obtain a set of input fields compatible with the wanted field and we solve the neuron problem. In quantum mechanics we make the same process where we have the basis quantum state for n qubits. The problem in quantum mechanics is to classify sets of functions by a quantum measure as projection of a vector into one particular state that is the key to recognize by one query the property of the function or the values of the functions. In this case we select one and only one Morphogenetic source among others in a way to know the researched property. We also explain the meaning of the entanglement and Morphogenetic synchronic process as the global property to eliminate local contradiction in hidden variables by the explanation of the Coker Specher theorems. At the end we can also use the wave basis functions to reverse the physical process that from the sources generate the field. So when we know the field we can came back to compute the Morphogenetic sources and at the end in a physical domain obtain the wanted field. All different applications put in evidence one method denoted projection method that in a global way collect all the local knowledge in fields that with linear aggregation can give the solution of our problems. The Morphogenetic sources are used in holographic process to simulate from two dimension the three dimension. We know that in the global morphogenetic field the information is diffuse in all the reference space so the information is robust to external noise we can also find the same information also in one part of the reference space. In conclusion the Morphogenetic computing we eliminate the stressing work of the creation of the codes and in the same time errors or contradictions are eliminate in the generation of the morpho by the Morphogenetic sources.

References

1. Metwaly, A.F., Rashad, M.Z., Omara, F.A., Megahed, A.A.: Architecture of multicast centralized key management scheme using quantum key distribution and classical symmetric encryption. Eur. Phys. J. Spec. Top. **223**(8), 1711–1728 (2014)
2. Metwaly, A., Rashad, M.Z., Omara, F.A., Megahed, A.A.: Architecture of point to multipoint QKD communication systems (QKDP2MP). In: 8th International Conference on Informatics and Systems (INFOS), Cairo, pp. NW 25–31. IEEE (2012)

3. Farouk, A., Omara, F., Zakria, M., Megahed, A.: Secured IPsec multicast architecture based on quantum key distribution. In: The International Conference on Electrical and Bio-medical Engineering, Clean Energy and Green Computing, pp. 38–47. The Society of Digital Information and Wireless Communication (2015)
4. Farouk, A., Zakaria, M., Megahed, A., Omara, F.A.: A generalized architecture of quantum secure direct communication for N disjointed users with authentication. Sci. Rep. **5**, 16080 (2014)
5. Wang, M.M., Wang, W., Chen, J.G., Farouk, A.: Secret sharing of a known arbitrary quantum state with noisy environment. Quantum Inf. Process. **14**(11), 4211–4224 (2015)
6. Naseri, M., Heidari, S., Batle, J., Baghfalaki, M., Gheibi, R., Farouk, A., Habibi, A.: A new secure quantum watermarking scheme. Optik-Int. J. Light Electron Opt. **139**, 77–86 (2017)
7. Batle, J., Ciftja, O., Naseri, M., Ghoranneviss, M., Farouk, A., Elhoseny, M.: Equilibrium and uniform charge distribution of a classical two-dimensional system of point charges with hard-wall confinement. Phys. Scr. **92**(5), 055801 (2017)
8. Geurdes, H., Nagata, K., Nakamura, T., Farouk, A.: A note on the possibility of incomplete theory (2017). arXiv preprint arXiv:1704.00005
9. Batle, J., Farouk, A., Alkhambashi, M., Abdalla, S.: Multipartite correlation degradation in amplitude-damping quantum channels. J. Korean Phys. Soc. **70**(7), 666–672 (2017)
10. Batle, J., Naseri, M., Ghoranneviss, M., Farouk, A., Alkhambashi, M., Elhoseny, M.: Shareability of correlations in multiqubit states: optimization of nonlocal monogamy inequalities. Phys. Rev. A **95**(3), 032123 (2017)
11. Batle, J., Farouk, A., Alkhambashi, M., Abdalla, S.: Entanglement in the linear-chain Heisenberg antiferromagnet $Cu(C_4H_4N_2)$ $(NO_3)_2$. Eur. Phys. J. B **90**, 1–5 (2017)
12. Batle, J., Alkhambashi, M., Farouk, A., Naseri, M., Ghoranneviss, M.: Multipartite non-locality and entanglement signatures of a field-induced quantum phase transition. Eur. Phys. J. B **90**(2), 31 (2017)
13. Nagata, K., Nakamura, T., Batle, J., Abdalla, S., Farouk, A.: Boolean approach to dichotomic quantum measurement theories. J. Korean Phys. Soc. **70**(3), 229–235 (2017)
14. Abdolmaleky, M., Naseri, M., Batle, J., Farouk, A., Gong, L.H.: Red-Green-Blue multi-channel quantum representation of digital images. Optik-Int. J. Light Electron Opt. **128**, 121–132 (2017)
15. Farouk, A., Elhoseny, M., Batle, J., Naseri, M., Hassanien, A.E.: A proposed architecture for key management schema in centralized quantum network. In: Handbook of Research on Machine Learning Innovations and Trends, pp. 997–1021. IGI Global (2017)
16. Zhou, N.R., Li, J.F., Yu, Z.B., Gong, L.H., Farouk, A.: New quantum dialogue protocol based on continuous-variable two-mode squeezed vacuum states. Quantum Inf. Process. **16**(1), 4 (2017)
17. Batle, J., Abutalib, M., Abdalla, S., Farouk, A.: Persistence of quantum correlations in a XY spin-chain environment. Eur. Phys. J. B **89**(11), 247 (2016)
18. Batle, J., Abutalib, M., Abdalla, S., Farouk, A.: Revival of Bell nonlocality across a quantum spin chain. Int. J. Quantum Inf. **14**(07), 1650037 (2016)
19. Batle, J., Ooi, C.R., Farouk, A., Abutalib, M., Abdalla, S.: Do multipartite correlations speed up adiabatic quantum computation or quantum annealing? Quantum Inf. Process. **15**(8), 3081–3099 (2016)
20. Batle, J., Bagdasaryan, A., Farouk, A., Abutalib, M., Abdalla, S.: Quantum correlations in two coupled superconducting charge qubits. Int. J. Mod. Phys. B **30**(19), 1650123 (2016)
21. Batle, J., Ooi, C.R., Abutalib, M., Farouk, A., Abdalla, S.: Quantum information approach to the azurite mineral frustrated quantum magnet. Quantum Inf. Process. **15**(7), 2839–2850 (2016)
22. Batle, J., Ooi, C.R., Farouk, A., Abdalla, S.: Nonlocality in pure and mixed n-qubit X states. Quantum Inf. Process. **15**(4), 1553–1567 (2016)
23. Metwaly, A.F., Rashad, M.Z., Omara, F.A., Megahed, A.A.: Architecture of Multicast Network Based on Quantum Secret Sharing and Measurement (2015)

Morphogenetic Sources in Quantum, Neural and Wave Fields: Part 2

G. Resconi, K. Nagata, O. Tarawneh and Ahmed Farouk

Abstract Neural network and quantum computer have the same conceptual structure similar to Huygens sources in the wave field generation. Any point of the space is a source with different intensity of waves that transport information in all the space where are superposed in a complex way to generate the wave field. In wave theory this sources are denoted Huygens sources. The morphogenetic field is the wave field generate by computed sources that are designed in a way to transform the original field in a wanted field to satisfy wanted property. The morphogenetic computation is this type of global computation by sources like Huygens sources that in parallel and synchronic way give us the designed field. So the intensity of the sources must be computed a priory before the morphogenetic effective computation in a way to have an entanglement of the sources that in the same time compute the field. If we cannot design the sources a priory and we want generate the field by a recursion process we enter easily in a deadlock state for which one source generate local wanted field that destroy the generation of another local field. So we have a contradiction between the action of different non entangled sources that cannot generate all the wanted field. In neural network we have the superposition of the input vectors in quantum mechanics we have the superposition of the states. In the neural network the intensity of the sources are the neural weights and the threshold. In quantum mechanics the intensity of the sources are the coefficients of the quantum states superposition. To design neural sources intensity (weights) we use the matrix of all possible inputs by which

G. Resconi
Department of Mathematics and Physics, Catholic University, I-25121 Brescia, Italy

K. Nagata (✉)
Department of Physics, Korea Advanced Institute of Science and Technology, Daejeon 34141, Korea
e-mail: ko_mi_na@yahoo.co.jp

O. Tarawneh
Information Technology Department, Al-Zahra College for Women, P.O.Box 3365, Muscat, Oman

A. Farouk
Computer Sciences Department, Faculty of Computers and Information, Mansoura University, Mansoura, Egypt

A.E. Hassanien et al. (eds.), *Quantum Computing: An Environment for Intelligent Large Scale Real Application*, Studies in Big Data 33,
https://doi.org/10.1007/978-3-319-63639-9_15

we can define all possible outputs. In the design neural network we cannot use the simple theory of input output but all the past or future input output are used. Space and time is not important in the design the network more important is to use the space of all possible input and output. The same in the quantum computer where we must design the unitary transformation for which only one wanted state coefficient is different from zero all the other coefficients are put to zero. In this way we can select among a huge possible states any one wanted state solution of our problem. In this scheme we include Deutsch problems, Berstein Varizani theorem and Nagata parallel function computing. The difference between quantum computer and neural network is that in quantum computer the basis is the oracle square matrix without any threshold and contradiction. In neural network the basis is a rectangular matrix of possible input with possible contradiction and threshold. So in neural network is necessary first to enlarge the basis in a way to solve with the minimum enlargement the contradiction and after use the threshold to reduce the complexity of the input basis. In the one step neural method we compute the parameters in one step as in quantum computer we use one query is used to generate the wanted result by a unitary matrix. To select wanted result in quantum computer and to obtain the wanted function in neural network, we use the projection operator method for non orthogonal states as oracle and inputs in quantum computer and neural network. Coker Specher theorem is revised in the light of the projection operator. In fact projection operator can select in a superposition one and only one element. Now when we have many basis with elements in common the local projection can enter in conflict with other connected basis projections. This put up in evidence that quantum computer and neural computer include contradiction or conflict. So before any computation we must solve the contradiction itself by the entangled projection method.

1 Introduction

Quantum computer has no contradiction because any time transformation (unitary transformation) of the Hilbert space the input and the output can be reversed. In quantum computer we use the projection operator into a space that include the wave function that Fot Coker Specher theorems quantum computer can have conflicts only for a set of connected set of basis. Neural computer has contradiction because we cannot reverse the input output transformation (projection operator). So the quantum computer and the neural computation are complementary one with the other. When we try to reduce the quantum computer to an ordinary computer or a neural computer we must introduce hidden variables that create contradictions shown in KS and Bell paradoxes. So quantum computer has no locality (entanglement) no direction in a particular space (in fact is all the universe that change in time but not an individual particle), All cells or states in the quantum computer are computed as one entity in a superposition state (coherent cells state as in laser). Probability to found the same cell is in any place different from zero (non locality). In the superposition state that cells (states) can interfere in a negative way to eliminate any probability to see the cell itself. In classical computer the superposition and interference are an intrinsic contradictory condition because how is the meaning to have the same particle or cell in two different positions or how is the meaning of two memory cells to have. By way of 2017, the improvement and growth of a real quantum computer is still in early stages but many poetical and theoretical experimentations were implemented by many research groups [1–23]. In conclusion the identity of one cell is not sure in the quantum computer.

2 Oracle in Quantum Computer and Projection Operator

Given the equation

$$Ax = y \tag{1}$$

We have the solution

$$\begin{array}{l} A^T Ax = A^T y \\ and \\ x = (A^T A)^{-1} A^T y \end{array} \tag{2}$$

When A is a square matrix with determinant different from zero we have

$$x = A^{-1}(A^T)^{-1}A^T y = A^{-1} y \tag{3}$$

In quantum mechanics given the Oracle matrix

$$A = \begin{bmatrix} 1 & 0 & 0 & 0 \\ 1 & 1 & 0 & 1 \\ 1 & 0 & 1 & 1 \\ 1 & 1 & 1 & 0 \end{bmatrix} = [1 \quad x \quad y \quad x \oplus y] \tag{4}$$

And

$$y = \begin{bmatrix} f(0,0) \\ f(1,0) \\ f(0,1) \\ f(1,1) \end{bmatrix} \tag{5}$$

We have

$$x = A^{-1} y = \frac{1}{2} \begin{bmatrix} 2f(0,0) \\ -f(0,0) + f(0,1) - f(1,0) + f(1,1) \\ -f(0,0) - f(0,1) + f(1,0) + f(1,1) \\ -f(0,0) + f(0,1) + f(1,0) - f(1,1) \end{bmatrix} \tag{6}$$

For which we have for $f(x, y) = 1, or f(x, y) = 0$ that are the constant function with value 1 or 0 we have

$$x = A^{-1}y = \frac{1}{2}\begin{bmatrix} 2f(0,0) \\ -f(0,0) + f(0,1) - f(1,0) + f(1,1) \\ -f(0,0) - f(0,1) + f(1,0) + f(1,1) \\ -f(0,0) + f(0,1) + f(1,0) - f(1,1) \end{bmatrix} = \begin{bmatrix} 1 \\ 0 \\ 0 \\ 0 \end{bmatrix} \tag{7}$$

And

$$Ax = 1\begin{bmatrix} 1 \\ 1 \\ 1 \\ 1 \end{bmatrix} + 0\begin{bmatrix} 0 \\ 1 \\ 0 \\ 1 \end{bmatrix} + 0\begin{bmatrix} 0 \\ 0 \\ 1 \\ 1 \end{bmatrix} + 0\begin{bmatrix} 0 \\ 1 \\ 1 \\ 0 \end{bmatrix} \tag{8}$$

In the Dirac symbolism we have

$$\psi = 1|00\rangle + 0|10\rangle + 0|01\rangle + 0|11\rangle \tag{9}$$

For which we have for $f(x, y) = x$

$$x = A^{-1}y = \frac{1}{2}\begin{bmatrix} 2f(0,0) \\ -f(0,0) + f(0,1) - f(1,0) + f(1,1) \\ -f(0,0) - f(0,1) + f(1,0) + f(1,1) \\ -f(0,0) + f(0,1) + f(1,0) - f(1,1) \end{bmatrix} = \begin{bmatrix} 0 \\ 1 \\ 0 \\ 0 \end{bmatrix} \tag{10}$$

And

$$Ax = 0\begin{bmatrix} 1 \\ 1 \\ 1 \\ 1 \end{bmatrix} + 1\begin{bmatrix} 0 \\ 1 \\ 0 \\ 1 \end{bmatrix} + 0\begin{bmatrix} 0 \\ 0 \\ 1 \\ 1 \end{bmatrix} + 0\begin{bmatrix} 0 \\ 1 \\ 1 \\ 0 \end{bmatrix} \tag{11}$$

For which we have for $f(x, y) = y$

$$x = A^{-1}y = \frac{1}{2}\begin{bmatrix} 2f(0,0) \\ -f(0,0) + f(0,1) - f(1,0) + f(1,1) \\ -f(0,0) - f(0,1) + f(1,0) + f(1,1) \\ -f(0,0) + f(0,1) + f(1,0) - f(1,1) \end{bmatrix} = \begin{bmatrix} 0 \\ 0 \\ 1 \\ 0 \end{bmatrix} \tag{12}$$

And

$$Ax=0\begin{bmatrix}1\\1\\1\\1\end{bmatrix}+0\begin{bmatrix}0\\1\\0\\1\end{bmatrix}+1\begin{bmatrix}0\\0\\1\\1\end{bmatrix}+0\begin{bmatrix}0\\1\\1\\0\end{bmatrix} \quad (13)$$

For which we have for $f(x,y)=x\oplus y$

$$x=A^{-1}y=\frac{1}{2}\begin{bmatrix}2f(0,0)\\-f(0,0)+f(0,1)-f(1,0)+f(1,1)\\-f(0,0)-f(0,1)+f(1,0)+f(1,1)\\-f(0,0)+f(0,1)+f(1,0)-f(1,1)\end{bmatrix}=\begin{bmatrix}0\\0\\0\\1\end{bmatrix} \quad (14)$$

And

$$Ax=0\begin{bmatrix}1\\1\\1\\1\end{bmatrix}+0\begin{bmatrix}0\\1\\0\\1\end{bmatrix}+0\begin{bmatrix}0\\0\\1\\1\end{bmatrix}+1\begin{bmatrix}0\\1\\1\\0\end{bmatrix} \quad (15)$$

The coefficients (0,0,0,1) are the Morphogenetic sources that select one and only one state. Because the coefficients

$$x=A^{-1}y=\frac{1}{2}\begin{bmatrix}2f(0,0)\\-f(0,0)+f(0,1)-f(1,0)+f(1,1)\\-f(0,0)-f(0,1)+f(1,0)+f(1,1)\\-f(0,0)+f(0,1)+f(1,0)-f(1,1)\end{bmatrix}=\begin{bmatrix}1\\0\\0\\0\end{bmatrix} \quad (16)$$

Are asymmetric we can change the coefficient $f(0,0)$ in a way to have the same symmetry in agreement with the other $f(0,0)$ coefficients. In fact when we substitute $f(0,0)$ with

$$-\frac{1}{2}(f(0,0)+f(1,0)+f(0,1)+f(1,1))+1 \quad (17)$$

We have

Proposition When $f(x,y)=1$ we have

$$-\frac{1}{2}(f(0,0)+f(1,0)+f(0,1)+f(1,1))+1=-1 \quad (18)$$

When $f(x,y)=x,y,x\oplus y$

We have

$$-\frac{1}{2}(f(0,0)+f(1,0)+f(0,1)+f(1,1))+1=0 \tag{19}$$

Proof For $f(x,y)=x$ we have

$$f(x,y)=\begin{bmatrix} f(0,0)\\ f(1,0)\\ f(0,1)\\ f(1,1)\end{bmatrix}=\begin{bmatrix}0\\1\\0\\1\end{bmatrix} \tag{20}$$

$$-\frac{1}{2}(1+1)+1=0 \tag{21}$$

For $f(x,y)=y$ we have

$$f(x,y)=\begin{bmatrix} f(0,0)\\ f(1,0)\\ f(0,1)\\ f(1,1)\end{bmatrix}=\begin{bmatrix}0\\0\\1\\1\end{bmatrix} \tag{22}$$

$$-\frac{1}{2}(1+1)+1=0 \tag{23}$$

For $f(x,y)=x\oplus y$ we have

$$f(x,y)=\begin{bmatrix} f(0,0)\\ f(1,0)\\ f(0,1)\\ f(1,1)\end{bmatrix}=\begin{bmatrix}0\\1\\1\\0\end{bmatrix} \tag{24}$$

$$-\frac{1}{2}(1+1)+1=0 \tag{25}$$

The new equivalent coefficients are

$$x=A^{-1}y=\frac{-1}{2}\begin{bmatrix} f(0,0)+f(1,0)+f(0,1)+f(1,1))\\ f(0,0)-f(0,1)+f(1,0)-f(1,1)\\ f(0,0)+f(0,1)-f(1,0)-f(1,1)\\ f(0,0)-f(0,1)-f(1,0)+f(1,1)\end{bmatrix}+\begin{bmatrix}-1\\0\\0\\0\end{bmatrix} \tag{26}$$

That can write in this way

$$x = \frac{-1}{2}\begin{bmatrix} 1 & 1 & 1 & 1 \\ 1 & -1 & 1 & -1 \\ 1 & 1 & -1 & -1 \\ 1 & -1 & -1 & 1 \end{bmatrix}\begin{bmatrix} f(0,0) \\ f(1,0) \\ f(0,1) \\ f(1,1) \end{bmatrix} + \begin{bmatrix} -1 \\ 0 \\ 0 \\ 0 \end{bmatrix} \tag{27}$$

To compensate the vector

$$\begin{bmatrix} -1 \\ 0 \\ 0 \\ 0 \end{bmatrix} \tag{28}$$

We expand this vector in the eight dimensional space

$$\begin{bmatrix} -1 \\ 0 \\ 0 \\ 0 \end{bmatrix} \Rightarrow \begin{bmatrix} 1 \\ -1 \\ 0 \\ 0 \\ 0 \\ 0 \\ 0 \\ 0 \end{bmatrix} \tag{29}$$

To restore wanted symmetry transformation we use the matrix W in this way

$$\begin{bmatrix} 1 & 1 & 0 & 0 & 0 & 0 & 0 & 0 \\ 1 & -1 & 0 & 0 & 0 & 0 & 0 & 0 \\ 1 & 1 & 0 & 0 & 0 & 0 & 0 & 0 \\ 1 & -1 & 0 & 0 & 0 & 0 & 0 & 0 \\ 1 & 1 & 0 & 0 & 0 & 0 & 0 & 0 \\ 1 & -1 & 0 & 0 & 0 & 0 & 0 & 0 \\ 1 & 1 & 0 & 0 & 0 & 0 & 0 & 0 \\ 1 & -1 & 0 & 0 & 0 & 0 & 0 & 0 \end{bmatrix}\begin{bmatrix} 1 \\ -1 \\ 0 \\ 0 \\ 0 \\ 0 \\ 0 \\ 0 \end{bmatrix} = \begin{bmatrix} 0 \\ 2 \\ 0 \\ 2 \\ 0 \\ 2 \\ 0 \\ 2 \end{bmatrix} \tag{30}$$

Now we expand the other part

$$\begin{bmatrix} f(0,0)+f(1,0)+f(0,1)+f(1,1)) \\ f(0,0)-f(0,1)+f(1,0)-f(1,1) \\ f(0,0)+f(0,1)-f(1,0)-f(1,1) \\ f(0,0)-f(0,1)-f(1,0)+f(1,1) \end{bmatrix} \Rightarrow \begin{bmatrix} 0 \\ f(0,0)+f(1,0)+f(0,1)+f(1,1) \\ 0 \\ f(0,0)-f(0,1)+f(1,0)-f(1,1) \\ 0 \\ f(0,0)+f(0,1)-f(1,0)-f(1,1) \\ 0 \\ f(0,0)-f(0,1)-f(1,0)+f(1,1) \end{bmatrix} \tag{31}$$

Now we expand the transformation W to obtain again a wanted symmetry

$$\begin{bmatrix} 1 & 1 & 1 & 1 & 1 & 1 & 1 & 1 \\ 1 & -1 & 0 & 0 & 0 & 0 & 0 & 0 \\ 1 & 1 & 0 & 0 & 0 & 0 & 0 & 0 \\ 1 & -1 & 0 & 0 & 0 & 0 & 0 & 0 \\ 1 & 1 & 0 & 0 & 0 & 0 & 0 & 0 \\ 1 & -1 & 0 & 0 & 0 & 0 & 0 & 0 \\ 1 & 1 & 0 & 0 & 0 & 0 & 0 & 0 \\ 1 & -1 & 0 & 0 & 0 & 0 & 0 & 0 \end{bmatrix} \begin{bmatrix} 0 \\ f(0,0)+f(1,0)+f(0,1)+f(1,1) \\ 0 \\ f(0,0)-f(0,1)+f(1,0)-f(1,1) \\ 0 \\ f(0,0)+f(0,1)-f(1,0)-f(1,1) \\ 0 \\ f(0,0)-f(0,1)-f(1,0)+f(1,1) \end{bmatrix}$$
$$= \begin{bmatrix} 4f(0,0) \\ -(f(0,0)+f(1,0)+f(0,1)+f(1,1)) \\ (f(0,0)+f(1,0)+f(0,1)+f(1,1)) \\ -(f(0,0)+f(0,1)+f(1,0)+f(1,1)) \\ (f(0,0)+f(1,0)+f(0,1)+f(1,1)) \\ -(f(0,0)+f(1,0)+f(0,1)+f(1,1)) \\ (f(0,0)+f(1,0)+f(0,1)+f(1,1)) \\ -f(0,0)+f(0,1)+f(1,0)+f(1,1) \end{bmatrix} \tag{32}$$

We continue our expansion of W

$$\begin{bmatrix} 1 & 1 & 1 & 1 & 1 & 1 & 1 & 1 \\ 1 & -1 & 1 & -1 & 1 & -1 & 1 & -1 \\ 1 & 1 & 0 & 0 & 0 & 0 & 0 & 0 \\ 1 & -1 & 0 & 0 & 0 & 0 & 0 & 0 \\ 1 & 1 & 0 & 0 & 0 & 0 & 0 & 0 \\ 1 & -1 & 0 & 0 & 0 & 0 & 0 & 0 \\ 1 & 1 & 0 & 0 & 0 & 0 & 0 & 0 \\ 1 & -1 & 0 & 0 & 0 & 0 & 0 & 0 \end{bmatrix} \begin{bmatrix} 0 \\ f(0,0)+f(1,0)+f(0,1)+f(1,1) \\ 0 \\ f(0,0)-f(0,1)+f(1,0)-f(1,1) \\ 0 \\ f(0,0)+f(0,1)-f(1,0)-f(1,1) \\ 0 \\ f(0,0)-f(0,1)-f(1,0)+f(1,1) \end{bmatrix}$$
$$= \begin{bmatrix} 4f(0,0) \\ -4f(0,0) \\ (f(0,0)+f(1,0)+f(0,1)+f(1,1)) \\ -(f(0,0)+f(0,1)+f(1,0)+f(1,1)) \\ (f(0,0)+f(1,0)+f(0,1)+f(1,1)) \\ -(f(0,0)+f(1,0)+f(0,1)+f(1,1)) \\ (f(0,0)+f(1,0)+f(0,1)+f(1,1)) \\ -f(0,0)+f(0,1)+f(1,0)+f(1,1) \end{bmatrix} \tag{33}$$

Again we have

$$\begin{bmatrix} 1 & 1 & 1 & 1 & 1 & 1 & 1 & 1 \\ 1 & -1 & 1 & -1 & 1 & -1 & 1 & -1 \\ 1 & 1 & -1 & -1 & 1 & 1 & -1 & -1 \\ 1 & -1 & 0 & 0 & 0 & 0 & 0 & 0 \\ 1 & 1 & 0 & 0 & 0 & 0 & 0 & 0 \\ 1 & -1 & 0 & 0 & 0 & 0 & 0 & 0 \\ 1 & 1 & 0 & 0 & 0 & 0 & 0 & 0 \\ 1 & -1 & 0 & 0 & 0 & 0 & 0 & 0 \end{bmatrix} \begin{bmatrix} 0 \\ f(0,0)+f(1,0)+f(0,1)+f(1,1) \\ 0 \\ f(0,0)-f(0,1)+f(1,0)-f(1,1) \\ 0 \\ f(0,0)+f(0,1)-f(1,0)-f(1,1) \\ 0 \\ f(0,0)-f(0,1)-f(1,0)+f(1,1) \end{bmatrix}$$

$$= \begin{bmatrix} 4f(0,0) \\ -4f(0,0) \\ (f(0,0)+f(1,0)+f(0,1)+f(1,1))-(f(0,0)-f(0,1)+f(1,0)-f(1,1))+f(0,0)+ \\ \qquad f(0,1)-f(1,0)-f(1,1)-(f(0,0)-f(0,1)-f(1,0)+f(1,1)) \\ -(f(0,0)+f(0,1)+f(1,0)+f(1,1)) \\ (f(0,0)+f(1,0)+f(0,1)+f(1,1)) \\ -(f(0,0)+f(1,0)+f(0,1)+f(1,1)) \\ (f(0,0)+f(1,0)+f(0,1)+f(1,1)) \\ -f(0,0)+f(0,1)+f(1,0)+f(1,1) \end{bmatrix}$$

$$= \begin{bmatrix} 4f(0,0) \\ -4f(0,0) \\ 4f(1,0) \\ -(f(0,0)+f(0,1)+f(1,0)+f(1,1)) \\ (f(0,0)+f(1,0)+f(0,1)+f(1,1)) \\ -(f(0,0)+f(1,0)+f(0,1)+f(1,1)) \\ (f(0,0)+f(1,0)+f(0,1)+f(1,1)) \\ -f(0,0)+f(0,1)+f(1,0)+f(1,1) \end{bmatrix} \tag{34}$$

For the complete expansion of W we have

$$\begin{bmatrix} 1 & 1 & 1 & 1 & 1 & 1 & 1 & 1 \\ 1 & -1 & 1 & -1 & 1 & -1 & 1 & -1 \\ 1 & 1 & -1 & -1 & 1 & 1 & -1 & -1 \\ 1 & -1 & -1 & 1 & 1 & -1 & -1 & 1 \\ 1 & 1 & 1 & 1 & -1 & -1 & -1 & -1 \\ 1 & -1 & 1 & -1 & -1 & 1 & -1 & 1 \\ 1 & 1 & -1 & -1 & -1 & -1 & 1 & 1 \\ 1 & -1 & -1 & 1 & -1 & 1 & 1 & -1 \end{bmatrix} \begin{bmatrix} 0 \\ f(0,0)+f(1,0)+f(0,1)+f(1,1) \\ 0 \\ f(0,0)-f(0,1)+f(1,0)-f(1,1) \\ 0 \\ f(0,0)+f(0,1)-f(1,0)-f(1,1) \\ 0 \\ f(0,0)-f(0,1)-f(1,0)+f(1,1) \end{bmatrix} = \begin{bmatrix} 4f(0,0) \\ -4f(0,0) \\ 4f(1,0) \\ -4f(1,0) \\ 4f(0,1) \\ -4f(0,1) \\ 4f(1,1) \\ -4f(1,1) \end{bmatrix} \tag{35}$$

And for

$$\begin{bmatrix} 1 & 1 & 1 & 1 & 1 & 1 & 1 & 1 \\ 1 & -1 & 1 & -1 & 1 & -1 & 1 & -1 \\ 1 & 1 & -1 & -1 & 1 & 1 & -1 & -1 \\ 1 & -1 & -1 & 1 & 1 & -1 & -1 & 1 \\ 1 & 1 & 1 & 1 & -1 & -1 & -1 & -1 \\ 1 & -1 & 1 & -1 & -1 & 1 & -1 & 1 \\ 1 & 1 & -1 & -1 & -1 & -1 & 1 & 1 \\ 1 & -1 & -1 & 1 & -1 & 1 & 1 & -1 \end{bmatrix} \begin{bmatrix} 0 \\ f(0,0)+f(1,0)+f(0,1)+f(1,1) \\ 0 \\ f(0,0)-f(0,1)+f(1,0)-f(1,1) \\ 0 \\ f(0,0)+f(0,1)-f(1,0)-f(1,1) \\ 0 \\ f(0,0)-f(0,1)-f(1,0)+f(1,1) \end{bmatrix} \frac{1}{2} = \begin{bmatrix} 2f(0,0) \\ -2f(0,0) \\ 2f(1,0) \\ -2f(1,0) \\ 2f(0,1) \\ -2f(0,1) \\ 2f(1,1) \\ -2f(1,1) \end{bmatrix} \tag{36}$$

and

$$\begin{bmatrix} 1 & 1 & 1 & 1 & 1 & 1 & 1 & 1 \\ 1 & -1 & 1 & -1 & 1 & -1 & 1 & -1 \\ 1 & 1 & -1 & -1 & 1 & 1 & -1 & -1 \\ 1 & -1 & -1 & 1 & 1 & -1 & -1 & 1 \\ 1 & 1 & 1 & 1 & -1 & -1 & -1 & -1 \\ 1 & -1 & 1 & -1 & -1 & 1 & -1 & 1 \\ 1 & 1 & -1 & -1 & -1 & -1 & 1 & 1 \\ 1 & -1 & -1 & 1 & -1 & 1 & 1 & -1 \end{bmatrix} (\begin{bmatrix} 0 \\ f(0,0)+f(1,0)+f(0,1)+f(1,1) \\ 0 \\ f(0,0)-f(0,1)+f(1,0)-f(1,1) \\ 0 \\ f(0,0)+f(0,1)-f(1,0)-f(1,1) \\ 0 \\ f(0,0)-f(0,1)-f(1,0)+f(1,1) \end{bmatrix} \frac{1}{2} + \begin{bmatrix} 1 \\ -1 \\ 0 \\ 0 \\ 0 \\ 0 \\ 0 \\ 0 \end{bmatrix} = \begin{bmatrix} 2f(0,0) \\ 2-2f(0,0) \\ 2f(1,0) \\ 2-2f(1,0) \\ 2f(0,1) \\ 2-2f(0,1) \\ 2f(1,1) \\ 2-2f(1,1) \end{bmatrix} \tag{37}$$

For which we have

$$\begin{bmatrix} 2f(0,0) \\ 2-2f(0,0) \\ 2f(1,0) \\ 2-2f(1,0) \\ 2f(0,1) \\ 2-2f(0,1) \\ 2f(1,1) \\ 2-2f(1,1) \end{bmatrix} = 2\begin{bmatrix} f(0,0) \\ 1-f(0,0) \\ f(1,0) \\ 1-f(1,0) \\ f(0,1) \\ 1-f(0,1) \\ f(1,1) \\ 1-f(1,1) \end{bmatrix} = 2\begin{bmatrix} 0\oplus f(0,0) \\ 1\oplus f(0,0) \\ 0\oplus f(1,0) \\ 1\oplus f(1,0) \\ 0\oplus f(0,1) \\ 1\oplus f(0,1) \\ 0\oplus f(1,1) \\ 1\oplus f(1,1) \end{bmatrix} \tag{38}$$

Full basis for three dimensions or ORACLE is

$$\begin{bmatrix} x & y & z \\ 0 & 0 & 0 \\ 1 & 0 & 0 \\ 0 & 1 & 0 \\ 1 & 1 & 0 \\ 0 & 0 & 1 \\ 1 & 0 & 1 \\ 0 & 1 & 1 \\ 1 & 1 & 1 \end{bmatrix}, x\oplus y = \begin{bmatrix} 0 \\ 1 \\ 1 \\ 0 \\ 0 \\ 1 \\ 1 \\ 0 \end{bmatrix}, x\oplus z = \begin{bmatrix} 0 \\ 1 \\ 0 \\ 1 \\ 1 \\ 0 \\ 1 \\ 0 \end{bmatrix}, y\oplus z = \begin{bmatrix} 0 \\ 0 \\ 1 \\ 1 \\ 1 \\ 1 \\ 0 \\ 0 \end{bmatrix}, x\oplus y\oplus z = \begin{bmatrix} 0 \\ 1 \\ 1 \\ 0 \\ 1 \\ 0 \\ 0 \\ 1 \end{bmatrix} \tag{39}$$

And the matrix in the projection operator is

$$A = \begin{bmatrix} 1 & 0 & 0 & 0 & 0 & 0 & 0 & 0 \\ 1 & 1 & 0 & 0 & 1 & 1 & 0 & 1 \\ 1 & 0 & 1 & 0 & 1 & 0 & 1 & 1 \\ 1 & 1 & 1 & 0 & 0 & 1 & 1 & 0 \\ 1 & 0 & 0 & 1 & 0 & 1 & 1 & 1 \\ 1 & 1 & 0 & 1 & 1 & 0 & 1 & 0 \\ 1 & 0 & 1 & 1 & 1 & 1 & 0 & 0 \\ 1 & 1 & 1 & 1 & 0 & 0 & 1 & 1 \end{bmatrix} \tag{40}$$

The more interesting ones are the Toffoli gate (or AND/NAND gate) and the OR/NOR gate. The operations of Toffoli and OR/NOR gates are, respectively,

$$\begin{aligned} &|s,x,y\rangle \Rightarrow |x,y\rangle|s\oplus f(x,y)\rangle \\ &where \\ &f(x,y) = x\wedge y \\ &or \\ &f(x,y) = x\vee y \end{aligned} \tag{41}$$

3 Projection Operator and Solution by One Step of Neural Network

For the projection operator Q the linear combination of the column vectors in (4) is different from Y but is at the minimum distance respect to Y itself. The theorem proof is based on the definition of projection that has the minimum distance between the projection vector QY and the original vector Y. We know that the output y of a neuron is given by the expression

$$y_j = f\left[\sum_i w_i A_{i,j} - \theta\right] \tag{42}$$

where the weights w_j are the Morphogenetic sources. The function f is the step function (actually known as the Heaviside function) for which we have

$$\begin{array}{ll} f(x) = 1 & x > 0 \\ f(x) = 0 & x \leq 0 \end{array} \tag{43}$$

The superposition of the inputs vectors is $\sum_i w_i A_{i,j}$ where we assume that the weights of the neuron can be computed by the projection of the designed function Y into the space of the input vectors A. Because QY is not equal to Y, we choose among possible Y the Y which projection in input A is similar to Y. In this case we can choose a threshold value for which y_j in (42) is equal to the Boolean vector of output Y. To compute the threshold we use the expression (44)

$$\theta = \frac{\min[(QY)Y] + \max[(QY)(1-Y)]}{2} \tag{44}$$

Among the values QY we choose the values for which Y = 1 for this set of values we pick up the minimum value. Now from QY values we choose the values for which Y = 0 and we pick up the maximum. In the example (6) we have.

For Y = 1 we have the values $V_1 = \left\{\frac{2}{3}\right\}$, for Y = 0 we have the values $V_2 = \left\{0, \frac{1}{3}, \frac{1}{3}\right\}$ so we have

$$Min(V_1) = \frac{2}{3}, \; Max(V_2) = \frac{1}{3} \tag{45}$$

So the threshold is $\theta = \frac{\frac{2}{3} + \frac{1}{3}}{2} = \frac{1}{2}$ so we have the neuron result by the projection method

With projection operator we find the strength of the weights and threshold or Morphogenetic sources. To obtain the designed output without iteration process. This method that use projection operator is denoted One Step Method. Given the function

$$Y = \begin{bmatrix} 0 \\ 1 \\ 1 \\ 0 \end{bmatrix} \tag{46}$$

We have not parameters w for which

$$QY = A\,w = w_1 \begin{bmatrix} 0 \\ 1 \\ 0 \\ 1 \end{bmatrix} + w_2 \begin{bmatrix} 0 \\ 0 \\ 1 \\ 1 \end{bmatrix} = w_1 A_1 + w_2 A_2 = y_j \approx \begin{bmatrix} 0 \\ 1 \\ 1 \\ 0 \end{bmatrix} \tag{47}$$

We cannot solve by projection operator and by neural network the Y that is the well known XOR gate that with the digital gates can write in this way

$$(x \wedge \neg y) \vee (\neg x \wedge y) = Y \tag{48}$$

At the question why we want use the neural network for this simple operator when we have the easy gate expression. The motivation is not for simple expression but for more complex Boolean function where is very difficult or impossible to implement by a physical system the wanted Boolean function. This is the classical code problem in digital computer.

4 Expansion of the One Step Method as Solution of Neural Contradictions

Here, we present an advanced method that we use to compute the neural network and hidden neurons. This method does not use recursion methods that present the problem of the convergences and local minimum to compute neural network parameters for designed Boolean function to implement as output of a neural network. The new method is denoted one step method that give two results. The first is to detect if a Boolean function can be solve or if is impossible to solve with one neuron. When the Boolean function can be solved without hidden neurons we have an algorithm that in one step compute neural parameters as weights and threshold. When the Boolean function cannot be solved we have a method to know how much hidden neurons are necessary to solve the functions. The number of the hidden neurons is at the maximum equal to the number of inputs and this avoid any exponential explosion. Given the colon space of inputs

$$A = \begin{bmatrix} a_{11} & a_{12} & \dots & a_{1m} \\ a_{21} & a_{22} & \dots & a_{2m} \\ \dots & \dots & \dots & \dots \\ a_{n1} & a_{n2} & \dots & a_{nm} \end{bmatrix} \tag{49}$$

We have that

$$QY = w_1 \begin{bmatrix} a_{11} \\ a_{21} \\ \cdots \\ a_{n1} \end{bmatrix} + w_2 \begin{bmatrix} a_{12} \\ a_{22} \\ \cdots \\ a_{n2} \end{bmatrix} + \ldots. + w_m \begin{bmatrix} a_{1m} \\ a_{2m} \\ \cdots \\ a_{nm} \end{bmatrix} \tag{50}$$

Now if we extend the colon space to adjoin the vector Y we have

$$QY = w_1 \begin{bmatrix} a_{11} \\ a_{21} \\ \cdots \\ a_{n1} \end{bmatrix} + w_2 \begin{bmatrix} a_{12} \\ a_{22} \\ \cdots \\ a_{n2} \end{bmatrix} + \ldots. + w_m \begin{bmatrix} a_{1m} \\ a_{2m} \\ \cdots \\ a_{nm} \end{bmatrix} + w_{m+1} Y = Y \tag{51}$$

We know that the projection operator compute the best parameters for which the distance between QY and Y assume the minimum value. Now in (51) appear the function Y at the right and at the left so with the projection operator the best weights w are

$$W = \begin{bmatrix} w_1 \\ \cdots \\ w_m \\ w_{m+1} \end{bmatrix} = \begin{bmatrix} 0 \\ \cdots \\ 0 \\ 1 \end{bmatrix} \tag{52}$$

In this case we have QY = Y and $QY - Y = 0$ that is minimum value of the distance between QY and Y

Example

$$A = \begin{bmatrix} 0 & 0 \\ 1 & 0 \\ 0 & 1 \\ 1 & 1 \end{bmatrix} \oplus \begin{bmatrix} 0 \\ 1 \\ 0 \\ 0 \end{bmatrix} = \begin{bmatrix} 0 & 0 & 0 \\ 1 & 0 & 1 \\ 0 & 1 & 0 \\ 1 & 1 & 0 \end{bmatrix} \tag{53}$$

So

$$w = (A^T A)^{-1} A^T Y = \begin{bmatrix} 0 \\ 0 \\ 1 \end{bmatrix} \tag{54}$$

And

$$Aw = A(A^T A)^{-1} A^T Y = QY = \begin{bmatrix} 0 \\ 1 \\ 0 \\ 0 \end{bmatrix} \tag{55}$$

This is the collapse theorem in projection operator. For the collapse theorem we have that when we enlarge the colon space in this way

$$QY = w_1 \begin{bmatrix} a_{11} \\ a_{21} \\ \dots \\ a_{n1} \end{bmatrix} + w_2 \begin{bmatrix} a_{12} \\ a_{22} \\ \dots \\ a_{n2} \end{bmatrix} + \dots + w_m \begin{bmatrix} a_{1m} \\ a_{2m} \\ \dots \\ a_{nm} \end{bmatrix} + w_{m+1}(AU - Y) = Aw + w_{m+1}(AU - Y) \tag{56}$$

where

$$U = \begin{bmatrix} 1 \\ 1 \\ \dots \\ 1 \end{bmatrix} \tag{57}$$

Because QY include Y in the projection operator the minimum value of QY—Y must be zero and QY = Y. the computed weights must be

$$w = \begin{bmatrix} 1 \\ \dots \\ 1 \\ -1 \end{bmatrix} \tag{58}$$

And

$$QY = \begin{bmatrix} a_{11} \\ a_{21} \\ \dots \\ a_{n1} \end{bmatrix} + \begin{bmatrix} a_{12} \\ a_{22} \\ \dots \\ a_{n2} \end{bmatrix} + \dots + \begin{bmatrix} a_{1m} \\ a_{2m} \\ \dots \\ a_{nm} \end{bmatrix} - AU + Y = AU - AU + Y = Y \tag{59}$$

We remark that a priory we have no idea of the weights. But with the computation of the expression

$$w = (A^T A)^{-1} A^T Y \tag{60}$$

The system compute the weights for which QY—Y assume the minimum value. Now A include Y so the minimum value must be QY = Y.

Example Given the colon space

$$A = \begin{bmatrix} 0 & 0 \\ 1 & 0 \\ 0 & 1 \\ 1 & 1 \end{bmatrix} \text{ and Y } = \begin{bmatrix} 0 \\ 1 \\ 0 \\ 0 \end{bmatrix} \text{ we have } AU - Y = \begin{bmatrix} 0 & 0 \\ 1 & 0 \\ 0 & 1 \\ 1 & 1 \end{bmatrix} \begin{bmatrix} 1 \\ 1 \end{bmatrix} - \begin{bmatrix} 0 \\ 1 \\ 0 \\ 0 \end{bmatrix} = \begin{bmatrix} 0 \\ 1 \\ 0 \\ 1 \end{bmatrix} + \begin{bmatrix} 0 \\ 0 \\ 1 \\ 1 \end{bmatrix} - \begin{bmatrix} 0 \\ 1 \\ 0 \\ 0 \end{bmatrix} = \begin{bmatrix} 0 \\ 0 \\ 1 \\ 2 \end{bmatrix} \tag{61}$$

Now we have for

$$A=\begin{bmatrix} 0 & 0 & 0 \\ 1 & 0 & 0 \\ 0 & 1 & 1 \\ 1 & 1 & 2 \end{bmatrix} \tag{62}$$

$$w=(A^TA)^{-1}A^TY=\begin{bmatrix} 1 \\ 1 \\ -1 \end{bmatrix} \tag{63}$$

And

$$QY=1\begin{bmatrix} 0 \\ 1 \\ 0 \\ 1 \end{bmatrix}+1\begin{bmatrix} 0 \\ 0 \\ 1 \\ 1 \end{bmatrix}-1\begin{bmatrix} 0 \\ 0 \\ 1 \\ 2 \end{bmatrix}=\begin{bmatrix} 0 \\ 1 \\ 0 \\ 0 \end{bmatrix}=Y \tag{64}$$

When we decompose the expression $AU-Y$ into a Boolean set of vectors B_k we have

$$AU-Y=a_1B_1+a_2B_2+\ldots..+a_kB_k \tag{65}$$

For

$$QY=\begin{bmatrix} a_{11} \\ a_{21} \\ \cdots \\ a_{n1} \end{bmatrix}+\begin{bmatrix} a_{12} \\ a_{22} \\ \cdots \\ a_{n2} \end{bmatrix}+\ldots.+\begin{bmatrix} a_{1m} \\ a_{2m} \\ \cdots \\ a_{nm} \end{bmatrix}-(AU-Y)$$
$$=Y=AU-(AU-Y)=AU-(a_1B_1+a_2B_2+\ldots+a_kB_k) \tag{66}$$

So the weights to obtain the minimum value QY = Y must be

$$w=\begin{bmatrix} 1 \\ \cdots \\ 1 \\ -a_1 \\ \cdots \\ -a_k \end{bmatrix} \tag{67}$$

Example

$$AU-Y=\begin{bmatrix} 0 & 0 \\ 1 & 0 \\ 0 & 1 \\ 1 & 1 \end{bmatrix}\begin{bmatrix} 1 \\ 1 \end{bmatrix}-\begin{bmatrix} 0 \\ 1 \\ 0 \\ 0 \end{bmatrix}=\begin{bmatrix} 0 \\ 0 \\ 1 \\ 2 \end{bmatrix}=1\begin{bmatrix} 0 \\ 0 \\ 1 \\ 1 \end{bmatrix}+1\begin{bmatrix} 0 \\ 0 \\ 0 \\ 1 \end{bmatrix} \tag{68}$$

So

$a_1=1, a_2=1$

And

$$QY = 1\begin{bmatrix}0\\1\\0\\1\end{bmatrix} + 1\begin{bmatrix}0\\0\\1\\1\end{bmatrix} - (1\begin{bmatrix}0\\1\\0\\1\end{bmatrix} + 1\begin{bmatrix}0\\0\\1\\1\end{bmatrix} - \begin{bmatrix}0\\1\\0\\0\end{bmatrix}) = 1\begin{bmatrix}0\\1\\0\\1\end{bmatrix} + 1\begin{bmatrix}0\\0\\1\\1\end{bmatrix} - (1\begin{bmatrix}0\\0\\1\\1\end{bmatrix} + 1\begin{bmatrix}0\\0\\0\\1\end{bmatrix}) = \begin{bmatrix}0\\1\\0\\0\end{bmatrix} \tag{69}$$

So the computed weights for (67) are

$$w = \begin{bmatrix}1\\1\\-1\\-1\end{bmatrix} \tag{70}$$

Proposition Given the number of the inputs the maximum number of new hidden inputs or new colons in A are equal to the number of inputs. When we adjoin a number of colons equal to the number of inputs the neural network rebuilt from inputs and weights exactly the same values of the designed function Y and we have not necessity to use the threshold value. Now we can check by the use of the threshold if it is possible to reduce the maximum number of hidden inputs or neurons in a way to obtain a more efficient neural network.

Example For the output function $Y = \begin{bmatrix}0\\1\\0\\0\end{bmatrix}$ we have two input so at the maximum we have two new colons (hidden neurons) given by the values

$$B_1 = \begin{bmatrix}0\\0\\1\\1\end{bmatrix}, B_2 = \begin{bmatrix}0\\0\\0\\1\end{bmatrix} \tag{71}$$

Because we have two inputs the maximum set of inputs are

$$A = \begin{bmatrix}0 & 0 & 0 & 0\\1 & 0 & 0 & 0\\0 & 1 & 1 & 0\\1 & 1 & 1 & 1\end{bmatrix} \tag{72}$$

For with any Boolean function with two inputs can be solved by the neural network. Now because the new colons or hidden neurons function are ordered we can begin with simple neuron without hidden neurons and move to introduce hidden neurons as new inputs.

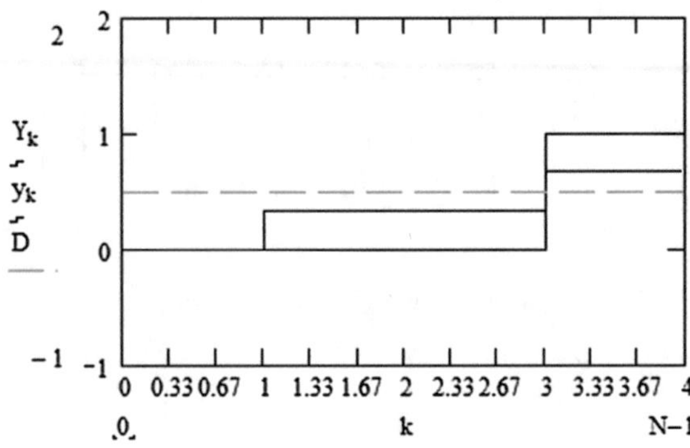

Fig. 1 Neuron outputs y for the four inputs 00, 10, 01, 11 and the function Y = (0 1 0 0)

Example For the function

$$Y = \begin{bmatrix} 0 \\ 1 \\ 0 \\ 0 \end{bmatrix} \tag{73}$$

No hidden neurons are necessary so the input are given by the matrix

$$A = \begin{bmatrix} 0 & 0 \\ 1 & 0 \\ 0 & 1 \\ 1 & 1 \end{bmatrix} \; w = \begin{bmatrix} \frac{2}{3} \\ -\frac{1}{3} \end{bmatrix}, and\, \theta = \frac{1}{2} \tag{74}$$

The solution is given in Fig. 1.

Now for $Y = \begin{bmatrix} 0 \\ 1 \\ 1 \\ 0 \end{bmatrix}$ we cannot solve this function for the neural contradiction without hidden neurons and new inputs.

So the inputs must be expanded in this way

$$UA - Y = \begin{bmatrix} 0 \\ 0 \\ 0 \\ 2 \end{bmatrix} \tag{75}$$

So we have

$$A = \begin{bmatrix} 0 & 0 & 0 & 0 \\ 1 & 0 & 0 & 0 \\ 0 & 1 & 0 & 0 \\ 1 & 1 & 1 & 1 \end{bmatrix} \tag{76}$$

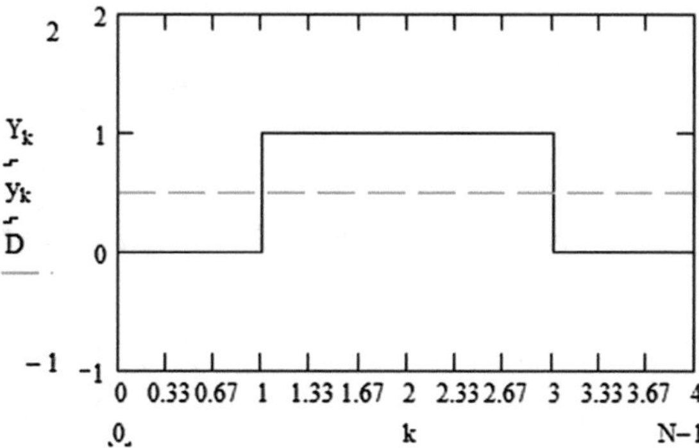

Fig. 2 Output y of neuron for the function Y with one hidden neuron

That can be reduce to

$$A = \begin{bmatrix} 0 & 0 & 0 \\ 1 & 0 & 0 \\ 0 & 1 & 0 \\ 1 & 1 & 1 \end{bmatrix} \tag{77}$$

and

$$w = (A^T A)^{-1} A^T Y = \begin{bmatrix} 1 \\ 1 \\ -2 \end{bmatrix}, A(A^T A)^{-1} A^T Y = QY = \begin{bmatrix} 0 \\ 1 \\ 1 \\ 0 \end{bmatrix} \tag{78}$$

And the solution is (Fig. 2).
So we have the neural network with one hidden (Fig. 3).

For $Y = \begin{bmatrix} 1 \\ 0 \\ 0 \\ 1 \end{bmatrix}$ we have

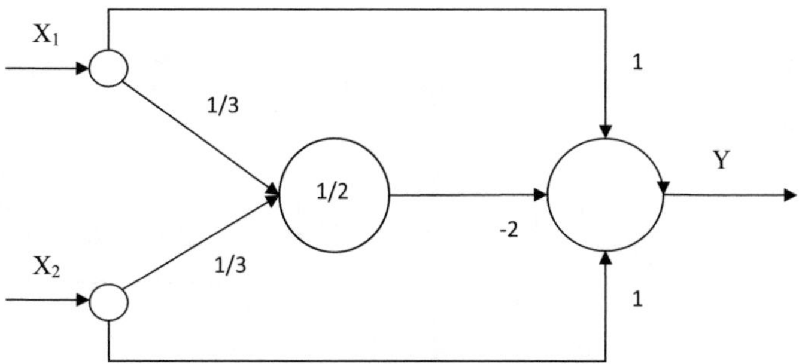

Fig. 3 Neural network with one hidden neuron to solve the function (0 1 1 0)

$$UA - Y = \begin{bmatrix} -1 \\ 1 \\ 1 \\ 1 \end{bmatrix} \tag{79}$$

So we have two possible hidden neuron for the two inputs so

$$A = \begin{bmatrix} 0 & 0 & 0 & 1 \\ 1 & 0 & 1 & 0 \\ 0 & 1 & 1 & 0 \\ 1 & 1 & 1 & 0 \end{bmatrix} \tag{80}$$

The third colon is for the values 1 and the fourth colon is for −1. Now we reduce the number of hidden neurons to obtain

$$A = \begin{bmatrix} 0 & 0 & 1 \\ 1 & 0 & 0 \\ 0 & 1 & 0 \\ 1 & 1 & 0 \end{bmatrix} \tag{81}$$

So we have

$$w = \begin{bmatrix} \frac{1}{3} \\ \frac{1}{3} \\ 1 \end{bmatrix}, \theta = \frac{1}{2} \tag{82}$$

And the result is (Fig. 4).

Digital Machine and Systems by neural network.

Given the machine with x the input, q the states and y the output

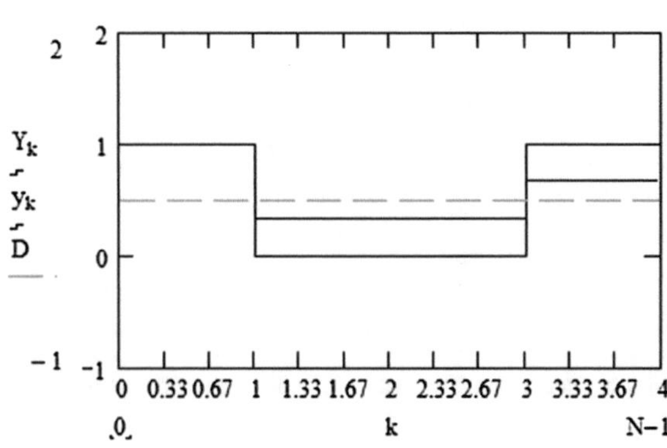

Fig. 4 Neural output y for the function Y = (1 0 0 1). The threshold is in *green*

$$\begin{bmatrix} q\backslash x & 0 & 1 & y \\ 1 & 3 & 6 & 0 \\ 2 & 3 & 4 & 1 \\ 3 & 2 & 5 & 0 \\ 4 & 5 & 2 & 0 \\ 5 & 6 & 3 & 0 \\ 6 & 5 & 1 & 0 \end{bmatrix} \tag{83}$$

With the code

$$\begin{bmatrix} 1 & \rightarrow & 000 \\ 2 & \rightarrow & 010 \\ 3 & \rightarrow & 100 \\ 4 & \rightarrow & 001 \\ 5 & \rightarrow & 111 \\ 6 & \rightarrow & 110 \end{bmatrix} \tag{84}$$

We have the transition state function

$$\begin{bmatrix} q\backslash x & 0 & 1 & y \\ 000 & 100 & 110 & 0 \\ 010 & 100 & 001 & 1 \\ 100 & 010 & 111 & 0 \\ 001 & 111 & 010 & 0 \\ 111 & 110 & 100 & 0 \\ 110 & 111 & 000 & 0 \end{bmatrix} \tag{85}$$

The system can be represented by the Boolean equations

$$\begin{cases} q_1(t+1) = \overline{q_1(t)}\,\bar{x} + \overline{q_2(t)}\,\overline{q_3(t)}\,x + q_1(t)q_2(t)(q_3(t) + \bar{x}) \\ q_2(t+1) = q_1(t)\,\bar{x} + xq_2(t)\overline{q_3(t)} + \overline{q_1(t)}\,\overline{q_2(t)}) \\ q_3(t+1) = q_1(t)q_2(t)\overline{q_3(t)}\,\bar{x} + \bar{x}q_1(t)\overline{q_2(t)} + q_2(t)\overline{q_1(t)} + q_3(t)\overline{q_1(t)}\,\bar{x} \end{cases} \tag{86}$$

Graphic image of the Boolean system for the first equation by elementary Boolean functions AND, OR, and NOT (Fig. 5).

Now we show that is possible to found a neural network that solve the previous system without the use AND, OR and NOT.

The initial states and the input form the neural input system or colon space A.

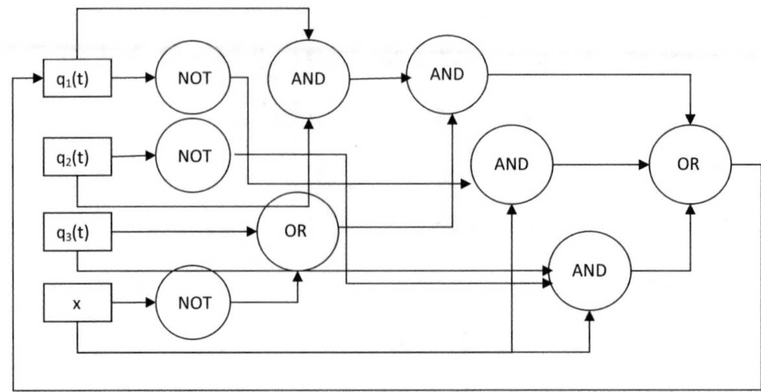

Fig. 5 AND, OR, NOT representation of the digital machine

$$A = \begin{bmatrix} q_1(t) & q_2(t) & q_3(t) & x \\ 0 & 0 & 0 & 0 \\ 0 & 1 & 0 & 0 \\ 1 & 0 & 0 & 0 \\ 0 & 0 & 1 & 0 \\ 1 & 1 & 1 & 0 \\ 1 & 1 & 0 & 0 \\ 0 & 0 & 0 & 1 \\ 0 & 1 & 0 & 1 \\ 1 & 0 & 0 & 1 \\ 0 & 0 & 1 & 1 \\ 1 & 1 & 1 & 1 \\ 1 & 1 & 0 & 1 \end{bmatrix} \text{the state output function is} \begin{bmatrix} 0 & 0 & 0 \\ 0 & 1 & 0 \\ 1 & 0 & 0 \\ 0 & 0 & 1 \\ 1 & 1 & 1 \\ 1 & 1 & 0 \end{bmatrix} \rightarrow \begin{bmatrix} 0 \\ 1 \\ 0 \\ 0 \\ 0 \\ 0 \end{bmatrix} \quad (87)$$

Neural network for first final state function

$$A = \begin{bmatrix} q_1 & q_2 & q_3 & x \\ 0 & 0 & 0 & 0 \\ 0 & 1 & 0 & 0 \\ 1 & 0 & 0 & 0 \\ 0 & 0 & 1 & 0 \\ 1 & 1 & 1 & 0 \\ 1 & 1 & 0 & 0 \\ 0 & 0 & 0 & 1 \\ 0 & 1 & 0 & 1 \\ 1 & 0 & 0 & 1 \\ 0 & 0 & 1 & 1 \\ 1 & 1 & 1 & 1 \\ 1 & 1 & 0 & 1 \end{bmatrix} \rightarrow \begin{bmatrix} 1 & 0 & 0 \\ 1 & 0 & 0 \\ 0 & 1 & 0 \\ 1 & 1 & 1 \\ 1 & 1 & 0 \\ 1 & 1 & 1 \\ 1 & 1 & 0 \\ 0 & 0 & 1 \\ 1 & 1 & 1 \\ 0 & 1 & 0 \\ 1 & 0 & 0 \\ 0 & 0 & 0 \end{bmatrix} = [\, q_1(t+1) \quad q_2(t+1) \quad q_3(t+1) \,] \quad (88)$$

$$G = AU - q_1(t+1) = \begin{bmatrix} -1 \\ 0 \\ 1 \\ 0 \\ 2 \\ 1 \\ 0 \\ 2 \\ 1 \\ 2 \\ 3 \\ 3 \end{bmatrix} \tag{89}$$

So we have four possible states $\{1, 2, 3, -1\}$ for which we can decompose G in this way.

That can be decomposed in this way

$$G = AU - q_1(t+1) = \begin{bmatrix} -1 \\ 0 \\ 1 \\ 0 \\ 2 \\ 1 \\ 0 \\ 2 \\ 1 \\ 2 \\ 3 \\ 3 \end{bmatrix} \rightarrow [g_1 \quad g_2 \quad g_3 \quad g_4] = \begin{bmatrix} 0 & 0 & 0 & 1 \\ 0 & 0 & 0 & 0 \\ 1 & 0 & 0 & 0 \\ 0 & 0 & 0 & 0 \\ 1 & 1 & 0 & 0 \\ 1 & 0 & 0 & 0 \\ 0 & 0 & 0 & 0 \\ 1 & 1 & 0 & 0 \\ 1 & 0 & 0 & 0 \\ 1 & 1 & 0 & 0 \\ 1 & 1 & 1 & 0 \\ 1 & 1 & 1 & 0 \end{bmatrix} \tag{90}$$

When at the colon space A we adjoin the four colons $[g_1 \quad g_2 \quad g_3 \quad g_4]$ we have the colon space

$$A = \begin{bmatrix} q_1 & q_2 & q_3 & x & g_1 & g_2 & g_3 & g_4 \\ 0 & 0 & 0 & 0 & 0 & 0 & 0 & 1 \\ 0 & 1 & 0 & 0 & 0 & 0 & 0 & 0 \\ 1 & 0 & 0 & 0 & 1 & 0 & 0 & 0 \\ 0 & 0 & 1 & 0 & 0 & 0 & 0 & 0 \\ 1 & 1 & 1 & 0 & 1 & 1 & 0 & 0 \\ 1 & 1 & 0 & 0 & 1 & 0 & 0 & 0 \\ 0 & 0 & 0 & 1 & 0 & 0 & 0 & 0 \\ 0 & 1 & 0 & 1 & 1 & 1 & 0 & 0 \\ 1 & 0 & 0 & 1 & 1 & 0 & 0 & 0 \\ 0 & 0 & 1 & 1 & 1 & 1 & 0 & 0 \\ 1 & 1 & 1 & 1 & 1 & 1 & 1 & 0 \\ 1 & 1 & 0 & 1 & 1 & 1 & 1 & 0 \end{bmatrix} = a_{j,k} \tag{91}$$

We remark that any permutation of the colons cannot change the computation of the neural network architecture.

$$A(A^TA)^{-1}Aq_1(t+1)=q_1(t+1) \tag{92}$$

And

$$(A^TA)^{-1}Aq_1(t+1)=W=\begin{bmatrix}1\\1\\1\\1\\-1\\-1\\1\\-1\end{bmatrix}, threshold=D=0.5 \tag{93}$$

So we have for the neural network $h(\sum_{j=1}^{8} w_j a_{j,k}-D)=q_{1,k}(t+1)$ where h(x) is the Heaviside function (Fig. 6).

That is the function $q_1(t+1)$. Now when we reduce the number of the colons that we adjoin we can have again the function $q_{1,k}(t+1)$ but not in a direct way but only by the Heaviside function. In fact the subsets $[g_2 \quad g_3 \quad g_4]$ of the four functions $[g_1 \quad g_2 \quad g_3 \quad g_4]$ can solve the Boolean function $q_1(t+1)$. In fact for the colon space

$$A=\begin{bmatrix} q_1 & q_2 & q_3 & x & g_2 & g_3 & g_4\\ 0&0&0&0&0&0&1\\ 0&1&0&0&0&0&0\\ 1&0&0&0&0&0&0\\ 0&0&1&0&0&0&0\\ 1&1&1&0&1&0&0\\ 1&1&0&0&0&0&0\\ 0&0&0&1&0&0&0\\ 0&1&0&1&1&0&0\\ 1&0&0&1&0&0&0\\ 0&0&1&1&1&0&0\\ 1&1&1&1&1&1&0\\ 1&1&0&1&1&1&0 \end{bmatrix}=a_{j,k} \tag{94}$$

We have

$$A(A^TA)^{-1}Aq_1(t+1)=y \tag{95}$$

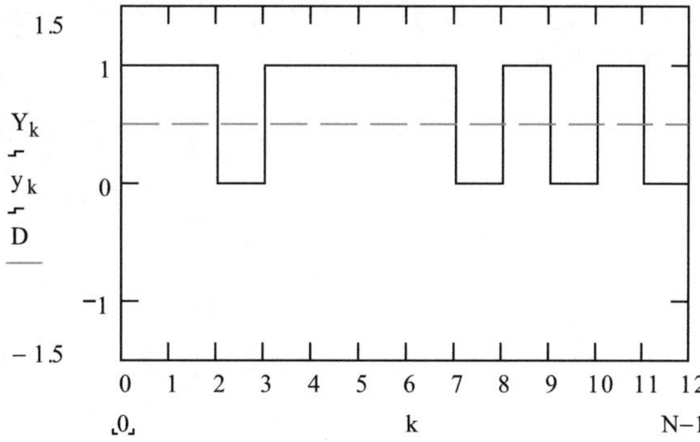

Fig. 6 Neural output of the function $q_1(t+1)$. We see that the neural output is equal to the function $q_1(t+1)$

And

$$(A^T A)^{-1} A q_1(t+1) = W = \begin{bmatrix} 0.197 \\ 1.004 \\ 1.084 \\ 0.736 \\ -1.615 \\ -0.364 \\ 1 \end{bmatrix}, threshold = D = 0.437 \tag{96}$$

In a graphic way we have (Fig. 7).
The function y is not equal to $q_1(t+1)$ but with the Heaviside function we have

$$h\left(\sum_{j=1}^{7} w_j a_{j,k} - D\right) = q_{1,k}(t+1) \tag{97}$$

To conclude the network we must solve the functions g_2, g_3, g_4.

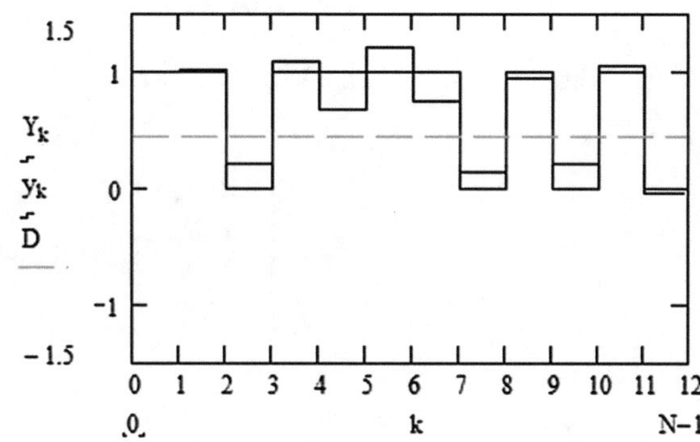

Fig. 7 The neuron output y is not equal to $q_1(t+1)$ but with the Heaviside function we can have at the neuron the same function $q_1(t+1)$ with less number of hidden neuron

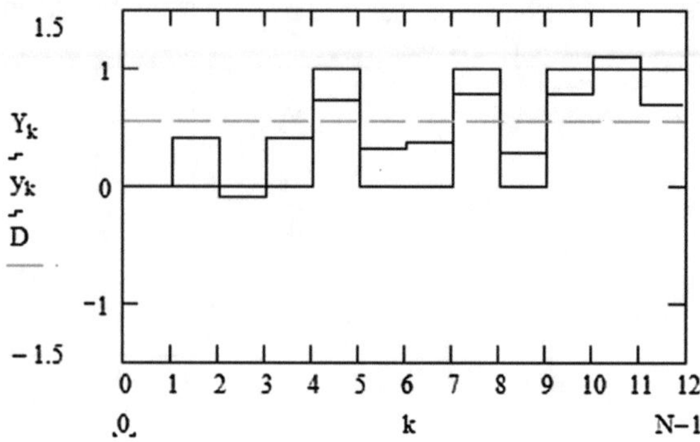

Fig. 8 Neural output y for the function g_2

Given the colon space

$$A = \begin{bmatrix} q_1 & q_2 & q_3 & \\ 0 & 0 & 0 & 0 \\ 0 & 1 & 0 & 0 \\ 1 & 0 & 0 & 0 \\ 0 & 0 & 1 & 0 \\ 1 & 1 & 1 & 0 \\ 1 & 1 & 0 & 0 \\ 0 & 0 & 0 & 1 \\ 0 & 1 & 0 & 1 \\ 1 & 0 & 0 & 1 \\ 0 & 0 & 1 & 1 \\ 1 & 1 & 1 & 1 \\ 1 & 1 & 0 & 1 \end{bmatrix} = a_{j,k} \tag{98}$$

The operators are

$$A(A^T A)^{-1} A g_2 = y \tag{99}$$

For which we have (Fig. 8)
And

$$W = \begin{bmatrix} -0.094 \\ 0.406 \\ 0.406 \\ 0.375 \end{bmatrix}, \theta = 0.547 \tag{100}$$

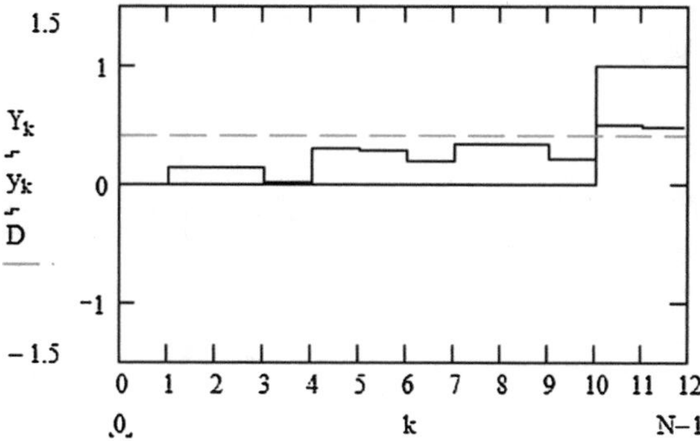

Fig. 9 Neural output y for the function g_3

For g_3 we have

$$A(A^TA)^{-1}Ag_3 = y \tag{101}$$

For which we have (Fig. 9)
And

$$W = \begin{bmatrix} 0.141 \\ 0.141 \\ 0.016 \\ 0.375 \end{bmatrix}, \theta = 0.398 \tag{102}$$

For g_4 we have (Figure 10)

$$A(A^TA)^{-1}Ag_4 = y \tag{103}$$

All the elements are equal to zero because the four inputs are all equal to zero. In this case no solution exist. But if we take the complementary function $Y = 1 - g_4$ we have (Fig. 11).

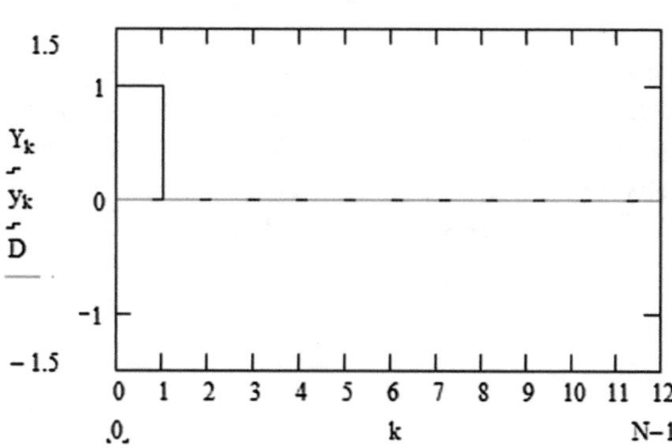

Fig. 10 Neural output for the function g_4

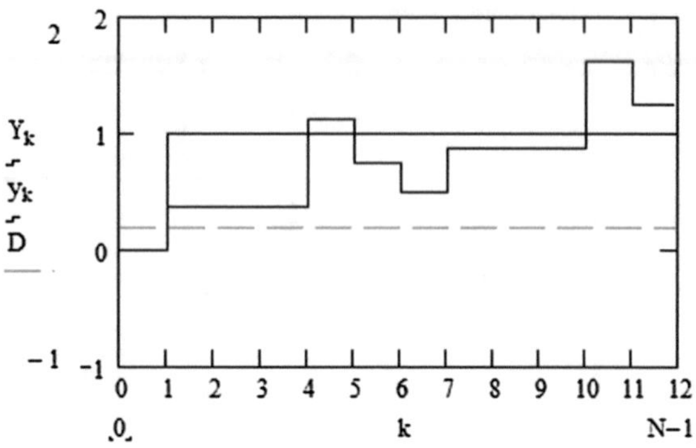

Fig. 11 Neural output for the function 1-g_4

And

$$W = \begin{bmatrix} 0.375 \\ 0.375 \\ 0.375 \\ 0.5 \end{bmatrix}, \theta = 0.188 \tag{104}$$

Because we have

$$(A^T A)^{-1} A(1 - g_4) = W \tag{105}$$

We have that for the complementary function g_4 we have

$$\begin{gathered} Y = 1 - g_4 \\ and \\ g_4 = 1 - Y \end{gathered} \tag{106}$$

And the neural network complementary function we have

$$y = 2\theta - A(A^T A)^{-1} A^T (1 - g_4) = 2\theta - AW \tag{107}$$

where W are the weights of the functions $1 - g_4$ that we can solve by the one step method in neural network and we can use the weights W to solve also the function g_4 by one step that was impossible to solve directly. The graph of the neural network is this (Fig. 12).

The neural network is (Fig. 13).

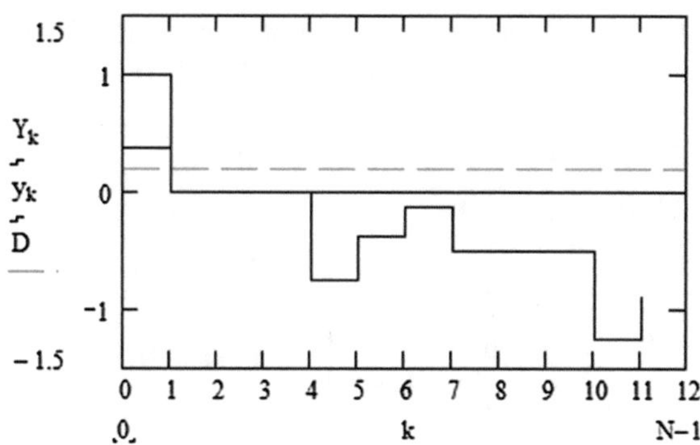

Fig. 12 Neural output for the function g_4

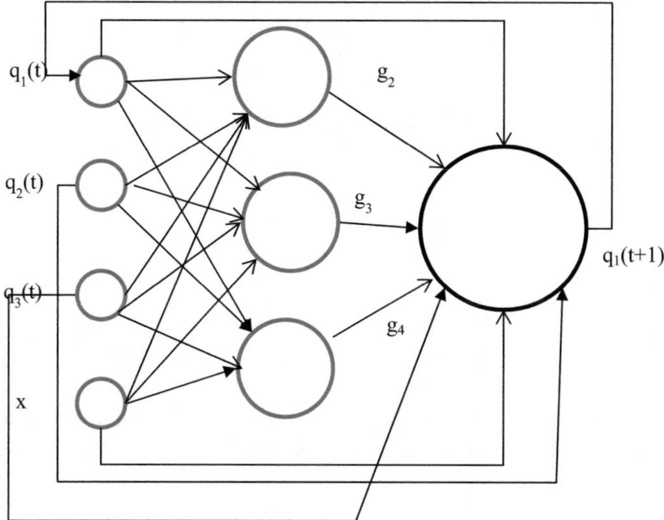

Fig. 13 Neural network for first bit $q_1(t+1)$ of the digital system

Coker Specher theorems and non locality in quantum and neural network due to multiple external information sharing and non monotonic logic.

Given the set of orthonormal references in quantum mechanics

$$
\begin{aligned}
&\mathrm{C1}=\begin{pmatrix}0&0&1&1\\0&0&1&-1\\0&1&0&0\\1&0&0&0\end{pmatrix}\quad \mathrm{C2}=\begin{pmatrix}0&0&1&1\\0&1&0&1\\0&0&1&-1\\1&0&0&0\end{pmatrix}\\
&\mathrm{C3}=\begin{pmatrix}1&1&1&0\\-1&-1&1&0\\1&-1&0&1\\-1&1&0&1\end{pmatrix}\quad \mathrm{C4}=\begin{pmatrix}1&1&1&0\\-1&1&0&1\\1&1&-1&0\\-1&1&0&-1\end{pmatrix}\\
&\mathrm{C5}=\begin{pmatrix}0&0&1&1\\0&1&0&0\\1&0&0&0\\0&0&1&-1\end{pmatrix}\quad \mathrm{C6}=\begin{pmatrix}1&1&1&0\\-1&1&0&1\\-1&1&0&-1\\1&1&-1&0\end{pmatrix}\\
&\mathrm{C7}=\begin{pmatrix}1&1&1&0\\1&1&-1&0\\-1&1&0&1\\1&-1&0&1\end{pmatrix}\quad \mathrm{C8}=\begin{pmatrix}1&-1&1&0\\1&1&0&1\\-1&1&1&0\\1&1&0&-1\end{pmatrix}\\
&\mathrm{C9}=\begin{pmatrix}1&-1&1&0\\1&1&0&1\\1&1&0&-1\\-1&1&1&0\end{pmatrix}
\end{aligned}
\tag{108}
$$

All the nine references include 18 four dimension vectors that we numerate from 1 to 18. For example we have for C1 the vectors

$$
1\equiv\begin{bmatrix}0\\0\\0\\1\end{bmatrix},2\equiv\begin{bmatrix}0\\0\\1\\0\end{bmatrix},3\equiv\begin{bmatrix}1\\1\\0\\1\end{bmatrix},4\equiv\begin{bmatrix}1\\-1\\0\\0\end{bmatrix}
\tag{109}
$$

So C1 can be represented by the set of vectors

$$
C1=\{1,2,3,4\}
\tag{110}
$$

We repeat the same for all the other orthonormal basis. We remark that the nine sets are not disjoin but has common elements as we show in this table

$$\begin{bmatrix} (1,2,3,4) & (1) & (3) & 0 & (2) & 0 & (4) & 0 & 0 \\ (1) & (1,5,6,7) & 0 & (7) & (5) & 0 & 0 & (6) & 0 \\ (3) & 0 & (3,8,9,10) & (8) & 0 & (9) & (10) & 0 & 0 \\ 0 & (7) & (8) & (7,8,11,12) & 0 & (11) & 0 & (12) & 0 \\ (2) & (5) & 0 & 0 & (2,5,13,14) & (14) & 0 & 0 & (13) \\ 0 & 0 & (9) & (11) & (14) & (9,11,14,15) & 0 & 0 & (15) \\ (4) & 0 & (10) & 0 & 0 & 0 & (4,10,16,17) & (16) & (17) \\ 0 & (6) & 0 & (12) & 0 & 0 & (16) & (6,12,16,18) & (18) \\ 0 & 0 & 0 & 0 & (13) & (15) & (17) & (18) & (13,15,,17,18) \end{bmatrix} \tag{111}$$

With the projection operator in the orthonormal basis C1 we select among the linear combination

$$\psi = c_1 \begin{bmatrix} 0 \\ 0 \\ 0 \\ 1 \end{bmatrix} + c_2 \begin{bmatrix} 0 \\ 0 \\ 1 \\ 0 \end{bmatrix} + c_3 \begin{bmatrix} 1 \\ 1 \\ 0 \\ 0 \end{bmatrix} + c_4 \begin{bmatrix} 1 \\ -1 \\ 0 \\ 1 \end{bmatrix} \tag{112}$$

For

$$A = \begin{bmatrix} 1 \\ -1 \\ 0 \\ 0 \end{bmatrix} \tag{113}$$

We have the projection operator (quantum measure) for C1

$$A(A^TA)^{-1}A^T\psi = AA^T\psi = |4\rangle\langle 4|(c_1|1\rangle + c_2|2\rangle + c_3|3\rangle + c_4|4\rangle) = c_4|4\rangle$$

$$c = (A^TA)^{-1}A^T\psi = A^T\psi = [1 \quad -1 \quad 0 \quad 0](c_1 \begin{bmatrix} 0 \\ 0 \\ 0 \\ 1 \end{bmatrix} + c_2 \begin{bmatrix} 0 \\ 0 \\ 0 \\ 1 \end{bmatrix} + c_3 \begin{bmatrix} 0 \\ 0 \\ 0 \\ 1 \end{bmatrix} + c_4 \begin{bmatrix} 0 \\ 0 \\ 0 \\ 1 \end{bmatrix}) = c_4 \tag{114}$$

Now for any orthonormal basis locally is possible to select one component and found the linear coefficient associate. The problem is to know if is possible to solve the same problem for all the orthonormal basis. The Coker Specher theorem show that local is possible to select one state by projection operator but globally is impossible for the far dependence of local basis with others as we show in the previous table. So when is true in one basis can be false in another basis for the connection of one basis with another. Given a basis external influence can destroy the local property as in the non monotonic logic. With the diagram of the connections in this figure (Fig. 14).

In basis 5 all the vectors are false so for the external action we cannot select one state by projection operator due to the false external action.

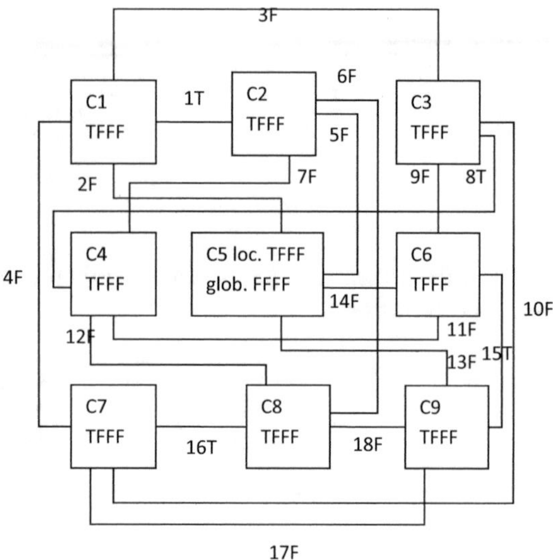

Fig. 14 Connection of nine basis one with common vector with the other in KS system

Explanation of the graph

Given the arrow

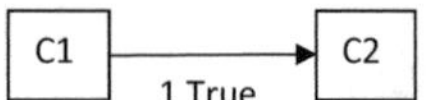

C1 and C2 are two set of vectors (basis) that had in common the vector v1 that has the value true. Any basis has one element that is selected that we denote a True all the other that are not selected are denoted as False. The basis C5 for the general rule must have one element that must be selected so for the local rule we have TFFF. Now for the selection of the other elements that are in common with C5 the selection is impossible. This generate contradiction between global and local projection to select one element.

5 Conclusion

In Turing Machine input output and states form a system. When system is in one state the input is transformed in output. Local and sequential process is the ordinary computation in Turing machine and in the language for the digital computer. The software is built to connect all different sequential computations (statements) in a way that the final result will be the solution of our problem. The problem with digital computer is to build a code that will be useful for our purpose. Also if the digital computer can solve a lot of problems when we compare the efficiency of the digital computer with the efficiency of the brain we remark a huge difference in the two type of computation. Brain can solve problems that digital computer cannot. But how is the explanation of this very high brain efficiency? When we study the brain we can see that is compose by elements denoted neurons. Any neurons receive a lot of inputs, make weighted linear combination, with threshold give its reply. Any input can be one or zero n inputs have all possible value of the Boolean vector with dimension n. For two inputs we have (0,0), (1,0), (0,1), (1,1) possible inputs. So for first input has this possible inputs 0101, the second inputs has the possible inputs 0011. The first input has the morphogenetic field 0101, the second has the morphogenetic field 0011. The weights are the Morphogenetic sources or strength of the fields. Given the wanted field in a Boolean form, we project this field into the inputs fields and we compute the weights or morphogenetic sources. The superposition of the input fields generate the neuron field that with the threshold is can be the wanted field. If the wanted field is incompatible with the input field we have a method to enlarge the number of inputs in a way to obtain a set of input fields compatible with the wanted field and we solve the neuron problem. In quantum mechanics we make the same process where we have the basis quantum state for n qubits. The problem in quantum mechanics is to classify sets of functions by a quantum measure as projection of a vector into one particular state that is the key to recognize by one query the property of the function or the values of the functions. In this case we select one and only one Morphogenetic source among others in a way to know the researched property. We also explain the meaning of the entanglement and Morphogenetic synchronic process as the global property to eliminate local contradiction in hidden variables by the explanation of the Coker Specher theorems. At the end we can also use the wave basis functions to reverse the physical process that from the sources generate the field. So when we know the field we can came back to compute the Morphogenetic sources and at the end in a physical domain obtain the wanted field. All different applications put in evidence one method denoted projection method that in a global way collect all the local knowledge in fields that with linear aggregation can give the solution of our problems. The Morphogenetic sources are used in holographic process to simulate from two dimension the three dimension. We know that in the global morphogenetic field the information is diffuse in all the reference space so the information is robust to external noise we can also find the same information also in one part of the reference space. In conclusion the Morphogenetic computing we eliminate the stressing work of the creation of the codes and in the same time errors or contradictions are eliminate in the generation of the morpho by the Morphogenetic sources.

References

1. Metwaly, A.F., Rashad, M.Z., Omara, F.A., Megahed, A.A.: Architecture of multicast centralized key management scheme using quantum key distribution and classical symmetric encryption. Eur. Phys. J. Spec. Top. **223**(8), 1711–1728 (2014)
2. Metwaly, A., Rashad, M.Z., Omara, F.A., Megahed, A.A.: Architecture of point to multipoint QKD communication systems (QKDP2MP). In: 8th International Conference on Informatics and Systems (INFOS), Cairo, pp. NW 25–31 (2012). IEEE
3. Farouk, A., Omara, F., Zakria, M., Megahed, A.: Secured IPsec multicast architecture based on quantum key distribution. In: The International Conference on Electrical and Bio-medical Engineering, Clean Energy and Green Computing, pp. 38–47 (2015). The Society of Digital Information and Wireless Communication
4. Farouk, A., Zakaria, M., Megahed, A., Omara, F.A.: A generalized architecture of quantum secure direct communication for N disjointed users with authentication. Sci. Rep. **5**, 16080 (2014)
5. Wang, M.M., Wang, W., Chen, J.G., Farouk, A.: Secret sharing of a known arbitrary quantum state with noisy environment. Quantum Inf. Process. **14**(11), 4211–4224 (2015)
6. Naseri, M., Heidari, S., Batle, J., Baghfalaki, M., Gheibi, R., Farouk, A., Habibi, A.: A new secure quantum watermarking scheme. Optik-Int. J. Light Electron Opt. **139**, 77–86 (2017)
7. Batle, J., Ciftja, O., Naseri, M., Ghoranneviss, M., Farouk, A., Elhoseny, M.: Equilibrium and uniform charge distribution of a classical two-dimensional system of point charges with hard-wall confinement. Phys. Scr. **92**(5), 055801 (2017)
8. Geurdes, H., Nagata, K., Nakamura, T., Farouk, A.: A note on the possibility of incomplete theory (2017). arXiv preprint arXiv:1704.00005
9. Batle, J., Farouk, A., Alkhambashi, M., Abdalla, S.: Multipartite correlation degradation in amplitude-damping quantum channels. J. Korean Phys. Soc. **70**(7), 666–672 (2017)
10. Batle, J., Naseri, M., Ghoranneviss, M., Farouk, A., Alkhambashi, M., Elhoseny, M.: Shareability of correlations in multiqubit states: optimization of nonlocal monogamy inequalities. Phys. Rev. A **95**(3), 032123 (2017)
11. Batle, J., Farouk, A., Alkhambashi, M., Abdalla, S.: Entanglement in the linear-chain Heisenberg antiferromagnet Cu (C 4 H 4 N 2) (NO 3) 2. Eur. Phys. J. B **90**, 1–5 (2017)
12. Batle, J., Alkhambashi, M., Farouk, A., Naseri, M., Ghoranneviss, M.: Multipartite non-locality and entanglement signatures of a field-induced quantum phase transition. Eur. Phys. J. B **90**(2), 31 (2017)
13. Nagata, K., Nakamura, T., Batle, J., Abdalla, S., Farouk, A.: Boolean approach to dichotomic quantum measurement theories. J. Korean Phys. Soc. **70**(3), 229–235 (2017)
14. Abdolmaleky, M., Naseri, M., Batle, J., Farouk, A., Gong, L.H.: Red-Green-Blue multi-channel quantum representation of digital images. Optik-Int. J. Light Electron Opt. **128**, 121–132 (2017)
15. Farouk, A., Elhoseny, M., Batle, J., Naseri, M., Hassanien, A.E.: A proposed architecture for key management schema in centralized quantum network. In: Handbook of Research on Machine Learning Innovations and Trends, pp. 997–1021 (2017). IGI Global
16. Zhou, N.R., Li, J.F., Yu, Z.B., Gong, L.H., Farouk, A.: New quantum dialogue protocol based on continuous-variable two-mode squeezed vacuum states. Quantum Inf. Process. **16**(1), 4 (2017)
17. Batle, J., Abutalib, M., Abdalla, S., Farouk, A.: Persistence of quantum correlations in a XY spin-chain environment. Eur. Phys. J. B **89**(11), 247 (2016)
18. Batle, J., Abutalib, M., Abdalla, S., Farouk, A.: Revival of bell nonlocality across a quantum spin chain. Int. J. Quantum Inf. **14**(07), 1650037 (2016)
19. Batle, J., Ooi, C.R., Farouk, A., Abutalib, M., Abdalla, S.: Do multipartite correlations speed up adiabatic quantum computation or quantum annealing? Quantum Inf. Process. **15**(8), 3081–3099 (2016)

20. Batle, J., Bagdasaryan, A., Farouk, A., Abutalib, M., Abdalla, S.: Quantum correlations in two coupled superconducting charge qubits. Int. J. Mod. Phys. B **30**(19), 1650123 (2016)
21. Batle, J., Ooi, C.R., Abutalib, M., Farouk, A., Abdalla, S.: Quantum information approach to the azurite mineral frustrated quantum magnet. Quantum Inf. Process. **15**(7), 2839–2850 (2016)
22. Batle, J., Ooi, C.R., Farouk, A., Abdalla, S.: Nonlocality in pure and mixed n-qubit X states. Quantum Inf. Process. **15**(4), 1553–1567 (2016)
23. Metwaly, A.F., Rashad, M.Z., Omara, F.A., Megahed, A.A.: Architecture of Multicast Network Based on Quantum Secret Sharing and Measurement (2015)

Part III
Quantum in Intelligent Applications

Quantum Inspired Evolutionary Algorithm in Load Frequency Control of Multi-area Interconnected Thermal Power System with Non-linearity

K. Jagatheesan, Sourav Samanta, Alokeparna Choudhury, Nilanjan Dey, B. Anand and Amira S. Ashour

Abstract Load Frequency Control (LFC) is an important issue in power system to maintain power system stability and quality of generated power supply during sudden load demand period. In order to overcome this issue, power systems are interconnected and secondary controllers are introduced to regulate the power system parameters within a specified limit during sudden load demand period. In this present work, three area single stage reheat thermal power systems are interconnected and each area comprises governor unit, reheated unit, turbine unit, Governor Dead Band (GDB), Generation Rate Constraint (GRC) non-linear components and boiler dynamics effect. One Percent (1%) Step Load Perturbation (SLP) is considered in thermal area 1 of the investigated power system. The Proportional-Integral-Derivative (PID) controller is introduced as a secondary controller. Since tuning of the controller gain values play a vital role, evolutionary

K. Jagatheesan
Department of EEE, Mahendra Institute of Engineering and Technology, Namakkal, Tamilnadu, India
e-mail: jaga.ksr@gmail.com

S. Samanta (✉)
Department of CSE, University Institute of Technology, Burdwan, Westbengal, India
e-mail: sourav.uit@gmail.com

A. Choudhury
Department of CS, St. Xavier's College Burdwan, Burdwan, Westbengal, India
e-mail: choudhury.am2011@gmail.com

N. Dey
Department of Information Technology, Techno India College of Technology, Kolkata, India
e-mail: neelanjan.dey@gmail.com

B. Anand
Department of EEE, Hindusthan College of Engineering and Technology, Coimbatore, Tamilnadu, India
e-mail: b_anand_eee@yahoo.co.in

A.S. Ashour
Faculty of Engineering, Department of Electronics and Electrical Communications Engineering, Tanta University, Tanta, Egypt
e-mail: amirasashour@yahoo.com

A.E. Hassanien et al. (eds.), *Quantum Computing: An Environment for Intelligent Large Scale Real Application*, Studies in Big Data 33,
https://doi.org/10.1007/978-3-319-63639-9_16

algorithms are introduced to tune the controller gain values. In the current work, the Genetic Algorithm (GA), Quantum Inspired Genetic Algorithm (QIGA) and Quantum Inspired Evolutionary Algorithm (QIEA) are proposed for tuning of controller gain values. The cumulative comparisons of the simulation result are clearly reported that QIGA and QIEA are more superior to the GA based PID controller performance in the same investigated power system in terms of time domain specification parameters.

Keywords Load frequency control · Interconnected power system · Genetic algorithm · Quantum inspired genetic algorithm · Quantum inspired evolutionary algorithm · Step load perturbation · Objective function · Optimal gain values · Time domain specification

1 Introduction

Recently, the power generating units are operated with more commitment due to enormous growth in the new technologies and the need of power regular supply. In this regard, the power generating units are interconnected through tie-line for satisfying load demand. The increase in the power system size leads to increase the complexity. The system responses are affected during any sudden load demand that may occur in any one of the interconnected power system. For regulating the system parameter within the specified limit, the secondary controller proper design becomes more crucial. In this regard, the PID controller is introduced as a secondary controller for regulating the power system operation during sudden load demand period. The suitable controller gain value selection is considered an essential issue to guarantee better controlled response during the sudden load period. So, several optimization techniques are implemented to optimize the secondary controller gain values in single/multi-area interconnected power system for regulating the system parameters and for maintaining quality of generated power supply during sudden load demand period.

Literature survey clearly shows that several evolutionary algorithms based optimization techniques are implemented to tune the controller gain values for solving the load frequency control/automatic generation control issue in single area and multi-area inter connected power system. Automatic Generation Control (AGC) of three area power generating interconnected hydrothermal power generating unit with Generation Rate Constraint (GRC) non linearity has been discussed in [1] by implementing Fuzzy Integral Double Derivative (FIDD) controller. The Bacterial Foraging (BF) optimization technique is utilized for optimizing the FIDD controller's gain values. The BF optimization technique tuned gain values based Fractional Order PID (FOPID) is implemented for AGC of multi-area interconnected thermal power generating unit under deregulated environment by considering GRC non linearity effect [2]. Two Degree-of-Freedom PID controller gain values are optimized by considering Integral Time Absolute Time error (ITAE), Integral Square Error (ISE) and Integral Time Square Error (ITSE) cost function based Differential Evolution (DE) tuned controller designed for solving LFC of two area interconnected thermal power system including Governor Dead Band (GDB) non linearity [3]. Imperialist Control Algorithm (ICA) optimization technique tuned PID controller has been implemented for solving LFC in nonlinear power system considering GRC non linearity [4]. Cuckoo Search (CS) optimization technique optimized PI controller has been implemented for solving LFC issue in two area interconnected power system. In addition, the power system has been equipped with GDB non linearity and Super Magnetic Energy Storage (SMES) unit for analyzing the system performance [5]. Fuzzy Logic Controller (FLC) has been designed for solving LFC issue in two area interconnected power system by considering GGRC and GDB non linearity effect. The FLC performance has been compared with conventional PI controller for measuring the supremacy of proposed FLC controller [6]. The FLC has been implemented into two area power system for solving AGC issue in two area interconnected and GRC non linearity equipped reheat thermal power system [7]. Adaptive LFC of two area interconnected power system (Hydrothermal) has been discussed [8] by considering GRC non linearity effect with adaptive implicit hybrid self tuning based controller. Furthermore, the artificial neural network (ANN) controller has been executed in two area interconnected thermal power system for analyzing AGC issue by considering GDB non linearity under deregulated environment [9]. The GRC, GDB, Boiler Dynamics (BD) and SMES unit equipped three area interconnected power system frequency have been stabilized by employing robust decentralized frequency stabilizer unit with integral controller [10]. The GDB non linearity effect is two area interconnected power system has been analyzed, where the system response at which the frequency and tie line power deviation has been considered for the analysis [11]. μ-synthesis based ANN technique has been implemented into LFC of interconnected power generating unit and GRC non linearity also has been considered for validating supremacy of proposed controller in [12]. GDB non linearity and BD equipped thermal power system is considered for investigation [13]. In addition, the integral controller and SMES energy unit were also considered for analysis. Lyapunov technique optimized parameters based reheat thermal power system has been considered for the investigation [14] with backlash non linearity effect for solving LFC in power system during sudden demand period. Continuous and discrete modes based AGC of two area interconnected thermal power generating unit with GRC non linearity has been proposed in [15]. The optimum selection based speed regulation parameters have been used for the AGC in two area interconnected reheat thermal power system by considering the non-linearity effect [16]. The Battery Energy Storage (BES) effect in LFC of interconnected power system with GDB and GRC non linearity has been discussed in [17]. The adaptive Controller has been designed and implemented into LFC of single area single area power system [18] considering GRC non linearity effect.

Three area interconnected thermal power system has been designed by considering GRC non linearity. In addition, the BF optimization technique has been implemented to optimizing I controller gain values in [19]. The different steam configuration implemented thermal and electric governor equipped hydro power systems are interconnected two area power system has been discussed for analyzing the AGC issue by considering several classical controllers in [20]. The effect of FLC in two area hydrothermal power generating unit has been studied by considering GDB, GRC non linearity and BD effect in [21]. Two area interconnected reheat thermal power system has been discussed by considering classical controller and GRC non linearity effect in [22]. Bacterial Foraging Optimization (BFO) technique optimized several classical controllers have been implemented into three area interconnected thermal power system with GRC non linearity effect [23]. AGC of two area interconnected reheat thermal power generating unit has been studied in [24] by considering GDB non linearity effect. Lyapunov technique is used to tune frequency bias parameter values to integral controller gain.

AGC control of three area interconnected hydro power generating unit with classical controller has been presented in [25]. The GRC non linearity effect has been considered and controller gain values are optimized by using Bacterial Foraging optimization technique [25]. Integral Square Error (ISE) cost function based controller has been designed into decentralized biased controller used LFC of interconnected power system in [26]. Cuckoo Search (CS) algorithm designed controller has been implemented in two area interconnected thermal power for solving AGC issue in [27]. Fractional Order PID (FOPID) controller has been designed for LFC of interconnected power system with GRC non linearity effect and dead zone effect in [28].The CS algorithm optimized FOPID controller has been implemented into three area interconnected thermal power system with GRC non linearity. The several Flexible Alternating Current Transmission System (FACTS) devices and 2-DOF controllers (2DOF-PI, 2DOF-PD, 2DOF-IDD, 2DOF-IDD) have been considered for the investigation [29]. Bat Algorithm (BA) optimized PD-PID cascade controller has been designed into AGC of three area interconnected thermal power system with GRC non linearity effect [30]. GRC and GDB non linearity equipped multi-source interconnected power generating unit has been considered for the investigation in [31] by considering Improved Particle Swarm Optimization (IPSO) technique based controller for solving LFC issue. Hybrid firefly algorithm and pattern search optimization designed PID controller has been included in AGC of multi-area interconnected power system [32] with GRC non linearity effect. Firefly Algorithm (FA) optimized controller gain values based PI/PID controllers have been implemented in three area unequal thermal power generating unit for solving LFC problem in power generating unit under sudden load demand [33]. Cuckoo Search (CS) algorithm optimized controller gain value based 2DOF controllers (2DOF-IDD, 2DOF-PI, 2DOF-PD) have been implemented into three area interconnected power system including GRC non linearity [34]. The Unequal three area interconnected thermal power system has been developed in [32] by considering time delay, BD and GRC non linearity effect for solving the AGC of power system. Thyristor Controller Series Compensator (TCSC) unit and SMES unit have been implemented multi-area multi-unit power system unit has been designed [35]. Discrete optimum integral controller

Table 1 Optimization technique related to LFC/AGC of single/multi-area interconnected power system

Year	Optimization techniques	References
1982	Lyapunov technique	[15]
1983	Continuous and discrete mode optimization	[16]
1989	Adaptive controller	[19]
2009	Fuzzy logic controller (FLC)	[22]
2015	Cuckoo search	[29]
2015	Bat inspired algorithm	[30]
2014	Firefly algorithm	[33]
2015	Cuckoo search	[34]
2015	Hybrid firefly algorithm	[32, 56]
1978	Parameter-plane technique	[48, 57]
1988	Optimal control theory	[37, 58]
2006	Artificial neural network (ANN)	[59]
2009	Genetic algorithm (GA)	[60]
2009	Stochastic particle swarm optimization (PSO)	[61]
2011	Genetic algorithm	[62]
2011	Bacterial foraging optimization algorithm (BFOA)	[63]
2011	Craziness based particle swarm optimization (CRAZYPSO)	[64]
2012	Artificial bee colony (ABC)	[65]
2013	Bacterial foraging (BF) technique	[66]
2013	Ant colony optimization	[67]
2014	Stochastic particle swarm optimization (SPSO)	[68]
2014	Ant colony optimization	[69]
2015	Ant colony optimization	[70, 71, 72]
2015	Beta wavelet neural network (BWNN)	[73]
2016	Particle swarm optimization	[74]
2016	Fuzzy logic controller	[75]
2016	Flower pollination algorithm (FPA)	[57, 76]
2016	Ant colony optimization	[58, 77]

performance has been compared with optimum PI controller in AGC of two area interconnected thermal power system [36] by including GRC non linearity effect. The optimization techniques used for solving AGC/LFC issue in interconnected power system are clearly depicted in Table 1.

In the current proposed work, the performance is evaluated for three area interconnected reheat thermal power system with Quantum Inspired Evolutionary Algorithm (QIEA), Quantum Inspired Genetic Algorithm (QIGA) and Genetic Algorithm (GA) optimized PID controller performance by using ITAE objective function. The thermal power system is considered for the analysis. Consequently, the main contribution of proposed research work is as follows:

(1) To design three areas reheat thermal power system with PID controller.
(2) To design the suitable secondary controller gain values based PID controller for the investigated thermal power system.
(3) To optimize the controller gain values by using QIEA, QIGA and GA optimization technique based ITAE based cost functions with 1% SLP in area 1.
(4) To compare the performance of different optimization technique based controller performance in the investigated power system response.

2 System Investigated

In the current proposed work, the LFC of three area interconnected reheat thermal power system is considered and presented. The considered three areas are equal size thermal power system and all three areas are interconnected through tie line. Transfer function model of three area interconnected power system shown in Fig. 1.

The power system parameters and their nominal values are given in Table 2. The each area of thermal power system comprises governor, reheater, turbine, power system (Generator) and speed regulator respectively. A one percent (1%) Step Load Perturbation (SLP) is considered in area 1 for analyzing the dynamic behaviors of power system with different optimization algorithm optimized controller performance during sudden load demand period. During nominal loading condition each power generating area takes care of its own load demand and keeps the power system parameter within the specified value.

During sudden load demand period of time performance of system affected in terms of time domain specification values (damping oscillation, large peak over and

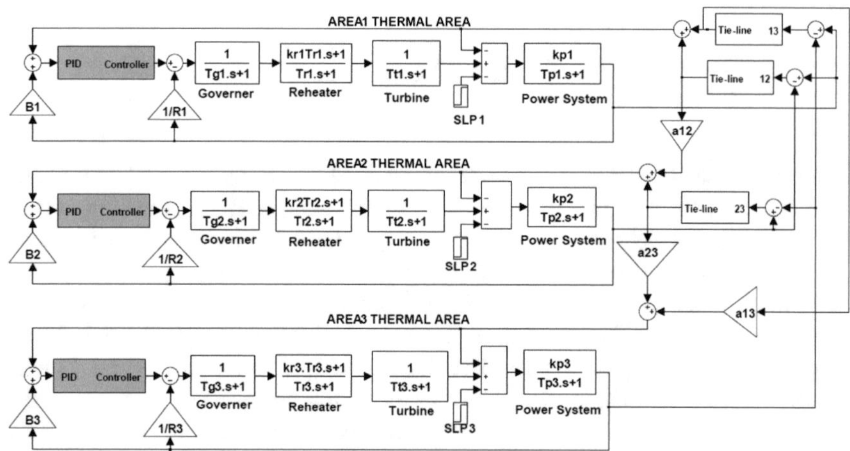

Fig. 1 Transfer function model of three area interconnected reheat thermal power system with PID controller

Table 2 Nominal values of power system parameter

Parameters	Symbol	Nominal value
Reheat time constant	$T_{t1} = T_{t2} = T_{t3}$	0.3 s
Speed governor time constant	$T_{g1} = T_{g2} = T_{g3}$	0.2 s
Self regulation for governor	$R_1 = R_1 = R_3$	2.4 Hz pu^{-1} MW
Frequency bias constant	$B_1 = B_1 = B_3$	0.425 puMW/Hz
Power system Gain	$K_{p1} = K_{p2} = K_{p3}$	120 Hz pu^{-1} MW
Power system time constant	$T_{p1} = T_{p2} = T_{p3}$	20 s
Tie line power coefficient	$T_{12} = T_{13} = T_{23}$	0.0707 MW rad^{-1}
Steam turbine reheat time constant	Tr1 = Tr2 = Tr3	10 s
Steam turbine reheat coefficient	Kr1 = Kr2 = Kr3	0.5

under shoot with large settling time). A proper design and implementation of the secondary controller are crucial for delivering good quality power supply to all consumers without any interruption. In order to conquer the above said issue PID controller is implemented as a secondary controller and discussed in the section "Control Strategy".

3 Control Strategy

In the current work, the PID controller is considered as a secondary controller for regulating power system operation during sudden load disturbance period. The transfer function of PID controller is given by:

$$G_{PID}(S) = K_p E(S) + \frac{K_i}{T_i S} E(S) + K_d E(S) \tag{1}$$

where, K_p, K_i and K_d indicate the Proportional, Integral and Derivative controller gain values; respectively, E(S) denotes the error signal. The structure of proposed PID controller is shown in Fig. 2. The input of controller is Area Control Error (ACE) and output of controller is the control signal ($delP_{ref}$), which is given to the power system as a reference signal. In the current work, three PID controllers are designed and implemented into three area interconnected reheat thermal power system. The PID controller consists of three basic terms such as integral, proportional and derivative terms, where, the K_p is proportional gain, K_i is integral gain, K_d is derivative gain, ACE is area control error and $delP_{ref}$ is control signal or reference signal.

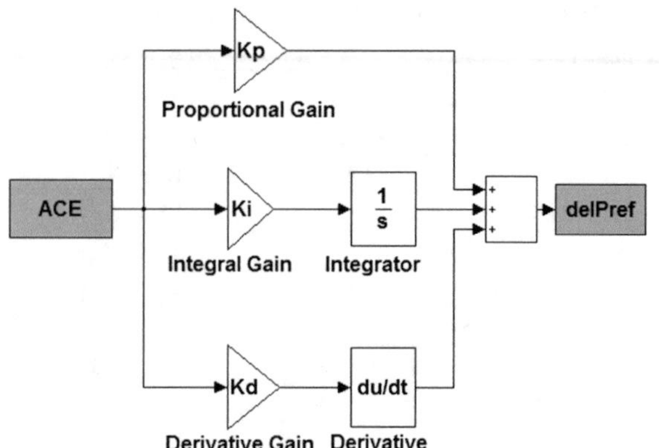

Fig. 2 Structure of PID controller

The input of controller ACE is given by:

$$ACE_i = B_i \Delta f_i + \Delta P_{tiei,j} \tag{2}$$

The output of controller dlP_{ref} is given by:

$$\Delta P_{refi} = -K_i \cdot ACE_i - \frac{K_P}{T_i} \int ACE_i - K_d T_d \frac{d}{dt} ACE_i \tag{3}$$

where, Δf_i is frequency deviation, ΔP_{tie} is tie line power deviation, B_i is frequency bias constant, ΔP_{refi} is reference signal given the *i*th area of the power of power system or control signal generated by PID controller, ACE is the value of area control error, K_p, K_i and K_d indicate the proportional, integral and derivative controller gain values; respective lyand T_i and T_d indicate the integration time and derivative time; respectively.

The performance and behavior of the proposed controller are mainly depending on the proper selection of controller parameters and choosing better optimal values. By using hit and trial method for selection of controller gain value is very tough and complex because it takes more time for selection of optimal value. The main aim of proposed research work is to implement evolutionary algorithm based optimization techniques. In the current work, the QIEA, QIGA and GA algorithms are implemented for optimizing controller gain values.

4 Quantum Inspired Algorithms

Quantum-Inspired Algorithms are a new class of hybrid algorithms, which are inspired by Quantum Computing. This hybrid concept generates many optimization algorithms, such as Quantum Inspired Evolutionary Algorithm (QIEA) and Quantum Inspired Genetic Algorithm (QIGA), which are actually a combined concept of some significant features from both Quantum Computing (QC) and Evolutionary Computing (EC). The classical approach of Evolutionary Computing includes a class of some optimization algorithms which are known as Evolutionary Algorithms (EA).

Evolutionary Algorithms (EAs) are basically a stochastic search and optimization method based on the principles of natural biological evolution. Compared to the traditional or conventional optimization methods (such as calculus-based), the EAs are considered more robust, can give global optimum solutions overcoming the problem of stuck in local optima and may be applied generally without domain-specific heuristics. The EA implements three major mechanisms from biological evolution, which are reproduction, mutation and natural selection. The EAs are characterized by the encoding of the solutions, and evaluation function for computing fitness of the representative solutions and other parameters as the population size, characteristic operators, and selection schemes. The three main-stream methods of evolutionary computation, which have been established before many years back are:

(i) Genetic Algorithm (GA) developed by [37–39],
(ii) Evolutionary Programming (EP) developed by [40], and
(iii) Evolution Strategies (ES) developed by [41, 42].

The main principles on which EAs operate are population of potential solutions, and the principle of survival of the fittest. EAs apply these to produce successively better approximate solutions.

Quantum Computing [43] is a branch of computer science concerning applications of unique quantum mechanical effects to solving computational problems. The two types of computations i.e., EC and QC are implemented by the algorithms named as Evolutionary Algorithm (EA) and Quantum Algorithm (QA); respectively. The Quantum Algorithms are mainly intended for quantum mechanical computers, but not for classical computers. In order to utilize the advantages of Quantum Computing in classical computers, researchers introduced the hybrid concept of computing algorithms, such as the QIEA and QIGA. For various optimization problems researchers use this new era of computation and it gives more efficient solution than any conventional evolutionary algorithms can give ever.

4.1 Evolutionary Computing

The population-based methods adopted concepts from biology and implemented them into automatic machine oriented computation. These set of methods are known as Evolutionary Computation (EC). It merges the concepts from population biology, genetics, and evolution. Evolutionary algorithms [39] which are heuristic search algorithms, do not always guarantee to provide the exact optimal solutions, but definitely they can find better optimal solutions within the less amount of time.

Evolutionary Algorithms, which are adaptive optimization algorithms and follow the process of natural selection and genetics [44], have been applied to a variety of optimization problems. A generic evolutionary algorithm consists of some operations, namely: initialization, selection, reproduction and replacement. Initialization means the seeding of initial population by using some suitable encoding scheme. The selection operator selects the individuals randomly or according to their fitness. In case of replacement, older individuals are replaced by new offspring's.

The basic generic evolutionary algorithm first constructs an initial population of some individuals, then iterates through three procedures. First, it assesses the fitness of all the individuals initialized. Second, it breeds a new population of offspring, based on the fitness information. Third, it recombines the fittest parents and children to form new generation population, and in this fashion the loop continues.

Algorithm of the generic classical Evolutionary Algorithm

```
begin
    Initialize random population;
    Evaluate the population;
    Select fittest solution and store it;
    Generation = 0;
        While () //termination criterion is not satisfied

            Evaluate the population;
            Select fitter chromosomes by survivor selection procedure;
            Upgrade the population with upgrade operators;
            Recombine the population with fittest solutions;
            Generation = Generation + 1;
        end
end
```

The first step in the above algorithm is to select a proper encoding technique and the population size. In order to get better optimal solution appropriate population size should be chosen, as large population increases the diversity amount.

4.2 Genetic Algorithm

Genetic algorithms are adaptive optimization algorithms that follow the process of natural selection and genetics [45]. They have been also applied to a variety of optimization problems, such as travelling salesman problem and knapsack problem. A generic GA consists of some operations, namely initialization, selection, reproduction and replacement which are same as EA in operation. Crossover and mutation operations are used to maintain balance between exploitation and exploration. Upon replacement, older individuals are replaced by new offspring's just like as in EA. The pseudo-code of the generic GA is as follows.

Algorithm: **Genetic Algorithm**

---

begin
  Assume initially random population;
  Assess the population;
  Generation = 0;
While ()
  Chose the superior chromosomes using the reproduction process;
  Crossover ;
  Select the best chromosomes ;
  Mutation ;
  Evaluate;
  Generation = Generation + 1;
end
end

In this pseudo-code, the optimal solution is obtained by using the proper population size. Typically, the the diversity increases by increasing the population size.

4.3 Quantum Computing

Generally, in computer chips, the information bits are physically stored as low and high voltages on the wires (i.e., 0 and 1 in binary). Inside Quantum Computing machines, the computations are performed based on the laws of quantum mechanics that can be described as the behaviour of the particle at the sub-atomic level. The single electron in this atom can either be in a ground state (the lowest energy level) or in an excited state (the high energy level). These two states can be used to encode the binary data bit values 0 and 1 (for low and high energy level respectively). In quantum physics the two basis states of electron are denoted by Dirac notations like, $|0\rangle$ (for ground state) and $|1\rangle$ (for excited state). Unlike classical physics, in quantum physics there comes one new concept of linear superposition principle along with these two basic states. The fundamental concept of superposition principle says that if a Quantum System can be present in one of the two basis states, $|0\rangle$ or $|1\rangle$, then it can also be in any linear combination of these two states, i.e., $\alpha_0|0\rangle\alpha_1|1\rangle$. This is called the superposition state, where the coefficient α_0 and α_1 are called the amplitudes of state $|0\rangle$ and $|1\rangle$; respectively, which give the probabilistic measure of the occurrence of state $|0\rangle$ and state $|1\rangle$; respectively.

4.4 Qubit Representation

The superposition $\alpha_0|0\rangle\alpha_1|1\rangle$ is the basic or the smallest unit of encoded information in quantum computers or quantum systems, which is called a Qubit. The representation of Qubit is given as follows:

$$|\psi\rangle = \alpha_0|0\rangle + \alpha_1|1\rangle \tag{4}$$

According to the superposition principle, the amplitudes α's are arbitrary complex numbers and the squares of their norms added up to 1, which is expressed by:

$$|\alpha_0|^2 + |\alpha_1|^2 = 1 \tag{5}$$

α's are probabilistic amplitude of the qubit that may exist in a state '0' or in state '1' and it satisfies the normalization condition. In quantum computer, the qubits are quaternary in nature. It has three states actually unlike the binary bits in classical computers. The states are state 0 or state 1 and the linear superposition of these two basic states. In order to measure the state of a qubit which is in a superposition state at present, the measurement process forces the qubit to change its present state from the superposition to any of the two basic states (i.e., state 0 or 1). Upon measurement the qubit changes its state towards state 0 or state 1 depending on the value of the probabilistic amplitude of it. A qubit is actually a vector in a vector space which is generally two dimensional and complex in nature. An example of a qubit state could be given as follows:

$$|\psi\rangle = \frac{1}{\sqrt{2}}|0\rangle + \frac{1}{\sqrt{2}}|1\rangle \tag{6}$$

The above example also satisfies the normalization condition (Eq. 5), in addition multiple qubits can be handled as in [46]. Suppose two qubits, which can be in one of the four computational basis states, 00, 01, 10, or 11. Similarly a pair of qubits can also exist in superposition of these four states as follows:

$$|\psi\rangle = \alpha_{00}|00\rangle + \alpha_{01}|01\rangle + \alpha_{10}|10\rangle + \alpha_{11}|11\rangle \tag{7}$$

The Normalization Condition can be expressed as:

$$|\alpha_{00}|^2 + |\alpha_{01}|^2 + |\alpha_{10}|^2 + |\alpha_{11}|^2 = 1 \tag{8}$$

Thus a qubit string (quantum chromosome), which is set of some individual qubit (quantum gene) can be represented as follows:

$$q = \begin{bmatrix} \alpha_{00} & \alpha_{01} \cdots & \alpha_{0m} \\ \alpha_{10} & \alpha_{11} \cdots & \alpha_{1m} \end{bmatrix} \quad (9)$$

where, an individual qubit is: $\begin{bmatrix} \alpha_0 \\ \alpha_1 \end{bmatrix}$; and m denotes the length of the quantum chromosome Q.

Quantum logic gates are used in quantum mechanics, which are basically quantum circuits. Quantum circuit is a quantum computation model, where computation is performed through a sequence of quantum gates. Quantum Gates are reversible i.e., like the inverter (NOT) gate, the output of which can be undone that means the input data can be obtained from the output one. Quantum gates are used for updating the qubits so that better optimal solution can be obtained and also the updated qubits can give better convergence rate if updating operation is performed properly. Q-gates mostly operate on one or two qubits, such as the classical logic gates. Thus, the gates are represented by 2 × 2 or 4 × 4 unitary matrices. This property ensures that any quantum gate U is always logically reversible.

4.5 Quantum Operator

For updating the qubit individual a quantum update operator is used, which in particular is a quantum rotational gate. A rotation gate $\mathrm{U}(\Delta\theta_i)$ is used to update the qubit individual as variation operator. The coefficients $(\alpha_{0i}, \alpha_{1i})$ of the ith qubit is updated as follows:

$$\begin{bmatrix} \alpha'_{0i} \\ \alpha'_{1i} \end{bmatrix} = \mathrm{U}(\Delta\theta_i) * \begin{bmatrix} \alpha_{0i} \\ \alpha_{1i} \end{bmatrix} \quad (10)$$

i.e., $\begin{bmatrix} \alpha'_{0i} \\ \alpha'_{1i} \end{bmatrix} = \begin{bmatrix} cos(\Delta\theta_i) & -sin(\Delta\theta_i) \\ sin(\Delta\theta_i) & cos(\Delta\theta_i) \end{bmatrix} * \begin{bmatrix} \alpha_{0i} \\ \alpha_{1i} \end{bmatrix}$, where, the rotation gate is given by: $\mathrm{U}(\Delta\theta_i) = \begin{bmatrix} cos(\Delta\theta_i) & -sin(\Delta\theta_i) \\ sin(\Delta\theta_i) & cos(\Delta\theta_i) \end{bmatrix}$

The algorithm of the qubit update procedure is given here in the following:

Procedure Update (q)

Begin
i← 0
while(i < m) **do**
Begin
i← i+1
Determine($\Delta\theta_i$) with the lookup table
Obtain $\left(\alpha'_{0i}\ ,\ \alpha'_{1i}\right)$ from the following:
if(q is located in the first/third quadrant)
$$\begin{bmatrix}\alpha'_{0i} & \alpha'_{1i}\end{bmatrix}^T = U(\Delta\theta_i)\begin{bmatrix}\alpha_{0i} & \alpha_{1i}\end{bmatrix}^T$$
else
$$\begin{bmatrix}\alpha'_{0i} & \alpha'_{1i}\end{bmatrix}^T = \mathrm{U}\,(-\Delta\theta_i)\begin{bmatrix}\alpha_{0i} & \alpha_{1i}\end{bmatrix}^T$$
end
q ←q'
end
end

In the above algorithm qubit individuals (q) in quantum chromosome are updated by applying Q-gates. By the update operation the updated qubit should satisfy the following normalization condition:

$$\left|\alpha'_0\right|^2 + \left|\alpha'_1\right|^2 = 1, \tag{11}$$

where, α'_0 and α'_1 are the values of the updated qubit. The following rotation gate is used as a Q-gate in hybrid quantum computing, such as:

$$\mathrm{U}(\Delta\theta i) = \begin{bmatrix} cos(\Delta\theta_i) & -sin(\Delta\theta_i) \\ sin(\Delta\theta_i) & cos(\Delta\theta_i) \end{bmatrix} \tag{12}$$

where, $\Delta\theta i$, $(i = 1, 2, \ldots, m)$ is a rotation angle of each qubit towards either 0 or 1 state depending on its sign.

4.6 *Quantum Inspired Evolutionary Algorithm*

The first proposed evolutionary algorithm based on the concepts and principles of quantum computing was presented in 1996 [45]. Other early research examples of quantum-inspired evolutionary algorithms with binary quantum representation based on qubits are due to [47]. In the past few years, several other works on quantum evolutionary algorithms have been also proposed [48–53].The complete pseudo-code of generic QIEA has been presented here. Here, Q(t) denotes the *t*th generation of a quantum population.

In quantum chromosomes, the genes states are observed during the phenotype creation. The observed population matrix is then used to evaluate the fitness of an individual. Afterwards, the quantum gates are used to update the qubit individuals.

4.7 *Quantum Inspired Genetic Algorithm*

Quantum Inspired Genetic Algorithm (QIGA) is a hybrid optimization algorithm enriched with the efficient computing characteristics from two important field of computation GA and Quantum Computing (QC). The bio-quantum-inspired optimization algorithm QIGA adopted the features from QC, such as qubit representation, superposition of state, quantum state measurement and the qubit updating.

Simultaneously, the GA adopted the features, such as initialization of population, selection, crossover and the mutation. QIGA uses the features, such as updating qubit, selection, crossover and mutation as functional operators of the algorithm to resample the population of solution candidates. The QIGA uses binary quantum chromosomes for representing solutions, which is encoded as:

$$q = \begin{bmatrix} \alpha_{00} & \alpha_{01} \cdots & \alpha_{0m} \\ \alpha_{10} & \alpha_{11} \cdots & \alpha_{1m} \end{bmatrix}$$

where, each column represents a binary quantum gene $|\Psi 1\rangle, \ldots, |\Psi m\rangle$. Hence, a state of the whole quantum population $Q = \{q_1, q_2, \ldots, q_N\}$ can be simply illustrated by a matrix of vectors, where $q_1, q_2, \ldots, q_N$ are binary quantum chromosomes. The complete pseudo-code of QIGA is as follows.

***Algorithm*: Quantum-Inspired Genetic Algorithm (QIGA)**

```
Begin
   t←0
Initialize Q(0)
...
   While not termination-criterion
      t←t+1

   Evaluate Q(t)
   Perform genetic operations on Q(t)

   end
end
```

In the preceding algorithm, the genes of all individuals in the quantum population Q(0) are initialized with linear superposition $\left(\sqrt{2}/2|0\rangle + \sqrt{2}/2|1\rangle\right)$, which results in sampling the whole search space with equal probability. During phenotype creation, states of all genes in quantum chromosomes are observed. Thus, the search space is sampled with respect to the probability distribution in quantum chromosomes as encoded. The evaluation of an individual's fitness is based on the observed population matrix P(t). The genetic operators applied in the algorithm are based on the quantum rotation gates [48]. Its rotate state vectors in the quantum gene state space [48]. In the above algorithm, the qubit individuals in a population of individuals Q (t) are updated by applying some quantum gates (Q-gates).

Recently, the fields of quantum-inspired genetic algorithms are one of the fastest growing areas of research work. Numerous extensions of the QIGA have been proposed in several fields and still growing on. Early research examples of quantum-inspired genetic algorithms with binary quantum representation based on qubits are presented in [45, 52–55].

5 The Proposed QIEA Based PID Controller Optimization Algorithm

For superior controlled performance during sudden load disturbance condition, the controller is designed and implemented. In addition, the controller performance is mainly depends on the proper selection of the controller gain values. In the current work, the PID controller value is optimized by using QIEA with considering Integral Time Absolute Error objective function. 1% SLP is applied into the area 1 of investigated power system. The proposed algorithm flow chart is shown in Fig. 3.

6 Result and Discussion

6.1 Design of GIEA Optimized PID Controller

In order to achieve superior response in any type of closed loop control response, a proper design of controller in practical situation is required. The response of controller is mainly depends on the proper selection of controller parameters. The controller parameters are obtained by proper selection of the objective function and

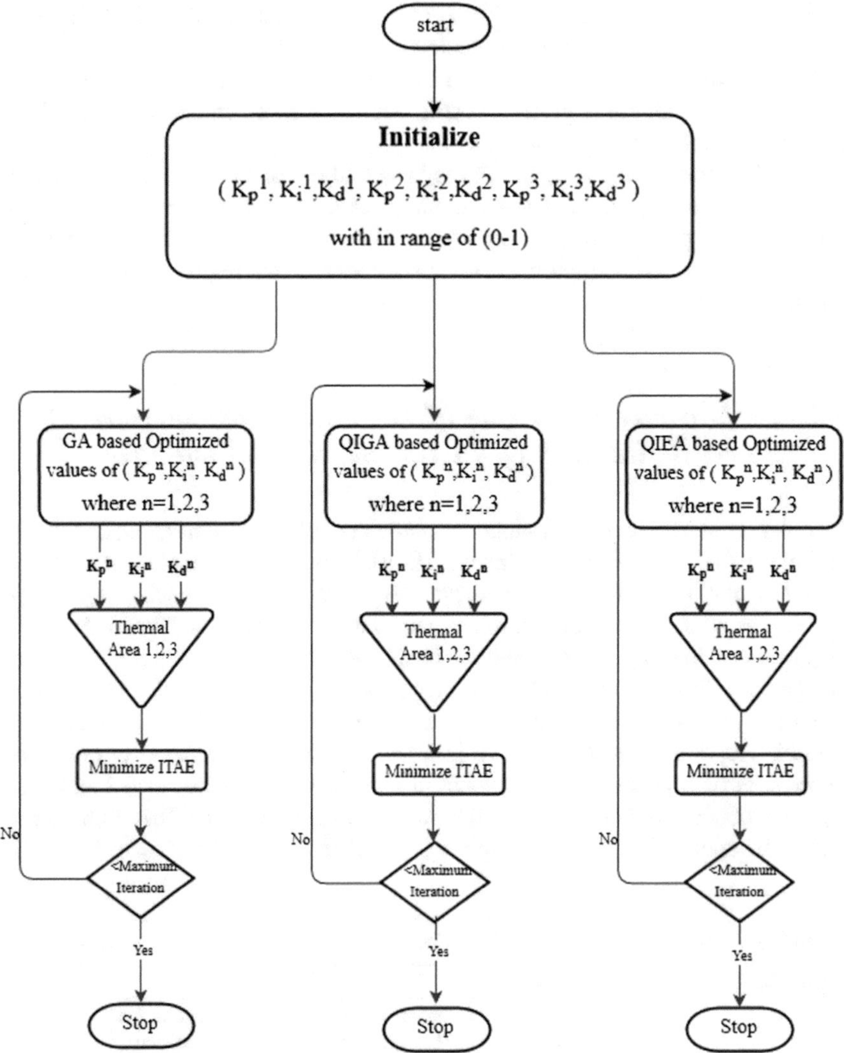

Fig. 3 Combined flow chart of proposed algorithm

the optimization technique for tuning the controller parameters. In the current work, the ITAE objective function is considered for the optimization PID controller parameters of Fig. 1.

$$J = ITAE = \int_{0}^{t} t \cdot |ACE_i| dt \tag{13}$$

The parameters gain values of PID controller optimized by using GA, QIGA and QIEA optimization algorithms are depicted in Table 3.

The execution time for optimizing PID controller gain values by using different optimization technique is given in Table 3 and bar plot comparisons also clearly shown in Fig. 4 as reported in Table 4.

Based on the numerical values in the table and bar plot comparison, it is clearly evident that the proposed QIEA technique takes lesser computing time to compute PID controller gain values compared to the QIGA and GA optimization technique. Furthermore, the GA computing time is shorter than that required with the QIGA optimization technique.

6.2 Performance Evaluation of Three Are Interconnected Power System with QIEA Optimized PID Controller

In order to examine the dynamic performance of evolutionary algorithm, the optimized PID controller performance 1% (0.01 pu) SLP is applied into thermal area 1 of three area interconnected thermal power system. The GA, QIGA and QIEA algorithm optimized controller gain values are tabulated in Table 3. The response obtained by using proposed QIEA optimized controller performance is compared with GA and QIGA controller based performance in the same investigated power system. The response comparisons clearly evident that the proposed algorithm optimized controller give fast settled response compared to other optimized controller performance as shown in Figs. 5, 6, 7, 8, 9, 10, 11, 12 and 13. In the figures, the solid lines show the response of proposed QIEA optimization technique optimized controller response, dotted lines give the response of QIGA

Table 3 PID controller gain values optimized by using different optimization algorithms

Controller gain		Technique		
		Genetic algorithm (GA)	Quantum inspired genetic algorithm (QIGA)	Quantum inspired evolutionary algorithm (QIEA)
Proportional gain (K_p)	K_{p1}	0.82362	0.90476	0.8739
	K_{p2}	0.80222	0.77778	0.53666
	K_{p3}	0.87107	0.85714	0.47116
Integral gain (K_i)	K_{i1}	0.96999	0.99999	0.7957
	K_{i2}	0.7065	0.53968	0.88856
	K_{i3}	0.61855	0.74603	0.79277
Derivative gain (K_d)	K_{d1}	0.22304	0.19048	0.18182
	K_{d2}	0.32176	0.31746	0.89443
	K_{d3}	0.10194	0.93651	0.35386

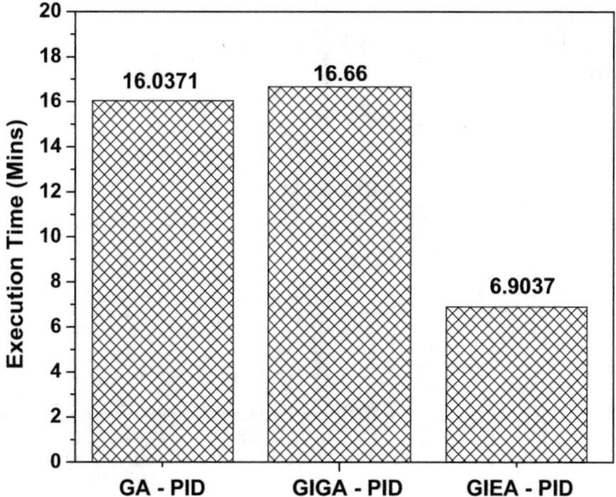

Fig. 4 Bar plot comparisons of execution time for different optimization algorithm

Table 4 Execution time for different optimization algorithm

Technique	GA–PID controller	QIGA–PID controller	QIEA–PID controller
Execution time (min)	16.0371	16.66	6.9037

optimized controller performance and dashed lines show the response of GA optimized controller response.

The numerical values of settling time for frequency deviation, tie line power deviation and area control error are clearly depicted in Tables 5, 6 and 7; respectively.

Based on the numerical values of settling time in Tables 5, 6 and 7, the bar plot is drawn. The bar plot compares the settling time in frequency and tie line power deviations, area control error as shown in Figs. 14, 15 and 16. It is clearly evident that the proposed QIEA optimized PID controller based controller response settled faster than the other optimization technique optimized controller response in the same investigated power system during sudden load demand period.

The preceding results depicts that the proposed QIEA technique optimized PID controller response provides fast settled response compared to the other optimized controller techniques in the same investigated power system during sudden load demand period.

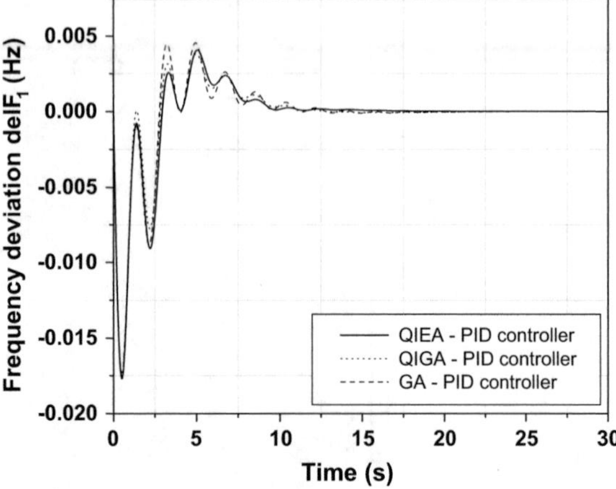

Fig. 5 Change in frequency in area 1 for 0.01 pu load disturbance in Thermal Area 1

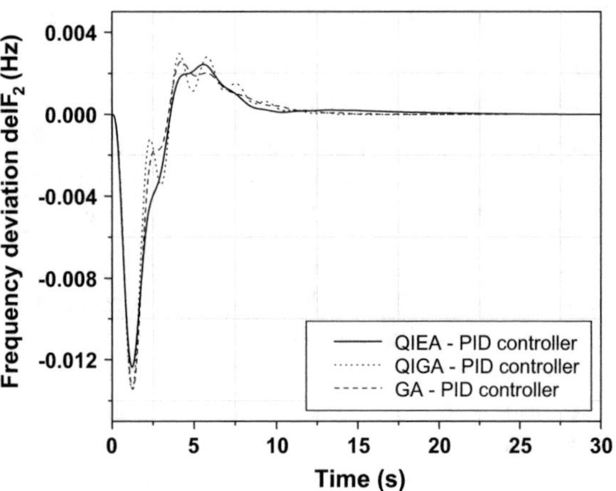

Fig. 6 Change in frequency in area 2 for 0.01 pu load disturbance in Thermal Area 1

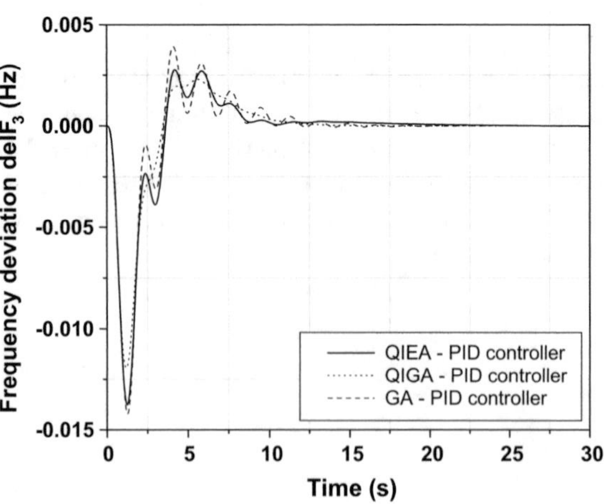

Fig. 7 Change in frequency in area 3 for 0.01 pu load disturbance in Thermal Area 1

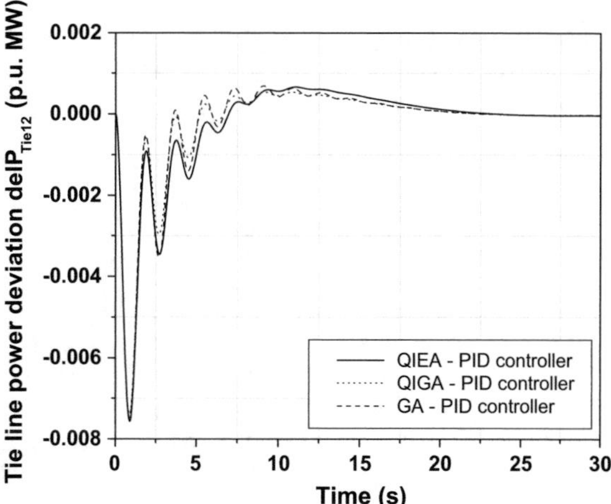

Fig. 8 Tie line power deviation in between area 1&2 for 0.01 pu load disturbance in Thermal Area 1

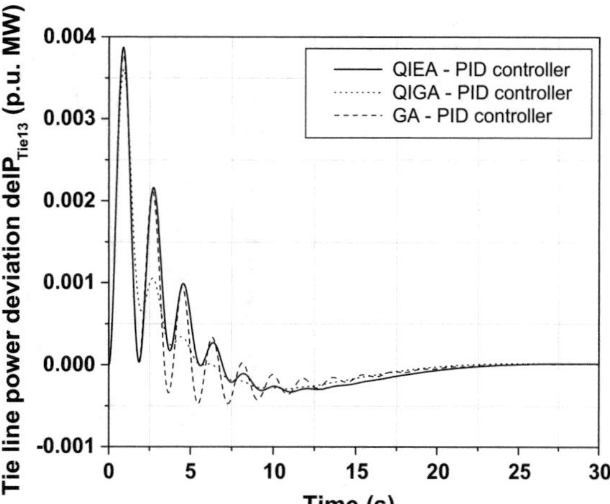

Fig. 9 Tie line power deviation in between area 1&3 for 0.01 pu load disturbance in Thermal Area 1

7 Conclusion

In the current work, the Quantum Inspired based optimization algorithms are used to optimize the PID controller gain values for solving load frequency control issue in multi-area interconnected thermal power system. The QIEA is proposed for optimizing the PID controller gain values by using ITAE objective function and applying 1% SLP in area 1. The supremacy of proposed controller performance is compared with quantum inspired genetic algorithm and genetic algorithm optimized controller performance in the same investigated power system. The computation

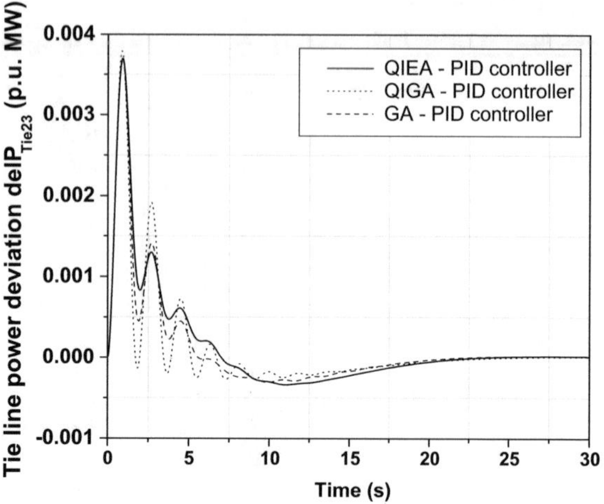

Fig. 10 Tie line power deviation in between area 2&3 for 0.01 pu load disturbance in Thermal Area 1

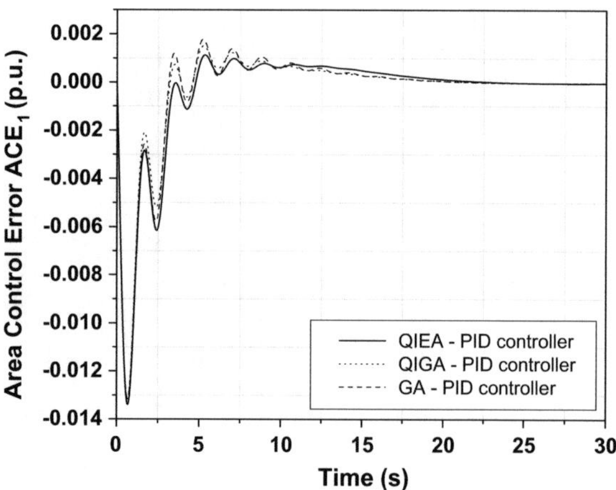

Fig. 11 Change in area control error in area 1 for 0.01 pu load disturbance in Thermal Area 1

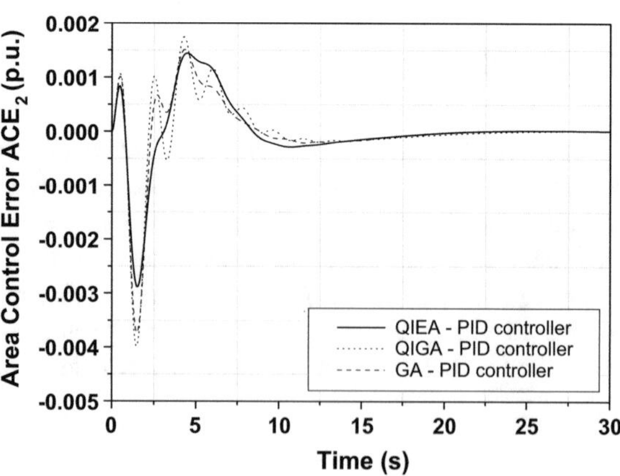

Fig. 12 Change in area control error in area 2 for 0.01 pu load disturbance in Thermal Area 1

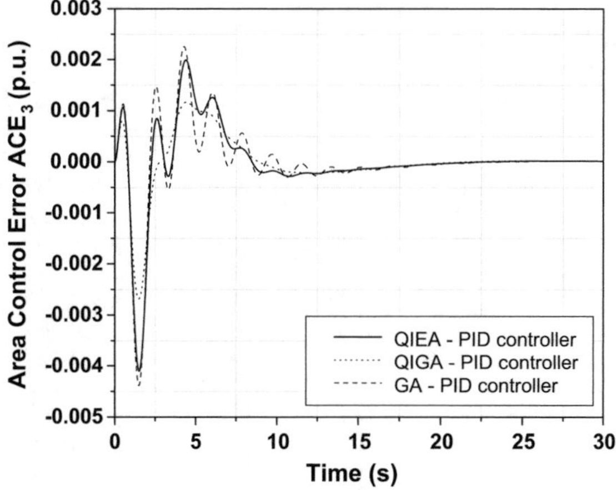

Fig. 13 Change in area control error in area 3 for 0.01 pu load disturbance in Thermal Area 1

Table 5 Settling time of frequency deviation for different optimization algorithm based controller

Response	ΔF_1 (Hz)			ΔF_2 (Hz)			ΔlF_3 (Hz)		
Technique	GA	QIGA	QIEA	GA	QIGA	QIEA	GA	QIGA	QIEA
Settling time (s)	20.05	18	17	20	17	16.5	22	17	16

Table 6 Settling time of tie line power deviation for different optimization algorithm based controller

Response	$\Delta Ptie_1$ (p.u. MW)			$\Delta Ptie_2$ (p.u. MW)			$\Delta Ptie_3$ (p.u. MW)		
Technique	GA	QIGA	QIEA	GA	QIGA	QIEA	GA	QIGA	QIEA
Settling time (s)	24	22	20.5	23	22.5	21	23	22	20

Table 7 Settling time of area control error deviation for different optimization algorithm based controller

Response	ACE_1 (p.u.)			ACE_2 (p.u.)			ACE_3 (p.u.)		
Technique	GA	QIGA	QIEA	GA	QIGA	QIEA	GA	QIGA	QIEA
Settling time (s)	24	23	21.5	26	25	23	28	23	22

time for computing controller gain value is clearly indicated that proposed QIEA controller takes lesser time for computation compare to QIGA and GA optimization technique. The performance of different optimization algorithm based controller performance clearly evident that proposed optimization technique based controller settled faster compared to other optimization technique based controller performance during 1% SLP in area 1. In addition, some response yield lesser peak over and under shoot frequency and tie line power deviation compared to QIGA and GA

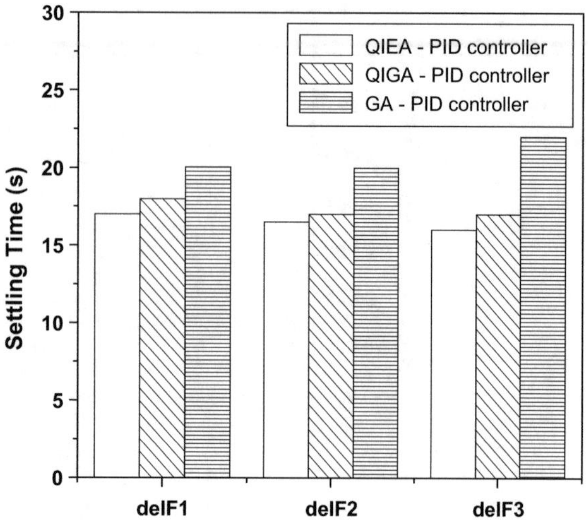

Fig. 14 Bar plot comparisons of settling time in frequency deviation with different algorithm optimized controller

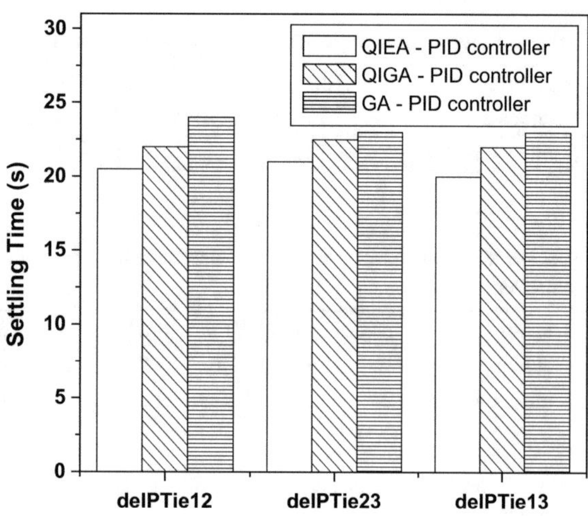

Fig. 15 Bar plot comparisons of settling time in tie line deviation with different algorithm optimized controller

based controller performance. Finally, it is proved that QIEA optimized PID controller maintain the system maintain and control the LFC issue in interconnected power system by maintain interconnected power system stability and delivering uninterrupted good quality power supply by keeping system parameter within the prescribed value during nominal and sudden load demand condition.

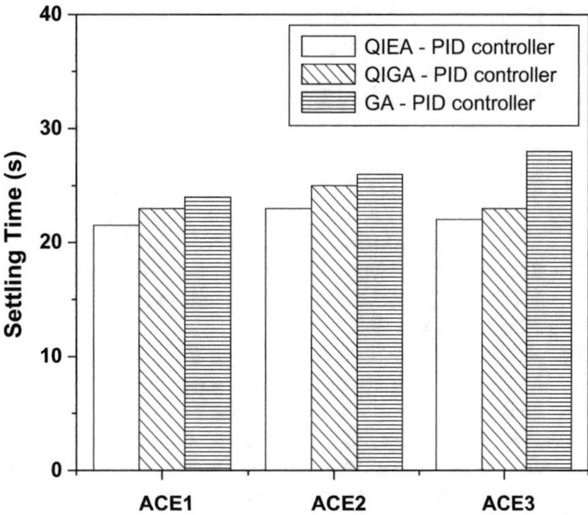

Fig. 16 Bar plot comparisons of settling time in area control error deviation with different algorithm optimized controller

References

1. Saikia, L.C., Sinha, N., Nanda, J.: Maiden application of bacterial foraging based fuzzy IDD controller in AGC of a multi-area hydrothermal system. Electr. Power Energy Syst. **45**, 98–106 (2013)
2. Debbarma, S., Saikia, L.S., Sinha, N.: AGC of a multi-area thermal system under deregulated environment using a non-integer controller. Electr. Power Syst. Res. **95**, 175–183 (2013)
3. Sahu, R.K., Panda, S., Rout, U.K.: DE optimized parallel 2-DOF PID controller for load frequency control of power system with governor dead-band nonlinearity. Electr. Power Energy Syst. **49**, 19–33 (2013)
4. Shabani, H., Vahidi, B., Ebrahimpour, M.: A robust PID controller based on Imperialist competitive algorithm for load frequency control of power systems. ISA Trans. **52**, 88–95 (2013)
5. Ramesh Kumar, S., Ganapathy, S.: Cuckoo Search optimization algorithm based load frequency control of interconnected power systems with GDB nonlinearity and SMES units. Int. J. Eng. Innovations **2**(12), 23–28 (2013)
6. Subha, S.: Load frequency control with fuzzy logic controller considering governor dead band and generation rate constraint non-linearities. World Appl. Sci. J. **29**(8), 1059–1066 (2014)
7. Nanda, J., Sakkaram, J.S.: Automatic generation control with fuzzy logic controller considering generation rate constraint. In: Proceedings of the 6th International Conference on Advances in Power System Control, Operation and Management, APSCOM 2003, Hong Kong, pp. 770–775 (2003)
8. Swain, A.K., Mohanty, A.K.: Adaptive load frequency control of an interconnected hydro thermal system considering generation rate constraint. J. Inst. Eng. (India) **76**, 109–114 (1995)
9. Parida, M., Nandha, J.: Automatic generation control of a hydro-thermal system in deregulated environment. In: Electrical Machines and Systems, 2005. ICEMS 2005. Proceedings of the Eighth International Conference, vol. 2, pp. 942–947 (2005)
10. Demiroren, A., Sengor, N.S., Lale Zeynelghi, H.: Automatic generation control by using ANN technique. Electr. Power Compon. Syst. **29**, 883–896 (2001)
11. Ngamroo, I.: Robust decentralized frequency stabilizers design for SMES taking into consideration system uncertainties. Electr. Power Energy Syst. **74**, 281–292 (2005)

12. Concordia, C., Kirchmayer, L.K., Szymanski, E.A.: Effect of speed-governor dead band on tie-line power and frequency control performance. IEEE Trans. Power Apparatus Syst. **76**(3), 429–434 (1957)
13. Shayeghi, H., Shayanfor, H.A.: Application of ANN technique based μ-synthesis to load frequency control of interconnected power and energy systems. Electr. Power Energy Syst. **28**, 503–511 (2006)
14. Tripathy, S.C., Balasubramaniam, R., Chandramohan Nair, P.S.: Effect of superconducting magnetic energy storage on automatic generation control considering governor dead and boiler dynamics. IEEE Trans. Power Syst. **7**(3), 1266–1273 (1992)
15. Tripathy, S.C., Hope, G.S., Malik, O.P.: Optimization of load-frequency control parameters for power systems with reheat steam turbines and governor dead band nonlinearity. IEE Proc. **129**(1), 10–16 (1982)
16. Nanda, J., Kothari, M.L., Satsangi, P.S.: Automatic generation control of an interconnected hydrothermal system in continuous and discrete modes considering generation rate constraints. IEE Proc. **130**(1), 17–27 (1983)
17. Hari, L., Kothari, M.L., Nanda, J.: Optimum selection of speed regulation parameters for automatic generation control in discrete mode considering generation rate constrains. IEE Proc. C **138**(4), 401–406 (1991)
18. Lu, C.-F., Liu, C.-C., Wu, C.-J.: Effect of battery energy storage system on load frequency control considering governor dead band and generation rate constraint. IEEE Trans. Energy Convers. **10**(3), 555–561 (1995)
19. Pan, C.T., Liaw, C.M.: An adaptive controller for power system load-frequency control. IEEE Trans. Power Syst. **4**(1), 122–128 (1989)
20. Nanda, J., Mishra, S., Sailkia, L.C.: Maiden application of bacterial foraging based optimization technique in multi area automatic generation control. IEEE Trans. Power Syst. **24**(2), 602–609 (2009)
21. Nanda, J., Mangla, A., Suri, S.: Some new findings on automatic generation control of an interconnected hydrothermal system with conventional controllers. IEEE Trans. Energy Convers. **21**(1), 187–194 (2006)
22. Anand, B., Jeyakumar, E.: Fuzzy Logic load frequency Control of hydro-Thermal system with non-Linearities. Int. J. Electr. Power Eng. **3**(2), 112–118 (2009)
23. Nanda, J., Saikia, L.C.: Comparison of performances of several types of classical controller in automatic generation control for an interconnected multi-area thermal system. In: Proceedings of 2008 Australasian Universities Power Engineering Conference (AUPEC'08), pp, 1–6 (2008)
24. Tripathy, S.C., Bhatti, T.S., Jha, C.S., Malik, O.P., Hope, G.S.: Sampled data automatic generation control analysis with reheat steam turbines and governor dead-band effects. IEE Trans. Power Apparatus Syst. **103**(5), 1045–1051 (1984)
25. Saikia, L.C., Bharali, A., Diixit, O., Malakar, T., Sharma, B., Kouli, S.: Automatic generation control of multi-area hydro system using classical controllers. In: 1st International Conference on Power and Energy in NERIST (ICPEN), 28–29 December 2012, Nirjuli, pp, 1–6 (2012)
26. Chidambaram, I.A., Velusami, S.: Decentralize biased controllers for load frequency control of interconnected power systems considering governor dead band non-linearity. In: IEEE Indicon 2005, Chennai, India, December 11–13, pp. 521–525 (2005)
27. Chaine, S., Tripathy, M.: Design of an optimal SMES for automatic generation control of two-area thermal power system using cuckoo search algorithm. J. Electr. Syst. Inf. Technol. (In Press) (2015)
28. Pan, I., Das, S.: Fractional-order load frequency control of interconnected power systems using chaotic multi-objective optimization. Appl. Soft Comput. **29**, 328–344 (2015)
29. Dash, P., Nidulsinha, L.C.S.: Comparison of performance of several FACTS devices using Cuckoo search algorithm optimized 2DOF controllers in multi-area AGC. Electr. Power Energy Syst. **65**, 316–324 (2015)

30. Dash, P., Saikia, L.C., Sinha, N.: Automatic generation control of multi area thermal system using bat algorithm optimized PD-PID cascade controller. Electr. Power Energy Syst. **68**, 364–372 (2015)
31. Zare, K., Hagh, M.T., Morsali, J.: Effective oscillation damping of an interconnected multi-source power system with automatic generation control and TCSC. Electr. Power Energy Syst. **65**, 220–230 (2015)
32. Sahu, R.K., Panda, S., Padhan, S.: A hybrid firefly algorithm and pattern search technique for automatic generation control of multi area power systems. Electr. Power Energy Syst. **64**, 9–23 (2015)
33. SarojPadhan, R.K., Sahu, S.P.: Application of firefly algorithm for load frequency control of multi-area interconnected power system. Electr. Power Compon. Syst. **42**(13), 1419–1430 (2014)
34. Dash, P., Saikia, L.C., Sinha, N.: Comparison of performances of several Cuckoo search algorithm based 2DOF controllers in AGC of multi-area thermal system. Electr. Power Energy Syst. **55**, 429–436 (2014)
35. Padhan, S., Sahu, R.K., Panda, S.: Automatic generation control with thyristor controlled series compensator including superconducting magnetic energy storage units. Ain Shams Eng. J. **5**, 759–774 (2014)
36. Kothari, M.L., Satsangi, P., Nanda, J.: Sampled-data automatic generation control of interconnected reheat thermal systems considering generation rate constraint. IEEE Trans. Power Apparatus Syst. **100** (5), 2334–2342 (1981)
37. Fraser, A.S.: Simulation of genetic systems by automatic digital computers. Aust. J. Biol. Sci. **10**, 484–491 (1957)
38. Bremermann, H.J.: Optimization through evolution and recombination in self organizing systems, pp. 93–106. In: Yovits, M.C., Jacobi, G.T., Goldstine, G.D. (eds.). Spartan, Washington, DC (1962)
39. Holland, J.H.: Adaptation in Natural and Artificial Systems. University Michigan Press, Ann Arbor(1975)
40. Fogel, L.J., Owens, A.J., Walsh, M.J.: Artificial Intelligence Through Simulated Evolution. Wiley, New York (1966)
41. Rechenberg, I.: Evolutions strategie: Optimierung technischer Systemenach Prinzipien der biologishen Evolution' Stuttgart. From-mann-Holzbog, Germany (1973)
42. Schwefel, H.-P.: Evolution and Optimum Seeking. Wiley, New York (1975)
43. Nielsen, M.A., Chuang, I.L.: Quantum Computation and Quantum Information. Cambridge University Press (2000)
44. Goldberg, D.E.: Genetic Algorithms in Search, Optimization, and Machine Learning. Addison Wesley Longman, (1989). ISBN 0-201-15767-5
45. Narayanan, A., Moore, M.: Quantum—inspired genetic algorithms. In: Proceedings of 1996 IEEE International Conference on Evolutionary Computation. Piscataway, pp. 61–66. IEEE Press, NJ (1996)
46. Hey, T.: Quantum computing: an introduction. Comput. Control Eng. J **10**(3), 105–112 (1999). IEEE Press, Piscataway, NJ
47. da Cruz, A.V.A., Pacheco, M.A.C., Vellasco, M.B.R., Barbosa, C.R.H.: Cultural operators for a quantum-inspired evolutionary algorithm applied to numerical optimization problems. In: Mira, J., Alvarez, J.R. (eds.) IWINAC (2), vol. 3562 of Lecture Notes in Computer Science, pp. 1–10. Springer (2005)
48. Sharma, M., Tyagi, S.: Novel knowledge based selective tabu initialization in genetic algorithm. Int. J. Adv. Res. Comput. Sci. Softw. Eng. **3**(5) (2013)
49. Han, K.-H., Kim, J.-H.: Quantum-inspired evolutionary algorithm for a class of combinatorial optimization. IEEE Trans. Evol. Comput. **6**(6) (2002); Nielsen, M.A., Chuang, I.L.: Quantum Computation and Quantum Information. Cambridge University Press (2010)
50. da Cruz, A.V.A., Vellasco, M.M.B., Pacheco, M.A.C.: Quantum-Inspired Evolutionary Algorithm for Numerical Optimization, pp. 2630–2637 (2006)

51. Fan, K., Brabazon, A., O'Sullivan, C., O'Neil, M.: Quantum-inspired evolutionary algorithms for financial data analysis. In: Evo Workshops, pp. 133–143 (2008)
52. Zhang, G., Jin, W., Li, N.: An improved quantum genetic algorithm and its application. Lecture Notes in Computer Science, pp. 449–452 (2003)
53. Zhao, S., Xu, G., Tao, T., Liang, L.: Real-coded chaotic quantum-inspired genetic algorithm for training of fuzzy neural networks. Comput. Math. Appl. **57**(11–12), 2009–2015 (2009)
54. Han, K.H., Kim, J.H.: Genetic quantum algorithm and its application to combinatorial optimization problem. In: Proceedings of the 2000 Congress on Evolutionary Computation, vol. 2, pp. 1354–1360. Citeseer (2000)
55. Nowotniak, R., Kucharski, J.: Building blocks propagation in quantum-inspired genetic algorithm. Sci. Bull. Acad. Sci. Technol. Automat. **14**, 795–810 (2010)
56. Naidu, K., Mokhlis, H., Bakar, A.H.A., Terzija, V., llias, H.A.: Application of firefly algorithm with online wavelet filter in automatic generation control of an interconnected power reheat thermal power systems. Electr. Power Energy Syst. **63**, 401–413 (2014)
57. Nanda, J., Kaul, B.L.: Automatic generation control of an interconnected power system. Proc. IEE **125**(5), 385–390 (1978)
58. Kothari, M.L., Nanda, J.: Application of optimal control strategy to automatic generation control of a hydrothermal system. IEE Proc. **135**(4), 268–274 (1988)
59. Demiroren, A., Zeynelgil, A.Z., Sengor, N.S.: The application of ANN technique to load frequency control for three-area power systems. In: IEEE Porto Power Tech Conference on 10–13th September, 2006, Portugal
60. Chidambaram, I.A., Paramasivam, B.: Genetic algorithm based decentralized controller for load-frequency control of interconnected power systems with RFB considering TCPS in the tie-line. Int. J. Electron. Eng. Res. **1**, 299–312 (2009)
61. Ebrahim, M.A., Mostafa, H.E., Gawish, S.A., Bendary, F.M.: Design of decentralized load frequency based-PID controller using stochastic particle swarm optimization technique. In: International Conference on Electric Power and Energy Conversion System, pp. 1–6 (2009)
62. Arivoli, A., Chidambaram, I.A.: Design of genetic algorithm (GA) based controller for load-frequency control of power systems interconnected with AC-DC tie-line. Int. J. Sci. Eng. Tech. **2**, 280–286 (2011)
63. Ali, E.S., Abd-Elazim, S.M.: Bacteria foraging optimization algorithm based load frequency controller for interconnected power system. Electr. Power Energy Syst. **33**, 633–638 (2011)
64. Gozde, H., Cengiz Taplamacioglu, M.: Automatic generation control application with craziness based particle swarm optimization in a thermal power system. Electr. Power Energy Syst. **33**, 8–16 (2011)
65. Gozde, H., Cengiz Taplamacioglu, M., Kocaarslan, I.: Comparative performance analysis of artificial bee colony algorithm in automatic generation control for interconnected reheat thermal power system. Electr. Power Energy Syst. **42**, 167–178 (2012)
66. Saikia, L.C., Sinha, N., Nanda, J.: Maiden application of bacterial foraging based fuzzy IDD controller in AGC of a multi-area hydrothermal system. Electr. Power Energy Syst. **45**, 98–106 (2013)
67. Omar, M., Solimn, M., Abdel Ghany, A.M., Bendary, F.: Optimal tuning of PID controllers for hydrothermal load frequency control using ant colony optimization. Int. J. Electr. Eng. Inform. **5**(3), 348–356 (2013)
68. Jagatheesan, K., Anand, B., Ebrahim, M.A.: Stochastic particle swarm optimization for tuning of PID controller in load frequency control of single area reheat thermal power system. Int. J. Electr. Power Eng. **8**(2), 33–40 (2014). ISSN: 1990-7958
69. Jagatheesan, K., Anand, B.: Automatic generation control of three area hydro-thermal power systems considering electric and mechanical governor with conventional and ant colony optimization technique. Adv. Nat. Appl. Sci. **8**(20), 25–33 (2014). ISSN: 1998-1090
70. Jagatheesan, K., Anand, B.: Performance analysis of double reheat turbine in multi -area AGC system using conventional and ant colony optimization technique. J. Electr. Electron. Eng. **15** (1), 1849–1854 (2015)

71. Jagatheesan, K., Anand, B., Dey, N.: Automatic generation control of thermal-thermal-hydro power systems with PID controller using ant colony optimization. Int. J. Serv. Sci. Manage. Eng. Technol. **6**(2), 18–34 (2015)
72. Jagatheesan, K., Anand, B., Dey, N., Ashour, A.S.: Artificial intelligence in performance analysis of load frequency control in thermal-wind-hydro power systems. Int. J. Adv. Comput. Sci. Appl. **6**(7), 203–212 (2015)
73. Francis, R., Chidambaram, I.A.: Optimized PI+load-frequency controller using BWNN approach for an interconnected reheat power system with RFB and hydrogen electrolyzer units. Electr. Power Energy Syst. **67**, 381–392 (2015)
74. Jagatheesan, K., Anand, B., Samanta, S., Dey, N., Ashour, A.S., Balas, V.E.: Particle swarm optimization based parameters optimization of PID controller for load frequency control of multi-area reheat thermal power systems. Int. J. Artif. Paradigm (Accepted for Publication) (2016)
75. Jagatheesan, K., Anand, B., Dey, N., Balas, V.E.: Load frequency control of hydro-hydro system with fuzzy logic controller considering non-linearity. In: World Conference on Soft Computing, Berkeley, May 22–25, 2016
76. Jagatheesan, K., Anand, B., Samanta, S., Dey, N., Santhi, V., Ashiur, A.S., Balas, V.E.: Application of flower pollination algorithm in load frequency control of multi-area interconnected power system with non-linearity. Neural Comput. Appl. (Accepted for Publication), 1–14 (2016)
77. Jagatheesan, K., Anand, B., Dey, N., Ashour, A.S.: Ant colony optimization algorithm based PID controller for LFC of single area power system with non-linearity and boiler dynamics. World J. Modeling Simul. **12**(1), 3–14 (2016)

Optimal Distributed Generation Allocation Using Quantum Inspired Particle Swarm Optimization

Morteza Nazari-Heris, Sajad Madadi, Mahmoud Pesaran Hajiabbas and Behnam Mohammadi-Ivatloo

Abstract Distributed Generation (DG), with respect to its ability in utilizing the alternative resources of energy, provides a promising future for power generation in electrical networks. Distributed generators contribution to the power systems includes improvement in energy efficiency and power quality to reliability and security. These benefits are only achievable with optimal allocation of distributed resources that considers the objective function, constraints, and employs a suitable optimization algorithm. In this chapter, a quantum inspired computational intelligence is exercised for the optimal allocation of distributed generators. The fact that most of power system optimization problems, when modelled accurately, are of non-convex and sometimes discrete nature has encouraged many researchers to develop optimization techniques to overcome such difficulties. The basic Particle Swarm Optimization (PSO) is one of the most favored optimization techniques with many attractive features. Early experiments of employing PSO in many applications in power systems have indicated its promising potential. Consequently, the more advanced alternatives of this algorithm such as Quantum behaved PSO (Q-PSO) may show the same or even better performance in power system problems. The aforementioned algorithm has already been employed for different optimization objectives in power systems such as: short-term non-convex economic scheduling, unit commitment problems, loss of power minimization, economic load dispatch, smart building energy management and power system operations. Nevertheless, the algorithm has never been used for optimal allocation of distributed generation units. In this chapter the above problem will be solved with a quantum behaved particle swarm optimization algorithm. The chapter will be started with an introduction to the optimal allocation of DG then the power system, including the DG units will be modeled. On the next step the Q-PSO will be adopted for the optimal allocation. Finally, the results and discussions will be presented.

M. Nazari-Heris · S. Madadi · M. Pesaran Hajiabbas (✉) · B. Mohammadi-Ivatloo
Smart Energy Systems Laboratory, Faculty of Electrical and Computer Engineering, University of Tabriz, Tabriz, Iran
e-mail: mpesaran@tabrizu.ac.ir

A.E. Hassanien et al. (eds.), *Quantum Computing: An Environment for Intelligent Large Scale Real Application*, Studies in Big Data 33,
https://doi.org/10.1007/978-3-319-63639-9_17

Keywords Particle swarm optimization (PSO) · Quantum inspired PSO · Optimization problem · Distributed generation (DG) allocation

1 Distributed Generation (DG) Allocation

Unlike the traditional generation, DG refers to a method in which a part of requested electric power is generated and delivered to customers with small generation units placed close to end users. It needs some changes in electric energy generation scheme. The DG is also addressed as dispersed generation, embedded generation or decentralized generation and as it most popular definition is an electrical source which is directly connected to the distribution network or placed on the customer site of meter [1].

Nowadays the technical developments are creating the opportunity of achieving enormous benefits from DGs in different field such as economical, technical and environmental [2–4]. The utilization of DGs in power systems improves the energy efficiency, power quality and security of the system. The integration of DGs in electrical networks is increased with respect to the advantages of DGs in terms of economic and security aspects [5]. Those advantages could be earned by optimal selection, sizing and placement of DGs with in a proper planning method. The optimal allocation of distributed resources aims to minimize the objective function considering equality and inequality constraints. The objective of optimal allocation of DGs in a power system is to select the proper size and site for the distributed generation units in the selected network.

There are technical and environmental restrictions in conventional power plants and stations expansion. Moreover unsecure fossil fuel market has led the electricity market toward new energy resources. Also, there are a number of incentives for encouraging network planners to use combined heat power (CHP) resources in distribution networks. Some of the issues which need to be considered for DG integration in distributed networks are: power losses, voltage control, reliability, stability, and fault level [6, 7]. It could be stated that DG installation in power networks changes the network characteristics [8, 9].

Distributed Generation Allocation (DGA) can be studied as a portion of Distributed Generation Planning (DGP). Since the same objectives, constraints and optimization approaches are common in either DGA or DGP, most of the studies, which focused on distributed generation planning and are related to this research, are reviewed in this section. There are a lot of methods have been employed in DGP according to the selected objectives and their respected operation constraints. Those methods can be categorized respecting to their approaches for optimization such as normal search methods, intelligent methods or fuzzy methods. A review and discussion on the objective functions and constraints, which are employed in DGP with their mathematical proper algorithms, can be summarized as follows.

Single or multi objective functions are considered to maximize the benefits of DG with respect to the equality and inequality constraints. Normally the real power loss is the base objective [10, 11]. Some other objectives may accompany this base objective: reactive power minimization [12], DG capacity maximization [13, 14] or maximizing social profits in a deregulated market [15, 16]. A variety of methods and models of objective selection have been used in DGP formulations such as: multi objective models, or comprehensive objective models [17]. There are also some studies on the influence of DG on supply system security and reliability which clarified that these parameters could be improved by the means of a proper DGP [18]. In [19, 20], the technical issues caused by the integration of variable DGs in distribution networks due to their time varying nature (e.g. wind turbines) are investigated using DGP approaches. A wide range of constraints has been selected, for optimal DGP fitness functions of either single and multi objective. These constraints can be categorized into two main classes: equality and inequality constraints. Equality constraints are normally referred to as conservation limits for power, while the most important inequality constraints can be listed as DG power limit, buses voltage limit, feeder thermal limit and transformer power limit. However inequality constraints may include voltage step [21], short circuit level and ratio (SCL and SCR) [22] and power exchange between areas [23].

Different studies are focused on utilization of optimization methods to solve the DG allocation problem. In [24], the authors integrated genetic algorithm (GA) and simulated annealing (SA) optimization technique to solve the optimal allocation of DGs in distribution networks. The authors studied the optimal size and location of DGs in the primary distribution network in [25], which aimed to minimize total power losses in the network. The authors compared several methods, including novel power loss sensitivity (NPLS), power stability index (PSI), and voltage stability index (VSI) for solving optimal DG allocation in [26]. The GA method is implemented in [27] in order to obtain optimal solution of the DG allocation problem in networks, which minimized the losses of the network and ensured an acceptable reliability level and voltage profile. Analytical and genetic algorithm methods are employed in [28] for minimizing the power losses of the system by optimal placement of DGs. In this reference, the active power of DGs, power factor and location of such units are considered. Cuckoo search algorithm (CSA) is applied in [29] for identifying the optimal size and location of DGs in the system, in which the search space is determined by the utilization of graph theory. Bacterial foraging optimization algorithm (BFOA) is employed in [30] to solve the optimal allocation of DGs considering minimization of power losses, reduction of operational costs and improvement of voltage stability.

The objective function of the DG allocation problem along with the electrical and operational constraints of the problem are provided in the following [31].

1.1 Common Objective for DGA

Most of the DGA studies have been done with the objective of real power loss minimization. Moreover, other objectives are also considered such as reactive power loss, voltage profile, the current reduction in weak lines, spinning reserve power and network MVA capacity. Normally the real power loss has been selected as a base objective index and other objectives have been used to form single or multi objective fitness functions for optimization. The most selected combinations are explained in the following sub-sections.

The selected objective function of the DG allocation problem is studied in this section. The objective function is based on power flow and voltage deviation calculation, which can be stated by the following equation:

$$Obj = w_P \times IP_L + w_Q \times IQ_L + w_V \times IV_{dev} \tag{1}$$

where, the indicators of the active and reactive power losses are demonstrated by IP_L and IQ_L. In addition, IV_{dev} shows the voltage deviation with respect to the original system. Weight factors (WF), which are between 0 and 1, are defined by w_P, w_Q, and w_V.

1.2 Operational and Electrical Constraints

The power flow constraints should be taken into account in the solution of optimal DG allocation problem, which can be written for bus i as:

$$P_i = V_i \sum_{j=1}^{N_{Bus}} Y_{ij} V_j cos(\delta_i - \delta_j - \gamma_{ij}) \tag{2}$$

$$Q_i = V_i \sum_{j=1}^{N_{Bus}} Y_{ij} V_j sin(\delta_i - \delta_j - \gamma_{ij}) \tag{3}$$

$$S_i = P_i + jQ_i \tag{4}$$

where, V_i is used to define the per unit magnitude of voltage with angle of δ_i at bus i. P_i and Q_i are the real and imaginary parts of injected power, respectively. In addition, the complex injected power is shown by S_i and the admittance of line ij is showm by Y_{ij}. The voltage of ith bus should be limited to the minimum and maximum values (V_i^{min}, V_i^{max}) by the following equations:

$$V_i^{min} \leq V_i \leq V_i^{max} \tag{5}$$

The upper bound of the generation capacity of the DG units should be considered as follows:

$$P_i^{DG} \leq P_{DG}^{max} \tag{6}$$

The obtained active and reactive power from the upstream network at bus i should be limited to the lower and upper bounds by the following equations:

$$P_{sub}^{min} \leq P_{sub} \leq P_{sub}^{max} \tag{7}$$

$$Q_{sub}^{min} \leq Q_{sub} \leq Q_{sub}^{max} \tag{8}$$

The power transferred between nodes i and j should be limited to the minimum and maximum line capacity as follows:

$$S_{ij}^{min} \leq S_{i,j} \leq S_{ij}^{max} \tag{9}$$

2 Particle Swarm Optimization (PSO)

The optimization techniques can be classified into two general categories to the mathematical and heuristic methods. The heuristic optimization techniques, which have experience-based characteristics, are defined as an efficient and quick method to provide optimal solutions [32]. Particle swarm optimization (PSO) is introduced as an effective tool for solving optimization problems with a series of equality and inequality constraints. PSO is a stochastic optimization technique, which uses a population of potential solutions to solve the problem. The population of PSO method is defined as swarm particles which are initialized by the random solutions.

Two main characteristics of each particle are the position and velocity, which are updated during the optimization process in order to obtain the optimal solution of the problem. The particles of PSO population are updated by employing optimization operators of the PSO. The PSO algorithm is based on shifting the particles toward the ones with a better solution [33]. The population is updated using the latest fitness values. The new velocity of each particle is altered based on its old value and the new position. The new velocity is assigned to each particle at the end of each iteration. As a result, the position of particles in new iteration is updated according to the present velocities of the particles and the its distance from the local and global best particles [34].

In the traditional PSO method, $X_i^k = (x_i1^k, x_i2^k, \ldots, x_iD^k)$ and $V_n^k = (v_i1^k, v_i2^k, \ldots, v_iD^k)$ are utilized to define the position and velocity vectors of the particles.

The position of each particle is updated considering the last velocity and best performance of that particle, as well as the best performance of the swarm overall, which can be formulated as:

$$V_i^{(k-1)} = w.V_i^k + r_1.c_1.(pbest_i^k - X_i^k) + r_2.c_2.(gbest^k - X_i^k) \tag{10}$$

$$x_i^{k+1} = x_i^k + v_i^{k+1} \tag{11}$$

where, the iteration counter of the PSO method is demonstrated by k. In addition, r_1 and r_2 are two random relaxation numbers, and the c_1 and c_2 are two learning positive constants and w is the inertia weight.

The PSO method is employed to solve a variety of power system problems in the area of scheduling problems including optimal scheduling of the hydrothermal systems [33], combined heat and power economic dispatch [35], optimal generation scheduling of hydro system [36], optimal generator maintenance scheduling [37], and optimal scheduling of micro-grids (MGs) in deregulated enviroments [38]. In addition, PSO algorithms is implemented on placement of power systems equipment in the network such as optimal placement of phasor measurement units (PMUs) in power networks [32], optimal capacitor placement [39], optimal placement of distributed generation in distribution systems [40], optimal switch placement [41] and optimal location of FACTS devices in power systems [42]. Moreover, optimal reactive power dispatch in power systems [43], and optimal sizing of hybrid energy systems [44] can be numerated as other applications of PSO method in the solution of power systems problems.

3 Quantum Particle Swarm Optimization (QPSO)

The PSO method has been altered in binary form to improve the algorithm performance in different applications. Mohan and Al-Kazemi have proposed to combine PSO with other nature inspired intelligence as well as quantum theory. Following variations have been suggested in [45]:

- Direct approach, in which the classical PSO algorithm is applied and the solutions are converted into bit strings using a hard decision decoding process.
- Bias vector approach, in which the velocitys update is randomly selected from the three parts in the right-hand side of (2), using probabilities depending on the value of the fitness function.
- Mixed search approach, where the particles are divided into multiple groups and each of them can dynamically adopt a local or a global version of PSO.

Furthermore, the quantum bit (Q-bit) is integrated to represent the linear superposition of binary solutions in a probabilistic manner [46, 47]. The results demonstrates quantum behaved PSO (QPSO) has faster and more efficient performance in comparison to the conventional PSO.

The proposed algorithm employs the quantum delta potential well approach to crawl around the previous best points [48] and integrates the mean best position to improve the particles full domain search ability [49]. The main difference between QPSO and the classic PSO is the iterative equation. In QPSO the velocity vectors for particles is not required and has fewer parameters to adjust, which make the implementation much easier [50].

QPSO is implemented in the variety of optimization problems in power systems due to its successful performance in solving the continues optimization problems. QPSO is employed in [51] for obtaining optimal generation scheduling of hydro and thermal units considering the minimization of operational cost and pollutant gas emissions. In [52], the authors introduced PSO based on quantum mechanics for solving Economic Dispatch (ED), in which the valve-point loading impact of the conventional thermal units is studied. The ED problem is solved in [53] by employing QPSO method considering ramp rate limitation of the generation units, prohibited operating zones (POZs), power transmission loss of the system, and quadratic objective functions. Additionally, QPSO has already been implemented for different optimization objectives in power systems such as unit commitment (UC) problem [20], electricity load forecasting [54, 55], smart building energy management [56], the loss of power minimization [57], and power quality monitor placement method [58].

The PSO algorithm is extended in [59] by using the quantum computing, which is introduced as quantum PSO (QPSO). QPSO differs from PSO in terms of iterative procedure and the particles velocity characteristics updating approach. In addition, the QPSO employs less number of adjusting parameters, which results in better performance of the optimization technique. QPSO does not distribute the particles randomly by using the wave function which is employed in order to determine the particle state. Moreover, the density function of the position of particle is presented by $|\varphi|^2$. In other words, each particle is defined by quantum bit and angle in QPSO. The analysis proves that the convergence of PSO method may be provided by the convergence of each particle to its local attractor [60], which is defined by $p_i = (p_{i,1}, p_{i,2}, \ldots, p_{i,D})$. $p_{i,j}$, which is the jth dimension of vector p_i, can be stated as follows:

$$p_{i,j}(t) = \varphi_j(t) \times pbest_{i,j}(t) + (1 - \varphi_j(t)) \times gbest_j(t) \tag{12}$$

where, $\varphi_j(t) = \frac{c_1 r_{1,j}}{c_1 r_{1,j} + c_2 r_{2,j}}$. consider that in PSO, the acceleration coefficients c_1 and c_2 are the same, $\varphi_j(t)$ will be a sequence of uniformly distributed random number between 0 and 1. In QPSO, it is supposed that the ith particle at jth population, moves in n-dimensional area with a δ potential well at $p_{i,j}(t)$. The particles will move according to the following equation, which is obtained by using Monte Carlo method [61]:

$$x_{i,j}(t+1) = p_{i,j}(t) + \alpha \times |C_j(t) - x_{i,j}(t)| \times ln(l/u) \qquad If \quad k \geq 0 \tag{13}$$

$$x_{i,j}(t+1) = p_{i,j}(t) - \alpha \times |C_j(t) - x_{i,j}(t)| \times ln(l/u) \qquad If \quad k \leq 0 \tag{14}$$

where, the generated random variables by a uniform probability distribution in range (0, 1) are demonstrated by u and k. In addition, α is contraction expansion coefficient,

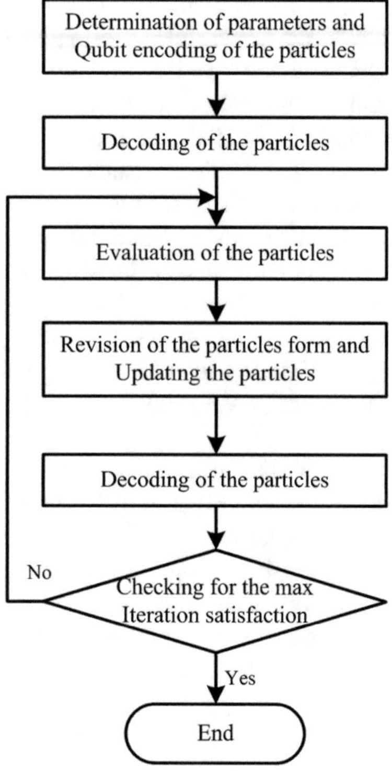

Fig. 1 The simplified flowchart of the QPSO method

which controls the convergence characteristics of the particle. $C_j(t)$ is defined as the mean best position and is formulated as follows:

$$C_j(t) = (1/m)\sum_{i=1}^{m} pbest_{i,j}(t) \tag{15}$$

The simplified flowchart of the QPSO method is demonstrated in Fig. 1.

4 Simulation Results

The QPSO method is employed in this chapter for obtaining optimal location of DGs in IEEE 30-bus test system in order to minimize the cost and power loss of the system. The IEEE 30-bus test system is demonstrated in Fig. 2. The QPSO method is employed to obtain the optimal site and the size of DGs in a multi objective aspect, where the WFs are employed in the objective function of the problem. The largest WF will be assigned to the smallest objective value, which ensures that the objective

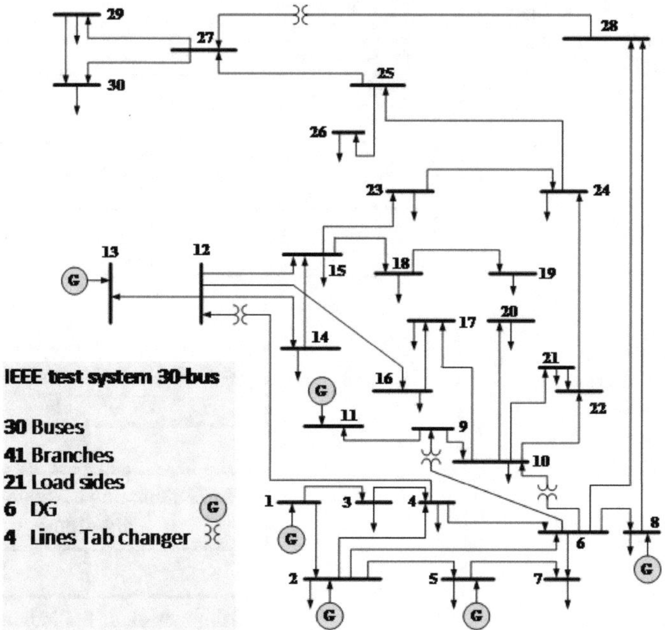

Fig. 2 IEEE 30-bus test system

values are equally effective on the objective function. The WFs for the test system are calculated by the following equations:

$$w_{p0} = \frac{1}{\sum_i^{DistBus}(IP_{Li})} \tag{16}$$

$$w_{q0} = \frac{1}{\sum_i^{DistBus}(IQ_{Li})} \tag{17}$$

$$w_{v0} = \frac{1}{\sum_i^{DistBus}(IV_{Li})} \tag{18}$$

$$W = w_{p0} + w_{q0} + w_{v0} \tag{19}$$

$$w_p = \frac{w_{p0}}{W}, w_q = \frac{w_{q0}}{W}, w_v = \frac{w_{v0}}{W}, \tag{20}$$

where, *DistBus* shows the candidate buses, which is able to install DG units.

The optimal location of the DGs and related sizes are obtained by employing the QPSO method in this chapter. The problem solution methodology is demonstrated in Fig. 2.

Table 1 The results of the DG optimization problem using conventional PSO

Iteration	62
Objective function	0.46468
Optimal buses	DG share
15	15%
19	15%
21	40%
22	10%
30	20%

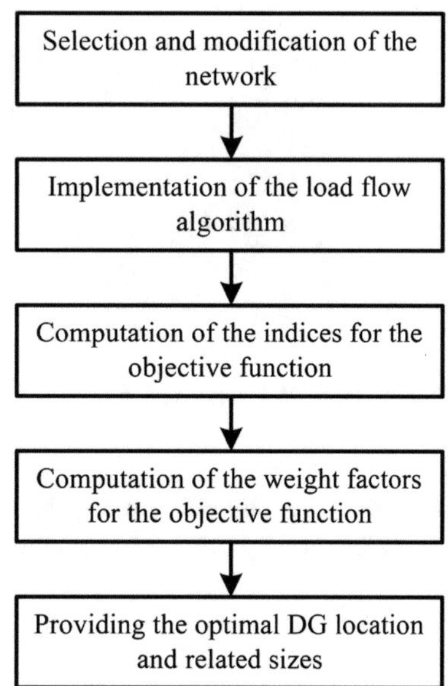

Fig. 3 The solution methodology of the DG allocation problem

The conventional PSO is applied to the DG allocation problem and the simulation results are demonstrated in Table 1. As seen in this table, the optimal locations of the DGs are buses 15, 19, 21, 22, and 30. The optimal value of the objective function, which is provided in iteration 62, is 0.46468 (Fig. 3).

In addition, the QPSO is applied to identify the optimal locations of DGs in IEEE 30-bus test system. The simulation results are reported in Table 2. As it is obvious in this table, the optimal buses of the DGs locations are buses 20, 21, 26, 27, and 30. The optimal value of the objective function by the application of QPSO method is equal to 0.460141, which is better than the conventional PSO method. The voltage profile is illustrated in Fig. 4. The voltage profile is improved by increasing the number of DG units. The total power losses for different DG allocation methods and without DG unit is shown in Fig. 5.

Table 2 The results of the DG optimization problem using QPSO

Iteration	51
Objective function	0.460141
Optimal buses	DG share
20	15%
21	50%
26	10%
27	15%
30	10%

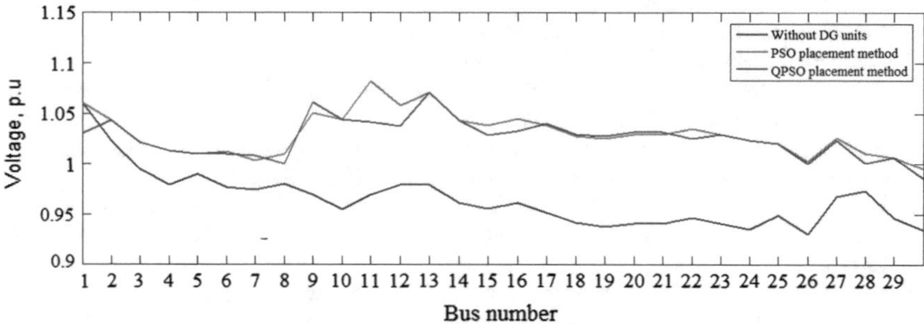

Fig. 4 The voltage profile for various number of DG units

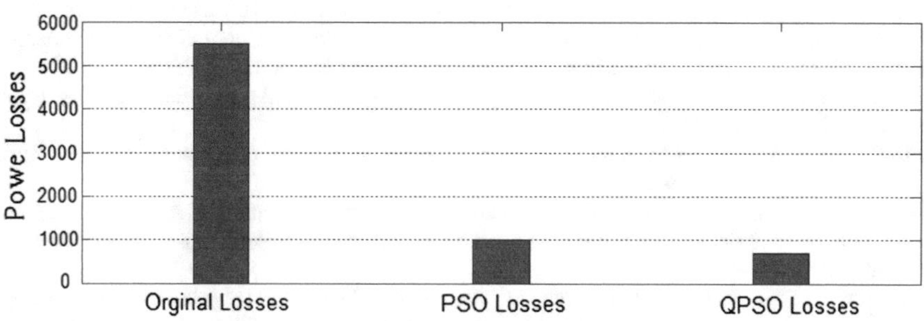

Fig. 5 The power losses for different DG allocation methods

5 Conclusion

In this chapter, the quantum particle swarm optimization (QPSO) is introduced as an efficient technique to obtain the optimal solution of an optimization problems. The QPSO method is employed for solving the optimal distributed generation (DG) allocation problem in power systems. The optimization problem is simulated in MATLAB environment and the optimal solutions are reported and analyzed. The simulation results of the conventional PSO and QPSO methods are compared for IEEE 30-bus test system. The comparison of the obtained results ensures high performance of the QPSO for the DG allocation problem. In addition, it is understood that the employment of QPSO in optimization problems with complicated constraints speeds up the convergence. The proposed QPSO not only shows improvement in the objective function of the problem comparing to the conventional PSO method, but also it is effective in voltage profile of the system.

References

1. Ackermann, T., Andersson, G., Söder, L.: Distributed generation: a definition. Electr. Power Syst. Res. **57**(3), 195–204 (2001)
2. Chiradeja, P., Ramakumar, R.: An approach to quantify the technical benefits of distributed generation. IEEE Trans. Energy Convers. **19**(4), 764–773 (2004)
3. El-Khattam, W., Salama, M.M.: Distributed generation technologies, definitions and benefits. Electr. Power Syst. Res. **71**(2), 119–128 (2004)
4. Pepermans, G., Driesen, J., Haeseldonckx, D., Belmans, R., Dhaeseleer, W.: Distributed generation: definition, benefits and issues. Energy Policy **33**(6), 787–798 (2005)
5. Hung, D.Q., Mithulananthan, N., Bansal, R.: An optimal investment planning framework for multiple distributed generation units in industrial distribution systems. Appl. Energy **124**, 62–72 (2014)
6. HA, M.P., Huy, P.D., Ramachandaramurthy, V.K.: A review of the optimal allocation of distributed generation: objectives, constraints, methods, and algorithms. Renew. Sustain. Energy Rev. (2016)
7. Walling, R., Saint, R., Dugan, R.C., Burke, J., Kojovic, L.A.: Summary of distributed resources impact on power delivery systems. IEEE Trans. Power Deliv. **23**(3), 1636–1644 (2008)
8. Ault, G.W., McDonald, J.R.: Planning for distributed generation within distribution networks in restructured electricity markets. IEEE Power Eng. Rev. **20**(2), 52–54 (2000)
9. Dugan, R.C., McDermott, T.E., Ball, G.J.: Planning for distributed generation. IEEE Ind. Appl. Mag. **7**(2), 80–88 (2001)
10. Rau, N.S., Wan, Y.-H.: Optimum location of resources in distributed planning. IEEE Trans. Power Syst. **9**(4), 2014–2020 (1994)
11. Padma Lalitha, M., Veera Reddy, V., Sivarami Reddy, N.: Application of fuzzy and abc algorithm for dg placement for minimum loss in radial distribution system. Iran. J. Electr. Electron. Eng. **6**(4), 248–257 (2010)
12. Popović, D., Greatbanks, J., Begović, M., Pregelj, A.: Placement of distributed generators and reclosers for distribution network security and reliability. Int. J. Electr. Power Energy Syst. **27**(5), 398–408 (2005)
13. Ochoa, L.F., Dent, C.J., Harrison, G.P.: Distribution network capacity assessment: variable dg and active networks. IEEE Trans. Power Syst. **25**(1), 87–95 (2010)
14. Dent, C.J., Ochoa, L.F., Harrison, G.P., Bialek, J.W.: Efficient secure ac opf for network generation capacity assessment. IEEE Trans. Power Syst. **25**(1), 575–583 (2010)
15. Kumar, A., Gao, W.: Optimal distributed generation location using mixed integer non-linear programming in hybrid electricity markets. IET Gener. Trans. Distrib. **4**(2), 281–298 (2010)
16. López-Lezama, J.M., Padilha-Feltrin, A., Contreras, J., Muñoz, J.I.: Optimal contract pricing of distributed generation in distribution networks. IEEE Trans. Power Syst. **26**(1), 128–136 (2011)
17. Ghosh, S., Ghoshal, S.P., Ghosh, S.: Optimal sizing and placement of distributed generation in a network system. Int. J. Electr. Power Energy Syst. **32**(8), 849–856 (2010)
18. Wang, D.T.-C., Ochoa, L.F., Harrison, G.P.: Dg impact on investment deferral: network planning and security of supply. IEEE Trans. Power Syst. **25**(2), 1134–1141 (2010)
19. Atwa, Y., El-Saadany, E., Salama, M., Seethapathy, R.: Optimal renewable resources mix for distribution system energy loss minimization. IEEE Trans. Power Syst. **25**(1), 360–370 (2010)

20. Khodr, H., Silva, M.R., Vale, Z., Ramos, C.: A probabilistic methodology for distributed generation location in isolated electrical service area. Electr. Power Syst. Res. **80**(4), 390–399 (2010)
21. Dent, C.J., Ochoa, L.F., Harrison, G.P.: Network distributed generation capacity analysis using opf with voltage step constraints. IEEE Trans. Power Syst. **25**(1), 296–304 (2010)
22. Keane, A., O'Malley, M.: Optimal allocation of embedded generation on distribution networks. IEEE Trans. Power Syst. **20**(3), 1640–1646 (2005)
23. Kim, J., Nam, S., Park, S., Singh, C.: Dispersed generation planning using improved hereford ranch algorithm. Electr. Power Syst. Res. **47**(1), 47–55 (1998)
24. Gandomkar, M., Vakilian, M., Ehsan, M.: A combination of genetic algorithm and simulated annealing for optimal dg allocation in distribution networks. In: Canadian Conference on Electrical and Computer Engineering, 2005, pp. 645–648. IEEE (2005)
25. Acharya, N., Mahat, P., Mithulananthan, N.: An analytical approach for dg allocation in primary distribution network. Int. J. Electr. Power Energy Syst. **28**(10), 669–678 (2006)
26. Murty, V., Kumar, A.: Optimal placement of dg in radial distribution systems based on new voltage stability index under load growth. Int. J. Electr. Power Energy Syst. **69**, 246–256 (2015)
27. Borges, C.L., Falcao, D.M.: Optimal distributed generation allocation for reliability, losses, and voltage improvement. Int. J. Electr. Power Energy Syst. **28**(6), 413–420 (2006)
28. Vatani, M., Alkaran, D.S., Sanjari, M.J., Gharehpetian, G.B.: Multiple distributed generation units allocation in distribution network for loss reduction based on a combination of analytical and genetic algorithm methods. IET Gener. Transm. Distrib. **10**(1), 66–72 (2016)
29. Nguyen, T.T., Truong, A.V., Phung, T.A.: A novel method based on adaptive cuckoo search for optimal network reconfiguration and distributed generation allocation in distribution network. Int. J. Electr. Power Energy Syst. **78**, 801–815 (2016)
30. Kowsalya, M., et al.: Optimal size and siting of multiple distributed generators in distribution system using bacterial foraging optimization. Swarm Evol. Comput. **15**, 58–65 (2014)
31. Pesaran, M., Mohd Zin, A.A., Khairuddin, A., Shariati, O.: Optimal sizing and siting of distributed generators by a weighted exhaustive search. Electr. Power Compon. Syst. **42**(11), 1131–1142 (2014)
32. Nazari-Heris, M., Mohammadi-Ivatloo, B.: Application of heuristic algorithms to optimal pmu placement in electric power systems: An updated review. Renew. Sustain. Energy Rev. **50**, 214–228 (2015)
33. Nazari-Heris, M., Mohammadi-Ivatloo, B., Gharehpetian, G.: Short-term scheduling of hydro-based power plants considering application of heuristic algorithms: a comprehensive review. Renew. Sustain. Energy Rev. **74**, 116–129 (2017)
34. Mahdi, F.P., Vasant, P., Rahman, M.M., Abdullah-Al-Wadud, M., Watada, J., Kallimani, V.: Quantum particle swarm optimization for multiobjective combined economic emission dispatch problem using cubic criterion function. In: 2017 IEEE International Conference on Imaging, Vision & Pattern Recognition (icIVPR), pp. 1–5. IEEE (2017)
35. Mohammadi-Ivatloo, B., Moradi-Dalvand, M., Rabiee, A.: Combined heat and power economic dispatch problem solution using particle swarm optimization with time varying acceleration coefficients. Electr. Power Syst. Res. **95**, 9–18 (2013)
36. Yuan, X., Wang, L., Yuan, Y.: Application of enhanced pso approach to optimal scheduling of hydro system. Energy Convers. Manag. **49**(11), 2966–2972 (2008)
37. Yare, Y., Venayagamoorthy, G.K., Aliyu, U.: Optimal generator maintenance scheduling using a modified discrete pso. IET Gener. Transm. Distrib. **2**(6), 834–846 (2008)
38. Mohammadi, M., Hosseinian, S., Gharehpetian, G.: Optimization of hybrid solar energy sources/wind turbine systems integrated to utility grids as microgrid (mg) under pool/bilateral/hybrid electricity market using pso. Solar Energy **86**(1), 112–125 (2012)
39. Yu, X.-M., Xiong, X.-Y., Wu, Y.-W.: A pso-based approach to optimal capacitor placement with harmonic distortion consideration. Electr. Power Syst. Res. **71**(1), 27–33 (2004)
40. Krueasuk, W., Ongsakul, W.: Optimal placement of distributed generation using particle swarm optimization. In: Proceedings of Power Engineering Conference in Australasian Universities, Australia. Citeseer (2006)

41. Moradi, A., Fotuhi-Firuzabad, M.: Optimal switch placement in distribution systems using trinary particle swarm optimization algorithm. IEEE Trans. Power Deliv. **23**(1), 271–279 (2008)
42. Saravanan, M., Slochanal, S.M.R., Venkatesh, P., Abraham, P.S.: Application of pso technique for optimal location of facts devices considering system loadability and cost of installation. In: The 7th International Power Engineering Conference, IPEC 2005, pp. 716–721. IEEE (2005)
43. Zhao, B., Guo, C., Cao, Y.: A multiagent-based particle swarm optimization approach for optimal reactive power dispatch. IEEE Trans. Power Syst. **20**(2), 1070–1078 (2005)
44. Moghaddas Tafreshi, S., Hakimi, S.: Optimal sizing of a stand-alone hybrid power system via particle swarm optimization (pso). In: International Power Engineering Conference, IPEC 2007, pp. 960–965. IEEE (2007)
45. Al-Kazemi, B., Mohan, C.: Discrete multi-phase particle swarm optimization. In: Information Processing with Evolutionary Algorithms, pp. 305–327. Springer (2005)
46. Yang, S., Wang, M.: A quantum particle swarm optimization. In: Congress on Evolutionary Computation, et al.: CEC2004, vol. 1, pp. 320–324. IEEE (2004)
47. Moore, P., Venayagamoorthy, G.K.: Evolving combinational logic circuits using a hybrid quantum evolution and particle swarm inspired algorithm. In: 2005 NASA/DoD Conference on Evolvable Hardware, 2005, pp. 97–102. Proceedings. IEEE (2005)
48. Sun, J., Feng, B., Xu, W.: Particle swarm optimization with particles having quantum behavior. In: Congress on Evolutionary Computation, CEC2004, vol. 1, pp. 325–331. IEEE (2004)
49. Sun, J., Xu, W., Feng, B.: A global search strategy of quantum-behaved particle swarm optimization. In: 2004 IEEE Conference on Cybernetics and Intelligent Systems, vol. 1, pp. 111–116. IEEE (2004)
50. Sun, J., Lai, C.-H., Wu, X.-J.: Particle Swarm Optimisation: Classical and Quantum Perspectives. CRC Press (2011)
51. Lu, S., Sun, C., Lu, Z.: An improved quantum-behaved particle swarm optimization method for short-term combined economic emission hydrothermal scheduling. Energy Convers. Manag. **51**(3), 561–571 (2010)
52. dos Santos Coelho, L., Mariani, V.C.: Particle swarm approach based on quantum mechanics and harmonic oscillator potential well for economic load dispatch with valve-point effects. Energy Convers. Manag. **49**(11), 3080–3085 (2008)
53. Jeong, Y.-W., Park, J.-B., Jang, S.-H., Lee, K.Y.: A new quantum-inspired binary pso: application to unit commitment problems for power systems. IEEE Trans. Power Syst. **25**(3), 1486–1495 (2010)
54. Wang, J., Liu, Z., Lu, P.: Electricity load forecasting based on adaptive quantum-behaved particle swarm optimization and support vector machines on global level. In: International Symposium on Computational Intelligence and Design, ISCID'08, vol. 1, pp. 233–236. IEEE (2008)
55. Tian, S., Tuanjie, L.: Short-term load forecasting based on rbfnn and qpso. In: Power and Energy Engineering Conference, APPEEC 2009. Asia-Pacific. IEEE (2009)
56. Badawy, R., Heßler, A., Albayrak, S., Hirsch, B., Yassine, A.: Quantum-inspired evolution for smart building energy management in future power networks. In: Eng Opt 2014, p. 226 (2014)
57. Esmin, A.A., Lambert-Torres, G., De Souza, A.Z.: A hybrid particle swarm optimization applied to loss power minimization. IEEE Trans. Power Syst. **20**(2), 859–866 (2005)
58. Ibrahim, A.A., Mohamed, A., Shareef, H., Ghoshal, S.P.: An effective power quality monitor placement method utilizing quantum-inspired particle swarm optimization. In: 2011 International Conference on Electrical Engineering and Informatics (ICEEI), pp. 1–6. IEEE (2011)
59. Sun, J., Fang, W., Palade, V., Wu, X., Xu, W.: Quantum-behaved particle swarm optimization with gaussian distributed local attractor point. Appl. Math. Comput. **218**(7), 3763–3775 (2011)
60. Clerc, M., Kennedy, J.: The particle swarm-explosion, stability, and convergence in a multidimensional complex space. IEEE Trans. Evol. Comput. **6**(1), 58–73 (2002)
61. Metropolis, N., Ulam, S.: The monte carlo method. J. Am. Stat. Assoc. **44**(247), 335–341 (1949)

A Boosting-Based Decision Fusion Method for Learning from Large, Imbalanced Face Data Set

Xiaohui Yuan, Mohamed Abouelenien and Mohamed Elhoseny

Abstract The acquisition of face images is usually limited due to policy and economy considerations, and hence the number of training examples of each subject varies greatly. The problem of face recognition with imbalanced training data has drawn attention of researchers and it is desirable to understand in what circumstances imbalanced data set affects the learning outcomes, and robust methods are needed to maximize the information embedded in the training data set without relying much on user introduced bias. In this article, we study the effects of uneven number of training images for automatic face recognition and proposed a boosting-based decision fusion method that suppresses the face recognition errors by training an ensemble with subsets of examples. By recovering the balance among classes in the subsets, our proposed multiBoost.imb method circumvents the class skewness and demonstrates improved performance. Experiments are conducted with four popular face data sets and two synthetic data sets. The results of our method exhibits superior performance in high imbalanced scenarios compared to AdaBoost.M1, SAMME, RUSboost, SMOTEboost, SAMME with SMOTE sampling and SAMME with random undersampling. Another advantage that comes with using subsets of examples is the significant gain in efficiency.

1 Introduction

Face recognition (FR) is an active research problem and many methods have been developed to improve the robustness and accuracy of the automatic process. Despite the great improvement, the application of the methods to many real-world scenarios

X. Yuan (✉) · M. Abouelenien
Department of Computer Science and Engineering, University of North Texas, Denton, Texas, USA
e-mail: xiaohui.yuan@unt.edu

M. Elhoseny
Faculty of Computers and Information, Mansoura University, Mansoura, Egypt

A.E. Hassanien et al. (eds.), *Quantum Computing: An Environment for Intelligent Large Scale Real Application*, Studies in Big Data 33,
https://doi.org/10.1007/978-3-319-63639-9_18

faces challenge of great variation in the number of training examples per human subject [1].

Imbalanced data set (IDS) is defined based on the ratio of the training data sizes [2]. When this ratio is much less (or greater) than one, the data set is considered imbalanced. In an IDS, the class with more examples is referred to as the majority class, and the other is the minority class. The uneven number of examples per class could pose an implicit bias to the learning process [3]. Without loss of generality, we use the ratio of the majority class (S_A) size and the minority class (S_I) size, denoted with $\beta = \frac{|S_A|}{|S_I|}$ and $\beta \geq 1$, in the rest of this article.

In the training of a FR algorithm, the number of images for each human subject is usually assumed to be equal. This is true in a controlled environment [4]. However, there are many applications, in which the acquisition of face images is constrained, and hence the number of training examples of each subject varies greatly. For example, images are taken from a terrorist in custody to provide extensive references for future recognition; whereas a large number of people only have a couple of such face images captured on occasions such as application for a driver license. The training images with abundant examples of some subjects, i.e., the majority classes, and much less number of examples of the others, i.e., the minority classes, exhibit the defining property of IDS.

The problem of face recognition with imbalanced training data has drawn attention of researchers and new methods are developed. Liu and Chen [1] incorporated a cost factor into the penalty function of Support Vector Machine (SVM). By assigning different costs to classes (i.e., subjects), the experiments demonstrated that the recognition of a person with less number of examples was improved. Lu and Tan [5] proposed a doubly weighted non-negative matrix factorization method to account for pairwise similarity of face samples within a class and a discriminant score of image pixels. The between sample weight was claimed to be a significant factor to improve the performance given imbalanced training set. Liu et al. [6] proposed an imbalanced SVM to deal with skewed class boundary in face detection. Similar to the method presented in [1], a cost factor was used to penalize the misclassification of the minority examples, i.e., the examples from the minority classes.

Despite the efforts devoted to the algorithm development for learning from IDS problem in FR, it is desirable to understand in what circumstances IDS affects the FR learning outcomes, and, hence, proper algorithmic remedies can be devised. Robust methods are needed to maximize the information embedded in the training data set without relying on user introduced bias. In this article, we analyze the effects of IDS to the performance of a face recognition system and propose a multi-class boosting method that suppresses the face recognition errors by training an ensemble of classifiers with subsets of examples. By recovering the balance among classes in the subsets, the proposed method circumvents the class skewness and demonstrates improved performance.

The rest of this article is organized as follows: Sect. 2 reviews the related work in multi-class boosting methods and ensemble for learning from imbalanced data.

Section 3 describes our proposed boosting-based, multi-class classification approach that takes advantage of data sampling and weight adjustment. Section 4 presents our experimental results and discussion. Section 5 concludes the paper.

2 Related Work

Boosting methods were designed to solve binary classification problems. Directly applying a boosting method to a multi-class problem, e.g., face recognition, is not straightforward. Intuitive solutions include translating a multi-class problem into several binary classification problems using one-against-all or one-against-one strategies [7]. Using one-against-all strategy, one model is constructed for each class; the one-against-one strategy constructs one model for each pair of classes [8]. Adaboost.MH [9] employed a hamming loss to represent the average error rate for the weak hypothesis over all the binary predictions. Using extra bits to encode class labels, the ensemble was able to tolerate mistakes made by a small number of classifiers [10]. Guruswami and Sahai extended AdaBoost.OC [11] and proposed AdaBoost.ECC [12] which replaced pseudo loss with a common measurement to evaluate the training error. An extension to AdaBoost was proposed by including the number of classes in the classifier weight [13]. The accuracy of each classifier only needs to be better than random guess (i.e., $1/K$). A generalization framework was developed by including an additive factor to the accuracy, which filled the theoretical gap of error relaxation for multi-class boosting [14].

In many real-world applications training data are usually uneven among classes. To address the problems of learning from IDS, one thrust of efforts focuses on using cost matrix. AdaUBoost [15] modified the weight updating rule and loss function such that the minority examples were emphasized with higher weights. A similar strategy was used in [16, 17] to boost multiple base-classifiers with asymmetric misclassification costs. Variations of cost-sensitive boosting was developed, each of which used a cost factor to modify examples' weights [16]. By assigning greater increment to the weight of a costly example and decreases less, repeatedly misclassified examples were allowed greater contributions to the learning process [18]. Joshi et al. [19] treated the misclassified minority examples and majority examples differently and proposed a confusion matrix-based weight to account for various difficulties in classifying rare classes.

Among the boosting methods for learning from IDS, sampling strategies have been heavily explored to create balanced training data sets. Chawla et al. [20] introduced SMOTEboost that generated synthetic minority examples using SMOTE strategy during training. Guo and Viktor [21] combined boosting and data generation and introduced the DataBoost-IM method, where hard-to-classify instances from both majority and minority classes were identified and used to generate synthetic examples. A similar idea of creating synthetic examples was also employed in E-Adsampling algorithms [22]. Compared to DataBoost-IM, which led to the creation of a large number of synthetic minority examples, E-Adsampling faced possible loss of the originally misclassified examples. Chen et al. [23] proposed RAMOBoost that ranked minority examples during boosting iteration and created synthetic minority examples based on a distribution function. Seiffert et al. proposed the RUSboost [24] that extended the AdaBoost methods by using random undersampling to select subsets of examples and demonstrated comparable performance to SMOTEboost. Galar et al. [25] proposed EUSBoost, a variation of RUSBoost, which employs an evolutionary undersampling approach to balance the data set.

Both cost embedding and sampling strategies improve binary classification using imbalanced training data. Extending to multi-class boosting, however, is not forthright [14]. In addition, in the application of face recognition, many state-of-the-art learning methods, e.g., LDA and Eigenface, produce stable classification results which abate the driving force of boosting strategy: diversity [26]. Our proposed multi-class boosting method addresses the problems of learning from imbalanced data and enabling employment of stable learners in the ensemble.

3 Multi-class Boosting for Learning from Imbalanced Face Data

In a multi-class classification problem, let the number of classes be K. The labels can then be encoded with values 1 and $-\frac{1}{K-1}$. For example, for an instance $\mathbf{x}_i$ that belongs to class 2, its label is expressed as $\{-\frac{1}{K-1}, 1, -\frac{1}{K-1}, \dots, -\frac{1}{K-1}\}^T$, where the value of the second component of the vector is 1 indicating that this example belongs to class 2 and the rest are $-\frac{1}{K-1}$.

Directly extending AdaBoost to address multi-class, imbalanced problems fails due to the stringent constraint on the performance of the weak learner [13, 27]. Given a multi-class data set, it is reasonable to assume equal probability for a random guess to label an instance to one of the K classes. Hence, the expected error is $1 - \frac{1}{K}$. The empirical error ϵ can then be expressed as follows:

$$\epsilon = \frac{1}{M} \sum_{j=1}^{K} (\sum_{i=1}^{M_j} \frac{K-1}{K}), \tag{1}$$

where M_j is the number of examples in class j and $M = \sum_j M_j$.

Provided with a multi-class, imbalanced data set, a classifier trained with an imbalanced data set could result in greater generalization error than a classifier trained with a balanced data set due to the dominating number of examples in the majority classes [8]. Following the same error minimization strategy, the classifier yields into the region of the minority class. Clearly, the cause of the suboptimal classifier is the uneven number of examples in the class overlap. Ideally, if the balance is restored in this region, the bias will diminish.

Another issue arises from stable learners that are frequently used in face recognition applications. For instance, Eigenface method constructs a subspace from the training examples and a face recognized by finding the nearest neighbor in the projected subspace. Hence, when a data set S is used to train such a face recognizer $f(\mathbf{x})$, the evaluation error over set S is close to zero. Knowing that the weight update is driven by error, we can expect little, if not zero, change in the next training round.

To address both issues of uneven data size induced bias and stable learner, we propose a multi-class boosting method (multiBoost.imb). Our method is presented in Algorithm 1. In multiBoost.imb, we introduce a perturbation strategy that selects a subset of examples from the majority classes according to the data distribution. The selected examples and the minority examples form a training set. Let $|S_I|$ denote the smallest class size. Following the data distribution w_i ($w_i : (\mathbf{x}_i, \mathbf{y}_i) \in S_A$), a subset of examples from each majority class S_A, denoted with S'_A, is randomly selected so that the size of this subset equals the size of the minority class, i.e., $|S'_A| = |S_I|$. The selected majority examples and the minority examples form a subset S' for training a weak learner:

$$S' = \{\cup S_{I_p}, \cup S'_{A_q}\} \text{ and } |S'_{A_q}| = |S_{I_p}|, \tag{2}$$

where p is the index of the minority classes and q is the index of the majority classes. S_{I_p} denotes the set of examples in the minority class p; S'_{A_q} denotes the subset of examples of the majority class q. The changing subset of training example ensures the construction of a group of diverse classifiers even with stable weak learners. In addition, the equal number of examples that represents all classes suppresses the influence of the IDS to the construction of a classifier f^t. The training process is repeated T times, which implies the number of base classifiers in the ensemble.

Unlike the training process, the entire data set S is used in the evaluation of each classifier f^t. This is necessary because not only the data distribution needs to be updated, but also the weight α^t to the classifier f^t has to be consistent to the overall performance of the learner. Without the knowledge of the underlying true data distribution and hence the overlapped regions among classes, empirical error is a reasonable metric.

The weight α^t determines how much a learner f^t contributes to the final decision as shown in Eq. (6). Given an IDS, the great empirical error of an unbiased learner results in a smaller weight assignment. In fact, as training continues, the examples within the overlapped regions are likely to have greater probabilities. To suppress possible over-weighting the biased classifier, an attenuation factor γ ($\gamma \geq 1$) is included in the weight calculation (see Eq. (4)), which are more likely to happen in the later stage of the training. Large γ subsides the impact of empirical error ϵ^t. When $\gamma = 1$ the weight calculation reduced to AdaBoost.M1 [9]; whereas when $\gamma = K - 1$ the weight becomes that of the SAMME algorithm in [13].

Assuming that classifiers are trained independently, the majority voting of an ensemble should lead to better results than using a single classifier [27]. This suggests that the weight of classifiers that perform better than random guessing should

Algorithm 1: MultiBoost.imb

Input: an imbalanced data set that consists of p minority classes and q majority classes: $S = \{\cup S_{I_p}, \cup S_{A_q}\}$.
Initialize the weight w_i for each $(\mathbf{x}_i, \mathbf{y}_i)$ with $\frac{1}{M}$.
for $t = 1, 2, \ldots, T$ **do**
    Construct a training set S' following Eq. (2).
    Train a classifier f^t using S' such that error is minimized.
    Compute error of f^t using the entire data set S:

$$\epsilon^t = \sum_{i=1}^{M} w_i^t [\![f^t(\mathbf{x}_i) \neq \mathbf{y}_i]\!] \tag{3}$$

    where $[\![\cdot]\!]$ is the indicator function that returns 1 if the argument is true.
    if $\epsilon^t \geq \frac{\gamma}{1+\gamma}$, then stop and set $T = t - 1$
    Compute the weight α^t for f^t:

$$\alpha^t = \log \frac{\gamma(1 - \epsilon^t)}{\epsilon^t} \tag{4}$$

    Update and normalize data distribution

$$w_i^{t+1} = \frac{w_i^t e^{-\frac{1}{2}\alpha^t \mathbf{y}_i f^t(\mathbf{x}_i)}}{W^t}, \tag{5}$$

    where $W^t = \sum_i w_i^t$.
end for
The ensemble $F(\mathbf{x})$ aggregates f^t by maximizing the weighted sum:

$$F(x) = \arg\max_k \left(\sum_{t=1}^{T} \alpha^t f^t(\mathbf{x})\right) \tag{6}$$

where $k \in [1, \ldots, K]$.

be positive. Hence, the ratio $\frac{\gamma(1-\epsilon)}{\epsilon}$ has to be greater than one. Following this assumption, the maximum acceptable error rate for a weak learner is bounded by

$$\epsilon < \frac{\gamma}{\gamma + 1}. \tag{7}$$

In contrast to AdaBoost, the inclusion of γ improves the error tolerance. If we relax our requirement of the error rate of the weak learners to be equivalent to that of the random guess, i.e.,

$$\epsilon = \frac{K - 1}{K}, \tag{8}$$

combining with Eq. (7) results in the upper bound for γ, i.e., $\gamma = K - 1$. Hence, the choice of γ lies in the range of $[1, K - 1]$.

The multiBoost.imb method updates data distribution following the exponential function as shown in Eq. (5). Given that the class label is encoded as a vector that consists of 1 and $\frac{-1}{K-1}$ [13], where the index of 1 indicates the class label, the dot product $\mathbf{y}_i f(\mathbf{x}_i)$ yields one of the following two values:

$$\mathbf{y}_i f(\mathbf{x}_i) = \begin{cases} \frac{K}{K-1} & \text{if } \mathbf{x}_i \text{ is correctly classified} \\ \frac{-K}{(K-1)^2} & \text{if } \mathbf{x}_i \text{ is misclassified} \end{cases} \tag{9}$$

It is clear that the update to w_i of a misclassified instance is smaller than that of a correctly classified instance. The gradually increased w_i of a misclassified instance is consistent with the relaxed constraint on the error rate. Hence, it prevents over emphasizing the large number of misclassified majority instances.

Given that the normalized data distribution w_i sums to one, we can express the sum of the data distribution as follows:

$$\begin{aligned} \sum_i w_i^{t+1} &= \sum_i w^t \frac{e^{-\alpha^t \mathbf{y}_i f^t(\mathbf{x}_i)}}{W^t} = \frac{1}{ZM} \sum_{i=1}^{M} \prod_{s=1}^{t} e^{-\alpha^s \mathbf{y}_i f^s(\mathbf{x}_i)} \\ &= \frac{1}{ZM} \sum_{i=1}^{M} e^{-\mathbf{y}_i \sum_{s=1}^{t} \alpha^s f^s(\mathbf{x}_i)} = \frac{1}{ZM} \sum_{i=1}^{M} e^{-\mathbf{y}_i f^*(\mathbf{x}_i)} = 1. \end{aligned}$$

where W^t is a normalization factor and $Z = \prod_{s=1}^{t} W^s$. Hence, the product of the normalization factor equals to the normalized sum of weight updates following AdaBoost:

$$Z = \frac{1}{M} \sum_{i=1}^{M} e^{-\mathbf{y}_i f^*(\mathbf{x}_i)} \tag{10}$$

where $f^*(\mathbf{x}_i) = \sum_{s=1}^{t} \alpha^s f^s(\mathbf{x}_i)$ is an intermediate ensemble.

When an instance is misclassified, i.e., $[\![f^t(\mathbf{x}_i) \neq \mathbf{y}_i]\!] = 1$, the function $e^{-\mathbf{y}_i f^*(\mathbf{x}_i)} > 1$. Together with Eq. (3), we have the upper bound of the error as the product of the normalization factors $\epsilon \leq \prod_s W^s$.

To find appropriate α, we minimize this error bound $\prod_s W^s$. Following the definition of W^t, we have

$$\prod_s W^s = \prod_s (\sum_i w_i^t e^{-\alpha^t \mathbf{y}_i f^t(\mathbf{x}_i)}) \tag{11}$$

Notice that $\mathbf{y}_i f^t(\mathbf{x}_i)$ results in two values as shown in Eq. (9), which is equivalent to $\frac{1}{2} f^t(\mathbf{x}_i) f^t(\mathbf{x}_i) \lambda (h^*(\mathbf{x}_i) - q)$, where $f^t(\mathbf{x}_i) f^t(\mathbf{x}_i) = \frac{K}{K-1}$ and q is a threshold. That is, the product of the true label and classification result is expressed as function $\frac{\lambda}{2}$ $(q - h^*(\mathbf{x}_i))$ that gives the following results:

$$\frac{\lambda}{2}(h^*(\mathbf{x}_i) - q) = \begin{cases} 1 & \text{if } \mathbf{x}_i \text{ is correctly classified} \\ \frac{-1}{(K-1)} & \text{if } \mathbf{x}_i \text{ is misclassified} \end{cases}$$

where h^* outputs 1 or -1 when $\mathbf{x}_i$ is classified correctly or incorrectly; λ is the inverse of the uninformative error rate, i.e., $\lambda = \frac{K}{K-1}$; q sets the threshold for deciding weight changes. Hence, combining with Eq. (8), we have

$$\mathbf{y}_i\mathbf{y}_i\lambda = f^t(\mathbf{x}_i)f^t(\mathbf{x}_i)\lambda = 1/\epsilon^2.$$

Based on the convexity of the exponential functions, the error upper bound in Eq. (11) is expressed as follows:

$$\prod_s w_i^t e^{-\alpha^t f^t(\mathbf{x}_i)f^t(\mathbf{x}_i)\frac{\lambda}{2}(h^*(\mathbf{x}_i)-q} = \prod_s w_i^t e^{-\alpha^t \frac{1}{\epsilon^2}(h^*(\mathbf{x}_i)-q)}$$
$$\leq \prod_s \sum_i w_i^t (h^*(\mathbf{x}_i) e^{-\alpha^t \frac{1}{\epsilon^2}(1-q)} + (1 - h^*(\mathbf{x}_i)) e^{\alpha^t \frac{1}{\epsilon^2} q}).$$

Taking the first derivative with respect to α^t and setting it to zero, we have the expression of α^t:

$$\alpha^t = \ln \frac{\gamma(1-\epsilon)}{\epsilon} \tag{12}$$

where $\gamma = \frac{1-q}{q}$.

Note that, in training a classifier, only a portion of examples from the majority classes are used. Hence, the error minimization is in the context of a subset of balanced training examples. However, the α given in Eq. (12) is subject to the entire training data set, which accounts for the entire data set.

4 Experiments and Discussion

4.1 Experiment Settings

We employed the Eigenface [28] and Fisherface [29] for face recognition and use four public face databases. The AT&T data set consists of 40 subjects with 10 images for each. The AR face database consists of 126 subjects (among which we used 50 to be consistent with the other two face data sets) and 11 images were cropped for each subject. The Yale database consists of 38 subjects and 65 images for each. LFW has more than 13,000 face images of over 5,000 subjects. The majority of the subjects has less than three images. In our experiments, we used the LFW images aligned with deep funneling method [30] and randomly selected 40 subjects, each of which has at least 20 images such that we can form different imbalance ratios and perform cross validation.

To simulate imbalanced training data, half of the classes (or subjects) were used as the majority classes, and the other half were treated as the minority classes. By re-sampling the data sets, we created training data with various imbalance ratios. The average performance of the leave-one-out cross validation serves as the baseline in our studies.

Cross-validation was used. Depending on the data set size, the number of folds varies. For example, AT&T database consists of 40 subjects. In the experiments of learning from imbalanced data set with imbalance ratio $\beta = 2$, five examples of each subject from 1 through 20 were randomly selected, and the other five were used as testing examples. Subjects 21 through 40 were treated as the majority classes and, based on the imbalance ratio, 10 examples were used for each subject in the training. The majority and the minority classes were switched in another experiment. Experiments were designed to reveal the effects of IDS with respect to the imbalance ratio and the difficulty of the problems. We focused on the evaluation of classifying the minority classes since that is the origin of most errors.

State-of-the-art methods were used in our comparison study. Extension of SMOTEboost for multi-class problem was developed based on AdaBoost.M2. In our initial study that follows this extension [20], it took more than 24 h to complete the training of one SMOTEboost ensemble of 10 base learners and the performance is no better than SMOTEboost using AdaBoost.M1 framework. Hence, the results reported in this comparison study for SMOTEboost is based on AdaBoost.M1. RUSboost, on the other hand, was developed for binary-class classification. For the comparison purpose, we extend it again following the spirit of AdaBoost.M1. We also limit our ensembles to 10 base learners due to great time expense for cross-validation.

4.2 *Performance Analysis*

In our performance analysis, we focus on classification of the minority classes under different imbalance ratios since the minority classes are usually the important ones in a FR problem. Table 1 summaries the average error rate (across all classes in each data set) for the testing scenarios. The standard deviation is included in the parenthesis. Since the baseline is the best result using leave-one-out cross-validation, there is only one baseline error for each data set. The bold face font highlights the best average performance in each case. Since the base learner recognizes a face based on the nearest face image in the training set the results represent Rank-1 recognition rate.

Among all scenarios, multiBoost.imb achieved 4 best performance out of 8 low imbalance ratio cases (four data sets with two different base learners) and 7 best performances out of 8 higher imbalance ratio cases. The maximum improvement as compared to the second best performance among all other methods is 12.8% in the high imbalance ratio cases and 8% in the low imbalance ratio cases. It is worth of

Table 1 The average error rate and standard deviation of multiBoost.imb with imbalanced face data sets

Data sets	S_i (β)	Average error rate (%) and standard deviation							
		Base	AdaBoost.M1	SAMME	multiBoost.imb	RUSboost	SMOTEboost	SAMME+RUS	SAMME+SMOTE
Eigenface									
AT&T	5 (2)	2.5	10.0 (16.8)	10 (16.8)	**8.8** (14.5)	9.5 (15.2)	**8.8** (14.2)	9.5 (15.2)	**8.8** (14.2)
	2 (5)	(6.3)	22.9 (18.3)	22.9 (18.3)	**18.1** (15.9)	18.5 (15.6)	20.9 (16.6)	19 (16.3)	20.9 (16.7)
AR	5 (2)	19.5	27.4 (29.5)	27.4 (29.5)	25.1 (29.1)	25.6 (28.7)	**25** (29.7)	25.6 (29.3)	**25** (29.7)
	2 (5)	(15.0)	49.4 (20.2)	49.6 (20.2)	**40.8** (19.7)	41.5 (21.5)	47.6 (20.6)	41.9 (19.6)	47.5 (20.5)
Yale	32 (2)	28.5	76.9 (7.2)	76.8 (7.2)	**75.1** (6.4)	75.7 (6.9)	76.2 (6.9)	75.2 (6.8)	76.3 (7.2)
	8 (8)	(10.2)	88.9 (3.5)	88.9 (3.5)	**82.4** (3.5)	84.1 (3.6)	88.1 (3.3)	84.9 (3.6)	88.2 (3.5)
LFW	5 (2)	83	90.5 (10.2)	90.5 (11.6)	88.6 (12.7)	89 (11.9)	81.8 (13.6)	88.4 (12.4)	**81.1** (13.2)
	2 (5)	(12.2)	95 (4.8)	95 (4.8)	**87.5** (7)	97.5 (15.8)	94.9 (4.8)	89.4 (6.2)	94.9 (4.8)
Fisherface									
AT&T	5 (2)	2	14.3 (20.6)	14.3 (20.6)	10.8 (18.2)	10.8 (17.6)	13.8 (22.4)	**9** (15)	13.8 (22.4)
	2 (5)	(4.1)	40.7 (22.9)	40.7 (22.9)	**16.3** (14)	**16.3** (13.8)	35.9 (22.6)	18.4 (14.7)	36.2 (22.2)
AR	5 (2)	2.8	30.8 (32.5)	30.8 (32.5)	18.6 (25.1)	**17.2** (25.9)	30.2 (29.2)	18.6 (26.3)	30.2 (29.2)
	2 (5)	(5.7)	46.3 (22.1)	46.3 (22.1)	**17.6** (16.7)	17.8 (16.9)	40.3 (21.1)	20.3 (17.1)	40 (22)
Yale	32 (2)	4.6	31.4 (13.1)	31.4 (13.1)	**28.1** (12)	29.9 (11.9)	29.6 (12.3)	35.5 (12.5)	29.7 (12.9)
	8 (8)	(2.5)	58.7 (8.4)	58.7 (8.4)	59.2 (7.3)	60.1 (7.3)	**50.7** (8.5)	63.5 (6.3)	50.8 (7.9)
LFW	5 (2)	66	84.5(16.4)	84.5(16.4)	**69.1** (21.3)	69.5 (21.2)	86.1 (16.8)	7.4 (20.2)	84.5 (16.4)
	2 (5)	(18.9)	97.5 (5.6)	97.5 (5.6)	**84.1** (11.9)	96.3 (15.8)	96.7 (6)	84.7 (11.5)	96 (5.7)

noting that with low imbalance ratio, i.e., $\beta = 2$, multiBoost.imb achieved highly competitive results against the baseline performance using both base learners.

Clearly, imbalance affects base learners differently. It is interesting to note that AT&T, AR, and Yale are fairly easy cases when Fisherface is applied to the balanced data sets. The average error rates of the baseline performance as reported in Table 1 are 2%, 2.8%, and 4.6% for AT&T, AR, and Yale, respectively. However, if the training data is imbalanced, ensembles using Fisherface as base learner degrade significantly in their performance.

It is evident that our proposed method multiBoost.imb achieves better results when the imbalance ratio is higher. In contrast to the low imbalance ratio, multiBoost.imb also achieved better performance with a couple of classes compared to the baseline. It is worth noting that RUSboost failed to create an effective classifier ensemble in the higher imbalance ratio with LFW data set. RUSboost mostly ignored all subjects except one in the case of LFW data set with $\beta = 5$. However, when we combined the random undersampling with SAMME, the performance improved greatly.

Among all other methods, the combination of SAMME with SMOTE sampling method yielded competitive performance in low imbalance ratio cases (4 out of 8 cases). It is interesting to note that SAMME with SMOTE exhibited even lower average error rate compared to the baseline with slightly greater STD. Although RUSboost and SMOTEboost have very close average error rate in high imbalance ratio with LFW data set, the performance of SMOTEboost is better than RUSboost.

4.3 The Attenuation Factor (γ)

One key parameter in our proposed method is the attenuation factor γ. Besides the benefit it introduces to relax the error upper bound for each learner and, hence, avoids early termination, it contributes a constant addition to the learner weight, which diversifies the ensemble. However, with larger γ, the ensemble becomes less likely to terminate due to the relaxed error upper bound. So it is important to identify appropriate γ.

We studied our method by varying γ value from 1 to $2K$, and the experiment using the combination of each data set and a γ value is repeated 6 times. Both KNN and decision trees are used as the base learner and the imbalance ratio include low ratio value, i.e., $\beta = 2$ and high ratio values ($\beta = 5$ for AT&T and AR and $\beta = 8$ for Yale). The average error rate of the ensemble with different base learner varies fairly greatly. Depending on the cases such difference can be up to 40%. Also, the imbalance ratios result in performance margin in the range of 10%. Hence we use the average error rate improvement as an indicator to reveal the trend of γ. In calculate the improvement rate, we use the error rate with $\gamma = 1$ as the base.

Figure 1 depicts the average error rate improvement. Because the number of classes in each data set differs, the range of each curve varies accordingly. The results with $\gamma = K$ is marked with an enlarged double-line symbol in red. When $\gamma = 1$,

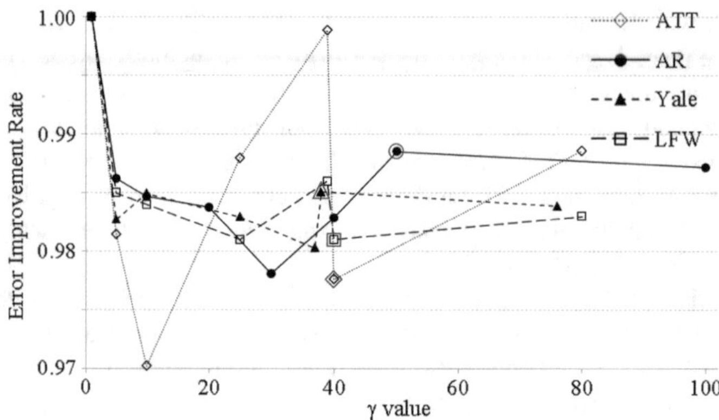

Fig. 1 The average error rate with respect to the attenuation factor. The double-line symbols highlight the average error rate with $\gamma = K$

multiBoost.imb degrades to AdaBoost.M1, and when $\gamma = K - 1$, multiBoost.imb becomes SAMME. It is clear that when $\gamma = 1$ all base learners and ensembles result in the greatest error rate. As γ increases, the error rate reduces. This trend continues even beyond $K - 1$. At $\gamma = 2K$, the error rate remains relatively small with little fluctuation from $\gamma = K - 1$. However, it is clear that the error rate gives suboptimal results when $\gamma = K - 1$. The best performance is mostly achieved when the attenuation factor is in the range of (1, K – 1). Clearly, an attenuation factor that is far greater than K – 1, however, does not result in significantly degraded performance. This is in part because the weight (or the contribution ratio) of the base learner becomes less dependent on the learner performance but the attenuation factor. That is, the ensemble becomes a collection of equally weighted weak learners trained with subsets of examples.

4.4 *Efficiency Analysis*

Table 2 presents the average training time. For SMOTEboost and SAMME with SMOTE sampling, we recorded both the sampling and learning times. Between two base learners, there is no significant difference in efficiency although Fisherface based methods took slightly longer time in comparison to Eigenface based methods.

Because less number of examples used to train each base learner, undersampling based methods took much shorter time to complete model creation. Data resampling takes time, which is trivial compared to the learning time, and we report the sampling and training tmes together. The maximum undersampling time is less than 5 s. Compared to AdaBoost.M1 and SAMME, multiBoost.imb, RUSboost, and SAMME with random undersampling used almost equivalent amount of time in the low imbalance ratio cases and about 50% less amount of time in the high imbalance ratio cases.

Table 2 The average training time in seconds

Data sets	β	AdaBoost.M1	SAMME	multiBoost.imb	RUSboost	SAMME+RUS	SMOTEboost		SAMME+SMOTE	
							Learning	Sampling	Learning	Sampling
Eigenface										
AT&T	2	27.59	26.04	25.93	25.92	17.14	35.14	1.62	33.85	3.5
	5	19.66	19.73	11.88	11.81	7.42	39.09	628.08	39	630.07
AR	2	48.07	43.95	48.48	48.66	31.61	65.61	2.94	70.3	3.02
	5	36.87	32.66	19.04	19.27	10.39	83.51	1026.04	64.69	1004.94
YALE	2	38.34	36.05	38.24	37.34	31.31	51.92	2.90	46.10	2.97
	8	30.17	32.34	13.96	13.75	8.83	44.64	372.58	43.35	368.02
LFW	2	68.71	68.9	50.57	52.18	37.87	91.07	2.93	96.29	3.31
	5	50.22	47.96	24.28	22.59	13.99	121.71	126.24	117.92	126.41
Fisherface										
AT&T	2	37.21	34.07	32.66	32.51	23.34	46.16	1.65	47.71	1.72
	5	27.57	27.56	18.43	18.45	13.06	51.13	629.73	58.46	641.88
AR	2	56.38	56.33	55.19	55.08	43.52	82.85	2.96	82.51	3.01
	5	44.05	45.08	27.49	27.36	18.78	82.07	1002.49	79.85	1005.34
YALE	2	56.67	55.62	52.91	55.08	46.99	62.52	3.02	58.36	3.01
	8	59.63	47.66	20.99	21.95	15.57	53.12	365.97	60.28	372.58
LFW	2	91.66	82.69	62.3	61.66	48.14	110.61	2.97	117.54	3.09
	5	56.74	59.33	31.55	28.92	20.67	140.91	126.01	126.05	126.17

The slight advantage in efficiency of undersampling based method comes from its simplicity. That is, random undersampling during training iterations is independent and no training history is consulted.

On the other hand, oversampling based methods, e.g., SMOTEboost and SAMME with SMOTE, clearly require a significant amount of time, especially when there is a high imbalance ratio, to resample the training set. The sampling time can take as high as 94% of the overall training time. Even without considering the sampling time used in SMOTEboost and SAMME with SMOTE, the learning time was much greater due to larger number of training examples.

Comparing time used for all face data sets, we can see that the increment of time in high imbalance ratio case is less dramatic. This is because each face example consists of more than ten thousands features. The high dimensionality requires significantly more time in training. When the imbalance ratio is low, this time for SMOTEboost is relatively small (in the order of 3 s). However, in the case of large imbalance ratio, the time cost to create new artificial examples is substantial for SMOTEboost. The exponential time increment makes SMOTEboost a less attractive method to handle highly imbalanced, high-dimensional problems.

5 Conclusion

In this article, we propose multiBoost.imb method that greatly improves the performance of face recognition to learn from imbalanced data without relying on user introduced bias. Experimental results demonstrated that the error rate of multiBoost.imb is consistently lower than that of AdaBoost.M1. The advantage becomes more apparent when the imbalance ratio enlarges. In some cases our method achieves even lower error rates beyond that of the baseline model, which is trained with all available examples. With our downsampling strategy, stable classifiers such as Eigenface can be employed as the base learner in an ensemble. Our studies of the attenuation factor show that the best performance is mostly achieved when the attenuation factor is within the range of $(1, K-1)$. An attenuation factor that is far greater than $K-1$, however, affected gently to the recognition performance.

The comparison study was conducted with respect to the state-of-the-art sampling-based ensemble methods for learning from multiclass, imbalanced data sets. When learning from balanced data sets or ones with low imbalance ratio, the performance of the compared methods is quite close. However, the improvement is substantial when the imbalance ratio is high. In contrast to the underampling based methods, e.g., RUSBoost, multiBoost.imb took much less number of training iterations to achieve a performance that took RUSboost many times more number of iterations to attain. Our efficiency analysis shows that multiBoost.imb and random undersampling based methods demonstrated similar efficiency given the same number of base learners, while oversampling based methods, e.g., SMOTEBoost, exhibited a inferior efficiency due to the excessive amount of time used to create and train with the additional synthetic examples. In the cases of large imbalance ratio, the extra time it took SMOTEboost to create the training set increases exponentially. This makes oversampling based methods less attractive ones to handle highly imbalanced, high-dimensional problems.

References

1. Liu, Y.-H., Chen, Y.-T.: Face recognition using total margin-based adaptive fuzzy support vector machines. IEEE Trans. Neural Netw. **18**(1), 178–192 (2007)
2. He, H., Edwardo, G.A.: Learning from imbalanced data. IEEE Trans. Knowl. Data Eng. **21**(9), 1263–1284 (2009)
3. Freund, Y., Schapire, R.E.: A short introduction to boosting. J. Jpn. Soc. Artif. Intell. **14**(5), 771–780 (1999)
4. Zhang, Y., Zhou, Z.-H.: Cost-sensitive face recognition. IEEE Trans. Pattern Anal. Mach. Intell. **32**(10), 1758–1769 (2010)
5. Lu, J., Tan, Y.-P.: A doubly weighted approach for appearance-based subspace learning methods. IEEE Trans. Inf. Forensic Secur. **5**(1), 71–78 (2010)
6. Liu, Y.-H., Chen, Y.-T., Lu, S.-S.: Face detection using kernel pca and imbalanced svm. In: Lecture Notes in Computer Science, International Conference on Natural Computation, vol. 4221, pp. 351–360 (2006)
7. Allwein, E.L., Schapire, R.E., Singer, Y.: Reducing multiclass to binary: a unifying approach for margin classifiers. J. Mach. Learn. Res. **1**, 113–141 (2000)
8. Freund, Y., Schapire, R.E.: A decision-theoretic generalization of on-line learning and an application to boosting. J. Comput. Syst. Sci. **55**(1), 119–139 (1997)
9. Schapire, R.E., Singer, Y.: Improved boosting algorithms using confidence-rated predictions. Machine Learning, pp. 80–91 (1999)
10. Dietterich, T.G., Bakiri, G.: Solving multiclass learning problems via error-correcting output codes. J. Artif. Intell. Res. **2**, 263–286 (1995)
11. Schapire, R.E.: Using output codes to boost multi-class learning problems. In: Proceedings of the 14th International Conference on Machine Learning, pp. 313–321 (1997)
12. Guruswami, V., Sahai, A.: Multiclass learning, boosting, and error-correcting codes. In: Proceedings of the 12th Annual Conference on Computational Learning Theory, pp. 145–155 (1999)
13. Zhu, J., Zou, H., Rosset, S., Hastie, T.: Multi-class adaboost. Stat. Interface **2**, 349–360 (2009)
14. Mukherjee, I., Schapire, R.E.: A theory of multiclass boosting. In: Proceedings of Twenty-Fourth Annual Conference on Neural Information Processing Systems (2010)
15. Karakoulas, G., Shawe-Taylor, J.: Optimizing classifiers for imbalanced training sets. In: Proceedings of the 1998 Conference on Advances in Neural Information Processing Systems II, pp. 253–259, Cambridge, MA, USA. MIT Press (1999)
16. Sun, Y., Kamel, M.S., Wong, A.K.C., Wang, Y.: Cost-sensitive boosting for classification of imbalanced data. Pattern Recogn. **40**, 3358–3378 (2007)
17. Wang, B.X., Japkowicz, N.: Boosting support vector machines for imbalanced data sets. In: Foundations of Intelligent Systems, pp. 38–47 (2008)
18. Fan, W., Stolfo, S.J., Zhang, J., Chan, P.K.: Adacost: misclassification cost-sensitive boosting. In: 16th International Conference on Machine Learning (1999)
19. Joshi, M.V., Kumar, V., Agarwal, R.C.: Evaluating boosting algorithms to classify rare classes: comparison and improvements. In: First IEEE International Conference on Data Mining, pp. 257–264 (2001)
20. Chawla, N.V., Lazarevic, A., Hall, L.O., Bowyer. K.W.: Smoteboost: improving prediction of the minority. In: Seventh European Conference on Principles and Practice of Knowledge Discovery in Databases, pp. 107–119 (2003)

21. Guo, H., Viktor, H.L.: Learning from imbalanced data sets with boosting and data generation: the databoost-im approach. SIGKDD Explor. **6**(1), 30–39 (2004)
22. Geiler, O.J., Hong, L., Yue-Jian, G.: An adaptive sampling ensemble classifier for learning from imbalanced data sets. In: International MultiConference of Engineers and Computer Scientists, vol. 1, March 2010
23. Chen, S., He, H., Garcia, E.A.: RAMOBoost: ranked minority oversampling in boosting. IEEE Trans. Neural Netw. **21**(10), 1624–1642 (2010)
24. Seiffert, C., Khoshgoftaar, T.M., Van Hulse, J., Napolitano, A.: RUSBoost: a hybrid approach to alleviating class imbalance. IEEE Trans. Syst. Man. Cybern. Part A Syst. Hum. **40**(1), 185–197 (2010)
25. Galar, M., Fernandez, A., Barrenechea, E., Francisco, H.: EUSBoost: enhancing ensembles for highly imbalanced data-sets by evolutionary undersampling. Pattern Recogn. **46**(12), 3460–3471 (2013)
26. Lu, J., Plataniotis, K.N., Venetsanopoulos, A.N., Li, S.Z.: Ensemble-based discriminant learning with boosting for face recognition. IEEE Trans. Neural Netw. **17**(1), 166–178 (2006)
27. Eibl, G., Pheiffer, K.-P.: Multiclass boosting for weak classifiers. J. Mach. Learn. Res. **6**, 189–210 (2005)
28. Turk, M., Pentland, A.: Eigenfaces for recognition. J. Cogn. Neurosci. **3**(1), 71–86 (1991)
29. Belhumeur, P., Hespanha, J., Kriegman, D.: Eigenfaces vs. fisherfaces: recognition using class specific linear projection. IEEE Trans. Pattern Anal. Mach. Intell. **19**(7), 711–720 (1997)
30. Huang, G.B., Mattar, M., Lee, H., Learned-Miller, E.: Learning to align from scratch. In: Advances in Neural Information Processing Systems (NIPS), Lake Tahoe, Nevada, United States, December 3–6, 2012

Automatic Construction of Aerial Corridor from Discrete LiDAR Point Cloud

Xiaohui Yuan, Dengchao Feng and Zejun Zuo

Abstract With the development of unmanned aerial systems (UASs), the lack of flight supervision mechanism and the related technical guidance in the airspace become a challenge for safety and privacy protection. In this paper, we present an automatic construction and visualization of airspace corridor from discrete LiDAR. In our method, DTM is generated with empirical decomposition method and the morphological operation and slope-based threshold, which provides an altitude-based upper zone in the space zoning. The detected non-ground objects and the boundary of the privacy-protected regions are used to construct the aerial corridor. To evaluate our proposed method, the ISPRS LiDAR datasets and a LiDAR dataset from Mustang Island were used. It was demonstrated that our proposed method improved the accuracy of delineation of the non-ground objects and improved the accuracy of DTM. Using DSM and DTM, the airspace is divided into the upper zone, safe zone, and takeoff/landing zone. The privacy sensitive regions were integrated into the zoning process and the route for UAS was planned automatically to avoid the private and restricted regions. The visualization technology was implemented to realize the construction of aerial corridor.

X. Yuan (✉)
Department of Computer Science and Engineering, University of North Texas, Denton, TX, USA
e-mail: Xiaohui.yuan@unt.edu

D. Feng
School of Electronic Engineering, North China Institute of Aerospace Engineering, Langfang, China

Z. Zuo
College of Information Engineering, Chinese University of Geosciences, Wuhan, China

A.E. Hassanien et al. (eds.), *Quantum Computing: An Environment for Intelligent Large Scale Real Application*, Studies in Big Data 33,
https://doi.org/10.1007/978-3-319-63639-9_19

1 Introduction

Unmanned aerial systems (UASs) is usually an unpiloted aircraft equipped with various sensors to carry out various dangerous and emergency missions. With the development of low altitude open-door policy, an increasing number of unmanned aircraft systems are produced and widely used in recreational activities, oil pipelines inspection, surveillance, etc., which complicates the low altitude airspace and causes risks for aircraft and ground objects. In order to improve the flight safety in low altitude airspace and realize the privacy protection, a series of research have been conducted. Most research focuses on the collision-avoidance system [1–4]. Soler et al. proposed a hybrid control method to avoid aircraft conflicts [5]. Another thrust of research is to regulate the low altitude airspace [6]. Feng et al. analyzed the influence of air environment on flight safety in low altitude airspace and explored the low altitude security alarm aeronautical chart visualization technology [7]. Sotiriou et al. proposed a trajectory conformance monitoring technology to increase the flight safety [8]. Chougdali et al. proposed a new model for aircraft landing scheduling using real-time algorithms scheduling [9]. Kim et al. proposed a construction of UAS traffic management to ensure safe management of the low-altitude UAS operation [10]. Foina et al. proposed an unmanned aerial traffic management solution using an air parcel model [11]. Fadlullah et al. proposed a dynamic trajectory control algorithm in UAV-aided network [12].

A key component in low altitude airspace regulation is to identify various regions on the ground [13, 14]. Yuan et al. proposed a spatially constrained, model-driven clustering method for water body delineation [15]. Feng et al. proposed the construction of aerial corridor for navigation of UASs in Class G airspace using LiDAR data [16]. Voss proposed a regulatory framework for small, unmanned aircraft to protecting privacy and property rights in the lowermost reaches of the atmosphere [17]. Kim et al. proposed a server-based real-time privacy protection scheme against video surveillance by unmanned aerial systems [18].

In this paper, we design an automatic aerial corridor construction method using discrete return LiDAR data to ensure flight safety and privacy protection. By estimating the noise magnitude in the digital surface model (DSM), Gaussian filters are used to suppress noise. Combined with empirical mode decomposition (EMD) method [19, 20] and morphological operations, the non-ground objects are detected and DTM is generated. The space zoning is designed according to the flight regulations, DSM, and DTM. The boundaries of the privacy protection region are identified as the restricted zone. The UASs route planning combines different flight modes in low altitude airspace. Using image inpainting method [21], the three-dimensional aerial corridors is generated.

The rest of this paper is organized as follows: Sect. 2 describes the construction of aerial corridor for safety and privacy protection and discusses DSM filtering, DTM generation, airspace zoning, and 3D visualization of the aerial corridor. Section 3 presents the experimental results and a comparison study. Section 4 concludes the paper with a summary of our method.

2 Aerial Corridor and Privacy Protection

To construct aerial corridor, the discrete LiDAR datasets are used to obtain the elevation of the ground objects to produce the digital terrain model (DTM). Figure 1 illustrates the flow chart of our proposed method for aerial corridor construction. In our method, a DSM is generated from LiDAR point cloud. The noise level of DSM is estimated and the corresponding filter based on noise level is applied to suppress noise. The filtered DSM is processed using empirical model decomposition and the morphological operations with threshold calculation to generate a DTM. The space zoning based on the DSM and DTM is designed under the guidance of low altitude flight regulations. To protect sensitive regions, private areas are marked in the space zoning and are taken into consideration in the route planning. According to the space delamination results and the privacy protection region, the route planning for UASs is designed to ensure flight safety and privacy protection. The visualization of the aerial corridor is achieved with inpainting [21] technology.

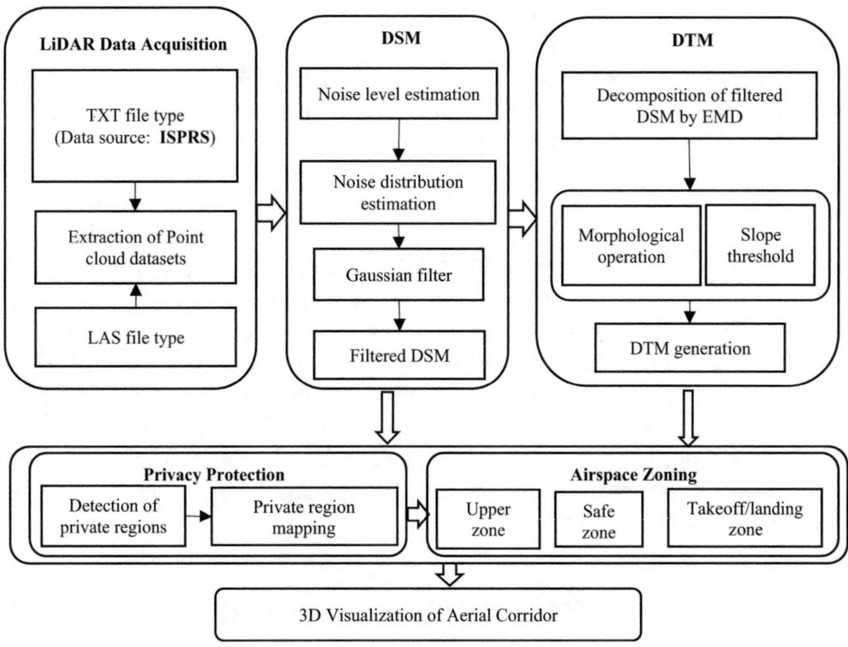

Fig. 1 Flow chart for the aerial corridor construction and visualization

2.1 Digital Terrain Model Construction

LiDAR datasets often contain various amounts of noise 20. The higher ground objects pose greater flight risk in low altitude airspace. Hence, methods such as median filter could induce loss of information. The noise level of DSM is estimated as follows 24:

$$\hat{\sigma}_n^2 = \lambda_{min}(\frac{1}{N}\sum_{i=1}^{N} y_i y_i^T) \tag{1}$$

$$y_i = x_i + n_{i,} \quad i = 1, 2, \ldots N \tag{2}$$

where $\hat{\sigma}_n^2$ is the estimated noise variance, x_i is the texture patch with the ith pixel at its center, N is the number of noise distribution area, λ_{min} is the minimum eigenvalue of the matrix of the noisy noise distribution area.

The Gaussian filter is produced based on the estimated noise level and performs filtering for each noisy texture patch of DSM. The scalar of the Gaussian filter is computed as follows:

$$M_{scalar} = 2\left[2\hat{\sigma}_n^2\right] + 1 \tag{3}$$

where $2\hat{\sigma}_n^2$ is a part of Gaussian kernel.

To generate DTM, the non-ground objects in the filter DSM need to be detected correctly. The filtered DSM is decomposed based on the empirical model decomposition method (EMD)19 as follows:

$$Y(i,j) = \sum_{l=1}^{L-1} X_l(i,j) + N(i,j) \tag{4}$$

where $Y(i,j)$ is the filtered DSM, $X_l(i,j)$ is the intrinsic mode functions (IMF), $N(i,j)$ is the residual signal and $L-1$ is the number of IMF in EMD. The decomposition is based on the extraction of energy associated with various intrinsic time scales, which will generate a collection of intrinsic mode functions (IMF) by a sifting process. In [20], the iterative decomposition of EMD depends on the IMF. That is, $X_l(i,j)$ is decomposed according to the maximum number of IMF. The morphological open operation is used in sifting process to obtain the residual signal $N(i,j)$, which is computed iteratively using morphological open operations based on a cost function F as follows:

$$F = \frac{\sum\left[Z(i,j) - Z_{avg}(i,j)\right]^2}{\sum Z(i,j)^2}, \quad F \in [0, 1] \tag{5}$$

where $Z(i,j)$ is the altitude of the filtered DSM with coordinates (i,j), $Z_{avg}(i,j)$ is the mean value of the altitude of the filtered DSM in the same position. When the cost is greater than a threshold, the iteration stops. The criterion for the residual matrix is as follows:

$$J = \frac{V_{th}}{V_{dsm} - V_{residue}} * \rho \tag{6}$$

where V_{th} is a preset threshold, V_{dsm} is the altitude, $V_{residue}$ is the altitude of the residue matrix, and ρ is a scaling factor ($\rho \in [0, 1]$). If $J \leq 1$, the point is marked as ground.

To obtain the non-ground objects, an elevation threshold is adopted [22] as follows:

$$R = R_{ini\_thr} + S_{slope} * C \tag{7}$$

$$S_{slope} = \sqrt[2]{\nabla X^2 + \nabla Y^2} \tag{8}$$

where R_{ini_thr} is the hard elevation threshold, S_{slope} is the slope of the ground, and C is the cell size. If the difference between DSM and DTM is greater than the threshold R, the corresponding cell in the matrix is marked as non-ground objects.

2.2 Space Zoning and Route Visualization for UASs

To construct the aerial corridor for UASs, the space delamination is designed according to the space height restriction in low altitude airspace, which includes the upper zone, safe zone, and takeoff/landing zone. The upper zone is the fixed zone (500 feet in U.S. for small UASs) above the ground. The shape of upper zone is the same as the DTM in the same region expect for the elevation value. The safety zone is between upper zone and takeoff/landing zone with a certain distance. takeoff/landing zone is the original earth surface with a natural obstacle and artificial objects, which follows DSM. Usually, takeoff/landing zone is more dangerous for UASs than other zones due to various non-ground objects, especially in the takeoff and landing stages.

Let $H_1(i,j)$ and $H_2(i,j)$ be the altitudes of the DSM and DTM at coordinates (i,j), respectively. Let $H_{constant}$ be the maximum permitted altitude for UASs in the low airspace. The altitudes of each zone are computed as follows:

$$H_{bottom\_zone}(i,j) = H_1(i,j) \tag{9}$$

$$H_{up\_zone}(i,j) = H_2(i,j) + H_{constant} \tag{10}$$

$$H_{safe\_zone}(i,j) \in \left[H_{bottom\_zone}(i,j), \mathrm{H}_{up\_zone}\right] \tag{11}$$

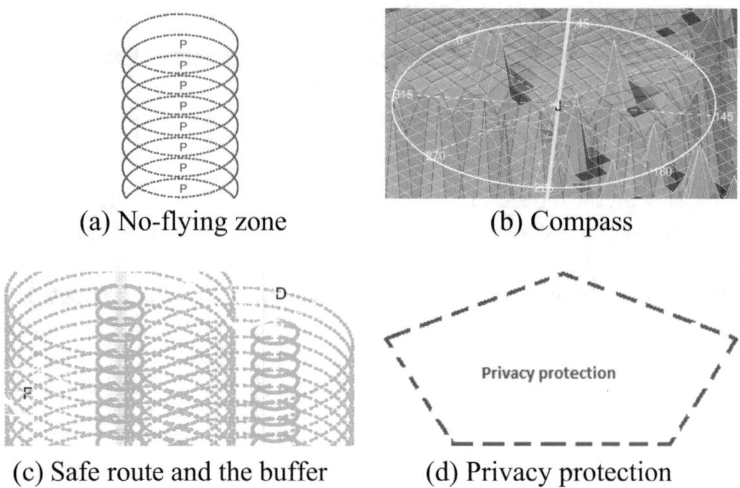

(a) No-flying zone (b) Compass

(c) Safe route and the buffer (d) Privacy protection

Fig. 2 The graphical icons used in the visualization

where $H_{bottom\_zone}(i,j)$ is the altitude of takeoff/landing zone, $H_{safe\_zone}(i,j)$ is the altitude range of safe zone and $H_{up\_zone}(i,j)$ is the altitude of upper zone.

The privacy invasion of UASs occurred frequently in recent years. Therefore, we need to mark regions that have privacy concerns. The contours of such regions are extracted and mapped. The marked boundary of privacy protection will be used to change the route of UASs in the aerial corridor to avoid passing by that forbidden zone without legal permission, which will decrease the illegal surveillance in that region.

In our visualization, we employed image inpainting method 21 to illustrate the three-dimensional aerial route, which includes three parts. The first part is the three-dimensional visualization of LiDAR point cloud datasets and space delamination, which include position (X, Y), altitude Z-value. The second part is the visualization of the three-dimensional DSM and DTM. The third part is to show the graphical icons in the model, which includes the obstruction, compass, restriction zone, prohibited zone, airport, etc. The fourth part is to display the flight paths, which include different flight altitude and topographic information. Some symbols on the route map are shown in Fig. 2. The visualized airspace model and the flight route are shown in Fig. 3. The yellow lines represent flight routes.

3 Experimental Results and Discussion

Our methods are implemented in Matlab 2015b. The LiDAR datasets used in our experiments include fifteen LiDAR datasets from the ISPRS with ground truth 22 and a LiDAR datasets of Mustang Island in Texas 23. The properties of these datasets are described in Tables 1 and 2.

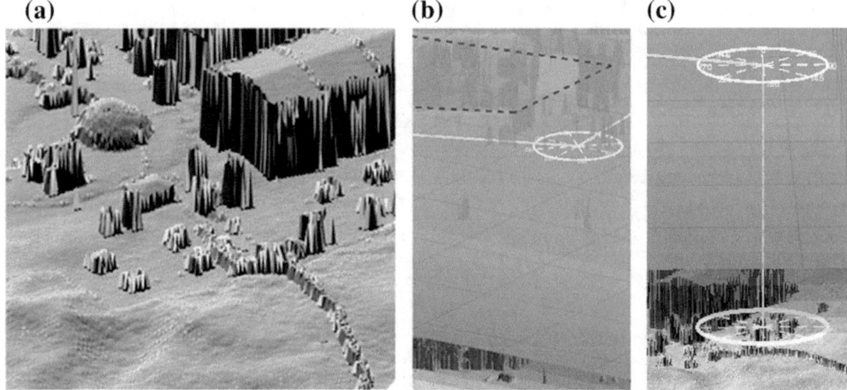

Fig. 3 Airspace model and flight path. **a** zoom-in view of DSM, **b** zoom-in view of safe zone, **c** zoom-in view of the takeoff/landing zone

Table 1 Features of the ISPRS dataset

ISPRS LiDAR datasets	Properties
Samp11, samp12	Steep slopes, mixture of vegetation and buildings on hillside, buildings on hillside, data gaps
Samp21, samp22, samp23, samp24	Large buildings, irregularly shaped buildings, road with bridge and small tunnel, data gaps
Samp31	Densely packed buildings with vegetation between them, building with eccentric roof (bottom left corner), open space with mixture of low and high features
Samp41, samp42	Railway station with trains (low density of terrain points), data gaps
Samp51, samp52, samp53, samp54	Steep slopes with vegetation, quarry, vegetation on river bank, data gaps
Samp61	Large buildings, road with embankment, data gaps
Samp71	Bridge, underpass, road with embankments, data gaps

To generate the accurate DTM, the following four methods, Thomas' method 25, Mongus' method 26, Özcan's method 20 and the proposed method are used to compare the performance of DTM. The parameters of the above methods are assigned in Table 3.

Kappa scores, error rate, skewness, and Kurtosis are adopted for objective evaluation of the DTM. The effect of the maximum numbers of IMF 20 in Özcan's method and the proposed method for Kappa scores will be discussed respectively in the experiments, which can be used to explain the reason for the above settings for the maximum number of IMF. For ISPRS fifteen standard datasets, the Kappa scores obtained by the above four methods are shown in Fig. 4.

As shown in Fig. 4a, the proposed method and Özcan's method obtained greater Kappa scores than the other two methods. For Samp22, Samp42, Samp51, Samp52,

Table 2 Features of the LiDAR dataset of Mustang Island in Texas

LiDAR data of Mustang Island	Properties
real_samp1 (Port Royal Ocean Resort and Conference Center)	Small buildings, Swimming pools with different shape, Bench, Fountains, Gallery, Tennis court
real_samp2 (Lost Colony Villas)	Small shrub, Residential area, Street lights, lane
real_samp3 (Mustang Island Conference center)	Buildings with different shapes, Lawn, Central air conditioning cooling tower, square
real_samp4 (Sandpiper Resort Condominiums)	Tall building, Swimming Pool, Cars parked in parking spaces, trees with small crown, lane

Table 3 Parameter used in the four methods

Thomas' method		Mongus' method		Özcan's method		Proposed method	
Cell size	1	b	0.25	Cell size	1	Cell size	1
Initial slope	0.2	k	0.05	Outlier threshold	5	Outlier threshold	5
Window size	[1, 16]	n		Iteration number	10	Iteration number	10
Max. threshold	0.45	Max. value of the filter	50	Max. window size	20	Max. window size	20
Elevation scaling factor	1.2			Max. threshold	4	Max. threshold	4
				Hard elevation threshold	0.6	Hard elevation threshold	0.6
				Logic value of Slope threshold	1	Logic value of slope threshold	1
				Max. number of IMF	1	Max. number of IMF	1
						Patch size	7
						Confidence interval	0.99
						Iteration for noise level estimation	3

Samp54 and Samp71, the Kappa Scores by the proposed method are 90.1381, 92.5285, 92.8452, 83.2073, 92.8114 and 90.6518 respectively, which is higher than the Kappa Scores by other three methods. For other samples, the proposed method also obtained high Kappa scores. Figure 4b shows both the proposed method and Özcan method can obtain good performance of total error rate than the other two methods. For Samp22, Samp42, Samp51, Samp52, Samp54 and Samp71, the total error rate by the proposed method is 4.1766, 3.1881, 2.3648, 3.2526, 3.5897 and 1.8089 respectively, which is lower than the total error rates by other three methods. For other samples, the proposed method also obtained low total error rates. From the comparison results of Kappa scores and the total error rates, it shows the

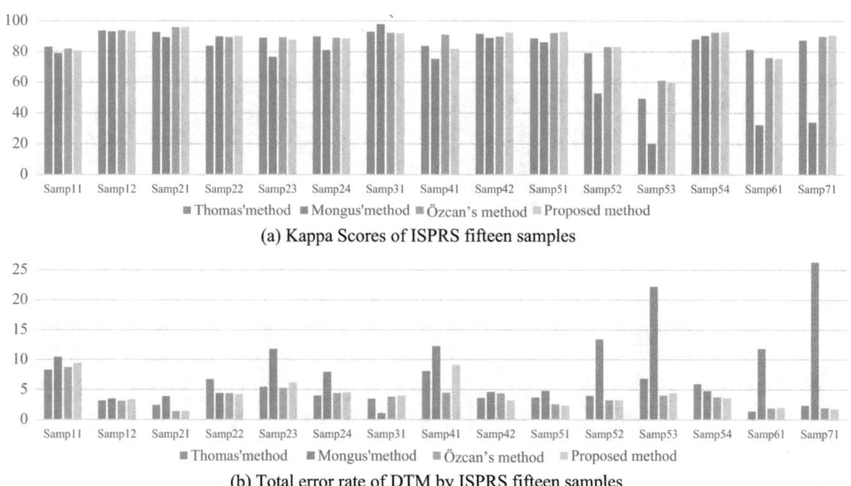

Fig. 4 The comparison results of DTM by ISPRS fifteen samples

proposed method can obtain the good performance for ISPRS datasets. The maximum number of IMF is very important for the performance of DTM. The relationship between the maximum number of IMF and the performance of DTM for Özcan's method and the proposed method are evaluated. The corresponding experiment results for different values of the maximum number of IMF are shown in Fig. 5.

Figure 5a shows the Kappa scores are shown a downward trend with the increase of the value of the maximum number of IMF, among which the best performance in ISPRS test samples can be obtained by comparison the Kappa scores when the maximum number of IMF is assigned 1 for Özcan's method. Figure 5b shows it is a downward trend for the Kappa scores of the generated DTM for ISPRS samples, among which the best performance of Kappa scores is usually obtained when the maximum number of IMF equals one. Figure 5c depicts the mean and standard deviation of Kappa scores of our proposed method and Özcan's method. Figure 5c shows the whole downward trends of mean values of Kappa scores with the increasing of the number of IMF. For the fifteen samples, when the maximum number of IMF increased from 1 to 6, the mean of Kappa scores in Özcan's method is 87.18235, 81.79479, 76.32071, 70.50371, 64.69338 and 59.75527 respectively. In the same condition, the mean of Kappa scores in the proposed method is 86.45674, 81.02425, 77.06374, 71.50672, 65.59312, and 60.03348. Both of the two methods showed that the Kappa scores descended gradually with the increasing number of maximum number of IMF. Therefore, when the maximum number of IMF equals one, the Kappa scores will be higher than other maximum numbers of IMF, which can explain the reason of the assigned 1 as the value of maximum number of IMF in Table 3. The estimated noise level and the detected noise distribution area by the proposed method are shown in

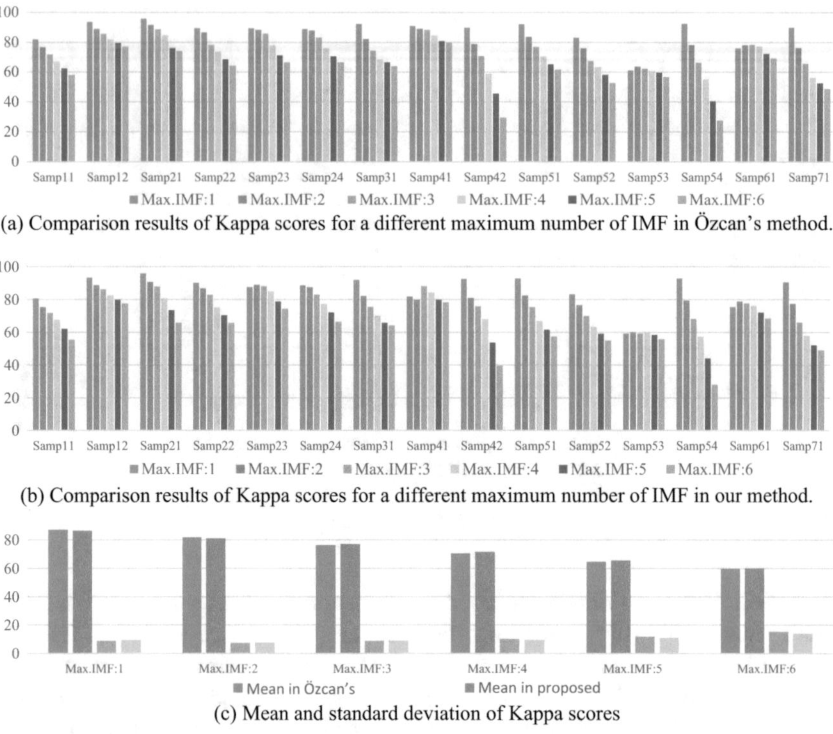

(a) Comparison results of Kappa scores for a different maximum number of IMF in Özcan's method.

(b) Comparison results of Kappa scores for a different maximum number of IMF in our method.

(c) Mean and standard deviation of Kappa scores

Fig. 5 Comparison results of Kappa scores in Özcan's method and the proposed method

Table 4 when the maximum number of IMF is assigned one. Both the methods achieved satisfactory results with IMF at one.

The following experiment using Mustang Island dataset compares the performance of DTM with IMF at one. The original DSM and the remote sensing images in the same regions of Mustang Island are shown in Fig. 6. The estimated noise level and the extracted noise distribution area by the proposed method for the real LiDAR datasets in Mustang Island are shown in Table 5. As shown in Table 5, the noise distribution areas in the four samples are detected according to the estimated noise level. For the real_samp1which will be used to construct the aerial corridor in the following experiments, the noise level is estimated to be 0.03 and the numbers of the noise distribution area are 158167. Correspondingly, the Gaussian filter is used to filter the 158167 noise area rather than the whole area of DSM, which can obtain more accurate filtered DSM than the median filter used for the whole DSM by Özcan's method. The four generated DTM by Özcan's method 20 and our method is shown in Fig. 7, in which the test sample is real_samp1, real_samp2, real_samp3 and real_samp4 respectively from left to right.

Compared the DTMs in Fig. 7a and b, it can be seen that more non-ground objects are detected in Fig. 7b, in which the non-objects can also be easily observed

Table 4 Noise level estimation of DSM for ISPRS fifteen samples by the proposed method

ISPRS	Noise level	Numbers of noise distribution area	ISPRS	Noise level	Numbers of noise distribution area
Samp11	0.33	9328.00	Samp42	0.12	8901.00
Samp12	0.24	16442.00	Samp51	0.05	43037.00
Samp21	0.15	5625.00	Samp52	0.04	41529.00
Samp22	0.10	14373.00	Samp53	0.04	81307.00
Samp23	0.17	10797.00	Samp54	0.07	3697.00
Samp24	0.14	2500.00	Samp61	0.02	39640.00
Samp31	0.22	11321.00	Samp71	0.02	33035.00
Samp41	0.08	6710.00			

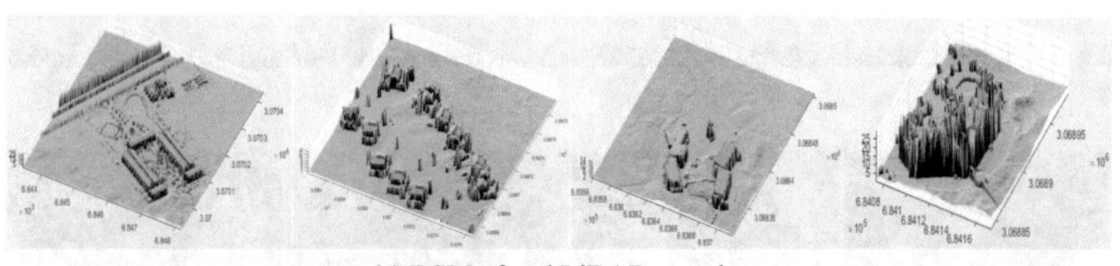

(a) DSM of real LiDAR samples

(b) Remote sensing images of selected areas in Mustang Island by google map

Fig. 6 DSM and the remote sensing images for Mustang Island of Texas

Table 5 The estimated noised level and the extracted patches by the proposed method

Mustang Island	Noise level	Numbers of noise distribution area	Mustang Island	Noise level	Numbers of noise distribution area
real_samp1	0.03	158167.00	real_samp3	0.03	12388.00
real_samp2	0.02	7373.00	real_samp4	0.04	4473.00

by the corresponding remote sensing image in Fig. 7b. According to the filtered DSM by noise level estimation and the generated DTM, the non-ground objects can be detected by the proposed method. The recognition results of a non-ground object by Özcan's method and the proposed method are shown in Fig. 8.

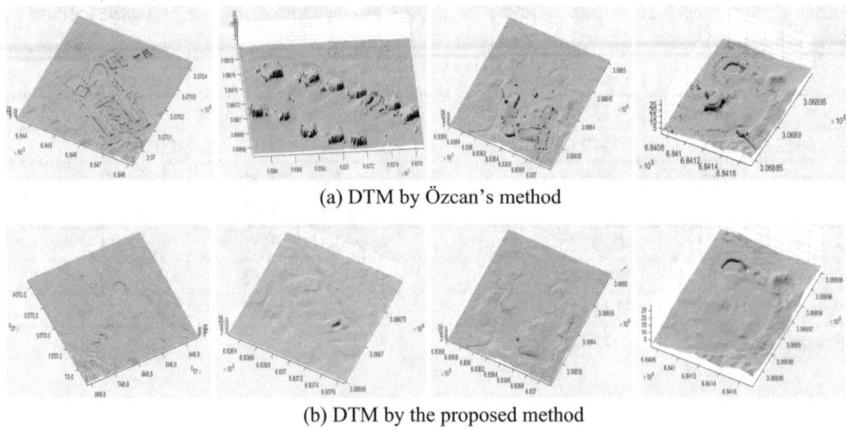

(a) DTM by Özcan's method

(b) DTM by the proposed method

Fig. 7 Comparison results for generated DTM between Özcan's method and the proposed method

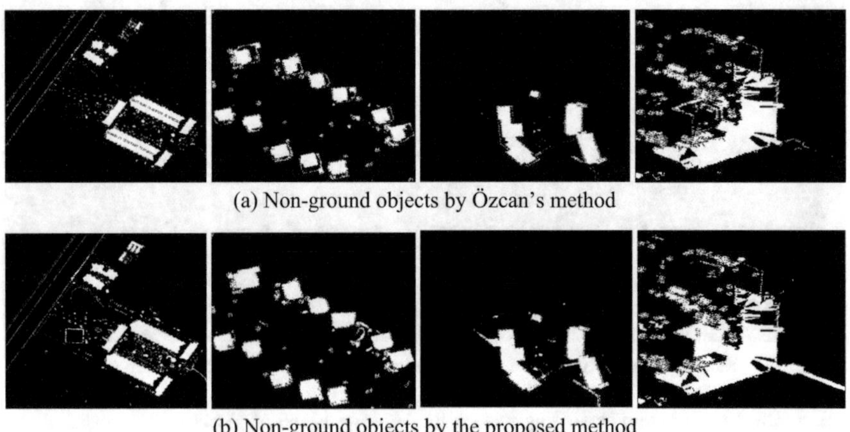

(a) Non-ground objects by Özcan's method

(b) Non-ground objects by the proposed method

Fig. 8 Comparison results for non-ground objects by Özcan's method and the proposed method

As shown in Fig. 8, more non-ground objects are obtained by the proposed method than Özcan's method, which shows the effectiveness of the proposed method. Unlikely the ISPRS dataset, the Mustang Island LiDAR dataset acquired has no reference. To estimate the performance of DTM, Skewness and Kurtosis are adopted. Figure 9 shows the proposed method can obtain lower Skewness values and Kurtosis values than Özcan's method, which shows that more ground objects can be detected by the proposed method. Figure 9a and b depict the skewness and kurtosis. The skewness value of the four samples (real_samp1, real_samp2, real_samp3, real_samp4) by the proposed method is 2.6303, 2.2056, 3.1926 and1.8916 respectively, which is lower than the skewness values (2.9041, 2.5997, 3.417 and 2.3185) by Özcan's method. The Kurtosis value of the four samples

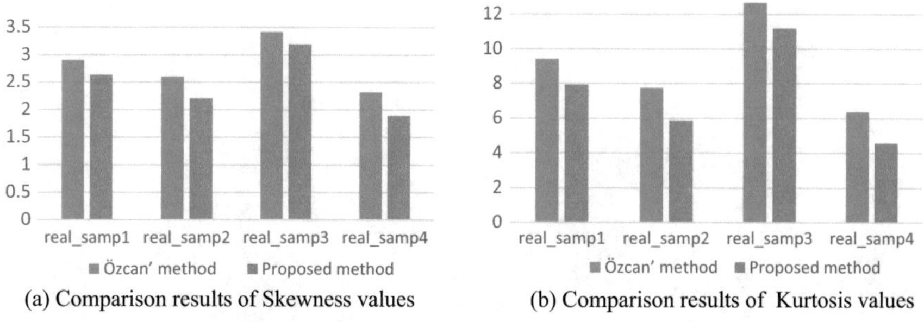

(a) Comparison results of Skewness values (b) Comparison results of Kurtosis values

Fig. 9 Comparison results of Skewness and Kurtosis values between Özcan's method and the proposed method

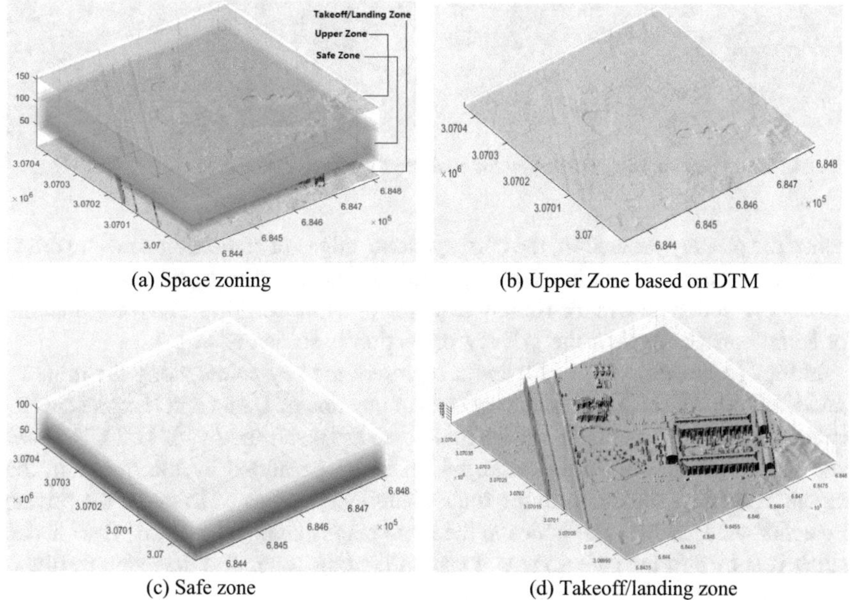

(a) Space zoning (b) Upper Zone based on DTM

(c) Safe zone (d) Takeoff/landing zone

Fig. 10 Space zoning for aerial corridor

(real_samp1, real_samp2, real_samp3 and real_samp4) by the proposed method is 7.9187, 5.8645, 11.192 and 4.578 respectively, which is also lower than the skewness values (9.4336, 7.7582, 12.6757 and 6.3757) by Özcan's method. The central limit theorem states that naturally measured samples will lead to a normal distribution. The non-ground object points may disturb the normal distribution. Low skewness and kurtosis values indicate the DTM approach to the normal distribution. Therefore, it is evidential that our method obtained accurate DTM consistently.

Figure 10 shows the visualization of space zoning result. Space is divided into three parts, namely upper zone, safe zone and takeoff/landing zone. In order to

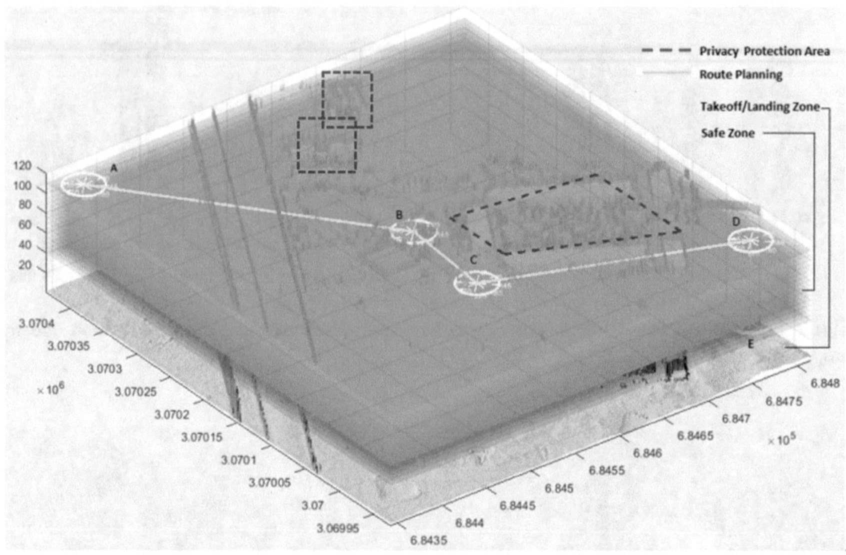

Fig. 11 Route planning for aerial corridor against privacy protection area

realize the privacy protection, the corresponding edges of restriction zone are drawn by the coordinates of the horizontal position in the DSM to avoid passing by the privacy protection area in the routing planning of aerial corridor. The route planning for aerial corridor against the privacy protection is shown in Fig. 11.

In Fig. 11, nodes A, B, C, D, and E represent the key points along the route for UASs, among which A represents the current position of UASs and E represents the landing location. If the nodes in the route map are consisted by A, B, D, E, it will pass by the privacy protection area, which is not permitted by the owner in that region. Therefore, the nodes in the route planning are adjusted to avoid the privacy protection area, namely the nodes in the route map is changed to A, B, C, D and E, which is shown in the safe zone and takeoff/landing zone. The detection results of privacy areas using real_samp1 are shown in Fig. 12.

Figure 12a shows the reference remote sensing image of real_samp1 with three privacy areas, which are marked by red rectangle respectively. Figure 12b is the objects recognition results by the proposed method, which can provide the border area for the aerial corridor to realize the privacy protection.

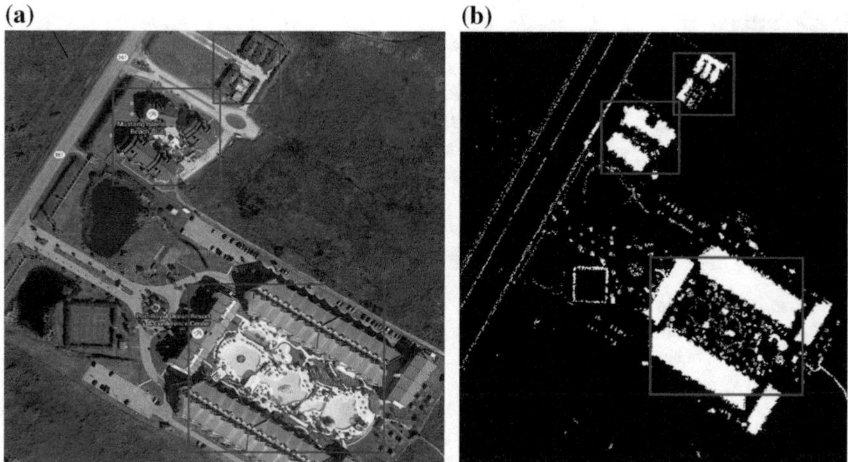

Fig. 12 Comparison of the results of privacy area detection between Özcan's method and the proposed method. **a** Remote sensing image of real_samp1 with three privacy areas. **b** Detection of privacy areas

4 Conclusion

In the paper, an automatic construction of aerial corridor from discrete LiDAR for UASs is proposed to ensure flight safety and privacy protection. The noise level estimation was used to suppress noise in the LiDAR data set. We analyze the maximum number of IMF based on the Kappa scores and the total error rates, and the results demonstrated that one was the most satisfactory choice for the maximum number of IMF. The DTM is generated by the decomposition of the filtered DSM based on EMD method and morphological operation.

As for the ISPRS dataset, it provides the reference values of ground and non-ground points. Both Kappa score and the total error rate were used to evaluate the proposed method and the state-of-the-art ones given reference dataset. With reference, Skewness and Kurtosis were used to evaluate the accuracy of DTM. In addition, satellite images were used as the reference to interpret the non-ground objects. It was demonstrated that our proposed method improved the accuracy of delineation of the non-ground objects and improved the accuracy of DTM. Using DSM and DTM, the airspace is divided into the upper zone, safe zone, and takeoff/landing zone. The privacy sensitive regions were integrated into the zoning process. Combined with the setting of privacy protection, the route planning of UAS can be adjusted automatically to avoid the private regions. Finally, the visualization technology was implemented to realize the construction of aerial corridor.

References

1. Insaurralde, C.C.: Architectural design for intelligent autonomy in unmanned aircraft. Proc. Digital Avion. Syst. 5C3-1–5C3-12 (2015)
2. Huang, W., Ding, W., Liu, C.: Multi-parametric programming approach for data link of UAS based on State Machine. In: Proceedings of Industrial Informatics-Computing Technology, Intelligent Technology, Industrial Information Integration, pp. 156–169 (2015)
3. Roelofsen, S., Gillet, D., Martinoli, A.: Reciprocal collision avoidance for quadrotors using onboard visual detection. In: Proceedings of Intelligent Robots and Systems, pp. 4810–4817 (2015)
4. Sereewattana, M., Ruchanurucks, M., Rakprayoon, P., Siddhichai, S., Hasegawa, S.: Automatic landing for fixed-wing UAV using stereo vision with a single camera and an orientation sensor: a concept. In: Proceedings of Advanced Intelligent Mechatronics, pp. 29–34 (2015)
5. Soler, M., Kamgarpour, M., Lioret, J., Lygeros, J.: A hybrid optimal control approach to fuel-efficient aircraft conflict avoidance. IEEE Trans. Intell. Transp. Syst. **99**, 1–13 (2016)
6. Feng, D., Yuan, X.: Advancement of safety corridor and emergency management visualization in low altitude airspace. J. Electron. Measur. Instrum. **30**(4), 485–495 (2016)
7. Feng, D., Yuan, X.: Advancement of security alarm aeronautical chart visualization in low altitude airspace. J. Electron. Measur. Instrum. **29**(3), 305–315 (2015)
8. Sotiriou, D., Kopsaftopoulos, F., Fassois, S.: An adaptive time-series probabilistic framework for 4-D trajectory conformance monitoring. IEEE Trans. Intell. Transp. Syst. **99**, 1–11 (2016)
9. Chougdali, S., Roudane, A., Mansouri, K., Yousfi, M., Qbadou, M.: New model for aircraft landing scheduling using real time algorithms scheduling. In: Proceedings of Intelligent Systems and Computer Vision, pp. 1–7 (2015)
10. Kim, Y., Jo, J., Shaw, M.: A light weight communication architecture for small UAS traffic management (SUTM). In: Proceedings of Integrated Communication Navigation, and Surveillance, pp. T4-1–T4-9 (2015)
11. Foina, A.G., Krainer, C., Sengupta, R.: An unmanned aerial traffic management solution for cities using an air parcel model. In: Proceeding of Unmanned Aircraft Systems, pp. 1235–1300 (2015)
12. Fadlullah, Z.M., Takaishi, D., Nishiyama, H., Kato, N., Miura, R.: A dynamic trajectory control algorithm for improving the communication throughput and delay in UAV-aided networks. IEEE Netw. **30**(1), 100–105 (2016)
13. Yuan, X., Hu, L., Buckles, B., Steinberg, L., Sarma, V.: An adaptive method for the construction of digital terrain model from Lidar data. In: Proceedings of Geoscience and Remote Sensing Symposium, pp. 828–830 (2008)
14. Sarma, V., Yuan, X., Hu, L., Yu, Y., Liu, X.: Road extraction from LiDAR using geometric and elevation features. Proceedings of the 2nd International conference on Earth Observation for Global Changes, pp. 25–29 (2009)
15. Yuan, X., Sarma, V.: Automatic urban water body detection and segmentation from sparse ALSM data via spatially constrained, model-driven clustering. IEEE Geosci. Remote Sens. Lett. **8**(1), 75–77 (2011)
16. Feng, D., Yuan, X.: Automatic construction of aerial corridor for navigation of unmanned aircraft systems in class G airspace using LiDAR. In: Proceedings SPIE 9828, Airborne Intelligence, Surveillance, Reconnaissance (ISR) Systems and Applications XIII, pp. 9828–18 (2016)
17. Voss, P.B.: Rethinking the regulatory framework for small unmanned aircraft: the case for protecting privacy and property rights in the lowermost reaches of the atmosphere. In: Proceedings Unmanned Aircraft Systems (ICUAS), pp. 173–178 (2013)
18. Kim, Y., Jo, J., Shrestha, S.: A server-based real-time privacy protection scheme against video surveillance by unmanned aerial systems. In: Proceedings Unmanned Aircraft Systems (ICUAS), pp. 684–692 (2014)

19. Huang, N.E., Shen, Z., Long, S.R., et al.: The empirical mode decomposition and the Hilbert spectrum for nonlinear and non-stationary time series analysis. Proc. R. Soc. Lond. A, 903–995 (1998)
20. Özcan, A.H., Ünsalan, C.: LiDAR data filtering and DTM generation using empirical mode decomposition. IEEE J. Sel. Topics Appl. Earth Obs. Remote Sens. 1–12, (2016)
21. Bertalmio, M., Guillermo, S., Caselles, V., Ballester, C.: Image inpainting. In: Proceedings of Computer Graphics and Interactive Techniques, pp. 417–424 (2000)
22. Thomas, J.P., Keith, C.C., William, A.M.: An improved simple morphological filter for the terrain classification of airborne LiDAR data. ISPRS J. Photogrammetry Remote Sens. **33**, 21–30 (2013)

SVD-DCT Based Medical Image Watermarking in NSCT Domain

Siddharth Singh, Rajiv Singh, Amit Kumar Singh and Tanveer J. Siddiqui

Abstract Medical images are of high importance and patient data must be kept confidential. In this chapter, we discuss a new hybrid transform domain technique for medical image watermarking and provide a detailed analysis of existing image watermarking methods. The proposed method uses a combination of nonsubsampled contourlet transform (NSCT), discrete cosine transform (DCT) and singular value decomposition (SVD) to achieve high capacity, robustness and imperceptibility. This method is non blind which requires cover image in receiver to extract watermarked image. Cover and watermark images are pre-processed in order to ensure accurate extraction of watermark. In this approach, we have considered medical images as cover and electronic patient record (EPR) is used as secret message. EPR message is embedded into selected sub band of cover image with selected gain factor so that there should be a good trade off among imperceptibility, robustness and capacity. NSCT increases hiding capacity and is more resistant to geometrical attacks. The combination of NSCT with DCT and SVD enhanced the perceptual quality and security of watermarked image. Experimental demonstration proved that the proposed method provides high robustness against geometrical and

S. Singh
Department of Electronics and Communication Engineering,
V. B. S. Purvanchal University, Jaunpur, Uttar Pradesh, India
e-mail: siddharthjnp@gmail.com

R. Singh (✉)
Department of Computer Science, Banasthali University,
Banasthali, Rajasthan, India
e-mail: jkrajivsingh@gmail.com

A.K. Singh
Department of Computer Science and Engineering, Jaypee University
of Information Technology, Waknaghat, Solan, Himachal Pradesh, India
e-mail: amit_245singh@yahoo.com

T.J. Siddiqui
Department of Electronics and Communication, University of Allahabad,
Allahabad Uttar Pradesh, India
e-mail: siddiqui.tanveer@gmail.com

A.E. Hassanien et al. (eds.), *Quantum Computing: An Environment for Intelligent Large Scale Real Application*, Studies in Big Data 33,
https://doi.org/10.1007/978-3-319-63639-9_20

signal processing attacks in terms of peak signal to noise ratio (PSNR) and correlation coefficient (CC).

Keywords Medical image watermarking · Nonsubsampled contourlet transform · Discrete cosine transform · Singular value decomposition · Transform domain image watermarking

1 Introduction

The advancement in multimedia systems and their wide use over the internet poses serious challenges to the information security. Multimedia watermarking provides solutions to the suspicious and malicious attacks on data privacy and security [1]. It includes the security of audio, image and video files. Watermarking, cryptography and steganography are the possible ways of data protection. Among these, watermarking is one of the most popular and found useful for copyright protection, content authentication, rightful ownership and secure communication [2–6]. In biomedical engineering, protection of medical data and images is crucial and their transmission to network should be made secured [7]. Also storage of electronic patient record (EPR) and medical images for small hospitals is of great concern. Furthermore, patient's disease should be kept confidential in medical reports accurately [8, 9]. One possible solution to keep medical data secure is digital image watermarking for this purpose, medical image is used as a cover image which reduces chance of tampering or manipulation. However, due to security threats to medical image security, development of medical data/image transmission algorithms is highly required.

In literature, a number of spatial and transform domain techniques have been proposed for image watermarking [10, 11]. However, transform domain techniques have gained great interest, as it provides high capacity and increased security [12–15]. Recently, hybrid techniques for image watermarking have become popular and several combinations of discrete wavelet transform (DWT), DCT and SVD have been proposed by many authors [16–20]. These methods have been found limited in hiding capacity and robustness. Therefore, in this chapter, we have focused on nonsubsampled contourlet transform (NSCT) and combined it with DCT and SVD for medical image watermarking. Since, NSCT decomposition yields a number of subbands, which increases the hiding capacity of the secret message. The combination of NSCT, DCT and SVD increases capacity along with robustness and imperceptibility which are desired for image watermarking. The proposed technique has been tested over a number of medical images and standard image data sets. To validate the performance of the proposed method, we tested it against geometrical and signal processing attack and compared with Singh et al. [20], Rosiyadi et al. [21], Srivastava and Saxena [22] and Tayal and Singh [23] in terms of peak signal to noise ratio (PSNR) and correlation coefficient (CC).

The rest of the chapter is organized as follows: related work is discussed in Sect. 2. In Sect. 3, brief details of NSCT, DCT and DWT are given. Section 4 explains the proposed method. Section 5, experimental results are discussed. Finally, Sect. 6 concludes the proposed work.

2 Related Work

Image watermarking techniques are very popular in literature and they differ in choice of cover media, recovery method and embedding domain. Image watermarking can be broadly divided into spatial domain [24], transform domain [25], spread spectrum techniques [26], intelligent and adaptive techniques [27]. In spatial domain, secret message is hidden directly in pixel values of cover image whereas in transform domain techniques embedding is done into transform coefficients obtained after forward transformation. Intelligent techniques can be used for optimally choose the transform coefficients for embedding data. The transform domain techniques are more robust against image processing attacks. Here, we provide a separate literature over spatial domain, transform domain and intelligent image watermarking techniques.

2.1 Spatial Domain Image Watermarking

In this approach, secret message bits are embedded directly by substituting the bits of the cover image. This method is computationally simple, straightforward and has large embedding capacity. Various methods for embedding in spatial domain are least significant bit (LSB) substitutions, patchwork, bit plane complexity segmentation, gray level modification, pixel value difference and quantization index modulation. Among all these techniques, LSB substitution is most popular. It involves random selection of a pixel of the cover image and replacement of its LSB with a message bit. This process is repeated until all data bits are inserted. LSB retains the image quality, as the changes in the values of LSBs during embedding will have least effect on stego image quality. LSB based watermarking achieve high embedding capacity and less robust against scaling, rotation, cropping and lossy compression attacks.

Patchwork technique is also used for embedding a message in a cover image [28]. A pseudorandom generator has been used to select two patches A and B. Patch A pixels are lightened whereas patch B are darkened. The changes of contrast in corresponding patches encodes one bit of embedding. The advantage of this technique is that the secret image is spread over the whole cover image, even if one patch is damaged, the other will persist. This method is not dependent on cover image and is more robust against JPEG compression operation.

Bit plane complexity segmentation (BPCS) watermarking is another spatial domain method which was proposed by Niimi et al. [29]. In this method concept of bit planes has been used for data hiding. Since in this method image blocks s divided into blocks, therefore, embedding capacity is increased and we get high quality stego image.

Potdar and Chang [30] proposed a data hiding method based on gray level modification. This method is used to transform data by modifying gray levels of pixels. It is based on concept of odd and even numbers to map data within cover image. This method has low computation complexity and high embedding capacity.

Another spatial domain watermarking technique for embedding secret message has been proposed on the basis of difference between values of two adjacent pixels [31]. Embedding process starts with dividing the cover image into non overlapping blocks of two adjacent pixels and pixel difference in each block has been calculated. The cover image contains large coefficient values in edge area and smaller coefficient values in smooth area. It means that edge area has large difference between adjacent pixels than smooth area in cover image. This method outperforms in term of imperceptibility as compared to LSB substitution method. Various method based on pixel value difference (PVD) are triway PVD, four pixel PVD etc. [32–34]. In quantization index modulation (QIM), embedding in cover image is done by first modulating the index of indices with hidden information and then quantizing cover signal with associated quantizer. It has higher embedding capacity and robustness against attacks. Thus, spatial domain watermarking methods achieve high embedding capacity and less robust against lossy compression and geometrical operations such as cropping, rotation, scaling etc.

2.2 Transform Domain Image Watermarking

Transform domain watermarking techniques provide better robustness and imperceptibility against signal and image processing attacks. Transform domain [35, 36] includes discrete fourier transform (DFT), discrete cosine transform (DCT) and DWT. DFT based watermarking techniques are not popular because it introduce round off errors that is ideal choice for data hiding applications. DCT was widely used by international data compression standards such as JPEG and MPEG file format. The use of DCT technique to hide information in digital media was introduced by Koch and Zhao [37] and Cox and Miller [38].

DCT based tools such as Jsteg [39], Outguess [40] and F5 [41] are used for image watermarking. F5 is further improved by Fan et al. [42]. Outguess spreads hidden data by selecting coefficients with the user-selected password [35].

Several image watermarking algorithms have been proposed [28] for data hiding in cover. For copyright protection, Barni et al. [43] has developed DCT based watermarking algorithm and embedded watermark in middle band DCT. This method enhances visual quality and showed robustness against image processing attacks. Another DCT based algorithm for copyright protection has been proposed by Hernanddez et al. [44]. This algorithm is blind and DCT was applied on cover image after dividing it into 8 × 8 blocks.

Statistical properties of DCT were utilized by Briassouli et al. [45]. They developed a blind watermark detector which used the Cauchy statistical model. Amin et al. [46] also proposed a statistically data hiding algorithm in DCT domain and improved security and robustness against noise addition, filter processing, sharpening, blurring and JPEG compression attacks. Histogram matching JPEG watermarking and JPEG watermarking were introduced by Noda et al. [47] using quantization index modulation (QIM) in DCT domain which was better than F5 in term of embedding rate and PSNR.

DCT based mod-4 blind watermarking method [48] supports both uncompressed and compressed images for data hiding. This method has been comparatively analyzed in terms of embedding capacity, PSNR, Universal Quality Index (Q) and steganalysis attacks. A lossless and reversible data hiding algorithm was proposed by Chang et al. [49] in DCT domain. Middle band DCT coefficients has been used to hide secret message and resulted in improved visual quality of stego image.

Lin et al. [50] have shown improvement in watermarking algorithm by utilizing low frequency DCT coefficients of the cover image. They used the concept of mathematical remainder which enabled robustness against JPEG compression. Singh et al. [51] proposed another DCT domain watermarking technique by embedding secret data in the middle frequency DCT coefficients. Among log normal, Pareto, Weibull and Gaussian distribution, they found that Weibull distribution is suitable for embedding data in middle band DCT coefficients. An adaptive DCT approach was proposed by Mali et al. [52] using energy threshold for selecting the embedding location in DCT domain. This method showed better robustness against compression, tampering, resizing, filtering and addition of Gaussian noise.

Wavelet transforms have been also used for image watermarking for data hiding [53]. Multi-resolution property of wavelet transform allows embedding data independently at different resolutions. Compression standards like JPEG2000 uses wavelet transforms. Therefore, wavelet transform are an obvious choice for data hiding.

Pixel wise masking technique was proposed by Barni et al. [54] for image watermarking. Watermark is hidden by modifying DWT coefficients of cover image which used HVS (Human visual system) characteristics for determining watermark strength. Kundur and Hatzinakos [55] proposed a robust watermarking technique using DWT based image fusion principle for copyright protection. Here again HVS model has been used to determine salient image component for embedding watermark. Kamstra and Heijmans [56] introduced a high capacity lossless reversible data embedding in wavelet domain and proposed least significant bit prediction and Sweldens lifting scheme. Liu et al. [57] proposed a new watermarking scheme for copyright protection and embedded watermark in difference value between the original cover and reference images.

Multiwavelet transform based watermarking algorithm has been proposed by Ghouti et al. [58] which is suitable for real time watermarking application. Lee et al.

[59] investigated a reversible image watermarking scheme in integer wavelet domain. In this approach, original image is divided into non overlapping blocks and watermark is embedded into high frequency integer wavelet coefficients using LSB substitution. Another integer wavelet based reversible data hiding algorithm was proposed by Peng et al. [60].

Lin et al. [61] proposed a blind image watermarking d based on the difference of wavelet coefficient quantization for copyright protection. Haar Digital wavelet transform (HDWT) based reversible data hiding scheme was proposed by Chan et al. [62]. Logo watermarking in wavelet domain was introduced by Bhatnagar et al. [63]. They obtained high quality of stego images and improved robustness against various intentional or unintentional attacks. Yeh et al. [64] proposed data hiding using wavelet bit plane for compressed images. In this method, bit planes of DWT coefficients have been used for message hiding using multistage encoding.

Numerous advance wavelet domain based image watermarking techniques has been proposed [65–70]. Contourlet transform [65] was an advancement over wavelet transform. The directionality and anisotropy properties of contourlet are suitable for watermarking. Sajedi et al. proposed a new adaptive contourlet transform based watermarking scheme that embeds data in specific cover image, selected by best cover image measures such as (PSNR) and number of modifications of cover image. They hide the secret data in contourlet coefficient through iterative embedding process to reduce the stego image distortion. Khalighi et al. [67] presented a new multiresolution image watermarking using contourlet transform. In the proposed method, contourlet transformed coefficients of host image are modified to embed the watermark. The quality analysis of the watermarked image was good in terms of visibility and PSNR values (average of 39.41 dB). The proposed method provides good visual quality and robustness against JPEG compression, addition noise, filtering and geometrical transformations; it is also suitable method for fingerprinting applications. Leung et al. [68] proposed six different non blind watermarking method based on curvelet transform and compared with existing watermarking scheme. They embedded different watermark in three different bands. Their method has better robustness against common image processing attacks. Mansouri et al. [70] proposed a new non blind SVD based image watermarking in complex wavelet transform (CWT) domain. They embedded modified singular values of CWT coefficients of cover image with singular value of watermark. The proposed method provides high robustness against common image processing and geometrical attack like cropping and rotation. Bhatnagar and Wu [69] introduced a novel biometric inspired watermarking based on fractional dual tree complex wavelet transform and singular value decomposition. The main advantage of using biometric based keys is that no one can get the key without original owner of biometric information. Their scheme provides high security and robustness against filtering, noise addition, JPEG compression, resizing, cropping, rotation, contrast adjustment and sharpen attack.

2.3 Intelligent and Adaptive Image Watermarking Techniques

Intelligent image watermarking are the applications of intelligent algorithms such as Genetic Algorithm (GA) and Particle Swarm optimization (PSO). These intelligent approaches are useful in optimizing embedding strength. GA based data hiding algorithm has been proposed by Shieh et al. [71] in DCT domain. Use of GA improved robustness against image processing attacks. PSO based method was introduced by Li and Wang [72] that modified JPEG quantization table and embed secret message bits in middle frequency DCT coefficient. This method also incorporated the concept of LSB substitution for data hiding.

Ishtiaq et al. [73] investigated an adaptive watermark strength selection algorithm using DCT and PSO. Like GA, PSO also optimized watermark strength in DCT domain and hence provided better robustness against addition of noise, low pass filtering, high pass filtering and median filtering, shifting and cropping operations. Wavelet based watermarking scheme using PSO has been proposed by Wang et al. [74] in order to improve imperceptibility, robustness and security. Model based image watermarking was introduced by Sallee [75] which allowed robust hiding of secret message without affecting statistical properties of cover image.

3 Theoretical Background

In this section, we provide a brief detail of the NSCT, DCT and SVD, and their usefulness in image watermarking.

3.1 Non Sub-sampled Contourlet Transform (NSCT)

Even though DWT and other hybrid combinations of DWT are popular, powerful, and familiar among techniques of watermarking, but it has its own limitations in capturing the directional information such as smooth contours and the directional edges of the image [76]. This problem is addressed by nonsubsampled contourlet transform (NSCT). NSCT offers directionality and anisotropy. NSCT methods perform much better than wavelet-based methods in images. Though, NSCT is a redundant transform [77] but have a very good property of shift invariance which will make our proposed algorithm more robust against various geometrical attacks. The redundancy property of NSCT is useful for accurate recovery of secret message, as no information loss occur on decomposition of watermark images [78, 79]. It contains two filter banks i.e. Non-down sampling pyramid Filter (NDSPF) and non-sub sampled directional filter bank (NSDFB) [77], shown in Fig. 1.

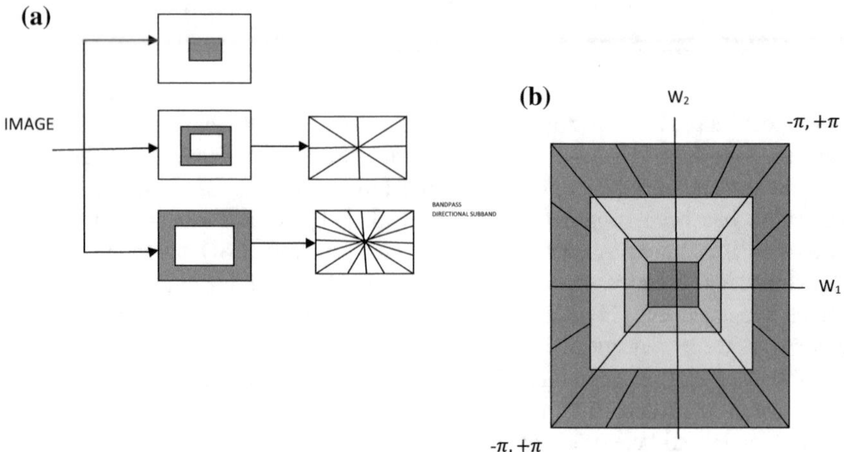

Fig. 1 **a** Non sub sampled filter bank structure, **b** The iterated frequency partitioning

3.2 Discrete Cosine Transform (DCT)

Discrete cosine transform (DCT) is a real domain transform which represents an image as coefficient of different frequencies of cosine which is basis vector for this transform [36, 80]. It works by separating an image into different blocks, such as low, high and middle frequency coefficients which makes it easier for embedding the watermark information into middle frequency band that will give an additional resistance with respect to the lossy compression techniques, while avoiding significant modification of the cover image. The DCT possess energy compaction property.

The forward transform equation is given below

$$b(u,v) = c(u)c(v)\sum_{x=0}^{N-1}\sum_{y=0}^{N-1} a(x,y)\cos\left(\frac{\pi u(2y+1)}{2N}\right) cos\left(\frac{\pi v(2y+1)}{2N}\right) \tag{1}$$

$$c(u) = \begin{cases} \frac{1}{\sqrt{N}} & if\ u=0 \\ \sqrt{\frac{2}{N}} & otherwise \end{cases} \tag{2}$$

where (x, y) is the intensity of the pixel in row x and column y of the image, and b (u, v) is the DCT coefficient in row u and column v of the DCT matrix.

3.3 Singular Value Decomposition

Any image of dimension $N \times N$ can be decomposed into two orthogonal matrices U *and* V and one diagonal matrix S [81] such that

$$M = USV^T \quad (3)$$

The major advantage of SVD for using it in watermarking technique is that a small variation in singular values does not affect visual perceptibility of image [78] and these singular values are unique in nature which makes any watermarking algorithm robust against geometrical attacks.

4 The Proposed Method

The proposed method of watermarking contains two stages: embedding and extraction followed by the combination of NSCT, DCT and SVD. The proposed method increases capacity, robustness and imperceptibility of watermarking without affecting the quality of cover image against various geometrical and signal processing attacks. We have considered NSCT because it captures the directional edges and smooth contours of medical images and is better than other wavelet transforms. Also, it decomposes cover image into six subbands, therefore, we have increased hiding capacity for data hiding.

4.1 Embedding Process

For proposed algorithm, the steps of the proposed method are given in Fig. 2.

Step 1: The cover image of size $N \times N$ and the watermark image of size $N/2 \times N/2$ are transformed into NSCT domain. For NSCT decomposition '1' level Laplacian pyramid filter followed by Directional filter Bank is used. After NSCT decomposition cover and watermark image is divided into 6 subbands {1, 1}, {1, 2}, {1, 3}{1, 1}, {1, 3}{1, 2}, {1, 4}{1, 1}, {1, 4}{1, 2}.

Step 2: Select any subband of NSCT decomposed cover and watermark image and then apply DCT and obtain DCT coefficients for the same.

Step 3: Compute SVD of DCT coefficients of cover image $[U_C\ S_C\ V_C^T]$ and also of watermark image $[U_W\ S_W\ V_W^T]$.

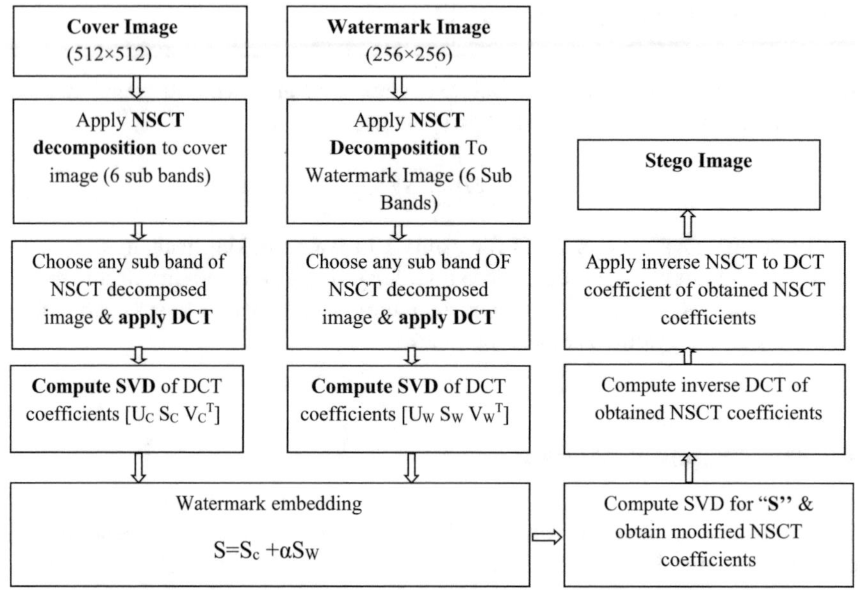

Fig. 2 The proposed embedding process

Step 4: The SVD coefficients are modified as:

$$S = S_c + \alpha\, S_W$$

where α is Gain factor which controls imperceptibility of watermark in cover image and robustness against attacks.

Step 5: Compute SVD of the modified coefficients "S" and obtain modified NSCT coefficients of cover image.

Step 6: Compute inverse DCT of the obtained modified NSCT coefficients and then apply inverse NSCT to get the watermarked image.

4.2 Recovery Process

The steps of recovery of embedded message are illustrated in Fig. 3.

Step 1: Perform NSCT decomposition on watermarked image and by this, watermarked image is divided into 6 sub bands {1, 1}, {1, 2}, {1, 3}{1, 1}, {1, 3}{1, 2}, {1, 4}{1, 1}, {1, 4}{1, 2}. The same band is selected which was selected in the embedding side.

Step 2: Applying DCT on subband of NSCT decomposed watermarked image and obtain DCT coefficients.

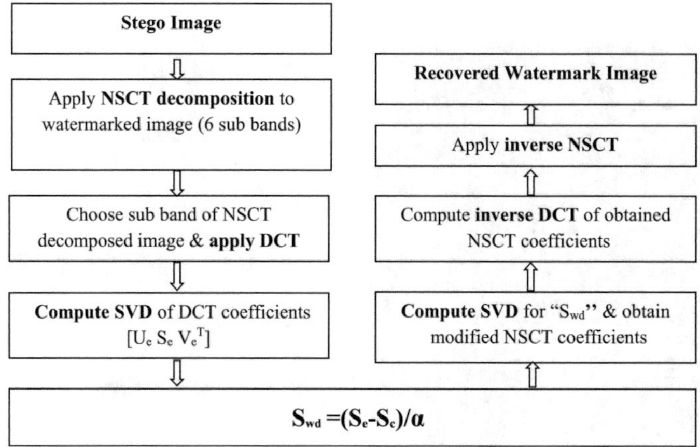

Fig. 3 The proposed recovery process

Step 3: Compute SVD of DCT coefficients $[U_e\ S_e\ V_e^T]$

Step 4: Performing the operation by modifying the SVD coefficients as:

$$S_{wd} = (S_e - S_c)/\alpha$$

where α is same gain factor which was used in the embedding side.

Step 5: Apply SVD to the "S_{wd}" and obtain the modified NSCT coefficients.

Step 6: Compute inverse DCT of the obtained modified NSCT coefficients and then apply inverse NSCT to get the Recovered watermark image.

5 Results and Discussion

We have performed experiment over various medical images such as brain MRI, chest CT, kidney stones, Hand X-ray images and other benchmark images. These images have been experimented for EPR and fingerprint watermarks. The robustness of the proposed method has been determined by PSNR and CC values. In the proposed method size of cover image is 512 × 512 and watermark image size of 256 × 256 has been selected. The overall performance of the proposed method depends on the size of cover and watermarked image, selection of subbands for data hiding.

Figure 4a, b shows brain MRI and watermark EPR images. For this pair of images, corresponding stego and recovered watermark image have been given in Fig. 4c, d. From Fig. 4, one can conclude that the visual quality of obtained stego

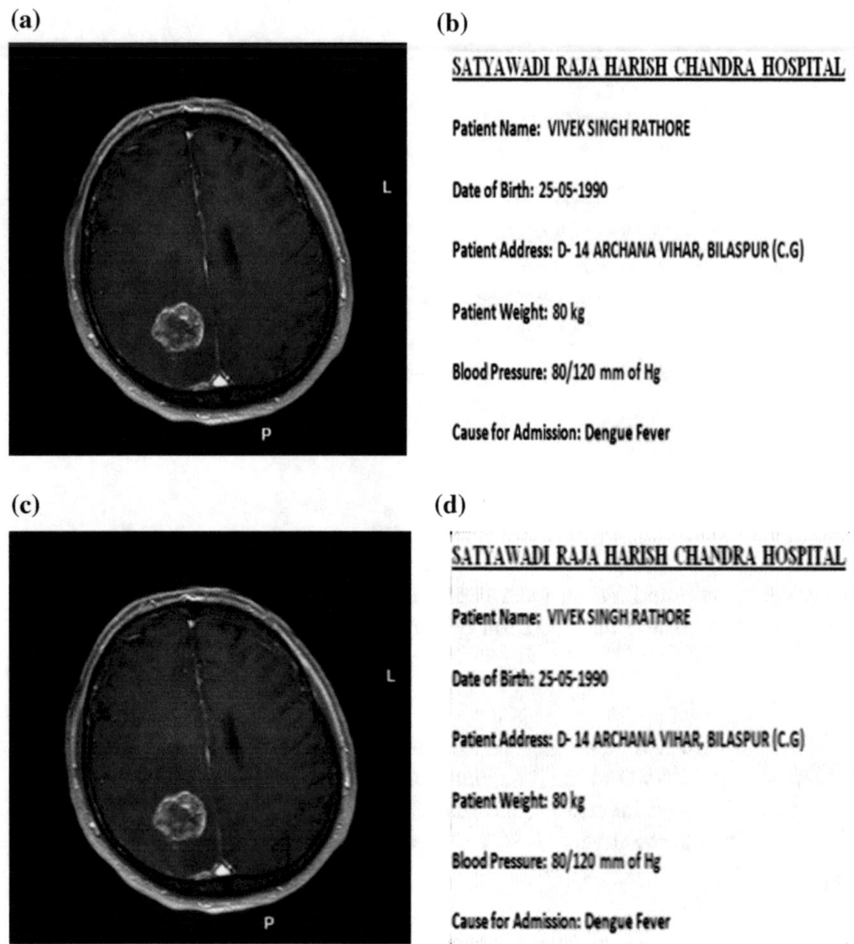

Fig. 4 **a** Original image, **b** watermark image, **c** stego image, and **d** recovered watermark image

and recovered watermark images are comparable to the original source images, which indicates the suitability of the proposed method.

PSNR and CC values for the proposed method have been determined at different values of gain factor. These values are given in Table 1. Referring Table 1, it is easily seen that by varying gain factor at different subbands the quality of the stego image and recovered watermark image improves or degrades. At gain factor 0.01 and 0.05, PSNR value is high but imperceptibility is low. Similarly, at higher gain factors imperceptibility is good but PSNR value goes down. Hence, there should be a good trade off between imperceptibility, robustness and capacity of embedding watermark into cover image. For these reasons, we have selected subband {1, 4}{1, 1} at gain factor = 0.1 in which CC is maximum i.e. 0.9990.

Table 1 Performance of proposed method in different subbands for different values of gain factor

Sub bands	Gain factor											
	0.01		0.05		0.10		0.50		1.0		5.0	
	PSNR	CC	PSNR	CC	PSNR	CC	PSNR	CC	PSNR	CC	PSNR	CC
{1, 2}	33.5393	0.9477	32.5943	0.9521	32.3171	0.9517	33.4348	0.9287	33.7641	0.9213	33.9209	0.9175
{1, 3}{1, 1}	31.3490	0.9903	25.3970	0.9971	21.8084	0.9981	23.8909	0.9943	24.5692	0.9932	24.4780	0.9934
{1, 3}{1, 2}	29.9157	0.9929	21.0559	0.9985	19.9164	0.9979	22.5656	0.9954	22.7016	0.9953	22.5782	0.9954
{1, 4}{1, 1}	**36.8402**	0.9646	26.1876	0.9958	19.4948	**0.9990**	20.5130	0.9966	21.1163	0.9961	20.6093	0.9966
{1,4}{1, 2}	34.8159	0.9770	20.7517	0.9989	17.2811	0.9989	20.0072	0.9969	20.0530	0.9969	19.6283	0.9972

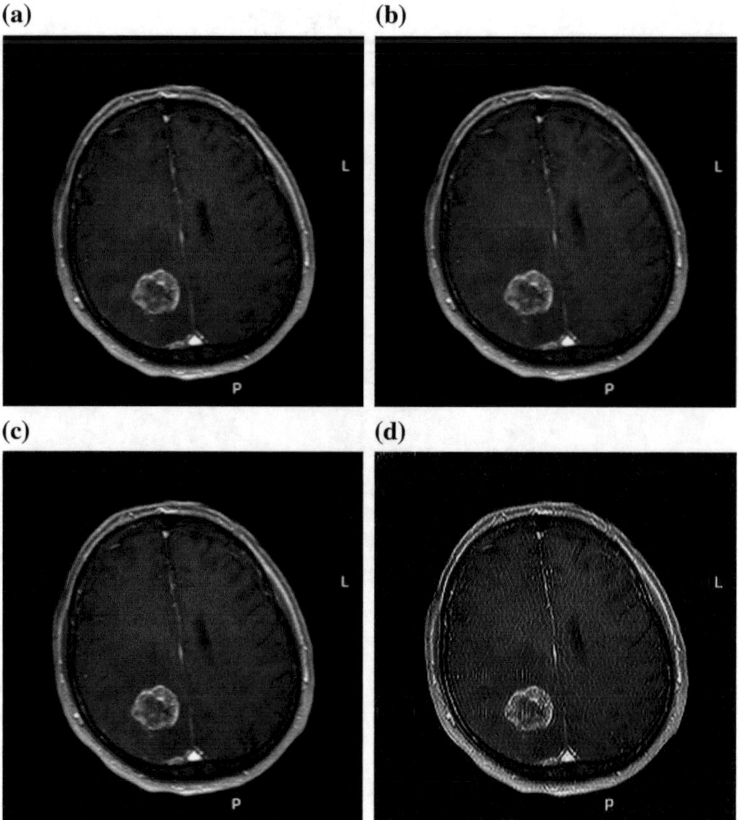

Fig. 5 **a** Brain MRI cover image, and **b–d** stego images with gain factor 0.01, 1.0 and 5.0 respectively

Furthermore, to validate the proposed method, we have shown stego and corresponding recovered watermarked images at different gain factors in Figs. 5 and 6. Here, again, we have found that the proposed method performs well for different gain factors. Moreover, we have performed experiments on standard Lena image as cover and fingerprint image as watermark, shown in Fig. 7. The quality of recovered watermark is comparable to the original message as given in Fig. 7. To test the robustness of the proposed algorithm, we have tested the proposed method against geometrical and signal processing attacks, namely average filter, Gaussian noise, Gaussian blur, Histogram equalization, Motion blur, Salt and pepper noise, resizing, rotation, speckle, Weiner filtering, Gaussian filtering, and Median filtering attacks. Recovered watermarks for Lena images against these attacks have been shown in Fig. 8. The results shown in Fig. 8 clearly indicate that the proposed method is robust against these attacks and the recovered watermarks are comparable to the original watermark. We have compared the performance of the proposed method in terms of PSNR and CC, shown in Table 2. Here, PSNR and CC values are given

(a)

SATYAWADI RAJA HARISH CHANDRA HOSPITAL

Patient Name: VIVEK SINGH RATHORE

Date of Birth: 25-05-1990

Patient Address: D- 14 ARCHANA VIHAR, BILASPUR (C.G)

Patient Weight: 80 kg

Blood Pressure: 80/120 mm of Hg

Cause for Admission: Dengue Fever

(b)

SATYAWADI RAJA HARISH CHANDRA HOSPITAL

Patient Name: VIVEK SINGH RATHORE

Date of Birth: 25-05-1990

Patient Address: D- 14 ARCHANA VIHAR, BILASPUR (C.G)

Patient Weight: 80 kg

Blood Pressure: 80/120 mm of Hg

Cause for Admission: Dengue Fever

(c)

SATYAWADI RAJA HARISH CHANDRA HOSPITAL

Patient Name: VIVEK SINGH RATHORE

Date of Birth: 25-05-1990

Patient Address: D- 14 ARCHANA VIHAR, BILASPUR (C.G)

Patient Weight: 80 kg

Blood Pressure: 80/120 mm of Hg

Cause for Admission: Dengue Fever

(d)

SATYAWADI RAJA HARISH CHANDRA HOSPITAL

Patient Name: VIVEK SINGH RATHORE

Date of Birth: 25-05-1990

Patient Address: D- 14 ARCHANA VIHAR, BILASPUR (C.G)

Patient Weight: 80 kg

Blood Pressure: 80/120 mm of Hg

Cause for Admission: Dengue Fever

Fig. 6 **a** EPR as watermark image, and **b–d** recovered watermark image with gain factors 0.01, 1.0 and 5.0 for Fig. 5b–d respectively

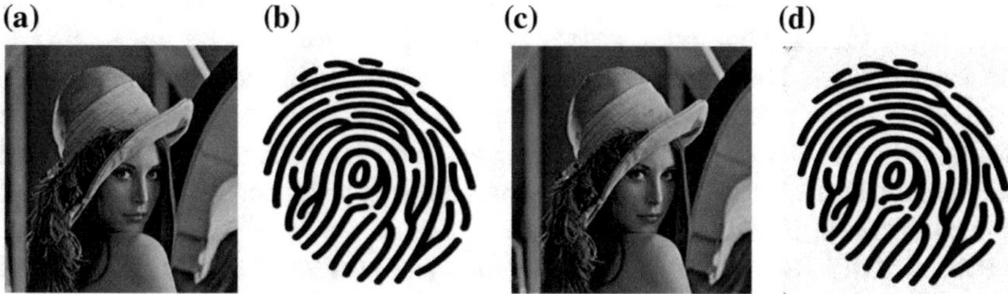

Fig. 7 **a** Cover image, **b** watermark image, **c** stego image, and **d** recovered watermark image

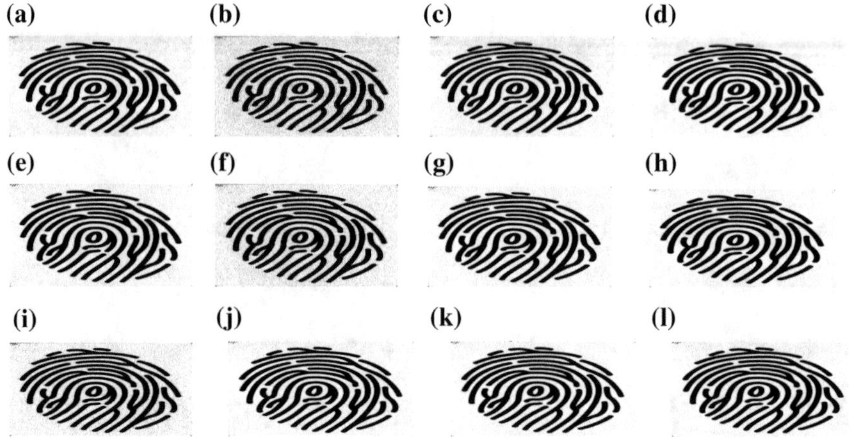

Fig. 8 Recovered watermarks from Lena image after attacks. **a** Average filter attack, **b** Gaussian noise attack (mean = 0.0 variance = 0.05), **c** Gaussian blur attack, **d** Histogram equalization attack, **e** Motion blur attack, **f** Pepper noise attack (density = 0.05), **g** resizing Attack (one half), **h** rotation attack (70), **i** Speckle noise attack (density = 0.05), **j** Weiner filtering [3 3], **k** Gaussian filtering [3 3], **l** Median filtering [3 3]

Table 2 Performance of proposed method for different cover image at gain factor = 0.1 for subband {1, 4}{1, 1}

Cover image	CC	PSNR
Brain MRI	0.9985	18.5950
Chest CT	0.9977	19.2730
Kidney stones	0.9988	20.1275
Hand X-Ray	0.9986	18.5183
Pancreas	0.9985	21.4953
Lena	0.9990	19.4948
Zelda	0.9983	18.6096

for different cover images at gain factor 0.1. From Table 2, it is clear that the proposed method possess high values of CC and PSNR. CC has been highest for Lena image and PSNR is highest for pancreas image. Further, Table 2 clearly shows that obtained PSNR and CC values are comparable, that can also be observed in Fig. 9. This figure shows the CC values for different cover images. Further, we have tested different cover images against JPEG compression, median filtering, Gaussian filtering, Weiner filtering, Gaussian noise, Salt and pepper noise, speckle noise, histogram equalization, resizing, Gaussian blur, motion blur, average filter and rotation attacks. The CC values of recovered watermarks at gain factor 0.1 have been shown in Table 3.

Referring this Table 3, the proposed method has been found suitable for different cover data including medical and other standard images. This can be easily verified from the Fig. 10 which shows the CC values of recovered watermark images against geometrical and image processing attacks.

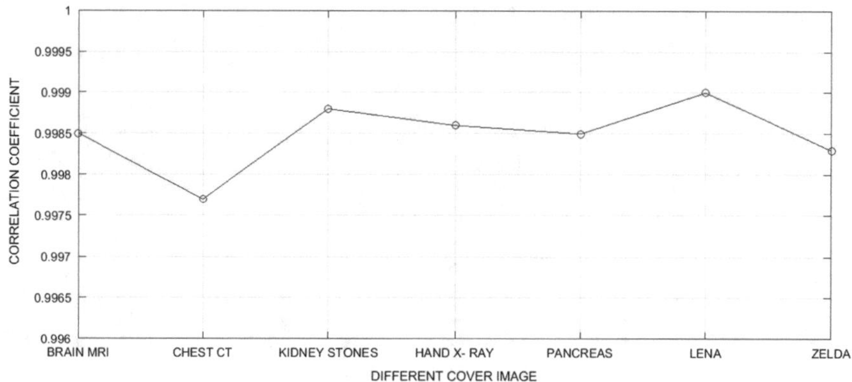

Fig. 9 Correlation coefficients for different cover images

Table 3 Performance of proposed method in terms of CC against different attacks at gain factor = 0.1 at subband {1, 4}{1, 1} for different cover image

Attacks	Brain MRI	Chest CT	Kidney stones	Hand X-Ray	Lena
JPEG compression (Qf-50)	0.9985	0.9969	0.9988	0.9985	0.9990
Median filtering [3 3]	0.9988	0.9953	0.9983	0.9988	0.9984
Gaussian filtering [3 3]	0.9989	0.9957	0.9983	0.9990	0.9983
Weiner filtering [3 3]	0.9989	0.9959	0.9983	0.9989	0.9983
Gaussian noise with mean = 0.0 variance = 0.05	0.9632	0.9608	0.9643	0.9743	0.9592
Salt and pepper noise (density = 0.05)	0.9685	0.9715	0.9700	0.9702	0.9758
Speckle noise (density = 0.05)	0.9955	0.9757	0.9943	0.9894	0.9840
Histogram equalization	0.9978	0.9982	0.9966	0.9985	0.9972
Resizing attack [1/2]	0.9987	0.9929	0.9970	0.9985	0.9966
Gaussian blur attack	0.9973	0.9913	0.9941	0.9971	0.9941
Motion blur attack	0.9978	0.9916	0.9947	0.9976	0.9951
Average filter attack	0.9975	0.9915	0.9943	0.9973	0.9944
Rotation attack (50)	0.9987	0.9937	0.9981	0.9977	0.9988

The proposed method is further compared against JPEG compression, median filtering, Gaussian noise, Salt and pepper noise and histogram equalization attacks with Singh et al. [20], Rosiyadi et al. [21], Srivastava and Saxena [22] and Tayal and Singh [23] and the CC values of these methods are given in Table 4. The dashed values in Table 4 indicate that the CC values are not available for the corresponding attacks. With reference to this Table 4, it is clear that the proposed method has the highest values of CC than the compared methods [20–23] except for JPEG compression in which CC value of Singh et al. [20] is higher than the

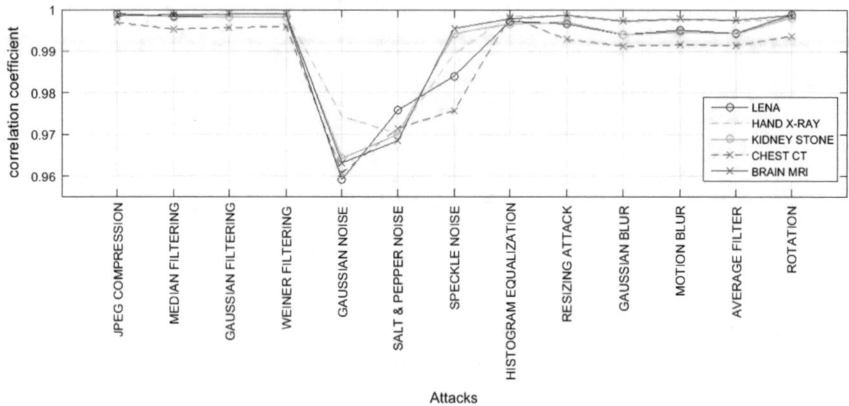

Fig. 10 Correlation coefficients of recovered watermark images against attacks

Table 4 Comparison of the proposed method Correlation coefficients

Attacks	Proposed method	Singh et al. [20]	Rosiyadi et al. [21]	Srivastava and Saxena [22]	Tayal and Singh [23]
JPEG compression (QF = 50)	0.9990	0.9994	0.1863	–	–
Median filtering [2 2]	0.9985	0.9752	0.4585	0.6019	–
Gaussian noise with mean = 0, variance = 0.01	0.9830	0.9754	–	0.6320	0.8893
Salt and pepper (density = 0.01)	0.9972	0.9952	–	–	0.9636
Histogram equalization	0.9972	0.9208	–	0.9123	0.9941

proposed method. Thus, from overall comparison of the proposed method, it can be concluded that the proposed method, i.e. combination of NSCT, DCT and SVD, provides high capacity, robustness and imperceptibility for medical images as well as for standard images.

6 Conclusions

In this chapter, we have proposed a new hybrid transform domain watermarking using NSCT, DCT and SVD for medical images. In this method, medical image is taken as cover and EPR as watermark and is embedded into higher frequency subband of cover image. By varying gain factor and subbands, quality of watermarked image and recovered watermark image changes in terms of PSNR, CC, and imperceptibility. Results demonstrated that the proposed combination provides high capacity to embed watermark and better imperceptibility. Furthermore, better robustness have been achieved against geometrical and signal processing attacks such as average filter, Gaussian noise, Gaussian blur, Histogram equalization, Motion blur, Salt and pepper noise, resizing, rotation, speckle, Weiner filtering, Gaussian filtering, and Median filtering attacks. The proposed method has been found superior to the existing image watermarking methods.

References

1. Potdar, V.M., Han, S., Chang, E.: A survey of digital image watermarking techniques. In: 2005 3rd IEEE International Conference on Industrial Informatics, INDIN'05, pp. 709–716. IEEE (2005)
2. Lin, S.D., Chen, C.F.: A robust DCT-based watermarking for copyright protection. IEEE Trans. Consum. Electron. **46**(3), 415–421 (2000)
3. Lu, C.S., Liao, H.Y.: Multipurpose watermarking for image authentication and protection. IEEE Trans. Image Process. **10**(10), 1579–1592 (2001)
4. Qi, X., Xin, X.: A quantization-based semi-fragile watermarking scheme for image content authentication. J. Vis. Commun. Image Represent. **22**(2), 187–200 (2011)
5. Liu, R., Tan, T.: An SVD-based watermarking scheme for protecting rightful ownership. IEEE Trans. Multimedia **4**(1), 121–128 (2002)
6. Celik, M.U., Sharma, G., Saber, E., Tekalp, A.M.: Hierarchical watermarking for secure image authentication with localization. IEEE Trans. Image Process. **11**(6), 585–595 (2002)
7. Cao, F., Huang, H.K., Zhou, X.Q.: Medical image security in a HIPAA mandated PACS environment. Comput. Med. Imaging Graph. **27**(2), 185–196 (2003)
8. Roy, S., Pal, A.K.: A robust blind hybrid image watermarking scheme in RDWT-DCT domain using Arnold scrambling. Multimedia Tools Appl. 1–40 (2016)
9. Ulutas, M., Ulutas, G., Nabiyev, V.V.: Medical image security and EPR hiding using Shamir's secret sharing scheme. J. Syst. Softw. **84**(3), 341–353 (2011)
10. Sreenivas, K., Prasad, V.K.: Fragile watermarking schemes for image authentication: a survey. Int. J. Mach. Learn. Cybern. 1–26 (2017)
11. Zheng, D., Liu, Y., Zhao, J., Saddik, A.E.: A survey of RST invariant image watermarking algorithms. ACM Comput. Surv. (CSUR) **39**(2), 5 (2007)
12. Meerwald, P., Uhl, A.: Survey of wavelet-domain watermarking algorithms. In: Photonics West 2001—Electronic Imaging, pp. 505–516. International Society for Optics and Photonics (2001)
13. Mousavi, S.M., Naghsh, A., Abu-Bakar, S.A.R.: Watermarking techniques used in medical images: a survey. J. Digital Imaging **27**(6), 714–729 (2014)
14. Al-Dmour, H., Al-Ani, A.: a medical image steganography method based on integer wavelet transform and overlapping edge detection. In: Neural Information Processing, pp. 436–444. Springer International Publishing (2015)
15. Navas, K.A., et al.: Lossless watermarking in JPEG2000 for EPR data hiding. In: 2007 IEEE International Conference on Electro/Information Technology (2007)
16. Singh, A.K., Dave, M., Mohan, A.: Hybrid technique for robust and imperceptible image watermarking in DWT–DCT–SVD domain. Natl. Acad. Sci. Lett. **37**(4), 351–358 (2014)
17. Rahimi, F., Rabbani, H.: A dual adaptive watermarking scheme in contourlet domain for DICOM images. Biomed. Eng. Online **10**(1), 53 (2011)
18. Singh, S., Rathore, V.S., Singh, R., Singh, M.K.: Hybrid semi-blind image watermarking in redundant wavelet domain. Multimedia Tools Appl. 1–25 (2017)

19. Singh, S., Rathore, V.S., Singh, R.: Hybrid NSCT domain multiple watermarking for medical images. Multimedia Tools Appl. 1–19 (2016)
20. Singh, A.K., Dave, M., Mohan, A.: Hybrid technique for robust and imperceptible multiple watermarking using medical images. Multimedia Tools Appl. **75**(14), 8381–8401 (2016)
21. Rosiyadi, D., et al.: A comparison between the hybrid using genetic algorithm and the pure hybrid watermarking scheme. Int. J. Comput. Theory Eng. **4**(3), 329 (2012)
22. Srivastava, A., Saxena, P.: DWT–DCT–SVD based semiblind image watermarking using middle frequency band. IOSR J. Comput. Eng. **12**(2), 63–66 (2013)
23. Tayal, A., Singh, A.: Choice of wavelet from wavelet families for DWT-DCT-SVD image watermarking. Choice **48**(17) (2012)
24. Su, Q., Chen, B.: Robust color image watermarking technique in the spatial domain. Soft Comput. 1–16 (2017)
25. Pandey, R., Singh, A.K., Kumar, B., Mohan, A.: Iris based secure NROI multiple eye image watermarking for teleophthalmology. Multimedia Tools Appl. **75**(22), 14381–14397 (2016)
26. Zarmehi, N., Aref, M.R.: Optimum decoder for multiplicative spread spectrum image watermarking with Laplacian modeling. ISC Int. J. Inf. Secur. **8**(2), 127–135 (2016)
27. Santhi, V.: Adaptive color image watermarking scheme using weibull distribution. In: Intelligent Techniques in Signal Processing for Multimedia Security, pp. 453–468. Springer International Publishing (2017)
28. Johnson, N.F., Jajodia, S.: Exploring steganography: seeing the unseen. IEEE Comput. **31**(2), 26–34 (1998)
29. Niimi, M., Noda, H., Kawaguchi, E.: A steganography based on region segmentation by using complexity measure. Trans. IEICE J81-D-II, 1132–1140 (1998)
30. Potdar, V.M., Chang, E.: Grey level modification steganography for secret communication. In: 2004 2nd IEEE International Conference on Industrial Informatics, INDIN'04, pp. 223–228. IEEE (2004)
31. Wu, D.C., Tsai, W.H.: A steganographic method for images by pixel-value differencing. Pattern Recogn. Lett. **24**(9), 1613–1626 (2003)
32. Yang, C.H., Weng, C.Y., Tso, H.K., Wang, S.J.: A data hiding scheme using the varieties of pixel-value differencing in multimedia images. J. Syst. Softw. **84**(4), 669–678 (2011)
33. Lee, Y.P., Lee, J.C., Chen, W.K., Chang, K.C., Su, J., Chang, C.P.: High-payload image hiding with quality recovery using tri-way pixel-value differencing. Inf. Sci. **191**, 214–225 (2012)
34. Liao, X., Wen, Q.Y., Zhang, J.: A steganographic method for digital images with four-pixel differencing and modified LSB substitution. J. Vis. Commun. Image Represent. **22**(1), 1–8 (2011)
35. Provos, N., Honeyman, P.: Detecting steganography content on the internet. CITI Technical Report, pp. 01–11 (2011)
36. Singh, S., Siddiqui, T.J.: Transform domain techniques for image steganography. Inf. Secur. Diverse Comput. Environ. 245–259 (2014)
37. Koch, E., Zhao, J.: Towards robust and hidden image copyright labeling. In: Proceeding of IEEE Workshop on Nonlinear Signal and Image Processing, Thessaloniki, Greece, pp. 452–455 (1995)
38. Cox, I.J., Miller, M.L.: U.S. Patent No. 6,108,434. Washington, DC: U.S. Patent and Trademark Office (2000)
39. Upham, D.: JPEG–Jsteg. http://zooid.org/~paul/crypto/jsteg
40. Provos, N.: Defending against statistical steganalysis. In: Proceedings of Tenth USENIX Security Symposium'01, pp. 323–335. Washington, DC (2001)
41. Westfeld, A.: F5-a steganographic algorithm: high capacity despite better steganlysis. In: Lecture Notes in Computer Science, vol. 2137, pp. 259–302. Springer, Berlin (2001)
42. Fan, L., Gao, T., Yang, Q., Cao, Y.: An extended matrix encoding algorithm for steganography of high embedding efficiency. Comput. Electr. Eng. **37**(6), 973–981 (2011)

43. Barni, M., Bartolini, F., Cappellini, V., Piva, A.: A DCT domain system for robust image watermarking. Signal Processing. In: European Association for Signal Processing (EURASIP), vol. 66(3) pp. 357–372 (1998)
44. Hernandez, J.R., Amado, M., Perez-Gonzalez, F.: DCT-domain watermarking techniques for still images: detector performance analysis and a new structure. IEEE Trans. Image Process. **9** (1), 55–68 (2000)
45. Briassouli, A., Tsakalides, P., Stouraitis, A.: Hidden messages in heavy-tails: DCT-domain watermark detection using alpha-stable models. IEEE Trans. Multimedia **7**(4), 700–715 (2005)
46. Amin, P.K. Liu, N., Subbalakshmi, K.P.: Statistically secure digital image data hiding. In: Proceedings of the IEEE 7th Workshop on Multimedia Signal Processing, Shanghai, pp. 1–4 (2005)
47. Noda, H., Niimi, M., Kawaguchi, E.: High-performance JPEG steganography using quantization index modulation in DCT domain. Pattern Recogn. Lett. **27**(5), 455–461 (2006)
48. Wong, K., Qi, X., Tanaka, K.: A DCT-based Mod4 steganographic method. Sig. Process. **87** (6), 1251–1263 (2007)
49. Chang, C.C., Lin, C.C., Tseng, C.S., Tai, W.L.: Reversible hiding in DCT-based compressed images. Inf. Sci. **177**(13), 2768–2786 (2007)
50. Lin, S.D., Shie, S.C., Guo, J.Y.: Improving the robustness of DCT-based image watermarking against JPEG compression. Comput. Stand. Interfaces **32**(1), 54–60 (2010)
51. Singh, H.V., Rai, S., Mohan, A., Singh, S.P.: Robust copyright marking using weibull distribution. Comput. Electr. Eng. **37**(5), 714–728 (2011)
52. Mali, S.N., Patil, P.M., Jalnekar, R.M.: Robust and secured image-adaptive data hiding. Digit. Signal Proc. **22**(2), 314–323 (2012)
53. Abdelwahab, A.A., Hassaan, L.A.: A discrete wavelet transform based technique for image data hiding. In: 2008 National Radio Science Conference, NRSC 2008, pp. 1–9. IEEE, March 2008
54. Barni, M., Bartolini, F., Piva, A.: Improved wavelet-based watermarking through pixel-wise masking. IEEE Trans. Image Process. **10**(5), 783–791 (2001)
55. Kundur, D., Hatzinakos, D.: Toward robust logo watermarking using multiresolution image fusion principles. IEEE Trans. Multimedia **6**(1), 185–198 (2004)
56. Kamstra, L., Heijmans, H.J.: Reversible data embedding into images using wavelet techniques and sorting. IEEE Trans. Image Process. **14**(12), 2082–2090 (2005)
57. Liu, J.L., Lou, D.C., Chang, M.C., Tso, H.K.: A robust watermarking scheme using self-reference image. Comput. Stan. Interfaces **28**(3), 356–367 (2006)
58. Ghouti, L., Bouridane, A., Ibrahim, M.K., Boussakta, S.: Digital image watermarking using balanced multiwavelets. IEEE Trans. Signal Process. **54**(4), 1519–1536 (2006)
59. Lee, S., Yoo, C.D., Kalker, T.: Reversible image watermarking based on integer-to-integer wavelet transform. IEEE Trans. Inf. Forensics Secur. **2**(3), 321–330 (2007)
60. Peng, F., Li, X., Yang, B.: Adaptive reversible data hiding scheme based on integer transform. Sig. Process. **92**(1), 54–62 (2012)
61. Lin, W.H., Horng, S.J., Kao, T.W., Fan, P., Lee, C.L., Pan, Y.: An efficient watermarking method based on significant difference of wavelet coefficient quantization. IEEE Trans. Multimedia **10**(5), 746–757 (2008)
62. Chan, Y.K., Chen, W.T., Yu, S.S., Ho, Y.A., Tsai, C.S., Chu, Y.P.: A HDWT-based reversible data hiding method. J. Syst. Softw. **82**(3), 411–421 (2009)
63. Bhatnagar, G., Jonathan, Wu, Q.M., Raman, B.: Robust gray-scale logo watermarking in wavelet domain. Comput. Electr. Eng. **38**(5), 1164–1176 (2012)
64. Yeh, H.L., Gue, S.T., Tsai, P., Shih, W.K.: Wavelet bit-plane based data hiding for compressed images. AEU-Int. J. Electr. Commun. (2013)
65. Sajedi, H., Jamzad, M.: ContSteg: contourlet-based steganography method. J. Wireless Sens. Netw. **1**(3), 163–170 (2009)
66. Sajedi, H., Jamzad, M.: Using contourlet transform and cover selection for secure steganography. Int. J. Inf. Secur. **9**(5), 337–352 (2010)

67. Khalighi, S., Tirdad, P., Rabiee, H.R.: A contourlet-based image watermarking scheme with high resistance to removal and geometrical attacks. EURASIP J. Adv. Signal Proces. **21** (2010)
68. Leung, H.Y., Cheng, L.M., Cheng, L.L.: Robust watermarking schemes using selective curvelet coefficients based on a HVS model. Int. J. Wavelets Multiresolut. Inf. Process. **8**(06), 941–959 (2010)
69. Bhatnagar, G., Wu, Q.J.: Biometrics inspired watermarking based on a fractional dual tree complex wavelet transform. Future Gener. Comput. Syst. **29**(1), 182–195 (2013)
70. Mansouri, A., Aznaveh, A.M., Azar, F.T.: SVD-based digital image watermarking using complex wavelet transform. Sadhana **34**(3), 393–406 (2009)
71. Shieh, C.S., Huang, H.C., Wang, F.H., Pan, J.S.: Genetic watermarking based on transform-domain techniques. Pattern Recogn. **37**(3), 555–565 (2004)
72. Li, X., Wang, J.: A steganographic method based upon JPEG and particle swarm optimization algorithm. Inf. Sci. **177**(15), 3099–3109 (2007)
73. Ishtiaq, M., Sikandar, B., Jaffar, M.A., Khan, A.: Adaptive watermark strength selection using particle swarm optimization. ICIC Express Lett. Int. J. Res. Surv. **4**(6) (2010)
74. Wang, Y.R., Lin, W.H., Yang, L.: An intelligent watermarking method based on particle swarm optimization. Expert Syst. Appl. **38**(7), 8024–8029 (2011)
75. Sallee, P.: Model-based steganography. In: Digital Watermarking, pp. 154–167. Springer, Berlin (2004)
76. Do, M.N., Vetterli, M.: The contourlet transform: an efficient directional multiresolution image representation. IEEE Trans. Image Process. **14**(12), 2091–2106 (2005)
77. Da Cunha, A.L., Zhou, J., Do, M.N.: The nonsubsampled contourlet transform: theory, design, and applications. IEEE Trans. Image Process. **15**(10), 3089–3101 (2006)
78. Chen, J., Lin, P.: A reliable enhanced watermarking based on NSCT and SVD. Adv. Inf. Sci. Ser. Sci. **5**(5), 629 (2013)
79. Narasimhulu, C.V., Prasad, K.S.: A robust watermarking technique based on nonsubsampled contourlet transform and SVD. Int. J. Comput. Appl. **16**(8) (2011)
80. Singh, S., Siddiqui, T.J.: A security enhanced robust steganography algorithm for data hiding. Int. J. Comput. Sci. Issues **9**(1), 131–139 (2012)
81. Kalman, D.: A singularly valuable decomposition: the SVD of a matrix. Coll. Math. J. **27**(1), 2–23 (1996)

The Utilization of Quantum Inspired Computational Intelligent in Power Systems Optimization

Mahmoud Pesaran Hajiabbas, Morteza Nazari-Heris, Sajad Madadi and Behnam Mohammadi-Ivatloo

Abstract Many areas in power systems require solving one or more nonlinear optimization problems. While analytical methods might suffer from slow convergence and the curse of dimensionality, heuristics-based intelligence can be an efficient alternative. This need is highlighted by technology advancements and bulk integration of the renewable energies in power grids. The Quantum inspired computational intelligence (QCI) techniques as a young discipline in computational intelligence field of research shows a promising future in optimization problems. The Quantum inspired computational intelligence (QCI) is known to effectively solve large-scale nonlinear optimization problems. This chapter will present a detailed overview of the quantum inspired computational intelligence and its variants in power systems optimization. Also, it provides a survey on the power system applications that have benefited from the powerful QCI as an optimization technique. For each application, technical details that are required for applying QCI, and the most efficient fitness functions are also discussed. In this chapter the definition, categorization and motivation for QCI employment in power systems will be elaborated. The major challenges and hinders for implementation will be discussed. The significance of this study is to present an overview on the applications of QCI in solving various power system problems in electrical engineering, which may be a useful resource for researchers to understand the state of art in QCI application in electric power systems to enable them for further explores. Current chapter will present the generalized introduction to the power systems optimization, the areas and categories and how their results and targets can be affected by optimization variables. Furthermore, the QCI methods for the power systems optimization will be presented.

Keywords Particle swarm optimization (PSO) · Quantum inspired PSO · Optimization problem · Distributed generation (DG) allocation

M. Pesaran Hajiabbas (✉) · M. Nazari-Heris · S. Madadi · B. Mohammadi-Ivatloo
Smart Energy Systems Laboratory, Faculty of Electrical and Computer Engineering, University of Tabriz, Tabriz, Iran
e-mail: mpesaran@tabrizu.ac.ir

A.E. Hassanien et al. (eds.), *Quantum Computing: An Environment for Intelligent Large Scale Real Application*, Studies in Big Data 33,
https://doi.org/10.1007/978-3-319-63639-9_21

1 Introduction to the Power Systems Optimization Techniques

The optimization problems associated with power systems traditionally are handled by conventional and mathematical approaches such as linear programming, nonlinear programming [1], branch and bound method [2], decomposition techniques [3], dynamic programing (DP) [4] and Lagrange multiplier [5]. The optimization algorithms, which normally are employed in power systems, are being discussed in this section in 4 main categories.

1.1 Classical and Conventional Algorithms

Linear programming (LP): Linear programming algorithms are employed in [6–8] to form an objective function by formulating linear equations and constraints. However, the LP method is only applicable on linear equation and constraints, but it has the ability to handle a large number of various operational limitations for power system including contingency constraints.

Nonlinear programming (NLP): The first step to solve a nonlinear programming problem is choosing the direction of search for iterative procedure. Nonlinear programming has been employed in many different ways for instance in [9], a first order method which is also referred as Generalized Reduced Gradient (GRG) has been utilized to solve OPF. In [8], nonlinear programming was also employed in its second order form to solve the second order partial derivative of power flow equations by a sequential quadratic programming together with Newton's method. In addition, the constraints were also applied using second order partial derivatives (The Hessian). The amount of resources at selected nodes was computed using the second order method to minimize the objective function (losses) in [10]. As stated in [11], the NLP integration in large power systems faces two major difficulties: first, depending on the starting point of the search procedure, different optimal results can be achieved. Second, the convergence of this method is guaranteed and it is not affected by starting point, but it can be very slow because of zig-zaging over the final solution.

Mixed Integer Nonlinear Programming (MINLP): A mixed integer nonlinear programming method has been employed to formulate a comprehensive optimization objective function for DG optimal siting and sizing in [12], and a General Algebraic Modeling System (GAMS) with integer decision variables containing 0 and 1 is integrated to formulate the model. A probabilistic planning method has been used in [13] to determine the optimal shares of different renewable units in hybrid DG units consist of wind, solar and biomass modules subjected to minimizing the yearly energy losses within the constraints limitations. MINLP has been implemented for formulating the problem respecting to the renewable DG sources uncertainties and load profile hourly variations. The optimal site and number of DGs operating in a hybrid electricity market have been determined using MINLP in [14], and the objective function has been modeled in GAMS with use of SNOPT solver. However, the method has some improvements in comparison to the NLP method but the difficulties for those methods still exist.

Analytical Algorithms (AA): In [15, 16], different analytical approaches are employed for power system optimization. In [17], the authors have minimized the objective function for optimal bus recognition. Acharya et al. [15] used loss approximation instead of accurate load flow calculation to identify the optimal DG installation site. In [16], the authors employed one of the most popular method for distribution load flow analysis that uses equivalent current injection. In this approach the matrices of Branch-Current to Bus-Voltage (BCBV) and Bus-Injection to Branch-Current (BIBC) are implemented in calculations. Hung et al. [18] improved the proposed method of [15] by developing a comprehensive formulation to recognize the optimum location and size of DGs.

Exhaustive Search (ES): Exhaustive search, which also known as brute force, direct search or generate and test, is a thorough test of target function with all possible input values. For discrete problems such as DGs optimal allocation, this method could be used. However, it is not an efficient solution method, but the results are always reliable because of all possibility checking [19]. The exhaustive search method has been employed for optimal DG allocation in [20].

Optimal Power Flow Based Algorithms (OPF): The optimal power flow algorithm has been employed in [21–26] for power system optimization problems. In [21], Maximum DG capacity and system available headroom has been identified considering voltage and thermal constraints, by implementation of optimal power flow under "reverse load-ability" method. In [22], demand bids are also considered in addition to the generation bids in a traditional optimal power flow approach to minimize the cost. The authors integrated GD units' goodness factor directly into the distribution system model [23]. DG units' incremental contribution to power losses (both active and reactive) is implemented as Incremental Loss Indices (ILI) in OPF framework. The study in [24], has been done by integrating voltage step constraints in to the OPF algorithm to identify the DG accommodation capacity of network. Authors in [25], used a one by one line outage contingency solution in OPF to recognize the maximum generation under the security limitations. Researchers in [26] implement the OPF approach to indicate the maximum capacity of a network which incorporating with variable DG due to huge load of iterative calculation in this method.

Continuation Power Flow (CPF): Analysis of power flow continuation and search for the bus with most sensitivity to the voltage collapse is employed in [27] to determine the optimal place for DG installation by determination of maximum loading or most sensitive bus to voltage collapse.

All above methods have limitations and difficulties in solving the non-linear non-convex problems. Hence, heuristic and meta-heuristic optimization methods are introduced to handle the computational burdens. Heuristic techniques facilitate some difficulties of conventional methods in solving the power system problems.

1.2 Intelligence Based Algorithms

The most famous heuristic optimization methods implemented to solve the power system problems are genetic algorithm, teaching learning based optimization (TLBO), simulated annealing, clonal selection algorithm (CSA), differential evolution (DE), particle swarm optimization (PSO), artificial bee colony (ABC), chemical reaction optimization (CRO), cultural algorithm (CA), cuckoo search (CS), shuffled frog leaping (SFL) algorithm, gravitational search (GS), group search optimization (GSO). Followings are some samples of heuristic methods integration in power system optimization problems.

Evolutionary Algorithms (EAs): In EA approaches, unlike the conventional optimization algorithms, the objective function and constraints do not need to be differentiated. The evolutionary programming algorithm was utilized for power system planning in [28]. Single and multi-objective functions could be integrated in this approach considering different constraints. However, the result accuracy and the convergence of these methods are the concerning points.

Simulated Annealing (SA): In [29], the SA has been selected to minimize the losses, emission, and contingency by optimal siting and sizing of DG units. The main weakness of this algorithm is its dependence on initial values and cooling parameters setting.

Tabu Search (TS): Network configuration and tap positions of Voltage Regulators (VRs) as well as the installation location, size and operation of Distribution Generation Resources (DGRs) and Reactive Power Sources (RPSs) are identified in [30] by using TS. The constraints violating selections are added to tabu list to avoid future forbidden selection. Nara et al. [31] introduced the coordination/decomposition technique and implemented the proposed approach along with TS to optimal DG siting and sizing respecting to the total loss minimization. The disadvantage of this method is the necessity of solving the regression model for any change in the initial weights to achieve the mean squared error.

Imperialist Competitive Method (ICM): The ICM has been employed for power system optimization in [32], while the Soroudi and Ehsan [33] integrated ICM by considering a combination of technical and economic objectives for DGP. The ICM is implemented in [34], for recognition of size and site of DG units for a distribution network containing the sensitive loads working in islanded mode. The implementation time of this algorithm is longer than other methods, however, due to the various step size the time is reasonable.

Fuzzy Set (FS) Based Algorithms: Both objective function and constraints of a power system are handled by fuzzy sets in [35–38]. Lalitha et al. [35] achieved the DG Suitability Index (DGSI) by modeling the Power Loss Index (PLI) and nodal voltage using fuzzy set method. In [36], a multi objective model composed by technical risk, economic risk, and monetary cost indices are modeled by a fuzzy set theory. A fuzzy set has been employed along with GA and goal programming to form a multi objective function in [37]. In [38], authors developed their own Adaptive Interactive Decision Making System (AIDMS) based on Bellman-Zadeh method to solve multi objective resource allocation problem. In [39, 40], the load uncertainties is modeled by a fuzzy set with respect to load and voltage constraints. In addition, the researchers in [41] implemented the effect of power system parameter uncertainty into the system model and handed out a better compromised solution while reducing the needed iteration time. The main disadvantage of the fuzzy based methods is that there is no correction step or factor and a wrong classification of variables may result in complete incorrect answers.

1.3 Nature Inspired Search Methods

The nature inspired optimization techniques are effective and useful tools to handle complex optimization problems. Such methods are capable of dealing with nonlinear and non-convex problems and associated equality and inequalities. In addition, the large computational time required to solve the complex optimization problems is shrunk by nature inspired optimization techniques. Nature inspired optimization techniques, which are also known as strong experience-based optimization methods, can handle the complex optimization problems of power systems. Such methods are effectively implemented to obtain the optimal solution of combinatorial optimization problems like neural network training, control system designing, and power system optimization problems. Couple of those methods are being covered and discussed in the following.

Genetic Algorithm (GA): The capability of GA in solving optimal DGP was shown in [42]. The service restoration under cold load pickup has been investigated in [43], using GA in a Multi-Objective MO model for DG implementation. The authors in [44–46], have studied DG planning impact on network reliability with integration of GA. In [47], a Non-Dominated Sorting Genetic Algorithm (NSGA) along with multi objective programming method is integrated to find the best and maximum implementation configuration of DWG with respect to the voltage and thermal constraints.

Particle Swarm Optimization (PSO): PSO has been utilized in different research areas for power systems optimization [48, 49]. The active power losses of distribution system has been minimized by optimal allocation and sizing of multi DG units using PSO algorithm including different load models in [50]. Lalitha et al. [35] optimally placed and sized DGs through a two step algorithm using a fuzzy approach and PSO to reduce the system power losses.

Ant Colony System (ACS): In most power system cases [51, 52], the ACS as an extension to Ant Colony Optimization (ACO) has handed out better results. ACS approach has been used as optimal positioning method for fixed re-closer or DGP to increase the reliability in [53]. The authors also proposed to integrate the algorithm for simultaneous allocation of both DGs and re-closers.

Bee Colony Optimization (BCO): While, The Artificial BCO is employed for investigation of transient performance of grid connected distributed generation in [54], it is implemented for optimizing the distribution network configuration considering loss reduction in [55]. Moreover, authors in [56, 57] employed the BCO for DGP concentrating on total system real power losses as the objective function. Sohi et al. [58] extended the objective function of BCO to include the line capacity improvement in addition to the loss reduction.

Cuckoo Search Optimization (CSO): The CSO has been employed for a combination of biomass and solar-thermal units considering loss reduction and voltage profile improvement in [59]. It also implemented in [60] for voltage profile improvement, which is expressed by two regulation and variation indexes, and power loss reduction for power system.

Firefly Method (FFM): The FFM has been employed in power system optimization aiming to minimize the active and reactive power losses, voltage profile improvement for various models of loads, line current, level of short circuit, and total absorbed apparent power of the network in [61]. The authors in [62], are also applied FFM to minimize the real power losses by optimal location and size determination of DGs in distribution network. Two major challenging issues in FFM are attractiveness and light intensity variation formulation. For the sake of simplicity, the attractiveness of a firefly is assumed to be determined by its lightness which in turn is derived by objective function formulation.

Nature inspired search algorithms have some disadvantages. All these methods are highly affected by their parameters which are selected as their operator constants. They may also be trapped in local optimal points in case of wrong initial value or parameter selection. They also could have unstable movements in finding the extremum point, hence, the convergence of all these algorithms is questionable too. In response to this issue, there are several studies on Hybrid Intelligent Algorithms.

1.4 Hybrid Intelligent Techniques

Generally, Hybrid Intelligent Algorithms (HIA) refers to the algorithms which are the combination of different artificial intelligent methods work in parallel or cascaded mode. There are various studies which have focused on different combination of existing meta-heuristic methods for distributed generation planning, including: Genetic-Tabu search (GATS) [63], Genetic-Particle Swarm Optimization (GAPSO) [64], Genetic-Optimal Power Flow (GAOPF) [65, 66], and Particle Swarm Optimization-Optimal Power Flow (PSOOPF) [67].

The GATS method has been employed for power system optimization in [63], the objective function was power losses when the harmonic power losses were included. A novel GAPSO method is presented in [64] for optimal size and site identification in DG by minimizing the power losses and improving the voltage regulation and stability. Harrison et al. [65] had illustrated the robustness of GAOPF for determining the installation sites of dedicated number of DGs. In [66], GA is employed for optimal positioning again and the OPF is implemented for minimizing the operation, maintenance, and network upgrade costs along with cost for load growth causing losses. A combination of discretized PSO and OPF is applied in [67] for optimal power system planning within a distribution network for site and size recognition of specified number of DGs.

There are couples of hybrid algorithms which are implemented in power system optimization problems and comprised Fuzzy Set in combination with one of the other intelligent algorithms. For instance Genetic-Fuzzy (GAFZ) is presented in [36, 37, 68], and Tabu-Fuzzy (TSFZ) is reported in [39]. The Genetic-Fuzzy algorithm is proposed as a solution for power system planning in [37]. Another multi objective model considering technical and economic risks together with planning, operation, and monetary cost index is proposed in [36]. The GAFZ combination is also applied in [68], the authors selected economic profit maximizing and loading margin of the system as the objectives and used the fuzzy set to combine them and form a single objective. In addition, the TS method is incorporated with a fuzzy set algorithm in [39] for optimal power system planning.

2 QPSO in Power System Problems

Quantum inspired computational intelligent can be defined as optimization techniques manipulated by quantum bits. Such algorithms are based on the interaction between quantum computing and evolutionary optimization programming methods. Quantum inspired methods can be classified into three main types comprising evolutionary-designed quantum algorithms (EDQAs), quantum evolutionary algorithms (QEAs), quantum-inspired evolutionary algorithms (QIEAs). The EDQAs methods are based on the automated synthesis of new quantum methods utilizing evolutionary methods. GA can be introduced as an example of EDQAs. The application of evolutionary methods in a quantum computation environment resulted in QEAs. Furthermore, QIEAs aims to produce new evolutionary methods by utilization of some techniques and basics of quantum intelligence. The QPSO algorithm as a sample of QEAs has been employed in several power system optimization problems.A general form of QPSO algorithm which is implemented in power system studies is presented as Algorithm 1.

$$diversity(S) = \frac{1}{|S|.|A|} \sum_{i=1}^{|S|} (\sum_{j=1}^{D} (x_{ij} - \overline{x_i})^2)^{\frac{1}{2}} \tag{1}$$

where S shows group. The scale of group is shown by $|S|$. $|A|$ is the length of the longest diagonals in searching space. The scale of problem is determined by D. x_{ij} is the value of particle i in the dimension of j. The average value of population set is determined by $\overline{x_i}$.

The main categories in which the quantum inspired particle swarm optimization has been selected as the optimization algorithm are covered and discussed in the following subsections. The main objectives and procedure for each category is also

Algorithm 1 Quantum-Behaved Particle Swarm Optimization

```
procedure AQPSO
    Define randomly population x_i
    Calculate mean optimum solution mbest
    Compute the diversity of group by (1)
    if diversity < d_low then
        α = α_r
    end if
    if diversity > d_high then
        α = α_a
    end if
    for i 1:population size(M) do
        if f(p_i) < f(x_i) then
            p_i = x_i
            p_g = argmin(p_i)
        end if
    end for
    for d 1:number of dimension(D) do
        f_i = rand(0, 1)
        p = f_i × p_id + (1 − f_i) × P_gd
        u = rand(0, 1)
        if rand(0, 1) > 0.5 then
            X_id = P − α × |mbest − x_id| × ln(frac1u
        else
            X_id = P + α × |mbest − x_id| × ln(frac1u
        end if
    end for
end procedure
```

elaborated. Moreover, the methodology is clarified using proper flowcharts. Furthermore, a conclusive review of each optimization study is presented. The power system optimization studies have been classified into four main categories as follows:

2.1 *Unit Commitment/Hydrothermal System Scheduling*

The objective of the unit commitment (UC) problem is to provide a unit commitment schedule for the generation units in power system aiming to minimize the operational cost of the units to meet the load demand. The UC problem is a decision making process, which is recognized as important activity of the system operators, in electrical energy markets. The UC problem should be solved considering a series of equality and inequality constraints including load balance, minimum down-time and maximum up-time of generation units, power flow equilibrium of systems, and the plants capacity limitations. Such problem is solved by utilizing different optimization methods such as Lagrangian relaxation [69], a combination of Lagrangian relaxation and particle swarm optimization [70], improved gravitational search algorithm [71], and gravitational search algorithm [72]. It should be highlighted that there always are some complications in solving the UC problems due to the uncertain nature of the renewable energy sources, uncertainties associated with load demand, and power market uncertainties.

In [73], a new model based on the enhanced particle swarm optimization is presented to determine the best solution for short term hydro thermal scheduling problem. This model considers the general hydro thermal scheduling constraints such as non-convex fuel cost function along with a variety of technical and operational constraints of hydro and thermal units. The major difference between conventional PSO and enhanced PSO is in their velocity operators. In the enhanced PSO, the particle position is changed with higher velocity at the early flights. This higher velocity leads to global search space coverage than the conventional PSO. However, the particle positions are changed slower in later flights to enhance the local search.

The environmental constraints motivate the researchers to implement such constraints in scheduling model. However, considering environmental constraints in power system planning model causes some problems. A type of environmental constraints is the pollutant emission, which is generally covered in hydro thermal scheduling problems. In [74], a modified quantum-behaved particle swarm optimization is employed to solve the short-term combined economic emission hydro thermal scheduling problem. The combined economic emission hydro thermal scheduling problem is modeled as a two objective problem including minimizing pollutant emission. Then, it is simplified into a single objective by price penalty factor. The global search ability of QPSO algorithm is increased by integrating the differential mutation operation. At last, the modified QPSO algorithm is implemented to identify the best schedule of hydro and thermal units. Moreover, the combined economic emission hydro thermal scheduling problem is solved by a improved QPSO in [74]. This paper presents the heuristic rules to meet some of the constraints, which are related to the water dynamic balance. In addition, the active power balance limitation is handled by a priority list.

The quantum-inspired binary PSO (BQPSO) is proposed to solve the binary PSO bugs for instance premature convergence, which are occurred in a problem with high constraints by [72]. The difference between QBPSO and PSO is the velocity update step elimination and the replacement of the Q-bit individual for the probabilistic representation. The QBPSO is employed to determine optimal solution of unit commitment problem.

2.2 Economic Dispatch

Economic Dispatch (ED) problem is defined as obtaining the optimal generation scheduling of the generations units to meet load demand in the scheduling time interval. The ED problem should be solved according to some operational and electrical constraints, which include valve-point loading effect of the conventional thermal units, load balance of the system, and generation capacity of the plants. The dispatch problem becomes more realistic considering the multi-objective economic-emission dispatch of the generation units, which is reported in some publications. The economic dispatch problem is solved using genetic algorithm (GA) [75], particle swarm optimization [76], and some other heuristic methods such as teaching learning-based optimization (TLBO) method or Cuckoo search algorithm [77].

A quantum mechanics theories inspired classical PSO method, using the harmonic oscillator potential well (HQPSO), is employed to solve the economic dispatch problems in [78]. The integrated harmonic oscillator is a very common potential distribution, which also is one of the most important model systems in quantum mechanics. The proposed HQPSO algorithm is applied on a 13-units test system with incremental fuel cost function that takes into account the valve-point loading effects and compared the results with that of classical PSO, QPSO, and other optimization algorithms. As reported, the proposed HQPSO has the lowest cost in comparison to all other methods. Later, QPSO algorithm is employed to solve ED problem aiming to minimize the generation cost considering the equality and inequality constraints in [79]. In the proposed, method the differential mutation is implemented in QPSO algorithm to improve the global search capability of the algorithm. To create a more practical model various nonlinear characteristics of the generator, such as non-smooth cost functions, prohibited operating zones, and ramp rate limits are considered. The performance of the proposed algorithm is verified against four other algorithms on three different power systems. The results depicted that the proposed algorithm acts better in terms of the convergence, robustness, and solution quality and solves the ED in more efficient and stable manner.

Following the research in the field of the QEAs and PSO, two definitions are introduced to the QEAs as: quantum bit and quantum rotation gate in [80]. The quantum bits as the smallest information units are employed to represent the particles probabilities which can form quantum strings from quantum bit individuals. In addition, the quantum rotation gate is adopted to drive quantum bits toward the local attractors and eventually achieving the global best. Furthermore, three definitions are introduced on immunology bases: individual affinity, individual concentration, and selection possibility. Those values are defined to facilitate the self-adaptive probability selection and chaotic sequences mutation integration, which in turn improves the algorithm search performance by preventing premature convergence and increasing population diversity. The proposed approach is applied on five standard benchmark functions to solve Economic Load Dispatch (ELD) problem in three power systems consisting of 3, 13, and 40 thermal units. The comparisons with other meta-heuristic methods such as the immune system algorithm (ISA), genetic algorithm (GA), evolutionary programming (EP), and other versions of particle swarm optimization (PSO) support the proposed methods efficient and reliable performance. In the same manner, Quantum-behaved particle swarm optimization algorithm is employed for ELD of power system in two different cases and achieved similar results [81].

In the most recent study, the quantum particle swarm optimization (QPSO) is utilized to solve a multi-objective combined economic emission dispatch (CEED) problem [82]. The problem is formulated by cubic criterion function and employing normalized max/max price penalty factor to form a single value objective function from two considered objectives. The proposed algorithm is applied on a 6-units power generation system and the results are compared against the Lagrangian relaxation method, classic PSO and simulated annealing (SA). Consequently, the effectiveness and robustness of QPSO method is approved and it has been suggested that the algorithm may be implemented in other power dispatch problems as well.

2.3 Load Forecasting

Load forecasting has a significant role in reliable operation and planning of power networks. A well-known load forecasting in power systems is short-term forecasting of the load demands which is used for power transfer and generation scheduling of the production units. The accurate load forecasting can considerably minimize the operational cost of the system and consequently results in cost savings of the system. Moreover, a series of tasks are related to the load forecasting including identification of the generation units capacity, fuel purchases, and power transfer between different sections of the system [83]. The researches in this area can be classified into four general types, which include (i) Long-term forecasting for the time intervals more than 1 year [84], (ii) Mid-term forecasting, which deals with time intervals between 1 week to 1 year [85], (iii) Short-term forecasting, which handles times between 1 and 168 h [86], and (iv) Very short-term forecasting tasks, which covers scheduling times less than 1 h [87].

Radial basis function (RBF) is a type of artificial neural networks which is generally used to solve the non-linear forecasting projects. The RBF network has a hidden layer of basis function or neurons. The output of each neurons calculation is based on distance between the neuron center and the input vector. Since, the tuning of the RBF neural network parameters is a challenge for researchers, the heuristic algorithms are applied to set these parameters. In [88], a new approach for training RBF neural network is proposed which is based on QPSO algorithm. In this algorithm, all network parameters are defined as individual particles, which can reach to optimal-adaptive values randomly throughout the search domain. Therefore, the parameters of RBF network can be quickly and accurately selected. More recently, the researchers have used the machine learning approaches to predict the time series. Support vector machine (SVM) is a common type of machine learning approaches. Generally forecasting models based on SVM have a good response however, SVM application can be improved by selection of the best learning parameters. In [89], a short-term load prediction model based on SVM with Adaptive Quantum-behaved Particle Swarm Optimization algorithm (AQPSO) is proposed. Then, a diversity-guided into the QPSO algorithm is employed to set the free parameters of SVM model automatically.

2.4 Utility Based Optimizations

An effective power quality monitor (PQM) placement method for voltage sag assessment by utilizing quantum-inspired particle swarm optimization technique has been studied in [90]. To handle the observability constraints with respect to the economic capability and sensitivity, the topological monitor reach area (TMRA) concept was implemented to give more flexibility to the search algorithm. Then, the multi-objective function was solved to obtain the optimal number and location of PQMs in power systems. The sag severity and monitor overlapping indices are combined to form the objective function for this study. The proposed approach solves the non-monitored fault and boundary issues on line segment at the boundaries. They have compared the optimization performance of the proposed technique against GA and standard binary PSO on the IEEE 69-bus and the 118-bus test systems to illustrate the algorithms efficiency. Finally, they have demonstrated that the quantum-inspired particle swarm optimization is the most effective technique in comparison to its counterparts.

A computational framework is developed for carbon tax and wind power uncertainty integration in economic dispatch (ED) in [91] where, the valve-point effects are also taken into account. The Weibull distribution and nonlinear wind power curve are employed to embed the stochastic probability of wind power in the model. The QEAs can crawl over the search domain with a smaller number of individuals and find the global solution within a less iterations. Hence, the QPSO is adopted to solve the revised ED strategy due to its stronger search ability and faster convergence. The simulation is carried out on a modified IEEE benchmark system which includes two wind farms and six thermal units. The applied wind data is the real wind speed data obtained from two meteorological stations in Australia.

The concept of smart building (home) energy management (SBEM) is based on the management of energy generation and consumption by households. This approach is derived from demand side management (DSM) which is trying to reshape the energy consumption pattern during peak hours for the end user. The quantum-inspired evolution algorithm along with power consumption controlling multi-agent system has been applied on the innovative SBEM systems to match the supply and demand in [92]. The problem is formulated as a multi-objective optimization for coordinating the multi-agent system operation to match the supply and demand. The model is solved by QEA and the effectiveness of the proposed algorithm performance has been tested using the models of household devices. The test result shows that the algorithm is able to minimize the simultaneous "ON" times of the consumption devices within home/building in the control procedure aiming to distribute the devices consumption and to minimize the consumption peaks.

3 Conclusion

In this chapter, a comprehensive review is prepared for the application of the quantum inspired evolutionary methods to power system problems. At first, the implementation of different optimization concepts is studied, which are classified to the conventional methods and heuristic optimization approaches. The quantum inspired evolutionary methods is then introduced as an effective method to handle the optimization problems. Different areas of the power system problems including economic dispatch (ED), unit commitment (UC), utility based optimization, and forecasting of the system parameters are studied. Different versions of the quantum inspired methods, which are employed as the solution, are discussed in this chapter. The highlighted points in this study are wished to help the researchers in the area of quantum inspired evolutionary methods and their applications in the power system problems.

References

1. Suo, M., Li, Y., Huang, G., Deng, D., Li, Y.: Electric power system planning under uncertainty using inexact inventory nonlinear programming method. J. Environ. Inform. **22**, (1) (2013)
2. Ding, T., Bo, R., Li, F., Sun, H.: A bi-level branch and bound method for economic dispatch with disjoint prohibited zones considering network losses. IEEE Trans. Power Syst. **30**(6), 2841–2855 (2015)
3. Abdolmohammadi, H.R., Kazemi, A.: A benders decomposition approach for a combined heat and power economic dispatch. Energy Convers. Manage. **71**, 21–31 (2013)
4. Tang, Y., He, H., Wen, J., Liu, J.: Power system stability control for a wind farm based on adaptive dynamic programming. IEEE Trans. Smart Grid **6**(1), 166–177 (2015)
5. Sandoval-Moreno, J., Besancon, G., Martinez, J.J.: Lagrange multipliers based price driven coordination with constraints consideration for multisource power generation systems. In: European Control Conference (ECC), pp. 1987–1992. IEEE (2014)
6. El-Ela, A.A., Allam, S.M., Shatla, M.: Maximal optimal benefits of distributed generation using genetic algorithms. Electr. Power Syst. Res. **80**(7), 869–877 (2010)
7. Keane, A., O'Malley, M.: Optimal allocation of embedded generation on distribution networks. IEEE Trans. Power Syst. **20**(3), 1640–1646 (2005)
8. Keane, A., O'Malley, M.: Optimal utilization of distribution networks for energy harvesting. IEEE Trans. Power Syst. **22**(1), 467–475 (2007)
9. Wu, F.F., Gross, G., Luini, J.F., Look, P.M.: A two-stage approach to solving large-scale optimal power flows. In: Power Industry Computer Applications Conference, PICA-79. IEEE Conference Proceedings, pp. 126–136. IEEE (1979)
10. Rau, N.S., Wan, Y.-H.: Optimum location of resources in distributed planning. IEEE Trans. Power Syst. **9**(4), 2014–2020 (1994)
11. Zhang, W., Li, F., Tolbert, L.M.: Review of reactive power planning: objectives, constraints, and algorithms. IEEE Trans. Power Syst. **22**(4), 2177–2186 (2007)
12. El-Khattam, W., Hegazy, Y., Salama, M.: An integrated distributed generation optimization model for distribution system planning. IEEE Trans. Power Syst. **20**(2), 1158–1165 (2005)
13. Atwa, Y., El-Saadany, E., Salama, M., Seethapathy, R.: Optimal renewable resources mix for distribution system energy loss minimization. IEEE Trans. Power Syst. **25**(1), 360–370 (2010)
14. Kumar, A., Gao, W.: Optimal distributed generation location using mixed integer non-linear programming in hybrid electricity markets. IET Gener. Transm. Distrib. **4**(2), 281–298 (2010)

15. Acharya, N., Mahat, P., Mithulananthan, N.: An analytical approach for dg allocation in primary distribution network. Int. J. Electr. Power Energy Syst. **28**(10), 669–678 (2006)
16. Gözel, T., Hocaoglu, M.H.: An analytical method for the sizing and siting of distributed generators in radial systems. Electr. Power Syst. Res. **79**(6), 912–918 (2009)
17. Wang, C., Nehrir, M.H.: Analytical approaches for optimal placement of distributed generation sources in power systems. IEEE Trans. Power Syst. **19**(4), 2068–2076 (2004)
18. Hung, D.Q., Mithulananthan, N., Bansal, R.: Analytical expressions for dg allocation in primary distribution networks. IEEE Trans. Energy Convers. **25**(3), 814–820 (2010)
19. Paar, C., Pelzl, J.: Understanding Cryptography: A Textbook for Students and Practitioners. Springer (2009)
20. Pesaran, M., Mohd Zin, A.A., Khairuddin, A., Shariati, O.: Optimal sizing and siting of distributed generators by a weighted exhaustive search. Electr. Power Compon. Syst. **42**(11), 1131–1142 (2014)
21. Harrison, G., Wallace, A.: Optimal power flow evaluation of distribution network capacity for the connection of distributed generation. IEE Proc. Gener. Transm. Distrib. **152**(1), 115–122 (2005)
22. Gautam, D., Mithulananthan, N.: Optimal dg placement in deregulated electricity market. Electr. Power Syst. Res. **77**(12), 1627–1636 (2007)
23. Algarni, A.A., Bhattacharya, K.: Disco operation considering dg units and their goodness factors. IEEE Trans. Power Syst. **24**(4), 1831–1840 (2009)
24. Dent, C.J., Ochoa, L.F., Harrison, G.P.: Network distributed generation capacity analysis using opf with voltage step constraints. IEEE Trans. Power Syst. **25**(1), 296–304 (2010)
25. Dent, C.J., Ochoa, L.F., Harrison, G.P., Bialek, J.W.: Efficient secure ac opf for network generation capacity assessment. IEEE Trans. Power Syst. **25**(1), 575–583 (2010)
26. Ochoa, L.F., Dent, C.J., Harrison, G.P.: Distribution network capacity assessment: Variable dg and active networks. IEEE Trans. Power Syst. **25**(1), 87–95 (2010)
27. Hedayati, H., Nabaviniaki, S.A., Akbarimajd, A.: A method for placement of dg units in distribution networks. IEEE Trans. Power Delivery **23**(3), 1620–1628 (2008)
28. Lai, L.L., Ma, J.: Application of evolutionary programming to reactive power planning-comparison with nonlinear programming approach. IEEE Trans. Power Syst. **12**(1), 198–206 (1997)
29. Sutthibun, T., Bhasaputra, P.: Multi-objective optimal distributed generation placement using simulated annealing. In: 2010 International Conference on Electrical Engineering/Electronics Computer Telecommunications and Information Technology (ECTI-CON), pp. 810–813. IEEE (2010)
30. Golshan, M.H., Arefifar, S.: Distributed generation, reactive sources and network-configuration planning for power and energy-loss reduction. IEE Proc. Gener. Transm. Distrib. **153**(2), 127–136 (2006)
31. Nara, K., Hayashi, Y., Ikeda, K., Ashizawa, T.: Application of tabu search to optimal placement of distributed generators. In: Power Engineering Society Winter Meeting, 2001, vol. 2, pp. 918–923. IEEE (2001)
32. Jahani, R., Hosseinzadeh, A., Hosseinzadeh, A., Abadi, M.M.: Ica-based allocation of dgs in a distribution system. Am. J. Sci. Res. **33**, 64–75 (2011)
33. Soroudi, A., Ehsan, M.: Imperialist competition algorithm for distributed generation connections. IET Gener. Transm. Distrib. **6**(1), 21–29 (2012)
34. Rahmatian, M., Ebrahimi, E., Ghanizadeh, A., Gharehpetian, G.: Optimal sitting and sizing of dg units considering islanding operation mode of sensitive loads. In: 2012 2nd Iranian Conference on Smart Grids (ICSG), pp. 1–5. IEEE (2012)
35. Lalitha, M.P., Reddy, V.V., Usha, V., Reddy, N.S.: Application of fuzzy and pso for dg placement for minimum loss in radial distribution system (2006)
36. Haghifam, M.-R., Falaghi, H., Malik, O.: Risk-based distributed generation placement. IET Gener. Transm. Distrib. **2**(2), 252–260 (2008)
37. Kim, K.-H., Lee, Y.-J., Rhee, S.-B., Lee, S.-K., You, S.-K.: Dispersed generator placement using fuzzy-ga in distribution systems. In: Power Engineering Society Summer Meeting, 2002 IEEE, vol. 3, pp. 1148–1153. IEEE (2002)

38. Ekel, P.Y., Martins, C., Pereira, J., Palhares, R.M., Canha, L.N.: Fuzzy set based multiobjective allocation of resources and its applications. Comput. Math. Appl. **52**(1–2), 197–210 (2006)
39. Ramírez-Rosado, I.J., Domínguez-Navarro, J.A.: Possibilistic model based on fuzzy sets for the multiobjective optimal planning of electric power distribution networks. IEEE Trans. Power Syst. **19**(4), 1801–1810 (2004)
40. Haghifam, M.-R., Malik, O.: Genetic algorithm-based approach for fixed and switchable capacitors placement in distribution systems with uncertainty and time varying loads. IET Gener. Transm. Distrib. **1**(2), 244–252 (2007)
41. El-Khattam, W., Bhattacharya, K., Hegazy, Y., Salama, M.: Optimal investment planning for distributed generation in a competitive electricity market. IEEE Trans. Power Syst. **19**(3), 1674–1684 (2004)
42. Silvestri, A., Berizzi, A., Buonanno, S.: Distributed generation planning using genetic algorithms. In: International Conference on Electric Power Engineering. PowerTech Budapest 99, p. 257. IEEE (1999)
43. Kumar, V., Gupta, I., Gupta, H.O.: Dg integrated approach for service restoration under cold load pickup. IEEE Trans. Power Delivery **25**(1), 398–406 (2010)
44. Borges, C.L., Falcao, D.M.: Optimal distributed generation allocation for reliability, losses, and voltage improvement. Int. J. Electr. Power Energy Syst. **28**(6), 413–420 (2006)
45. Teng, J.-H., Luor, T.-S., Liu, Y.-H.: Strategic distributed generator placements for service reliability improvements. In: Power Engineering Society Summer Meeting, IEEE, vol. 2, pp. 719–724. IEEE (2002)
46. Popović, D., Greatbanks, J., Begović, M., Pregelj, A.: Placement of distributed generators and reclosers for distribution network security and reliability. Int. J. Electr. Power Energy Syst. **27**(5), 398–408 (2005)
47. Ochoa, L.F., Padilha-Feltrin, A., Harrison, G.P.: Time-series-based maximization of distributed wind power generation integration. IEEE Trans. Energy Convers. **23**(3), 968–974 (2008)
48. Del Valle, Y., Venayagamoorthy, G.K., Mohagheghi, S., Hernandez, J.-C., Harley, R.G.: Particle swarm optimization: basic concepts, variants and applications in power systems. IEEE Trans. Evol. Comput. **12**(2), 171–195 (2008)
49. AlRashidi, M.R., El-Hawary, M.E.: A survey of particle swarm optimization applications in electric power systems. IEEE Trans. Evol. Comput. **13**(4), 913–918 (2009)
50. El-Zonkoly, A.: Optimal placement of multi-distributed generation units including different load models using particle swarm optimization. Swarm Evol. Comput. **1**(1), 50–59 (2011)
51. Gomez, J., Khodr, H., De Oliveira, P., Ocque, L., Yusta, J., Villasana, R., Urdaneta, A.: Ant colony system algorithm for the planning of primary distribution circuits. IEEE Trans. Power Syst. **19**(2), 996–1004 (2004)
52. Vlachogiannis, J.G., Hatziargyriou, N.D., Lee, K.Y.: Ant colony system-based algorithm for constrained load flow problem. IEEE Trans. Power Syst. **20**(3), 1241–1249 (2005)
53. Wang, L., Singh, C.: Reliability-constrained optimum placement of reclosers and distributed generators in distribution networks using an ant colony system algorithm. IEEE Trans. Syst. Man Cybern. Part C (Appl. Rev.) **38**(6), 757–764 (2008)
54. Chatterjee, A., Ghoshal, S., Mukherjee, V.: Artificial bee colony algorithm for transient performance augmentation of grid connected distributed generation. In: International Conference on Swarm, Evolutionary, and Memetic Computing, pp. 559–566. Springer (2010)
55. Rao, R.S., Narasimham, S., Ramalingaraju, M.: Optimization of distribution network configuration for loss reduction using artificial bee colony algorithm. Int. J. Electr. Power Energy. Syst. Eng. **1**(2), 116–122 (2008)
56. Lalitha, M.P., Reddy, N.S., Reddy, V.V.: Optimal dg placement for maximum loss reduction in radial distribution system using abc algorithm. Int. J. Rev. Comput. **3**, 44–52 (2010)
57. Hussain, I., Roy, A.K.: Optimal distributed generation allocation in distribution systems employing modified artificial bee colony algorithm to reduce losses and improve voltage profile. In: 2012 International Conference on Advances in Engineering, Science and Management (ICAESM), pp. 565–570. IEEE (2012)

58. Sohi, M.F., Shirdel, M., Javidaneh, A.: Applying bco algorithm to solve the optimal dg placement and sizing problem. In: Power Engineering and Optimization Conference (PEOCO), 5th International, pp. 71–76. IEEE (2011)
59. Noroozian, R., Molaei, S., et al.: Determining the optimal placement and capacity of dg in intelligent distribution networks under uncertainty demands by coa. In: 2012 2nd Iranian Conference on Smart Grids (ICSG), pp. 1–8. IEEE (2012)
60. Moravej, Z., Akhlaghi, A.: A novel approach based on cuckoo search for dg allocation in distribution network. Int. J. Electr. Power Energy Syst. **44**(1), 672–679 (2013)
61. Saravanamutthukumaran, S., Kumarappan, N.: Sizing and siting of distribution generator for different loads using firefly algorithm. In: 2012 International Conference on Advances in Engineering, Science and Management (ICAESM), pp. 800–803. IEEE (2012)
62. Sulaiman, M.H., Mustafa, M.W., Azmi, A., Aliman, O., Rahim, S.A.: Optimal allocation and sizing of distributed generation in distribution system via firefly algorithm. In: Power Engineering and Optimization Conference (PEDCO) Melaka, Malaysia, IEEE International, pp. 84–89. IEEE (2012)
63. Gandomkar, M., Vakilian, M., Ehsan, M.: A genetic-based tabu search algorithm for optimal dg allocation in distribution networks. Electr. Power Compon. Syst. **33**(12), 1351–1362 (2005)
64. Moradi, M.H., Abedini, M.: A combination of genetic algorithm and particle swarm optimization for optimal dg location and sizing in distribution systems. Int. J. Electr. Power Energy Syst. **34**(1), 66–74 (2012)
65. Harrison, G.P., Piccolo, A., Siano, P., Wallace, A.R.: Hybrid ga and opf evaluation of network capacity for distributed generation connections. Electr. Power Syst. Res. **78**(3), 392–398 (2008)
66. Naderi, E., Seifi, H., Sepasian, M.S.: A dynamic approach for distribution system planning considering distributed generation. IEEE Trans. Power Delivery **27**(3), 1313–1322 (2012)
67. Gomez-Gonzalez, M., López, A., Jurado, F.: Optimization of distributed generation systems using a new discrete pso and opf. Electr. Power Syst. Res. **84**(1), 174–180 (2012)
68. Akorede, M.F., Hizam, H., Aris, I., Ab Kadir, M.: Effective method for optimal allocation of distributed generation units in meshed electric power systems. IET Gener. Transm. Distrib. **5**(2), 276–287 (2011)
69. Zeynal, H., Hui, L.X., Jiazhen, Y., Eidiani, M., Azzopardi, B.: Improving lagrangian relaxation unit commitment with cuckoo search algorithm. In: 2014 IEEE International Conference on Power and Energy (PECon), pp. 77–82. IEEE (2014)
70. Yu, X., Zhang, X.: Unit commitment using lagrangian relaxation and particle swarm optimization. Int. J. Electr. Power Energy Syst. **61**, 510–522 (2014)
71. Ji, B., Yuan, X., Chen, Z., Tian, H.: Improved gravitational search algorithm for unit commitment considering uncertainty of wind power. Energy **67**, 52–62 (2014)
72. Ji, B., Yuan, X., Li, X., Huang, Y., Li, W.: Application of quantum-inspired binary gravitational search algorithm for thermal unit commitment with wind power integration. Energy Convers. Manage. **87**, 589–598 (2014)
73. Jadoun, V.K., Gupta, N., Niazi, K.R., Swarnkar, A.: Enhanced particle swarm optimization for short-term non-convex economic scheduling of hydrothermal energy systems. J. Electr. Eng. Technol. **10**(5), 1940–1949 (2015)
74. Lu, S., Sun, C., Lu, Z.: An improved quantum-behaved particle swarm optimization method for short-term combined economic emission hydrothermal scheduling. Energy Convers. Manage. **51**(3), 561–571 (2010)
75. Haghrah, A., Nazari-Heris, M., Mohammadi-Ivatloo, B.: Solving combined heat and power economic dispatch problem using real coded genetic algorithm with improved mühlenbein mutation. Appl. Thermal Eng. **99**, 465–475 (2016)
76. Sun, J., Palade, V., Wu, X.-J., Fang, W., Wang, Z.: Solving the power economic dispatch problem with generator constraints by random drift particle swarm optimization. IEEE Trans. Ind. Inform. **10**(1), 222–232 (2014)
77. Nazari-Heris, M., Mohammadi-Ivatloo, B., Gharehpetian, G.: Short-term scheduling of hydro-based power plants considering application of heuristic algorithms: a comprehensive review. Renew. Sustain. Energy Rev. **74**, 116–129 (2017)

78. dos Santos Coelho, L., Mariani, V.C.: Particle swarm approach based on quantum mechanics and harmonic oscillator potential well for economic load dispatch with valve-point effects. Energy Convers. Manage. **49**(11), 3080–3085 (2008)
79. Sun, J., Fang, W., Wang, D., Xu, W.: Solving the economic dispatch problem with a modified quantum-behaved particle swarm optimization method. Energy Convers. Manage. **50**(12), 2967–2975 (2009)
80. Meng, K., Wang, H.G., Dong, Z., Wong, K.P.: Quantum-inspired particle swarm optimization for valve-point economic load dispatch. IEEE Trans. Power Syst. **25**(1), 215–222 (2010)
81. Zhisheng, Z.: Quantum-behaved particle swarm optimization algorithm for economic load dispatch of power system. Expert Syst. Appl. **37**(2), 1800–1803 (2010)
82. Mahdi, F.P., Vasant, P., Rahman, M.M., Abdullah-Al-Wadud, M., Watada, J., Kallimani, V.: Quantum particle swarm optimization for multiobjective combined economic emission dispatch problem using cubic criterion function. In: 2017 IEEE International Conference on Imaging, Vision & Pattern Recognition (icIVPR), pp. 1–5. IEEE (2017)
83. Sheikhan, M., Mohammadi, N.: Neural-based electricity load forecasting using hybrid of ga and aco for feature selection. Neural Comput. Appl. **21**(8), 1961–1970 (2012)
84. Ardakani, F., Ardehali, M.: Novel effects of demand side management data on accuracy of electrical energy consumption modeling and long-term forecasting. Energy Convers. Manage. **78**, 745–752 (2014)
85. Gonzalez-Romera, E., Jaramillo-Moran, M.A., Carmona-Fernandez, D.: Monthly electric energy demand forecasting based on trend extraction. IEEE Trans. Power Syst. **21**(4), 1946–1953 (2006)
86. Nazari-Heris, M., Abapour, S., Mohammadi-Ivatloo, B.: Optimal economic dispatch of fc-chp based heat and power micro-grids. Appl. Thermal Eng. **114**, 756–769 (2017)
87. Amjady, N.: Short-term hourly load forecasting using time-series modeling with peak load estimation capability. IEEE Trans. Power Syst. **16**(4), 798–805 (2001)
88. Tian, S., Tuanjie, L.: Short-term load forecasting based on rbfnn and qpso. In: Power and Energy Engineering Conference, APPEEC 2009. Asia-Pacific, pp. 1–4. IEEE (2009)
89. Wang, J., Liu, Z., Lu, P.: Electricity load forecasting based on adaptive quantum-behaved particle swarm optimization and support vector machines on global level. In: International Symposium on Computational Intelligence and Design, ISCID'08, vol. 1, pp. 233–236. IEEE (2008)
90. Ibrahim, A.A., Mohamed, A., Shareef, H., Ghoshal, S.P.: An effective power quality monitor placement method utilizing quantum-inspired particle swarm optimization. In: 2011 International Conference on Electrical Engineering and Informatics (ICEEI), pp. 1–6. IEEE (2011)
91. Yao, F., Dong, Z.Y., Meng, K., Xu, Z., Iu, H.H.-C., Wong, K.P.: Quantum-inspired particle swarm optimization for power system operations considering wind power uncertainty and carbon tax in australia. IEEE Trans. Ind. Inform. **8**(4), 880–888 (2012)
92. Badawy, R., Heßler, A., Albayrak, S., Hirsch, B., Yassine, A.: Quantum-inspired evolution for smart building energy management in future power networks. In: EngOpt2014, p. 226 (2014)

Printed in the United States
By Bookmasters